# *Contributions to Algebra*

*A Collection of Papers Dedicated to Ellis Kolchin*

J. Miller

# *Contributions to Algebra*

*A Collection of Papers Dedicated to Ellis Kolchin*

---

*Edited by*

HYMAN BASS
*Department of Mathematics*
*Columbia University*
*New York, New York*

PHYLLIS J. CASSIDY
*Department of Mathematics*
*Smith College*
*Northampton, Massachusetts*

JERALD KOVACIC
*Department of Mathematics*
*Brooklyn College*
*Brooklyn, New York*

ACADEMIC PRESS New York San Francisco London 1977
*A Subsidiary of Harcourt Brace Jovanovich, Publishers*

ACADEMIC PRESS, INC.
111 Fifth Avenue, New York, New York 10003

*United Kingdom Edition published by*
ACADEMIC PRESS, INC. (LONDON) LTD.
24/28 Oval Road, London NW1

LIBRARY OF CONGRESS CATALOG CARD NUMBER:

**Library of Congress Cataloging in Publication Data**

Main entry under title:

Contributions to algebra.

Includes bibliographies.
1. Algebra–Addresses, essays, lectures.
2. Kolchin, Ellis Robert, Date
I. Kolchin, Ellis Robert, Date II. Bass, Hyman, Date III. Cassidy, Phyllis.
IV. Kovacic, Jerald.
QA155.C66 512 76-45980
ISBN 0-12-080550-2

PRINTED IN THE UNITED STATES OF AMERICA

*A number of friends, colleagues, and students of Ellis Kolchin have joined together to present this volume to him on the occasion of his sixtieth birthday. We believe we can give no better expression of our high regard than by this joint effort, for the papers reflect in their varied content his work and influence. He shaped the course of differential algebra and brought it into the mainstream of mathematics. He initiated the modern study of linear algebraic groups by departing from the Lie tradition and submitting the theory to the discipline of algebraic geometry. But it is not only as a researcher that we honor him. Our warm regards go also to the teacher, whose kindness and deep love for mathematics enable him to give unreservedly of his time and creative insights. And most of all we toast the man, whose warm heart and loyalty make him a true friend.*

*Happy birthday, Ellis!*

# *Contents*

## On finite projective groups
*Richard Brauer*



## Unipotent differential algebraic groups
*Phyllis J. Cassidy*



## Solutions in the general solution
*Richard M. Cohn*



## Folk theorems on elliptic equations
*Jean Dieudonné*



## Limit properties of stochastic matrices
*Samuel Eilenberg and Alex Heller*

## On the hyperalgebra of a semisimple algebraic group
*J. E. Humphreys*



## A notion of regularity for differential local algebras
*Joseph Johnson*



## The Engel–Kolchin theorem revisited
*Irving Kaplansky*



## Prime differential ideals in differential rings
*William F. Keigher*



## Constrained cohomology
*J. Kovacic*

## On elementary, generalized elementary, and liouvillian extension fields

*Maxwell Rosenlicht and Michael Singer*



## Derivations and valuation rings

*A. Seidenberg*



## On theorems of Lie–Kolchin, Borel, and Lang

*Robert Steinberg*



## A differential-algebraic study of the intrusion of logarithms into asymptotic expansions

*Walter Strodt*



## A "theorem of Lie–Kolchin" for trees

*J. Tits*

## Regular elements in anisotropic tori

*F. D. Veldkamp*

## *List of contributors*

*Numbers in parentheses indicate the pages on which the authors' contributions begin.*

*Hyman Bass* (1), Department of Mathematics, Columbia University, New York, New York

*Lipman Bers* (25), Department of Mathematics, Columbia University, New York, New York

*Lenore Blum* (37), Department of Mathematics, Mills College, Oakland, California, and University of California, Berkeley, California

*Richard Brauer*† (63), Department of Mathematics, Harvard University, Cambridge, Massachusetts

*Phyllis J. Cassidy* (83), Department of Mathematics, Smith College, Northampton, Massachusetts

*Richard M. Cohn* (117), Rutgers University, The State University of New Jersey, New Brunswick, New Jersey

*Jean Dieudonné* (129), Université de Nice, Nice, France

*Samuel Eilenberg* (135), Department of Mathematics, Columbia University, New York, New York

*J. Fogarty* (151), Department of Mathematics, University of Massachusetts, Amherst, Massachusetts

*P. X. Gallagher* (157), Department of Mathematics, Columbia University, New York, New York

*Howard Garland* (165), Department of Mathematics, Yale University, New Haven, Connecticut

*Harish-Chandra* (175), The Institute for Advanced Study, Princeton, New Jersey

*Alex Heller* (135), Department of Mathematics, The Graduate Center of CUNY, New York, New York

† Deceased.

*G. Hochschild* (183), Department of Mathematics, University of California, Berkeley, California

*J. E. Humphreys* (203), Department of Mathematics and Statistics, University of Massachusetts, Amherst, Massachusetts

*Joseph Johnson* (211), Department of Mathematics, Rutgers University, New Brunswick, New Jersey

*Irving Kaplansky* (233), Department of Mathematics, The University of Chicago, Chicago, Illinois

*William F. Keigher* (239), Department of Mathematics, University of Tennessee, Knoxville, Tennessee

*Jerald Kovacic* (251), Department of Mathematics, Brooklyn College, City University of New York, Brooklyn, New York

*Masatake Kuranishi* (267), Department of Mathematics, Columbia University, New York, New York

*James Lepowsky* (165), Department of Mathematics, Yale University, New Haven, Connecticut

*Hideyuki Matsumura* (279), Department of Mathematics, Nagoya University, Nagoya, Japan

*P. Norman* (151), Department of Mathematics, University of Massachusetts, Amherst, Massachusetts

*Takashi Ono* (295), Department of Mathematics, The Johns Hopkins University, Baltimore, Maryland

*Frans Oort* (305), Mathematisch Instituut, Amsterdam, The Netherlands

*Charles F. Osgood* (321), Naval Research Laboratory, Washington, D.C.

*R. J. Proulx*† (157), Department of Mathematics, Columbia University, New York, New York

*Maxwell Rosenlicht* (329), Department of Mathematics, University of California, Berkeley, California

*A. Seidenberg* (343), Department of Mathematics, University of California, Berkeley, California

*Michael Singer* (329), Department of Mathematics, North Carolina State University, Raleigh, North Carolina

*Robert Steinberg* (349), Department of Mathematics, University of California, Los Angeles, California

*Walter Strodt* (355), Department of Mathematics, St. Lawrence University, Canton, New York

*J. Tits* (377), Collège de France, Paris, France

*F. D. Veldkamp* (389), Department of Mathematics, University of Utrecht, Utrecht, The Netherlands

† Present address: Department of Mathematics, Wellesley College, Wellesley, Massachusetts.

# *Preface*

Although the articles in this volume are in the main devoted to commutative algebra, linear algebraic group theory, and differential algebra, the diversity of subjects covered—complex analysis, algebraic $K$-theory, logic, stochastic matrices, differential geometry, . . . —is a reflection of Ellis Kolchin's wide-ranging mathematical curiosity. His deep and abiding interest has always been in the application of the powerful and clarifying techniques of algebra to problems in the theory of differential equations. Following the tradition set by Joseph Fels Ritt (1893–1951), the founding father of differential algebra, his desire has been to remove the algebraic aspects of differential equations from analysis. It was in the course of applying the Ritt theory to the classical Picard–Vessiot theory that he became one of the pioneers of linear algebraic group theory. In this volume we celebrate the influence that Kolchin's work on the Galois theory of differential fields has had on the development of differential algebra and linear algebraic group theory.

In 1948 appeared Kolchin's beautiful paper "Algebraic matric groups and the Picard–Vessiot theory of homogeneous linear ordinary differential equations."[1] The intervening 30 years have witnessed the birth of a new subject, linear algebraic group theory, as algebraists developed a lively interest in the groups *ipsorum causa*.[2] However, just as finite group theory grew out of the Galois theory of polynomial equations, the theory of linear algebraic groups has its origins in the theory of differential equations.

[1] *Ann. of Math.* **49** (1948), 1–42.

[2] See Armand Borel, "Linear Algebraic Groups." Benjamin, New York and Amsterdam, 1969.

During the last years of the previous century, Émile Picard and Ernest Vessiot developed a Galois theory of homogeneous linear ordinary differential equations.[3] The role of the Galois group of such an equation is played by a group of linear substitutions with constant coefficients of a fundamental system of solutions. The kinds of analytic operation needed to obtain a fundamental system are characterized, as in the Galois theory, by the structure of the quotient groups in a composition series for the Galois group. In particular, Vessiot proved that the equation is "solvable by quadratures" if and only if its group is solvable. Unfortunately, the proof is marred by a lack of precision in the use of the words "solvable by quadratures," as Kolchin points out in his Picard–Vessiot paper, since one meaning is used in the sufficiency part and another in the necessity part.

The essential ingredient in the Picard–Vessiot theory is the fact that the Galois group of a homogeneous linear ordinary differential equation is an algebraic group. For Picard and Vessiot, the force of this was to enable them to make full use of the transcendental machinery of the Lie theory. This resulted in the dependence of the Picard–Vessiot theory, in essence algebraic, on the analytic theory of Lie groups. Furthermore, the relentlessly local approach of the classical Lie theory did not allow consideration of such global properties as connectedness and component of the identity, which can be described by abstract group criteria.[4] As Kolchin remarks in his Picard–Vessiot paper:

> the literature seems to be devoid of any basic theory of algebraic matric groups as such. For lack of such a theory, these groups, when encountered on a large scale (as in the Picard–Vessiot theory) have been treated as special cases of Lie groups. As a result, the generally brilliant theory of Picard and Vessiot suffered on the one hand from the lack of rigor of the early theory of Lie groups, and on the other hand from being too intimately bound up with the analytic point of view of the Lie theory, thereby obscuring the algebraic nature of the subject matter.[5]

During the 50 years following the seminal work of Picard and Vessiot, the Galois theory of differential equations was largely neglected. Departing from the Lie tradition, Kolchin developed an autonomous theory of algebraic matric groups. A guiding principle of his work was to weave together the group properties and the algebro-geometric properties of the

[3] See Vessiot's account in "Méthodes d'integration élémentaires," Encyclopedie des sciences mathematiques pures at appliquées, t. II, f. 1 (1910), 58–170.

[4] E. R. Kolchin, On certain concepts in the theory of algebraic matric groups, *Ann. of Math.* **49** (1948), 774–789.

[5] Kolchin, *loc. cit.* p. 2.

algebraic groups in order to obtain proofs for arbitrary characteristic and with a minimum of heavy machinery.

In his early papers on the Galois theory, Kolchin developed concepts that are basic to linear algebraic group theory, emphasizing the theory of solvable groups. A nonsingular matrix over an algebraically closed field can be written uniquely as a commuting product of a diagonalizable matrix and a unipotent matrix (the so-called semisimple and unipotent parts). Kolchin proved a most important property, characteristic of algebraic matric groups, that any such group contains the semisimple and unipotent parts of its elements. He goes on to show that a commutative algebraic matric group splits uniquely into a direct product of algebraic subgroups $T$ and $U$, which are simultaneously reducible to diagonal form and special triangular form, respectively. $T$ consists of the semisimple parts and $U$ of the unipotent parts of the matrices in the group.

Since a commutative group is solvable, this global "Jordan decomposition" was, of course, suggested by the famous Lie–Kolchin theorem: *A necessary and sufficient condition that a set of matrices be reducible to triangular form is that the set be contained in the underlying manifold of a connected solvable algebraic matric group*. The proof, which is beautiful in its simplicity and is carried through for arbitrary characteristic, has recourse neither to infinitesimal transformations nor to powerful tools of algebraic geometry.[6] Armand Borel, in his paper "Groups linéaires algébriques,"[7] extended Kolchin's decomposition theorem from commutative groups first to nilpotent groups, and then, by replacing direct product with semidirect product, to arbitrary connected solvable algebraic matric groups.

Throughout his work, Kolchin's primary interest is in algebraic groups as representations of transformation groups acting on the solution sets of differential equations. In order to free these Galois groups as much as possible from the particular matric representation and even from the particular rational structure, Kolchin tries to give abstract group-theoretic criteria whenever possible. He defines an algebraic group to be anticompact if it contains no element of finite order greater than 1 not divisible by the characteristic of the base field. For a matric group $G$, the property of being anticompact is equivalent to the condition that $G$ be composed of unipotent matrices ($G$ is unipotent) and, if the base field has positive characteristic $p$, to the condition that each of the elements of $G$ have order a power of $p$. Using Burnside's theorem, Kolchin then proved a global analogue of Engel's theorem, valid for arbitrary characteristic: *Every an-*

[6] See Robert Steinberg's paper, On theorems of Lie–Kolchin, Borel, and Lang, in this volume.

[7] *Ann. of Math.* **64** (1956), 20–82.

*ticompact algebraic matric group is nilpotent and is reducible to special triangular form.*

In his definition of quasicompact algebraic group, Kolchin characterizes, at least in the case of connected matric groups, diagonal groups (algebraic tori). An algebraic group containing no nontrivial anticompact algebraic subgroup is called quasicompact. This definition, which is not and indeed cannot be given solely in the terms of the abstract group structure, can be so given in the case of a connected group of given dimension. Kolchin goes on to show that the condition on an algebraic matric group to be quasicompact is equivalent to the requirement that every matrix in the group be diagonalizable. He then proves that a connected quasicompact algebraic matric group is diagonalizable, hence is, in particular, commutative.

Kolchin's Picard–Vessiot paper marks the beginning of a Galois theory of differential fields rather than of differential equations. The first definition of abstract differential field was given by Vessiot in an effort to find the correct "domain of rationality" for the Galois theory of differential equations. A differential field is a field $\mathfrak{F}$ of characteristic 0 on which is defined a derivation operator. In the classical case, the field is an extension of the field of rational functions of a complex variable $z$ and the operator is $d/dz$. We assume that the field $\mathfrak{C}$ of constants of $\mathfrak{F}$ is algebraically closed. A differential field extension $\mathfrak{G}$ of $\mathfrak{F}$ is called a Picard–Vessiot extension if the field of constants of $\mathfrak{G}$ is $\mathfrak{C}$ ($\mathfrak{G}$ has no new constants) and $\mathfrak{G}$ is obtained by adjoining to $\mathfrak{F}$ a fundamental system of solutions of a homogeneous linear ordinary differential equation with coefficients in $\mathfrak{F}$ and their derivatives.[8]

In the language of Lie theory, the set of solutions of the equation "admits" the general linear group $GL(n, \mathfrak{C})$ as a transitive transformation group. The problem is to describe the subgroup that represents the Galois group. Using the concept of Picard–Vessiot extension, this is easily done. For any differential field extension $\mathfrak{G}$ of $\mathfrak{F}$, let $G(\mathfrak{G}/\mathfrak{F})$ be the group of differential automorphisms $\sigma$ of $\mathfrak{G}$ over $\mathfrak{F}$ (thus $\sigma$ leaves fixed the elements of $\mathfrak{F}$ and commutes with the derivation operator). Kolchin proves that when $\mathfrak{G}$ is a Picard–Vessiot extension, $G(\mathfrak{G}/\mathfrak{F})$ is "abundant," i.e., for any intermediate differential field $\mathfrak{F}_1$ and any element $\alpha$ in $\mathfrak{G}$ not in $\mathfrak{F}_1$, there is a differential automorphism $\sigma$ of $\mathfrak{G}$ over $\mathfrak{F}_1$ such that $\sigma\alpha \neq \alpha$. This normality property of a Picard–Vessiot extension was stated in the classical case

[8] The existence of a fundamental system of solutions introducing no new constants was established by Kolchin in Existence theorems connected with the Picard–Vessiot theory of homogeneous linear ordinary differential equations, *Bull. Amer. Math. Soc.* **54** (1948), 927–932.

(and when $\mathfrak{F}_1 = \mathfrak{F}$) by Picard,[9] "Toute fonction rationnelle de $x$ et d'un système fondamental $y_1, y_2, \ldots, y_n$ et leurs dérivées, qui reste invariable par les substitutions du groupe $G$, est une fonction rationnelle de $x$." The subgroup $G$ of $GL(n, \mathfrak{C})$ representing $G(\mathfrak{G}/\mathfrak{F})$ relative to $y_1, \ldots, y_n$ is the set of all matrices $(a_{ij})$ such that there is a differential automorphism $\sigma$ of $\mathfrak{G}$ over $\mathfrak{F}$ with $\sigma y_j = \Sigma\, a_{ij} y_i$. Kolchin shows that $G$ is an algebraic subgroup of $GL(n, \mathfrak{C})$ whose dimension equals the transcendence degree of $\mathfrak{G}$ over $\mathfrak{F}$, and that the usual assignments define a Galois correspondence between the set of all intermediate differential fields $\mathfrak{F}_1$ and the set of all algebraic subgroups of $G$. $\mathfrak{F}_1$ is a normal extension of $\mathfrak{F}$, in the sense that $G(\mathfrak{F}_1/\mathfrak{F})$ is abundant, if and only if $G(\mathfrak{G}/\mathfrak{F}_1)$ is a normal subgroup of $G(\mathfrak{G}/\mathfrak{F})$. When this is the case, $G(\mathfrak{F}_1/\mathfrak{F})$ is isomorphic to $G(\mathfrak{G}/\mathfrak{F}) / G(\mathfrak{G}/\mathfrak{F}_1)$.[10] In particular, the identity component of $G(\mathfrak{G}/\mathfrak{F})$ is the Galois group of $\mathfrak{G}$ over the relative algebraic closure of $\mathfrak{F}$ in $\mathfrak{G}$. The finiteness of the quotient group in this case reflects the fact that the relative algebraic closure is a finite Galois extension of $\mathfrak{F}$.

The notion of Picard–Vessiot extension enabled Kolchin to define "solvable by quadratures" precisely. An extension $\mathfrak{G}$ of $\mathfrak{F}$ is called liouvillian if $\mathfrak{G}$ contains no new constants and is obtained by combinations of certain allowable operations: adjoining algebraic elements, integrals, and exponentials of integrals.[11] Kolchin proved that a Picard–Vessiot extension is contained in a liouvillian extension if and only if the identity component of its Galois group is solvable. It is contained in a liouvillian extension obtained by adjoining integrals (exponentials of integrals) alone if and only if its Galois group is representable by an anticompact (quasicompact) algebraic matric group.

In his papers "Galois theory of differential fields,"[12] "On the Galois theory of differential fields,"[13] and "Abelian extensions of differential fields,"[14] Kolchin describes a class of differential fields, which includes all Picard–Vessiot extensions, for which there is a viable Galois theory. He calls $\mathfrak{G}$ strongly normal over $\mathfrak{F}$ if $\mathfrak{G}$ itself contains no new constants, and if $\sigma$ is a differential isomorphism of $\mathfrak{G}$ over $\mathfrak{F}$ into an extension of $\mathfrak{F}$ then the field compositum $\mathfrak{G}\sigma\mathfrak{G}$ is generated over $\mathfrak{G}$ by constants, i.e., $\sigma$ fails to be

[9] É. Picard, "Traité d'analyse," t. III, ch. 17, Gauthier-Villars, Paris, 1896.

[10] The fact that the quotient group is a linear group is proved in a joint paper with C. Chevalley, Two proofs of a theorem on algebraic groups, *Proc. Amer. Math. Soc.* **2** (1951), 126–134.

[11] See also the paper, On elementary, generalized elementary, and liouvillian extension fields, by Maxwell Rosenlicht and Michael Singer in this volume.

[12] *Amer. J. Math.* **75,** (1953), 753–824.

[13] *Ibid.* **77** (1955), 868–894.

[14] *Ibid.* **82** (1960), 779–790.

an automorphism only by the introduction of constants. We may state Kolchin's fundamental theorem about strongly normal extensions $\mathfrak{G}$ of $\mathfrak{F}$ as follows: The relative automorphism group $G(\mathfrak{G}/\mathfrak{F})$ is representable by the group of $\mathfrak{C}$-rational points of an algebraic group $G$.[15] There is a Galois correspondence between the set of intermediate differential fields and the set of all $\mathfrak{C}$-subgroups of $G$. He goes on to show that $\mathfrak{G}$ is Picard–Vessiot over $\mathfrak{F}$ if and only if $G(\mathfrak{G}/\mathfrak{F})$ is linear, i.e., is representable by an algebraic matric group. In the classical case, $\mathfrak{G}$ is obtained by adjoining abelian functions if and only if $G(\mathfrak{G}/\mathfrak{F})$ is an abelian variety. Finally, using Chevalley's structure theorem for algebraic groups[16] Kolchin is able to characterize the strongly normal extensions $\mathfrak{G}$ of $\mathfrak{F}$ by means of a tower of differential fields $\mathfrak{F} \subset \mathfrak{F}^{\circ} \subset \mathfrak{E}^{\circ} \subset \mathfrak{G}$, with $G(\mathfrak{F}^{\circ}/\mathfrak{F})$ finite, $G(\mathfrak{E}^{\circ}/\mathfrak{F}^{\circ})$ representable by an abelian variety, $G(\mathfrak{G}/\mathfrak{E}^{\circ})$ representable by an algebraic matric group. In the classical case, this says that $\mathfrak{G}$ can be obtained by adjoining finitely many algebraic functions, then finitely many abelian functions, and finally a fundamental system of solutions of a homogeneous linear ordinary differential equation.

Ellis Kolchin placed Picard–Vessiot theory in its natural setting, the rich and elegant Ritt theory of algebraic differential equations. Although he considers himself first and foremost a differential algebraist, his work had its origins in another branch of mathematics—Lie theory—and opened up new paths in yet another—linear algebraic group theory. We have here limited ourselves to a discussion of his early work and as a result have barely touched upon his extensive and deep contributions to differential algebra. Indeed Ellis Kolchin *is* differential algebra. He has shaped the course of the subject and his influence is felt in every aspect of it. The articles in this volume, a number of which evolved from the "continuing differential algebra seminar," jointly and severally reflect this influence.

[15] This result was proved independently by H. Matsumura in Automorphism groups of differential fields and group varieties, *Mem. Coll. Sci. Univ. Kyoto, Ser. A* **28** (1954), 283–292.

[16] See Maxwell Rosenlicht, Some basic theorems of algebraic groups, *Amer. J. Math.* **78** (1956), 401–443.

# *Quadratic modules over polynomial rings*

*Hyman Bass*

*Columbia University*

This paper is dedicated in warm friendship to Ellis Kolchin.

## Introduction

Let $A = F[X_1, \ldots, X_d]$ be a polynomial algebra in $d$ variables over a field $F$ of characteristic $\neq 2$. Let $S = S(X_1, \ldots, X_d)$ be a symmetric $r$ by $r$ matrix over $A$ whose determinant is a nonzero constant in $F$. We consider the problem:

Is $S$ equivalent to a matrix with coefficients in $F$; more precisely, is there a $U$ in $GL_r(A)$ such that $US{}^tU = S_0$, where the t denotes transpose, and $S_0 = S(0, \ldots, 0)$? A theorem of Harder (see [12]) implies that the answer is affirmative for $d = 1$. A theorem of Karoubi [11] implies that the answer is "stably" affirmative. However, Parimala [13] has recently given a remarkable counterexample showing that the answer is negative for $F = \mathbb{R}$, $d = 2$, and $r = 4$. She raised the question of whether the answer is affirmative for $F = \mathbb{C}$. We prove here that the answer is affirmative whenever $F$ is algebraically closed and $d \leq 3$.† The proof uses Karoubi's theorem but not Harder's, thus giving a new proof of Harder's theorem in this special case. The proof also uses a general Witt Cancellation Theorem, proved in [4]. Because of the generality of its setting, the latter theorem and its proof are far

† *Note added in proof*: M. S. Raghunathan has just found an independent proof of this valid for all $d$, in a new paper, "Principal bundles on affine space."

more technical than the case of it needed here. It seemed worthwhile therefore to give the direct proof of the version used here. This is further justified by the fact that we obtain a sharpening of the theorem under the assumption that 2 is invertible.

The effect of the Cancellation Theorem plus Karoubi's theorem is to reduce our problem to one where $r$ is small (in fact, $r \leq 2d$). By using non-abelian étale cohomology, and the special isomorphisms of $\mathrm{Spin}_r$ with other classical groups for small $r$, we can solve the problem affirmatively for $r \leq 6$ (and all $d$) using the fact that the Brauer group $\mathrm{Br}(A)$ coincides with the "cohomological Brauer group" $H^2_{\mathrm{et}}(\mathrm{spec}(A), \mathbb{G}_m)$ (see [10]), away from $p = \mathrm{char}(F)$. This had been proved by Auslander–Goldman–Grothendieck [10] for $d \leq 2$, and recently in general by R. Hoobler.

## 1. Definitions, and background of the problem

*All rings here are understood to be commutative.* Let $A$ be a ring and let $E$ be an $A$-module. We call $E$ an *A-space* if it is a finitely generated projective $A$-module. Let $\varphi\colon E \times E \to A$ be a bilinear form on $E$. It induces two homomorphism from $E$ to $E^* = \mathrm{Hom}_A(E, A)$, and we say $\varphi$ is *invertible* (or nonsingular) if these are isomorphisms. We call $(E, \varphi)$ an *inner product* (resp., *symplectic*) *A-module* if $\varphi$ is symmetric (resp., alternating, i.e., $\varphi(x, x) = 0$ for all $x \in E$). We call $(E, Q)$ a *quadratic A-module* if $Q\colon E \to A$ is a quadratic form on $E$, i.e., $Q(ax) = a^2Q(x)$ and $\varphi_Q(x, y) = Q(x + y) - Q(x) - Q(y)$ is a (symmetric) bilinear form on $E$. An inner product (resp., symplectic $A$-module $(E, \varphi)$ is called an inner *product* (resp., *symplectic*) *A-space* if $E$ is an $A$-space and $\varphi$ is invertible. A quadratic $A$-module $(E, Q)$ is called a *quadratic A-space* if $(E, \varphi_Q)$ is an inner product $A$-space.

Let $f\colon A \to B$ be a ring homomorphism. If $\varphi$ is a bilinear form on $E$, we write $B \otimes_A (E, \varphi)$ for $(B \otimes_A E, B \otimes_A \varphi)$. If $Q$ is a quadratic form on $E$, then $B \otimes_A(E, Q) = (B \otimes_A E, B \otimes_A Q)$ is characterized by the formula $(B \otimes_A Q)(b \otimes x) = b^2f(Q(x))$ for $b \in B$, $x \in E$ (cf. [7, §3, No. 4 Prop. 3]). A *B-module* is said to be *extended from A* if it is isomorphic to $B \otimes_A E$ for some $A$-module $E$. One similarly defines the notion of an *inner product* (or *symplectic*, or *quadratic*) *B-module extended from A*.

We shall investigate here the following quadratic variations on Serre's problem, denoted $(\mathrm{S})_{\mathrm{L}}$.

Let $F$ be a field, and let $A = F[X_1, \ldots, X_d]$, a polynomial algebra in $d$ variables over $F$.

$(\mathrm{S})_{\mathrm{L}}$ Is every $A$-space extended from $F$?
$(\mathrm{S})_{\mathrm{Sp}}$ Is every symplectic $A$-space extended from $F$?

$(\mathrm{S})_{\mathrm{IP}}$ Is every inner product $A$-space extended from $F$?
$(\mathrm{S})_{\mathrm{Q}}$ Is every quadratic $A$-space extended from $F$?

The first problem has recently been solved by Quillen and Suslin (see [14]):

***Quillen–Suslin Theorem*** *The answer to* $(\mathrm{S})_{\mathrm{L}}$ *is affirmative, i.e. A-spaces are free.*

***Corollary*** *The answer to* $(\mathrm{S})_{\mathrm{Sp}}$ *is affirmative.*

This is an immediate consequence of the Quillen–Suslin theorem and the next proposition. For any ring $A$, $A$-module $E$, and $\varepsilon = \pm 1$, define the *hyperbolic* module $H_\varepsilon(E) = (E \oplus E^*, \varphi)$ by $\varphi((x, y), (x', y')) = \langle x, y'\rangle + \varepsilon\langle x', y\rangle$, where $\langle x, y\rangle = y(x)$ for $x \in E$ and $y \in E^*$. This is an inner product (resp., symplectic) $A$-module for $\varepsilon = 1$ (resp., $\varepsilon = -1$). It is an inner product (resp., symplectic) $A$-space if $E$ is an $A$-space. We similarly define the *hyperbolic quadratic A-module* $H_{\mathrm{Q}}(E) = (E \oplus E^*, Q_E)$ by $Q_E((x, y)) = \langle x, y\rangle$; it is a quadratic $A$-space if $E$ is an $A$-space. If $f\colon A \to B$ is a ring homomorphism then $B \otimes_A H_\varepsilon(E) \cong H_\varepsilon(B \otimes_A E)$ and $B \otimes_A H_{\mathrm{Q}}(E) \cong H_{\mathrm{Q}}(B \otimes_A E)$ if $E$ is an $A$-space. (See [4, Ch. I].)

***Proposition*** *Let A be a ring without nontrivial idempotents. Suppose that each A-space of* rank $\geq 2$ *has a direct summand isomorphic to A (e.g. that all A-space are free, or that A is noetherian of dimension* 1, *say a Dedekind ring). Every symplectic A-space is isomorphic to* $H_{-1}(A^n)$ *for some* $n \geq 0$.

Let $(E, \varphi)$ be a symplectic $A$-space. The proposition is well known when $A$ is a field, so it follows that the rank $r$ of $E$ is even. If therefore $E \neq 0$, we have $r \geq 2$ so, by hypothesis, $E$ contains a unimodular element $e$ (i.e. $e$ is a basis for a direct summand of $E$). Since $\varphi$ is invertible there is an $f \in E$ such that $\varphi(e, f) = 1$. Then $H = Ae + Af$ is a hyperbolic plane ($\cong H_{-1}(A)$) in $E$, hence an orthogonal direct summand of $(E, \varphi)$ (cf. [4, Ch. I, 3.2 and 4.10.2]). The proposition thus follows by induction on $r$, applied to the orthogonal complement to $H$.

*Remark* This proposition is only a slight variant of [6, Prop. 2].

We now come to problems $(\mathrm{S})_{\mathrm{IP}}$ and $(\mathrm{S})_{\mathrm{Q}}$.

*Counterexamples in characteristic 2.* Let $F$ be a field of characteristic 2, and let $A = F[T]$, where $T$ is an indeterminate. The quadratic $A$-space $(A^2, Q)$, where $Q(x, y) = x^2 + xy + Ty^2$ is not extended from $F$, as can be seen from its Arf invariant, which is the class of $T$ modulo $\{a^2 - a \mid a \in A\}$. Similarly the inner product $A$-space $(A^2, \varphi)$, where $\varphi$ has matrix $\left(\begin{smallmatrix} T & 1 \\ 1 & 0 \end{smallmatrix}\right)$, is not extended from $F$, for otherwise it would be a hyperbolic plane (specialize $T$

to 0), whereas $\varphi$ is not alternating. These examples show that both $(S)_{IP}$ *and* $(S)_Q$ *have negative responses if* $\text{char}(F) = 2$. One might still ask whether a quadratic $A$-space whose Arf invariant comes from $F$ is extended from $F$. Part of $(S)_{IP}$ is salvaged by the following result of Harder (see [12, Th. 13.4.3]).

***Harder's Theorem*** *Let $F$ be a field, let $T$ be an indeterminate, let $A = F[T]$, and let $(E, \varphi)$ be an inner product $A$-space which is anisotropic (i.e. $\varphi(x, x) \neq 0$ for all $x \neq 0$ in $E$). Then $(E, \varphi)$ is extended from $F$.*

***Corollary*** *Let $F$ be a field of characteristic $\neq 2$. Then the answer to $(S)_{IP}$ and $(S)_Q$ is affirmative for $d = 1$.*

Indeed, when 2 is invertible the notions of symmetric bilinear form $\varphi$ and quadratic form $Q$ are equivalent; e.g. the relation $\varphi_Q(x, x) = 2Q(x)$ determines $Q$ in terms of $\varphi_Q$. Thus $(S)_{IP}$ and $(S)_Q$ are equivalent problems when $\text{char}(F) \neq 2$. Secondly, if $A$ is a principal ideal domain containing $\frac{1}{2}$ then every quadratic $A$-space is the orthogonal direct sum of an anisotropic one and a hyperbolic one, $H_Q(A^n)$ for some $n$. Indeed, if $(E, Q)$ is a quadratic $A$-space and if $Q(e) = 0$ for some $e \neq 0$ in $E$, we may assume $e$ is unimodular (since $A$ is a PID) and, so find an $f' \in E$ such that $\varphi_Q(e, f') = 1$. Setting $f = f' - Q(f')e$, we have $Q(f) = 0$, so that $e$ and $f$ span a hyperbolic plane $H \cong H_Q(A)$, and $(E, \varphi)$ thus has $H$ as an orthogonal direct summand. The claim now follows by induction on the rank of $E$. Applying these observations to quadratic spaces over $A = F[T]$ when $\text{char}(F) \neq 2$ we see, in view of Harder's theorem, that they are all extended from $F$, thus proving the corollary.

Let $A$ be a ring. We call $A$-modules $E$, $E'$ *stably isomorphic* if $E \oplus A^n \cong E' \oplus A^n$ for some $n \geq 0$. We call quadratic $A$-modules $(E, Q)$, $(E', Q')$ *stably isomorphic* if $(E, Q) \oplus H_Q(A^n) \cong (E', Q') \oplus H_Q(A^n)$ for some $n \geq 0$.

***Karoubi's ("Homotopy Invariance") Theorem*** *Let $A$ be a ring in which 2 is invertible, and let $B = A[T_1, \ldots, T_d]$, a polynomial algebra in $d$ variables over $A$. Let $(E, Q)$ be a quadratic $B$-space. If $E$ is stably extended from $A$, then $(E, Q)$ is stably extended from $A$.*

"Stably extended from A" signifies "stably isomorphic to an object extended from $A$." For convenience of the reader we shall give the proof, in Section 5, of this theorem, which is really a special case of what Karoubi proves. Karoubi's theorem plus the Quillen–Suslin theorem show that the response to $(S)_Q$ is "stably affirmative" when $\text{char}(F) \neq 2$. These results plus Harder's make the following example quite remarkable.

***Parimala's Example*** [13] *There is an anisotropic quadratic space of rank 4 over* $\mathbb{R}[X, Y]$ *which is not extended from* $\mathbb{R}$.

Parimala suggested, however, that $(S)_Q$ might still have an affirmative response for $F = \mathbb{C}$. We prove this here for $d \leq 3$.

***Theorem 1*** *Let* $F$ *be an algebraically closed field of characteristic* $\neq 2$. *Every quadratic space over* $F[X, Y, Z]$ *is extended from* $F$.

Our proof requires only that $F$ be separably algebraically closed. Our methods should be useful also in the study of quadratic spaces over the affine algebras of smooth algebraic surfaces.

## 2. Reduction by the Cancellation Theorem to low ranks

Let $A$ be a ring. We say a quadratic $A$-module $(E, Q)$ has *Witt index* $\geq r$, denoted

$$\mathrm{WI}(E, Q) \geq r,$$

if $(E, Q)$ contains an orthogonal direct summand isomorphic to $H_Q(L)$, where $L$ is an $A$-space of rank $\geq r$.

If $u \in A$, we denote by $\langle u \rangle_Q$ the quadratic $A$-module $(A, Q_u)$ defined by $Q_u(x) = ux^2$ for $x \in A$. The associated bilinear map, $(x, y) \mapsto 2uxy$, is invertible if and only if $2u \in A^\times$ (the group of units of $A$).

***Cancellation Theorem*** *Suppose the maximal ideal space* $\max(A)$ *is noetherian of dimension* $\leq d$ *(e.g. that* $A$ *is a noetherian ring of Krull dimension* $\leq d$*). Let* $(E, Q)$ *be a quadratic* $A$*-module, and assume either of conditions* I *or* II:

(I) $\mathrm{WI}((E, Q) \oplus H_Q(A)) \geq d + 2$.

(II) $(E, Q) \oplus H_Q(A)$ *has an orthogonal direct summand isomorphic to* $H_Q(L) \oplus \langle u \rangle_Q$, *where* $2u \in A^\times$ *and* $L$ *is an* $A$*-space of rank* $\geq d + 1$ *such that* $GL(P)$ *acts transitively on the set of unimodular elements in* $P$.

*Then if* $(E, Q)$ *is stably isomorphic to a quadratic* $A$*-module* $(E', Q')$ *we have* $(E, Q) \cong (E', Q')$.

***Corollary 1*** *Let* $A$ *be a semilocal ring. If* $2 \in A^\times$ *then stably isomorphic quadratic* $A$*-spaces are isomorphic. In general, stably isomorphic quadratic* $A$*-spaces, one of which contains a hyperbolic plane, are isomorphic.*

In fact the hypotheses of the corollary furnish those of the Cancellation Theorem with $d = 0$, once one makes the obvious reduction to the case where $A$ has no nontrivial idempotents. For case II one uses the easily proved fact that, if $2 \in A^\times$, quadratic $A$-spaces are diagonolizable, i.e., direct sums of $\langle u \rangle_Q$'s with $u \in A^\times$. This follows from the corresponding result modulo rad($A$), and application of Nakayama's lemma.

***Corollary 2*** ("Rank Reduction Theorem") *Let $F$ be a field of characteristic $\neq 2$, and let $A = F[X_1, \ldots, X_d]$, a polynomial algebra in $d$ variables. Let $(E, Q)$ be a quadratic $A$-space. Put $(E_0, Q_0) = F \otimes_A (E, Q)$, where we identify $F$ with $A/\sum AX_i$, and $(E', Q') = A \otimes_F (E_0, Q_0)$. Then $(E, Q)$ and $(E', Q')$ are stably isomorphic, they are locally isomorphic over $A_m$ for each maximal ideal $\mathfrak{m}$ of $A$, and, if* $\mathrm{WI}(E_0, Q_0) \geq d$ *and* $\dim E_0 \geq 2d + 1$, *they are isomorphic.*

In fact the Quillen–Suslin theorem implies that $E \cong E'$, so we can apply Karoubi's theorem to conclude that $(E, Q)$ and $(E', Q')$ are stably isomorphic whence (Corollary 1) locally isomorphic. If $\mathrm{WI}(E_0, Q_0) \geq d$ and $\dim E_0 \geq 2d + 1$ then $(E', Q')$ satisfies condition II of the Cancellation Theorem (again using the Quillen–Suslin theorem for the transitivity condition), so $(E, Q) \cong (E', Q')$.

***Corollary 3*** *Suppose, in Corollary 2, that $F$ is algebraically closed. If* $\mathrm{rank}(E) \geq 2d + 1$ *then* $(E, Q) \cong (E', Q')$.

Indeed, since $\dim E_0 \geq 2d + 1$ and $F$ is algebraically closed, we must have $\mathrm{WI}(E_0, Q_0) \geq d$, so Corollary 3 results from Corollary 2.

We shall prove the Cancellation Theorem in Section 4. Under hypothesis I, the Cancellation Theorem is a special case of [4, Ch. IV, Cor. (3.6)].

*Remark 1* The linear Cancellation Theorem [2, Ch. IV, (3.5)] is a companion to Serre's stability theorem [2, Ch. IV, Th. 2.5] saying, when $\max(A)$ is a noetherian space of dimension $\leq d$, that $A$-spaces of rank $\geq d + 1$ contain unimodular elements. It is striking that one apparently knows no analogue of Serre's result for quadratic $A$-spaces. One might conjecture that a quadratic $A$-space $(E, Q)$ contains a hyperbolic plane if $(E, Q)$ is "sufficiently isotropic." One reasonable sense to give the latter expression is that, for each maximal ideal $\mathfrak{m}$ of $A$, the Witt index of $(E_\mathfrak{m}, Q_\mathfrak{m})$ is $\geq d + 1$. In this form the conjecture is true, and easily proved, if $A$ is a Dedekind domain. However it is false in general, even if $A$ is semilocal, as the following example shows. Let $B$ be an integral domain such that $\bar{B} = B/\mathrm{rad}(B)$ is a product $K_0 \times K_1$ of two fields. Define $A$ by the cartesian square

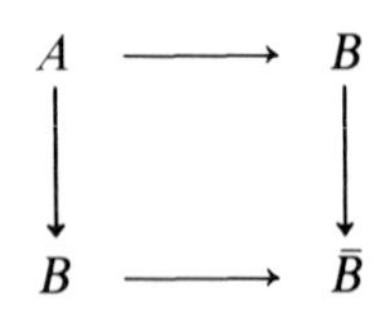

We can build a quadratic $A$-space $(E, Q)$ from the cartesian square

$$\begin{array}{ccccc} (E, Q) & \longrightarrow & & & H_Q(B) \\ \downarrow & & & & \downarrow \\ H_Q(B) & \longrightarrow & H_Q(\bar{B}) & \xrightarrow{\sigma} & H_Q(\bar{B}) \end{array}$$

where the matrix of $\sigma$, relative to the decomposition $\bar{B} = K_0 \times K_1$, is $(\begin{pmatrix} 1 & 0 \\ 0 & 1 \end{pmatrix}, \begin{pmatrix} 0 & 1 \\ 1 & 0 \end{pmatrix})$. Then $A$ is a semilocal ring with no nontrivial idempotents, and $(E, Q)$ and $H_Q(A)$ are locally isomorphic, but not isomorphic. These assertions are not difficult to verify. On the other hand I know of no counterexample to the above conjecture when $A$ is an integral domain. Another possibility is to weaken the conjecture by assuming the Witt index of $(E, Q)$ is large not just locally, but semilocally.

*Remark 2* Just as the Linear Cancellation Theorem has been sharpened, in the case of polynomial rings, by Suslin, Swan, Vaserštein, ... (cf. [6]), there should be an analogous sharpening of the above quadratic version.

## 3. Quadratic spaces of low rank

Throughout this section $A$ denotes an integral domain with field of fractions $K$.

The inner product $A$-spaces $(L, \varphi)$ of rank 1 are stable under tensor product. This gives to the set $\mathrm{Discr}(A)$ of their isomorphism classes $[L, \varphi]$ the structure of a group of exponent 2, which fits into an exact sequence

$$A^\times \xrightarrow{2} A^\times \xrightarrow{f} \mathrm{Discr}(A) \xrightarrow{g} \mathrm{Pic}(A) \xrightarrow{2} \mathrm{Pic}(A), \tag{1}$$

where $g[L, \varphi] = [L]$, and where, for $u \in A^\times$, $f(u) = \langle u \rangle = [A, \varphi_u]$, where $\varphi_u(x, y) = uxy$ (cf. [5, (2.1)]). If $(E, \varphi)$ is an inner product $A$-space of rank $r$ then $(\det E, \det \varphi) = (\Lambda^r E, \Lambda^r \varphi)$ is one of rank 1, whose class we denote

$$\det(E, \varphi) \in \mathrm{Discr}(A).$$

(See [5, (2.1)], or [12, (2.4)], where $\mathrm{Discr}(A)$ is denoted $Q(A)$.) It is proved in [5, (2.6)] that

$$\mathrm{Discr}(A) \to \prod_{\mathfrak{m}} \mathrm{Discr}(A_{\mathfrak{m}}) = \prod_{\mathfrak{m}} A_{\mathfrak{m}}^{\times}/A_{\mathfrak{m}}^{\times 2} \tag{2}$$

is injective, where $\mathfrak{m}$ varies over all maximal ideals of $A$, and that

$$\mathrm{Discr}(A) \to \mathrm{Discr}(K) = K^\times/K^{\times 2} \tag{3}$$

is injective if $A$ is integrally closed.

*Example 3.1* Let $P$ be an $A$-space of rank $r$. Then det $H_1(P) = \langle(-1)^r\rangle$. This is immediate when $P = A^r$, and follows in general by localizing and using the injectivity of (2).

Let $(E, Q)$ be a quadratic $A$-space of rank $r$. We shall attempt to describe such $(E, Q)$'s when $r$ is small. First note that if $r$ is odd then we must have $2 \in A^\times$. For otherwise there is a homomorphism from $A$ to a field $k$ of characteristic 2, and then $k \otimes_A (E, \varphi_Q)$ is a symplectic $k$-space of dimension $r$, whence $r$ would have to be even. Secondly we recall that, when $2 \in A^\times$, $(E, Q)$ and $(E, \varphi_Q)$ determine each other. It follows that, if $2 \in A^\times$, the quadratic $A$-spaces of rank 1 are parametrized, up to isomorphism, by Discr$(A)$.

*Example 3.2* If $A$ is factorial (a UFD) and $2 \in A^\times$ then every quadratic $A$-space of rank 1 is isomorphic to $\langle u\rangle_Q$ for some $u \in A^\times$, determined uniquely modulo $A^{\times 2}$. This follows because Pic$(A) = 0$ when $A$ is factorial. This applies notably when $A = F[X_1, \ldots, X_d]$ with $F$ a field of characteristic $\neq 2$. Then $A^\times = F^\times$ and we see that all quadratic $A$-space of rank 1 are extended from $F$.

***Proposition 3.3*** *Let* $(E, Q)$ *be a quadratic* $A$*-space of rank* 2. *Consider the conditions:*

(a) $(E, Q) \cong H_Q(L)$ *for some* $A$*-space* $L$ *of rank* 1.
(b) $(E, Q)$ *is stably isomorphic to* $H_Q(L)$ *for some* $A$*-space* $L$ *of rank* 1.
(c) $(E, Q)$ *is locally isomorphic to* $H_Q(A)$ *at each maximal ideal of* $A$.
(d) $K \otimes_A (E, Q) \cong H_Q(K)$. (Recall that $K$ is the field of fractions of $A$.)
(e) $\det(E, \varphi_Q) = \langle -1\rangle$

*Conditions* (a)–(c) *are equivalent and imply* (d) *and* (e). *If* $2 \in A^\times$ *they are equivalent to* (e). *If* $A$ *is integrally closed then* (d) *implies* (e).

The implications (a) ⇒ (b) and (c) ⇒ (d) are obvious; (a) ⇒ (e) follows from Example 3.1, and (b) ⇒ (c) follows from Corollary 1 of the Cancellation Theorem (Section 2). To prove (c) ⇒ (a), note first that $K \otimes_A (E, Q) \cong H_Q(K)$ contains exactly two isotropic lines (where $Q$ vanishes); let $L$ and $L'$ denote their intersections with $E \subset K \otimes_A E$. It suffices to show that $E = L \oplus L'$. But this results from its validity locally, where $L_\mathfrak{m}$ and $L_\mathfrak{m}'$ must be *the* two isotropic lines of which $(E_\mathfrak{m}, Q_\mathfrak{m}) \cong H_Q(A_\mathfrak{m})$ is the direct sum.

To prove (e) ⇒ (c) when $2 \in A^\times$ we may assume $A$ is local, hence $E$ is free. We can choose a basis $e, f$ of $E$ so that the matrix $\left(\begin{smallmatrix} a & b \\ b & c \end{smallmatrix}\right)$ of $\varphi_Q$ satisfies $ac - b^2 = -1$. Write $x \cdot y = \varphi_Q(x, y)$ for $x, y \in E$. Since $(b + 1) - (b - 1) = 2 \in A^\times$ and $A$ is local we can choose $t = \pm 1$ so that $u = 2(b + t) \in A^\times$. Let $e' = (b + t)e - af$ and $f' = -ce + (b + t)f$. Then

$$e' \cdot e' = (b + t)^2 a - 2a(b + t)b + a^2 c = a(b + t)^2 - 2(b + t)b + (b^2 - 1)$$

$$= 0,$$

and similarly $f' \cdot f' = 0$. Further

$$\begin{aligned} e' \cdot f' &= -(b+t)ac - (b+t)ac + ((b+t)^2 + ac)b \\ &= -2(b+t)(b^2-1) + ((b+t)^2 + (b^2-1))b = 2(b+t) = u. \end{aligned}$$

Hence $e', u^{-1}f'$ is a hyperbolic basis for $(E, \varphi_Q)$; since $2 \in A^\times$ it follows that $(E, Q) \cong H_Q(A)$.

*Remark* The implication just proved is contained in [3, Prop. (5.1)] from which we have adapted the above proof.

Finally, the implication (d) $\Rightarrow$ (e) when $A$ is integrally closed follows from the injectivity of (3). This proves the proposition.

**Corollary** *Suppose $A$ is a polynomial algebra $F[X_1, \ldots, X_d]$ where $F$ is a field of characteristic $\neq 2$. Let $(E, Q)$ be a quadratic $A$-space of rank $r$, let $(E_0, Q_0) = F \otimes_A (E, Q)$, and let $(E', Q') = A \otimes_F (E_0, Q_0)$. If $r = 1$ then $(E, Q) \cong (E', Q')$. If $r = 2$ and if $(E_0, Q_0) \cong H_Q(F)$ (this is automatic if $F$ algebraically closed) then $(E, Q) \cong (E', Q') \cong H_Q(A)$.*

The case $r = 1$ was noted above (Example 3.2). In case $r = 2$ the assertion follows from the proposition ((b) $\Rightarrow$ (a)) since Karoubi's theorem (see Section 1) implies that $(E, Q)$ and $(E', Q') \cong H_Q(A)$ are stably isomorphic.

*Remark* Parimala has pointed out that $(E, Q) \cong (E', Q')$ always when $r = 2$. The corollary above implies this is so after extension to the separable closure $F_s$ of $F$. Therefore the $(E, Q)$'s as above are classified by the galois cohomology set $H^1(G, O_2(F_s[X_1, \ldots, X_d]))$ where $G = \mathrm{Gal}(F_s/F)$ and $O_2(B) = \mathrm{Aut}(H_Q(B))$ for any ring $B$. The point now is that $O_2(F_s[X_1, \ldots, X_d]) = O_2(F_s)$, whence the result.

To treat quadratic $A$-spaces of higher ranks we shall use non-abelian étale cohomology (see [9] or [8, Ch. III]). If $G$ is a sheaf of groups on the étale topology on spec$(A)$, we write

$$H^i(G) = H^i_{\mathrm{et}}(\mathrm{spec}(A), G), \tag{4}$$

the $i$th étale cohomology with coefficients in $G$. If $G$ is non-abelian then this is defined only for $i = 0, 1$, and $H^1(G)$ is only a pointed set, not a group.

**Low Rank Theorem** *Let $F$ be an algebraically closed field of characteristic $\neq 2$, and let $A = F[X_1, \ldots, X_d]$, a polynomial $F$-algebra in $d$ variables. Every quadratic $A$-space of rank $\leq 6$ is extended from $F$.*

Combining this theorem with Corollary 3 of the Cancellation Theorem, we obtain the following corollary, which contains our main result, Theorem 1 at the end of Section 1.

**Corollary** *If $d \leq 3$ then every quadratic $A$-space is extended from $F$.*

To prove the Low Rank Theorem consider, for each integer $r \geq 0$, the quadratic $A$-space $(A^r, Q_r)$ defined as follows: If $r = 2s$ is even,

$$Q_{2s}(x_{-s}, \ldots, x_{-1}, x_1, \ldots, x_s) = x_1 x_{-1} + \cdots + x_s x_{-s}.$$

If $r = 2s + 1$ is odd,

$$Q_{2s+1}(x_{-s}, \ldots, x_{-1}, x_0, x_1, \ldots, x_s) = -x_0{}^2 + Q_{2s}(x_{-s}, \ldots, x_s).$$

Thus $(A^{2s}, Q_{2s}) \cong H_Q(A^s)$, *and* $(A^{2s+1}, Q_{2s+1}) \cong H_Q(A^s) \oplus \langle -1 \rangle_Q$. Let $O_r$ denote the (orthogonal) group scheme over spec$(A)$ of automorphism of $(A^r, Q_r)$; its points $O_r(B)$ in an $A$-algebra $B$ form the orthogonal group $O(B \otimes_A (A^r, Q_r))$.

Let $(E, Q)$ be a quadratic $A$-space of rank $r$. Let $(E_0, Q_0) = F \otimes_A (E, Q)$ and $(E', Q') = A \otimes_F (E_0, Q_0)$. Since $F$ is algebraically closed, quadratic $F$-spaces of the same dimension are isomorphic, so we have $(E', Q') \cong (A^r, Q_r)$. By Karoubi's theorem $(E, Q)$ and $(E', Q')$ are stably isomorphic, so Corollary 1 to the Cancellation Theorem implies that $(E, Q)$ and $(A^r, Q_r)$ are locally isomorphic in the Zariski topology on spec$(A)$, a fortiori also in the étale topology. Now the quadratic $A$-spaces isomorphic to $(A^r, Q_r)$ locally in the étale topology on spec$(A)$ are parametrized up to isomorphism, by the pointed set $H^1(O_r)$ $(= H^1_{\text{et}}(\text{spec}(A), O_r))$. (See [8, Ch. III, §5, No. 2].) Thus our aim is to prove that $H^1(O_r) = 0$.

We have the following standard sequences which, since $2 \in A^\times$, are exact in the étale topology on spec$(A)$:

$$1 \longrightarrow SO_r \longrightarrow O_r \xrightarrow{\det} \mu_2 \longrightarrow 1, \tag{5}$$

$$1 \longrightarrow \mu_2 \longrightarrow \text{Spin}_r \longrightarrow SO_r \longrightarrow 1, \tag{6}$$

$$1 \longrightarrow \mu_2 \longrightarrow \mathbb{G}_m \xrightarrow{2} \mathbb{G}_m \longrightarrow 1. \tag{7}$$

The following lemma records some results of Auslander and Goldman [1], Grothendieck [10], and Hoobler [15].

***Lemma 3.4*** *Let* $p = \text{char}(F)$.

(a) *The Brauer group* Br$(A)$ *is a p-group if* $p > 0$, *and* Br$(A) = 0$ *if* $p = 0$.

(b) *The canonical injection from* Br$(A)$ *to* $H^2(\mathbb{G}_m)$ *is bijective if* $d \leq 2$, *and an isomorphism away from p-torsion in general.*

Assertion (a) follows, by induction on $d$, from [1, Th. 7.2 and Props. 7.6 and 7.7]. Assertion (b) follows from [10, II, Cor. 2.2] for $d \leq 2$, and from [15] for large $d$.

***Lemma 3.5*** $H^1(\mu_2) = 0$ *and* $H^2(\mu_2) = 0$.

Identifying $H^1(\mathbb{G}_m)$ with $\mathrm{Pic}(A)$, the cohomology exact sequence of (7) takes the form

$$1 \to \mu_2(A) \to A^\times \xrightarrow{2} A^\times \to H^1(\mu_2) \to \mathrm{Pic}(A) \xrightarrow{2} \mathrm{Pic}(A)$$
$$\to H^2(\mu_2) \to H^2(\mathbb{G}_m) \xrightarrow{2} H^2(\mathbb{G}_m) \to \cdots. \tag{8}$$

(In fact the first line of (8) is just the exact sequence (1) above, once we identify $H^1(\mu_2)$ with $\mathrm{Discr}(A)$.) Since $A = F[X_1, \ldots, X_d]$ is factorial, $\mathrm{Pic}(A) = 0$, and since $F$ is algebraically closed, $A^\times$ $(= F^\times) = A^{\times 2}$, whence $H^1(\mu_2) = 0$, and

$$H^2(\mu_2) = \mathrm{Ker}(H^2(\mathbb{G}_m) \xrightarrow{2} H^2(\mathbb{G}_m)).$$

In view of Lemma 3.4 and the fact that $\mathrm{char}(F) \neq 2$ we conclude that

$$H^2(\mu_2) = \mathrm{Ker}(\mathrm{Br}(A) \xrightarrow{2} \mathrm{Br}(A)) = 0.$$

Consider now the cohomology sequence of (5) (see [9, Ch. III, Prop. 3.3.1], or [8, Ch. III, §4, 5.1]):

$$\cdots \to H^1(SO_r) \to H^1(O_r) \to H^1(\mu_2).$$

Since $H^1(\mu_2) = 0$ (Lemma 3.5) the map $H^1(SO_r) \to H^1(O_r)$ is surjective, so it suffices to prove that $H^1(SO_r) = 0$. Next consider the cohomology sequence of (6) [9, Ch. IV, 4.2, 10]:

$$\cdots \to H^1(\mu_2) \to H^1(\mathrm{Spin}_r) \to H^1(SO_r) \to H^2(\mu_2).$$

Since $H^2(\mu_2) = 0$ (Lemma 3.5) we see that $H^1(\mathrm{Spin}_r) \to H^1(SO_r)$ is surjective, so it suffices to show that $H^1(\mathrm{Spin}_r) = 0$. Now for small values of $r$ the spin representations give canonical isomorphisms:

| | Dynkin diagrams |
|---|---|
| $\mathrm{Spin}_2 \cong \mathbb{G}_m$ | |
| $\mathrm{Spin}_3 \cong SL_2 \cong Sp_2$ | ○ |
| $\mathrm{Spin}_4 \cong SL_2 \times SL_2 \cong Sp_2 \times Sp_2$ | ○ ○ |
| $\mathrm{Spin}_5 \cong Sp_4$ | ○═○ |
| $\mathrm{Spin}_6 \cong SL_4$ | ○—○—○ |

The proof of the Low Rank Theorem is thus completed by the following lemma.

***Lemma 3.6*** *For all* $r \geq 1$, $H^1(GL_r) = 0$, $H^1(SL_r) = 0$, *and* $H^1(Sp_{2r}) = 0$.

Indeed, $H^1(GL_r)$ classifies projective $A$-modules of rank $r$ [8, Ch. III, §5, 1.2], so $H^1(GL_r) = 0$ by the Quillen–Suslin theorem. Similarly, $H^1(Sp_{2r})$ classifies symplectic $A$-modules of rank $2r$, so the corollary to the Quillen–Suslin theorem implies that $H^1(Sp_{2r}) = 0$. The cohomology sequence of the exact sequence

$$1 \longrightarrow SL_r \longrightarrow GL_r \xrightarrow{\det} \mathbb{G}_m \longrightarrow 1$$

contains the exact sequence

$$GL_r(A) \xrightarrow{\det} A^\times \longrightarrow H^1(SL_r) \longrightarrow H^1(GL_r) = 0$$

which shows that $H^1(SL_r) = 0$, whence the lemma.

*Remark* The proof of Lemma 3.6 uses no assumptions about the field $F$.

## 4. Proof of the Cancellation Theorem

Let $A$ be a ring. If $(E, Q)$ is a quadratic $A$-module we shall write

$$x \cdot y = \varphi_Q(x, y)$$

for $x, y \in E$. If $X$ is a subset of $E$ we put

$$X^\perp = \{y \in E \mid x \cdot y = 0\},$$

and $x^\perp = \{x\}^\perp$ if $x \in E$. If $Q(x) \in A^\times$ the *symmetry*

$$\sigma_x(y) = y - ((y \cdot x)/Q(x))x$$

sends $x$ to $-x$ and fixes the elements of $x^\perp$; $\sigma_x$ belongs to the (orthogonal) group $O(E, Q)$ of automorphisms of $(E, Q)$. For each $x, y \in E$ define $\tau_{x,y} \in \mathrm{End}_A(E)$ by

$$\tau_{x,y}(z) = z + (z \cdot x)y - (z \cdot y)x - (z \cdot x)Q(y)x; \tag{1}$$

note that if $\sigma \in O(E, Q)$ we clearly have

$$\sigma\tau_{x,y}\sigma^{-1} = \tau_{\sigma x, \sigma y}\,. \tag{2}$$

***Proposition 4.1*** (a) *If* $x, y \in E$ *satisfy*

$$Q(x) = 0 \qquad \text{and} \qquad x \cdot y = 0, \tag{3}$$

*then* $\tau_{x,y} \in O(E, Q)$.

(b) *Let* $x \in E$ *be isotropic, i.e.* $Q(x) = 0$. *Then* $y \mapsto \tau_{x,y}$ *is a homomorphism from* $x^\perp$ *to* $O(E, Q)$. *We have* $\tau_{ax,y} = \tau_{x,ay}$ *for* $a \in A$, $y \in x^\perp$, *and* $\tau_{x,x} = 1$.

(c) *Let $L$ be a totally isotropic submodule of $E$ (i.e. $Q$ vanishes on $L$). Then $(x, y) \mapsto \tau_{x,y}$ is an alternating bilinear map from $L \times L$ to $O(E, Q)$. If $y \in E$ and $L \subset y^\perp$ then, for $x, x' \in L$, we have $\tau_{x',y} \circ \tau_{x,y} = \tau_{x+x',y} \circ \tau_{x,Q(y)x'}$.*

Assuming (3), we have

$$Q(\tau_{x,y}(z)) - Q(z) = z \cdot ((z \cdot x)y - (z \cdot y)x - (z \cdot x)Q(y)x) \\ + Q((z \cdot x)y - (z \cdot y)x - (z \cdot x)Q(y)x) \\ = -(z \cdot x)^2 Q(y) + (z \cdot x)^2 Q(y) = 0,$$

whence (a). Suppose $Q(x) = 0$ and $y, y' \in x^\perp$. Let $z' = \tau_{x,y}(z)$. Then

$$\tau_{x,y'}(\tau_{x,y}(z)) = z' + (z' \cdot x)y' - (z' \cdot y')x - (z' \cdot x)Q(y')x,$$

which, since $z' \cdot x = z \cdot x$, and $z' \cdot y' = z \cdot y' + (z \cdot x)(y \cdot y')$, equals

$$z' + (z \cdot x)y' - (z \cdot y')x - (z \cdot x)(y \cdot y')x - (z \cdot x)Q(y')x \\ = z + (z \cdot x)(y + y') - (z \cdot y + y')x - (z \cdot x)qx,$$

where $q = Q(y) + y \cdot y' + Q(y') = Q(y + y')$. Thus $\tau_{x,y'} \circ \tau_{x,y} = \tau_{x,y+y'}$. The formulas $\tau_{ax,y} = \tau_{x,ay}$ and $\tau_{x,x} = 1$ are obvious, whence (b). Suppose $x, x'$, $y \in E$ satisfy $Q(x) = Q(x') = x \cdot x' = x \cdot y = x' \cdot y = 0$. Let $z' = \tau_{x,y}(z)$. Then

$$\tau_{x',y}(\tau_{x,y}(z)) = z' + (z' \cdot x')y - (z' \cdot y)x' - (z' \cdot x')Q(y)x',$$

which, since $z' \cdot x' = z \cdot x'$ and $z' \cdot y = z \cdot y + (z \cdot x)(y \cdot y)$, equals

$$z' + (z \cdot x')y - (z \cdot y)x' - (z \cdot x)(y \cdot y)x' - (z \cdot x')Q(y)x' \\ = z + (z \cdot x + x')y - (z \cdot y)(x + x') - w,$$

where

$$w = (z \cdot x)Q(y)x + (z \cdot x)2Q(y)x' + (z \cdot x')Q(y)x' \\ = (z \cdot x + x')Q(y)(x + x') + Q(y)((z \cdot x)x' - (z \cdot x')x).$$

Since $\tau_{x+x',y}$ fixes $x$ and $x'$ it follows that $\tau_{x',y} \circ \tau_{x,y} = \tau_{x+x',y} \circ \tau_{x,Q(y)x'}$. This gives the last assertion of (c), and, when $Q(y) = 0$, the first assertion as well.

The elements $\tau_{x,y}$, where $x, y$ satisfy (3), are called *transvections*. If $Y \subset x^\perp$ we write $\tau_{x,Y}$ for $\{\tau_{x,y} | y \in Y\}$. If $L$ is as in 4.1(c) we write $\tau_{L,L} = \{\tau_{x,y} | x, y \in L\}$, and $\tau_{L,y} = \{\tau_{x,y} | x \in L)$ if $L \subset y^\perp$.

Call $e, f \in E$ a *hyperbolic pair* if

$$Q(e) = 0 = Q(f), \qquad \text{and} \qquad e \cdot f = 1. \tag{4}$$

These conditions signify that the map $A^2 \mapsto E$, $(x, y) \mapsto xe + yf$, is a morphism $H_Q(A) \to (E, Q)$, whence $e, f$ is a basis for a hyperbolic plane

$H = Ae \oplus Af$ in $E$, and $E = H \oplus H^{\perp}$ (cf. [4, Ch. I, Prop. 3.2]). We call $e \in E$ a *hyperbolic element* of $(E, Q)$ if it belongs to a hyperbolic pair.

If $x \in E$ then

$$O(x) = O_E(x) = \{\langle x, y \rangle \mid y \in E^*\} \tag{5}$$

is an ideal of $A$, and $O(x) = A$ if and only if $x$ is a basis for a free direct summand of $E$, in which case we say $x$ is *unimodular* in $E$. We call $x$ *isotropic* if $Q(x) = 0$. Condition (4) shows that the hyperbolic elements are unimodular and isotropic. Suppose, conversely, that $e \in E$ is unimodular and isotropic. Then $\langle e, y \rangle = 1$ for some $y \in E^*$, so, if $\varphi_Q$ is invertible, $e \cdot f' = 1$ for some $f' \in E$, and then $e$ and $f = f' - Q(f')e$ form a hyperbolic pair. Thus, *when $\varphi_Q$ is invertible, the hyperbolic elements of $(E, Q)$ are exactly the unimodular isotropic elements of $(E, Q)$.*

***Proposition 4.2*** *Let $U$ denote the set of hyperbolic elements of $(E, Q)$.*

(a) *If $e \in U$ then $\tau_{e, e^{\perp}}$ acts transitively on the set of $f \in E$ such that $e, f$ is a hyperbolic pair.*

(b) *Let $G$ be a subgroup of $O(E, Q)$. Suppose $G$ acts transitively on $U$ and $\tau_{e, e^{\perp}} \subset G$ for some $e \in U$. Then $\tau_{e, e^{\perp}} \subset G$ for every $e \in U$, and $G$ acts transitively on the set of hyperbolic pairs and the set of hyperbolic planes in $(E, Q)$.*

Let $e, f$ and $e, f'$ be hyperbolic pairs. Write $H = Ae \oplus Af$ and $f' = f_H + g$ with $g \in H^{\perp}$. Then

$$\begin{aligned}\tau_{e, -g}(f') &= f' + (f' \cdot e)(-g) - (f' \cdot -g)e - (f' \cdot -g)Q(-g)e \\ &= f' - g + ce\end{aligned}$$

for some $c \in A$, so $f'' = \tau_{e, -g}(f') \in H$, say $f'' = ae + bf$. Since $\tau_{e, -g}(e) = e$ we have $1 = e \cdot f'' = b$. Further $0 = Q(f'') = a$, so $f'' = f$. This proves (a). To prove (b) suppose $e, e' \in U$ and $\tau_{e, e^{\perp}} \subset G$. By assumption $\sigma(e) = e'$ for some $\sigma \in G$, so $G$ contains $\sigma\tau_{e, e^{\perp}}\sigma^{-1} = \tau_{e', e'^{\perp}}$. Let $e, f$ and $e', f'$ be hyperbolic pairs. There is a $\sigma \in G$ such that $\sigma e' = e$. Then there is a $\tau \in \tau_{e, e^{\perp}} \subset G$ such that $\tau\sigma f' = f$, so $\tau\sigma \in G$ carries $e', f'$ to $e, f$. Hence $G$ acts transitively on the sets of hyperbolic pairs and planes in $(E, Q)$.

We shall say that $(E, Q)$ satisfies the *Witt Cancellation condition* if, for any quadratic $A$-module $(E', Q')$ and any quadratic $A$-space $(E_0, Q_0)$, we have

$$(E, Q) \oplus (E_0, Q_0) \cong (E', Q') \oplus (E_0, Q_0) \Rightarrow (E, Q) \cong (E', Q').$$

***Proposition 4.3*** *The following conditions on the quadratic $A$-module $(E, Q)$ are equivalent.*

(1) *For all $n \geq 0$, $(E, Q) \oplus H_Q(A^n)$ satisfies the Witt cancellation condition.*

(2) *For all* $n > 0$, $O((E, Q) \oplus H_Q(A^n))$ *acts transitively on the set of hyperbolic planes in* $(E, Q) \oplus H_Q(A^n)$.

(3) *For all* $n > 0$, $O((E, Q) \oplus H_Q(A^n))$ *acts transitively on the set of hyperbolic elements in* $(E, Q) \oplus H_Q(A^n)$.

The equivalence of (2) and (3) follows from Proposition 4.2*b*. To prove (1) $\Rightarrow$ (2) write $H_Q(A^n) = H_Q(A^{n-1}) \oplus H$ and let $H'$ be a hyperbolic plane in $(E, Q) \oplus H_Q(A^n)$. To find an isometry carrying $H'$ onto $H$ it suffices to show that $H'^{\perp}$ is isomorphic to $H^{\perp} = (E, Q) \oplus H_Q(A^{n-1})$. But this follows from (1). To prove (2) $\Rightarrow$ (1) suppose that $(E, Q) \oplus H_Q(A^n) \oplus (E_0, Q_0)$ is isomorphic to $(E', Q') \oplus (E_0, Q_0)$, where $(E', Q')$ is a quadratic $A$-module and $(E_0, Q_0)$ is a quadratic $A$-space. By [4, Ch. I, Cor. (4.9)], $(E_0, Q_0)$ is an orthogonal direct summand of $H_Q(A^m)$ for some $m \geq 0$, so we can assume without loss that $(E_0, Q_0) = H_Q(A^m)$. We prove that $(E, Q) \oplus H_Q(A^n) \cong (E', Q')$ by induction on $m$. The assertion is evident if $m = 0$; if $m > 0$ write $H_Q(A^m) = H_Q(A^{m-1}) \oplus H$ with $H \rightarrow H_Q(A)$. Using the given isomorphism as an identification, we have

$$(E, Q) \oplus H_Q(A^n) \oplus H_Q(A^{m-1}) \oplus H = (E', Q') \oplus H_Q(A^{m-1})' \oplus H'.$$

By (2) there is an isometry carrying $H'$ to $H$, hence inducing an isomorphism from $(E', Q') \oplus H_Q(A^{m-1})'$ to $(E, Q) \oplus H_Q(A^n) \oplus H_Q(A^{m-1})$. The conclusion now follows by induction on $m$.

In view of Proposition 4.3, the Cancellation Theorem follows from the next result.

***Theorem 4.4*** *Assume that the maximal ideal space* $\max(A)$ *is noetherian of dimension d. Let* $(F, R)$ *be a quadratic A-module which decomposes as*

$$(F, R) = (E', Q') \oplus (E, Q)$$

*where* $(E, Q)$ *satisfies* I *or* II:

(I) $(E, Q) = H_Q(L)$, *where L is an A-space of rank* $\geq d + 2$.

(II) $2 \in A^{\times}$ *and* $(E, Q) = H_Q(L) \oplus \langle u \rangle_Q$, *where L is an A-space of rank* $\geq d + 1$, *and* $u \in A^{\times}$; *further,* $GL(L)$ *acts transitively on the set of unimodular elements of L.*

*Then* $O(F, R)$ *acts transitively on the set of hyperbolic planes in* $(F, R)$.

We have $F = E' \oplus E$ and $E = L \oplus L^* $ (case I) or $L \oplus L^* \oplus Ag$ (case II) with $Q(g) = u$. If $x \in F$ write $x = x_{E'} + x_E$ with $x_{E'} \in E'$ and $x_E \in E$, and similarly write $x_E = x_L + x_{L*}$ (case I) or $x_E = x_L + x_{L*} + x_g g$ (case II). Since rank $L \geq d + 1$ (in both cases), a theorem of Serre (see [2, Ch. IV, Th. 2.5]) implies that $L$ contains a unimodular element $e$. Choosing $f \in L^*$ so that $e \cdot f = 1$ we obtain a hyperbolic pair $e, f$ of $H_Q(L)$. In view of Proposition 4.3, the theorem will be proved if any hyperbolic element of $(F, R)$ can be transformed to $e$ by an element of $O(F, R)$. We shall identify $O(E, Q)$ with the subgroup of $O(F, R)$ fixing the elements of $E'$.

Let $x \in F$.

*Assertion 1* If $x_E$ is unimodular, there is a $\sigma \in O(E, Q)$ such that $\sigma(x)_L$ is unimodular.

This is a consequence of Theorem 4.8; we admit it for the present proof.

*Assertion 2* If $x$ is a hyperbolic element of $(F, R)$ there is a $\sigma \in O(F, R)$ such that $\sigma(x) \in E$.

Suppose first that $x_E$ is unimodular. Then, by Assertion 1, we can further arrange that $x_L$ is unimodular. We can then find $y \in L$ so that $x_L \cdot y = -1$, whence

$$\begin{aligned}\tau_{y, x_{E'}}(x) &= x + (x \cdot y)x_{E'} - (x \cdot x_{E'})y - (x \cdot y)Q(x_{E'})y \\ &= x - x_{E'} + ay\end{aligned}$$

for some $a \in A$. Then $\tau_{y, x_{E'}}(x) = x_E + ay \in E$, thus proving our assertion. We reduce the general case to the preceding one by constructing a $\tau \in O(F, R)$ such that $\tau(x)_E$ is unimodular. To do this consider a $y \in F$ such that $x$, $y$ is a hyperbolic pair; then $1 = x \cdot y = a + x_E \cdot y_E$, where $a = x_{E'} \cdot y_{E'}$. We see then that $A = Aa + O(x_E)$. Putting

$$J = \begin{cases} O(x_{L^*}) & \text{(case I)} \\ O(x_{L^*} + x_g g) & \text{(case II)} \end{cases}$$

we then have $A = Aa + O(x_L) + J$. Since rank $L \geq d + 1$ (in both cases) it follows from [2, Ch. IV, Th. 3.1] that there is a $z \in L$ such that

$$O(x_L - az) + J = A. \tag{6}$$

Consider now $x' = \tau_{z, y_{E'}}(x) = x + (x \cdot z)y_{E'} - (x \cdot y_{E'})z - (x \cdot z)Q(y_{E'})z$. We have $x_L' = x_L - az - (y_L \cdot z)Q(y_{E'})z$, $x'_{L^*} = x_{L^*}$, and, in case II, $x_g' = x_g$. It follows that $O(x_L') + O('_{L^*}) = O(x_L - az) + O(x_{L^*})$, whence, in view of (6) $O(x_E') = $ A. This proves Assertion 2.

By virtue of Assertion 2 the theorem now follows from:

*Assertion 3* Suppose $x \in E$ is a hyperbolic element. Then there is a $\sigma \in O(E, Q)$ such that $\sigma(x) = e$ (the above chosen unimodular element of $L$).

By Assertion 1 we may assume that $x_L$ is unimodular. Now $GL(L)$ acts transitively on the set of unimodular elements in $L$; this is our hypothesis in case II, and in case I it follows from [2, Ch. IV, Th. 3.4]. Recall that we have chosen a hyperbolic pair $e, f$ with $e \in L$ and $f \in L^*$. Choose $\alpha \in GL(L)$ so that $\alpha(x_L) = e$. The automorphism $H(\alpha) = \alpha \oplus \alpha^{*-1}$ belongs to $O(H_Q(L))$ (cf. [4, Ch. I, 4.7]). Replacing $x$ by $H(\alpha)(x)$ we may therefore assume that $x_L = e$. It follows that $x \cdot f = 1$, so $x, f$ is a hyperbolic pair. Now Proposition 4.2a implies that there is a $y \in f^\perp$ such that $\tau_{f, y}(x) = e$. This proves Theorem 4.4, modulo Theorem 4.8.

For the rest of this section $(E, Q)$ will denote one of the following types of quadratic $A$-spaces, where $L$ denotes an $A$-space:

(I) $(E, Q) = H_Q(L)$,
(II) $(E, Q) = H_Q(L) \oplus \langle u \rangle_Q$, with $2u \in A^\times$.

Let $\langle u \rangle_Q = (Ag, Q_u)$ with $Q_u(g) = u$; then $E = L \oplus L^* \oplus Ag$ in case II. We define a subgroup

$$EO(E, Q) \subset O(E, Q)$$

as follows. In case I it is the group generated by all transvections $\tau_{e, e^\perp} = \{\tau_{e, x} \mid x \in e^\perp\}$, where $e$ varies over all unimodular elements of $L$ and of $L^*$. In case II it is the group generated by all transvections $\tau_{x, g}$ with $x \in L$ or $x \in L^*$, i.e. by $\tau_{L, g} \cup \tau_{L^*, g}$.

***Lemma 4.5*** *Suppose we are in case* II.

(a) *Let $e \in L$ be unimodular and choose $f \in L^*$ so that $e \cdot f = 1$. Then $\omega = \tau_{e, g} \circ \tau_{f, u^{-1}g} \circ \tau_{e, g} \in EO(E, Q)$ and $\omega(e) = -u^{-1}f$, $\omega(f) = -ue$, $\omega(g) = -g$.*

(b) *$EO(H_Q(L))$, viewed as a subgroup of $O(E, Q)$ fixing $g$, is contained in $EO(E, Q)$.*

In the free module with basis $e, f, g$, the formula in (a) corresponds to the matrix equation

$$\begin{pmatrix} 1 & -u & -2u \\ 0 & 1 & 0 \\ 0 & 1 & 1 \end{pmatrix} \begin{pmatrix} 1 & 0 & 0 \\ -u^{-1} & 1 & -2 \\ u^{-1} & 0 & 1 \end{pmatrix} \begin{pmatrix} 1 & -u & -2u \\ 0 & 1 & 0 \\ 0 & 1 & 1 \end{pmatrix}$$

$$= \begin{pmatrix} 0 & -u & 0 \\ -u^{-1} & 0 & 0 \\ 0 & 0 & -1 \end{pmatrix}.$$

The calculation is left to the reader.

To prove (b), let $e, f$ be as in (a), so that $H = Ae \oplus Af$ is a hyperbolic plane, and, if $L = Ae \oplus M$, then $H_Q(L) = H \oplus H_Q(M)$. We then have $e^\perp = L \oplus M^*$. Since $x \mapsto \tau_{e, x}$ is a homomorphism from $e^\perp$ to $O(H_Q(L))$ we have $\tau_{e, e^\perp} = \tau_{e, L} \circ \tau_{e, M^*}$. If $x \in e^\perp$ we have (Proposition 4.1c),

$$\tau_{e, g} \circ \tau_{x, g} = \tau_{e+x, g} \circ \tau_{e, ux}. \tag{7}$$

When $x \in L$ this shows that $\tau_{e,ux} \in EO(E, Q)$, whence $\tau_{e,L} \subset EO(E, Q)$. Suppose $x \in M^*$. Note that $\omega$, in part (a), fixes $x$, so that

$$\omega\tau_{e,ux}\omega^{-1} = \tau_{\omega e,\omega ux} = \tau_{-u^{-1}f,ux} = \tau_{f,-x} \in \tau_{f,L^*},$$

and, just as above for $\tau_{e,L}$, we have $\tau_{f,L^*} \subset EO(E, Q)$. The lemma now follows easily from these observations.

***Lemma 4.6*** *Let $A \to A'$ be a surjective ring homomorphism. The natural homomorphism $EO(E, Q) \to EO(A' \otimes_A (E, Q))$ is surjective, provided that, in case* I, *every unimodular of $A' \otimes_A L$ is the image of a unimodular element of L. The latter condition holds if* $\max(A')$ *is a noetherian space of dimension $d'$,* $\operatorname{rank}(L) \geq d' + 2$, *and L contains a unimodular element.*

Use a prime to denote the image of an object under the base change $A' \otimes_A$. In case II, $\tau_{F',g'}$ is the image of $\tau_{F,g}$ for $F = L$ or $L^*$, so the lemma is clear in case II. In case I let $e'$ be unimodular in $L'$ or $L'^*$, say in $L'$. By assumption $e'$ is the image of some unimodular $e \in L$. It follows easily from the unimodularity of $e$ that $(e^\perp)' = (e')^\perp$, whence the surjectivity of $\tau_{e,e^\perp} \mapsto \tau_{e',(e')^\perp}$. Since the hypothesis on lifting unimodular elements of $L'$ easily implies the same property for $L'^*$, this proves the surjectivity in case I.

To show the liftability of unimodular elements, suppose $L$ has one, say $e$. Then we can write $L = Ae \oplus M$ and $L' = A'e' \oplus M'$. Let $E(Ae, M) \subset GL(L)$ and $E(A'e', M') \subset GL(L')$ be the subgroups defined as in [2, Ch. IV, p. 182], where it is shown (Proposition 3.3) that the homomorphism $E(Ae, M) \to E(A'e', M')$ is surjective. Let $f' \in L'$ be a unimodular element that we wish to lift. Since $\operatorname{rank}_A(L) \geq d' + 2$ the same follows for $\operatorname{rank}_{A'}(L')$, so [2, Ch. IV, Th. 3.4] implies that $\sigma'e' = f'$ for some $\sigma' \in E(A'e', M')$. If $\sigma \in E(Ae, M)$ lifts $\sigma'$ then $f = \sigma e$ is the required unimodular lifting of $f'$. This proves Lemma 4.6.

***Lemma 4.7*** *Suppose $A = A_1 \times \cdots \times A_n$ is a product of rings. Let $(E, Q) = (E_1, Q_1) \times \cdots \times (E_n, Q_n)$ be the corresponding decomposition of $(E, Q)$. Then*

$$EO(E, Q) = EO(E_1, Q_1) \times \cdots \times EO(E_n, Q_n).$$

Clearly $O(E, Q)$ is the product of the $O(E_i, Q_i)$'s, so it suffices to observe that $EO(E, Q)$ is generated by elements in the factors $EO(E_i, Q_i)$. In case I, this follows because, if $e$ is unimodular in $L$ or $L^*$, then $\tau_{e,e^\perp}$ is generated by the $\tau_{e,x}$, where $x$ varies over additive generators of $e^\perp$. In case II, write $L = L_1 \times \cdots \times L_n$. If $x \in L_i$ and $x' \in L_j$ with $i \neq j$ then $\tau_{x,g}\tau_{x',g} = \tau_{x+x',g}\tau_{x,ux'}$ (Proposition 4.1c), and clearly $\tau_{x,ux'} = 1$. Thus we see that $\tau_{L,g}$ is generated by the $\tau_{L_i,g} \subset EO(E_i, Q_i)$, and similarly for $\tau_{L^*,g}$. This proves the lemma.

***Theorem 4.8*** *Assume that* $\max(A)$ *is a noetherian space of dimension* $d$. *Let* $(E, Q)$ *be as above, and assume that the rank of* $L$ *is* $\geq d+2$ *in case* I, *and* $\geq d+1$ *in case* II. *Let* $x$ *be a unimodular element of* $E$. *Then there is a* $\sigma \in EO(E, Q)$ *such that the component of* $\sigma(x)$ *in* $L$ *is unimodular.*

We have $E = L \oplus L^*$ (I), or $L \oplus L^* \oplus Ag$ (II). For any $y \in E$ we accordingly write $y = y_L + y_{L*}$ (I) or $y = y_L + y_L + y_g g$ (II), with $y_L \in L$, $y_{L*} \in L$, $y_g \in A$. The conclusion of the theorem is that $\sigma(x)_L$ is unimodular. Since $\operatorname{rank}(L) \geq d+1$ (in both cases), Serre's theorem [2, Ch. IV, Th. 2.5] implies that $L$ contains a unimodular element $e$. Choose $f \in L^*$ so that $e \cdot f = 1$, hence $H = Ae \oplus Af$ is a hyperbolic plane, and we can write $H_Q(L) = H \oplus H_Q(M)$, where $L = Ae \oplus M$. We extend the notation above by writing $y_L = y_e e + y_M$ and $y_{L*} = y_f f + y_{M*}$.

*Assertion 1* If $A$ is a field there is a $\sigma \in EO(E, Q)$ such that $\sigma(x)_e \neq 0$.

If $x_e \neq 0$ we can take $\sigma = 1$, so suppose $x_e = 0$. Consider case II first. Let $a \in A$ and $x' = \tau_{e, ag}(x)$. Then $x_e' = -2x_g ua - x_f ua^2$. Since $\operatorname{char}(A) \neq 2$ (whence $\operatorname{card}(A) \geq 3$), we can choose $a$ so that $x_e' \neq 0$ provided that $x_g \neq 0$ or $x_f \neq 0$. Otherwise we have $x \in H_Q(M)$ so $M \neq 0$ and $\dim L \geq 2$. This case (thanks to Lemma 4.5b) is subsumed in case I, which we now treat. Let $y \in e^\perp$ and $x' = \tau_{e, y}(x)$. Then $x_e' = -x \cdot y - x_f Q(y)$. If this always vanishes then, taking $y \in L$ or $y \in M^*$, so that $Q(y) = 0$, we see that $x \in (e^\perp)^\perp = Ae$, so $x_e \neq 0$. This proves Assertion 1.

*Assertion 2* Let $S$ be a finite subset of $\max(A)$ meeting each irreducible component of $\max(A)$, and put $J = \bigcap_{\mathfrak{m} \in S} \mathfrak{m}$. There is a $\sigma \in EO(E, Q)$ such that $A\sigma(x)_e + J = A$.

Passing from $A$ to $A' = A/J$, and using Lemma 4.6, we can reduce to the case when $J = 0$. Then $A$ is a finite product of fields, so by Lemma 4.7 we can reduce to the case when $A$ is a field, whereupon the problem is solved by Assertion 1.

By virtue of Assertion 2, we may henceforth assume that

$$Ax_e + J = A. \tag{8}$$

We shall write

$$(E_1, Q_1) = H^\perp = (Ae \oplus Af)^\perp,$$

which is $H_Q(M)$ (case I) or $H_Q(M) \oplus \langle u \rangle_Q$ (case II), and $E = Ae \oplus Af \oplus E_1$. For $y \in E$ we write $y = y_e e + y_f f + y_{E_1}$, as usual.

*Assertion 3* There is a $y \in M$ such that $z = \tau_{e, y}(x)$ satisfies $Az_e + J = A$ and $Az_e + O(z_{E_1}) = A$.

Since $x$ is unimodular, we have

$$A = O(x) = Ax_e + Ax_f + O(x_M) + O(x_{M*}) \qquad (+Ax_g \text{ in case II}).$$

Let $I = Ax_e + O(x_{M*})$ $(+Ax_g$ in case II), so that

$$A = Ax_f + O(x_M) + I. \tag{9}$$

Let $A' = A/I$, and use a prime to denote reduction modulo $I$. Then $x_M{}'$ in the $A'$-module $M'$ satisfies $A'x_f{}' + O(x_M{}') = A'$, by (9). Since $x_e \in I$ it follows from (8) (cf. Assertion 2) that $d' = \dim\max(A') < \dim\max(A) \leq d$. We have $\operatorname{rank}_{A'}(M') = \operatorname{rank}_A(L) - 1 \geq d \geq d' + 1$, so it follows from [2, Ch. IV, Th. 3.1] that there is a $y' \in M'$ such that $O(x_M{}' + x_f{}'y') = A'$. Since $x_e \in I$ it follows from (8) that $I + J = A$. Hence we can lift $y' \in M'$ to an element $y \in JM$, and we then have

$$O(x_M + x_f y) + I = A. \tag{10}$$

We claim $y$ meets our requirements. In fact let

$$\begin{aligned} z = \tau_{e,y}(x) &= x + (x \cdot e)y - (x \cdot y)e \\ &= (x_e - (x_{M*} \cdot y))e + x_f f + (x_M + x_f y) + x_{M*} \\ &\qquad (+ x_g g \text{ in case II}). \end{aligned}$$

Then

$$z_e = x_e - (x_M \cdot y) \equiv x_e \bmod J$$

since $y \in JM$, so (8) implies that $Az_e + J = A$. We have $z_{E_1} = z_M + x_{M*}(+x_g g$ in case II) and $z_e = x_e \bmod O(x_{M*}) \subset O(z_{E_1})$. Thus

$$Az_e + O(z_{E_1}) = Ax_e + O(z_{E_1}) = O(z_M) + I.$$

Since $z_M = x_M + x_f y$ it follows from (10) that $Az_e + O(z_{E_1}) = A$, thus proving Assertion 3.

*Proof of Theorem 4.8* Let $J$ be as in Assertion 2. By virtue of Assertion 3 we may assume that

$$Ax_e + J = A \tag{11}$$

and

$$Ax_e + O(x_{E_1}) = A. \tag{12}$$

We shall prove the theorem by induction on $d$. If $d = 0$ then $A$ is semilocal, $J = \operatorname{rad}(A)$, and (11) implies that $x_e \in A^\times$, in which case the theorem is proved. If $x_e \notin A^\times$ let $A' = A/Ax_e$ and use a prime to denote reduction modulo $Ax_e$. Condition (11) and the definition of $J$ imply that

$$d' = \dim \max(A') < \dim \max(A) \leq d.$$

In view of (12) therefore we can apply our induction hypothesis to the unimodular element $x'_{E_1}$ in the quadratic $A$-space $(E_1{}', Q_1{}')$. The result is an element $\sigma' \in EO(E_1{}', Q_1{}')$ such that $\sigma'(x'_{E_1})_{M'}$ is unimodular. Since $\text{rank}_A(L) \geq d + 2$ in case I we have $\text{rank}_{A'}(M') \geq d' + 2$ in case I and $M$ then contains a unimodular element, by [2, Ch. IV, Th. 2.5]. It follows therefore from Lemma 4.6 that $\sigma'$ lifts to an element $\sigma$ of $EO(E_1, Q_1)$; we identify $O(E_1, Q_1)$ with the subgroup of $O(E, Q)$ fixing $e$ and $f$. If $x' = \sigma(x)$ we have $O(x_L{}') = O(x_e e + \sigma(x_{E_1})_M) = Ax_e + O(\sigma(x_{E_1})_M) = A$, as we see by reducing modulo $Ax_e$. This proves Theorem 4.8.

## 5. Proof of Karoubi's theorem

Karoubi's theorem (in Section 1) clearly follows from the following formulation.

***Theorem*** *Let $A$ be a ring in which 2 is invertible. Let $B = A[T]$, where $T$ is an indeterminate, and let $(E, \varphi)$ be an inner product $B$-space. If $E$ is stably extended from $A$ then $(E, \varphi)$ is stably extended from $A$.*

*Remark* The invertibility of 2 is used in two ways. One is to know that $\varphi$ is "even," i.e. $\varphi = \psi + {}^t\psi$ for some bilinear form $\psi$, where ${}^t\psi(x, y) = \psi(y, x)$; it suffices to take $\psi = \varphi/2$. However, it is conceivable that evenness of $\varphi$ suffices in the theorem, even if 2 is not invertible. The other, more serious, use of $\frac{1}{2}$ is to construct square roots of unipotent elements, a step we do not now know how to bypass.

We wish to show that, by adding extended inner product spaces to $(E, \varphi)$, we can make $(E, \varphi)$ extended. Let $(E_0, \varphi_0) = A \otimes_B (E, \varphi)$ and $(E', \varphi') = B \otimes_A (E_0, \varphi_0)$. Then $E$ and $E'$ are stably isomorphic. Adding a hyperbolic space $H_1(B^n)$ we can arrange that $E \cong E'$, and so identify the two modules. Choosing an $A$-space $(E_1, \varphi_1)$ so that $(E_0, \varphi_0) \oplus (E_1, \varphi_1) = H_1(A^m)$ we can further add $B \otimes_A (E_1, \varphi_1)$, and so arrange that $E = B^{2m}$. Identify the inner products on the free modules $B^p$ with their matrices relative to the standard basis. Then we can express $\varphi$ as a polynomial

$$\varphi = \varphi_0 + \varphi_1 T + \cdots + \varphi_d T^d$$

where each $\varphi_i$ is a symmetric $2m$ by $2m$ matrix over $A$. We wish to make $\varphi$ equivalent to $\varphi_0$, by a combination of transformations of the form $\varphi \mapsto \sigma\varphi{}^t\sigma$ with $\sigma \in GL_{2m}(B)$ (t denotes transpose), and

$$\varphi \mapsto \begin{pmatrix} \varphi & 0 \\ 0 & h_p \end{pmatrix}, \qquad \text{where} \quad h_p = \begin{pmatrix} 0 & I_p \\ I_p & 0 \end{pmatrix}$$

is the matrix of $H_1(B^p)$. Taking $p = 2m$ and

$$\sigma = \begin{pmatrix} I & \alpha & \beta \\ 0 & I & 0 \\ 0 & 0 & I \end{pmatrix} \qquad \text{(all entries are of size } 2m\text{)},$$

we have

$$\bar{\varphi} = \sigma \begin{pmatrix} \varphi & 0 \\ 0 & h_{2m} \end{pmatrix} {}^t\sigma = \begin{pmatrix} \varphi' & \beta & \alpha \\ {}^t\beta & 0 & I \\ {}^t\alpha & I & 0 \end{pmatrix},$$

where $\varphi' = \varphi + \beta\,{}^t\alpha + \alpha\,{}^t\beta$. Since $\varphi$ is even $(2 \in A^\times)$ we can write $\varphi_d = \psi + {}^t\psi$ for some matrix $\psi$ of ring $2m$. If $d > 1$ we can put $\alpha = \psi T^{d-1}$ and $\beta = -TI$, and so arrange that $\bar{\varphi}$ has degree $< d$ in $T$. Thus we can reduce to the case where $d = 1$. Then we have $\varphi = \varphi_0 + \varphi_1 T = \varphi_0(I + vT)$, where $= \varphi_0^{-1}\varphi_1$. Since $\tau = I + vT$, like $\varphi$, is invertible, $v$ must be nilpotent (its inverse $\sum_i (-vT)^i$ must be a polynomial). Since $\frac{1}{2} \in A$ we can form the polynomial

$$s(\tau) = \sum_{i \geq 0} \binom{-\frac{1}{2}}{i} (vT)^i,$$

such that $s(\tau)^2 = \tau^{-1}$. Since $\varphi_0$ and $\varphi_0\tau$ are symmetric we have $\varphi_0\tau = {}^t\tau\varphi_0$, and hence

$${}^ts(\tau)\varphi s(\tau) = {}^ts(\tau)(\varphi_0\tau)s(\tau) = \varphi_0 s(\tau)\tau s(\tau) = \varphi_0 .$$

This concludes the proof.

## References

1. M. Auslander and O. Goldman, The Brauer group of a commutative ring, *Trans. Amer. Math. Soc.* **97** (1960), 367–409.
2. H. Bass, "Algebraic $K$-Theory." Benjamin, New York, 1968.
3. H. Bass, Modules which support non-singular forms, *J. Algebra* **13** (1969), 246–252.
4. H. Bass, "Unitary algebraic $K$-theory," *Proc. Battelle Conf. Algebraic K-Theory* **3**, 57–265 (Lect. Notes in Math. **343**). Springer-Verlag, Berlin, 1973.
5. H. Bass, Clifford algebras and spinor norms over a commutative ring, *Amer. J. Math.* **96** (1974), 156–206.
6. H. Bass, Libération des modules projectifs sur certains anneaux de polynômes, *Sem. Bourbaki* (juin 1974), exposé 448.
7. N. Bourbaki, Formes sesquilinéaires et formes quadratiques, "Algèbre," Chap. 9. Hermann, Paris 1959.
8. M. Demazure and P. Gabriel, "Groupes algébriques," Vol. 1. North-Holland Publ., Amsterdam, 1970.
9. J. Giraud, "Cohomologie non-abélienne" (Grund. Math. Wiss. Eing. **179**). Springer-Verlag, Berlin, 1971.

10. A. Grothendieck, Le groupe de Brauer, I, II, III, "Dix exposés sur la cohomologie des schemas," pp. 46–188. North-Holland Publ., Amsterdam, 1968.
11. M. Karoubi, Periodicité de la $K$-théorie hermitienne, *Proc. Battelle Conf. Algebraic K-Theory* **3**, 301–411 (Lect. Notes in Math. **343**). Springer-Verlag, Berlin, 1973.
12. M. Knebusch, Grothendieck- und Wittringe von nichtausgearteten symmetrischen Bilinearformen, *Sitz. Heidelberg. Akad. Wiss., Math. -Naturwiss. Kl.* **3** (1970), 90–157.
13. S. Parimala, Failure of a quadratic analogue of Serre's conjecture, *Bull. Amer. Math. Soc.* (197 ).
14. D. Quillen, Projective modules over polynomial rings, *Invent. Math.* (1976).
15. R. Hoobler, A cohomological interpretation of Brauer groups (to appear.)

The author is grateful to S. Parimala, C. S. Seshadri, and R. Sridharan for helpful discussions of this work. I owe to Seshadri, in particular, the suggestion to use cohomological methods for small $r$. The author further thanks the Tata Institute of Fundamental Research for its support and hospitality during this research.

# *The action of the universal modular group on certain boundary points*

*Lipman Bers*

*Columbia University*

The subject of this note is the action of the universal modular group $\mathrm{Mod} = \mathrm{Mod}(1)$ on the boundary $\partial T = \partial T(1)$ of the universal Teichmüller space $T = T(1)$. We shall show that there are infinitely many points $\varphi \in \partial T$ such that, for every $\gamma \in \mathrm{Mod}$, $\gamma(\varphi)$ can be defined, in a natural way, and $\gamma(\varphi) \in \partial T$. The meaning of the terms will be recalled below. The reader is referred to my report [5] for a survey of Teichmüller space theory and for references to proofs, and to Ahlfors' lectures [1] or to the monograph by Lehto and Virtanen [7] for the needed information on quasiconformal mappings.

**A.** Every quasiconformal automorphism $\omega$ of the upper half plane $U = \{z = x + iy \in \mathbb{C}, y > 0\}$ is the restriction to $U$ of a topological automorphism of $U \cup \hat{R} = U \cup R \cup \{\infty\}$, which we denote by the same letter. We call $\omega$ *normalized* if it fixes 0, 1, and $\infty$, *equivalent to the identity*, if it fixes every $x \in R$, *equivalent* to another quasiconformal automorphism $\omega_1$ of $U$ if $\omega_1 \circ \omega^{-1}$ is equivalent to the identity. The equivalence class of $\omega$ will be denoted by $[\omega]$. The *universal Teichmüller space* $T = T(1)$ is the quotient of the group of all normalized quasiconformal selfmappings of $U$ over the (normal) subgroup of selfmappings equivalent to the identity.

The (Teichmüller) *distance* between two elements $[w_1]$ and $[w_2]$ of $T$ is

$$\delta([w_1], [w_2]) = \inf \log K(\omega_1 \circ \omega_2^{-1}), \qquad \omega_1 \in [w_1], \quad \omega_2 \in [w_2].$$

Here $K(f)$ denotes the *dilatation* of the quasiconformal mapping $f$. The above definition makes $T$ into a complete metric space but *not* into a topological group.

**B.** Every quasiconformal selfmapping $\omega$ of $U$ induces an isometric selfmapping $\omega_*$ of $T$ defined by

$$\omega_*([w]) = [\alpha \circ w \circ \omega^{-1}], \tag{1}$$

where $\alpha$ is a real Möbius transformation (conformal selfmapping of $U$) chosen so that $\alpha \circ w \circ \omega^{-1}$ fixes 0, 1, $\infty$. Note that $\omega_*$ depends only on $[\omega]$, that $\mathrm{id}_* = \mathrm{id}$ and that $(\omega_1 \circ \omega_2)_* = (\omega_1)_* \circ (\omega_2)_*$. The *universal modular group* $\mathrm{Mod} = \mathrm{Mod}(1)$ is the group of all $\omega_*$.

If $\omega$ is normalized, we have $\alpha = \mathrm{id}$ in (1), for all $w$. In this case $\omega_*$ is a *right translation* in $T$. The group of all right translations will be denoted by $\mathrm{Mod}_t$. If $\omega$ is a real Möbius transformation, $\omega_*$ is called a *rotation*; the group of rotations will be denoted by $\mathrm{Mod}_r$. Observe that

$$\mathrm{Mod} = \mathrm{Mod}_r \cdot \mathrm{Mod}_t = \mathrm{Mod}_t \cdot \mathrm{Mod}_r, \qquad \mathrm{Mod}_t \cap \mathrm{Mod}_r = 1 = \{\mathrm{id}\}.$$

**C.** Let $B$ denote the Banach space of holomorphic functions $\varphi(z)$ defined in the lower half plane $L = \{z = x + iy \in \mathbb{C},\ y < 0\}$ with

$$\|\varphi\| = \|\varphi\|_B = \sup |y^2 \varphi(z)| < +\infty. \tag{2}$$

The complex conjugate of $B$ is dual, via the scalar product

$$\iint_{y<0} y^2 \psi(z) \overline{\varphi(z)}\, dx\, dy,$$

to the Banach space $A$ of holomorphic functions $\psi(z)$, $z \in \mathrm{L}$, with norm

$$\iint_{y<0} |\psi(z)|\, dx\, dy < +\infty.$$

A sequence $\{\varphi_j\} \subset B$ *converges weakly* to a $\varphi \in B$ if $\|\varphi_j\| = O(1)$ and $\lim \varphi_j(z) = \varphi(z)$, $z \in L$.

For $\varphi \in B$ we denote by $W_\varphi(z)$ the solution of the differential equation

$$\{W, z\} = \varphi,$$

subject to the initial condition

$$W(z) = (z + i)^{-1} + O(|z + i|), \qquad z \to -i. \tag{3}$$

Here $\{W, z\}$ denotes the *Schwarzian derivative*:

$$\{W, z\} = \frac{W'''(z)}{W'(z)} - \frac{3}{2}\frac{W''(z)^2}{W'(z)^2}.$$

It is well known that

$$W_\varphi(z) = \frac{\eta_1(z)}{\eta_2(z)},$$

where $\eta_1$ and $\eta_2$ are solutions of the linear differential equation

$$2\eta''(z) + \varphi(z)\eta(z) = 0,$$

subject to the initial conditions

$$\eta_1(-i) = \eta_2'(-i) = 1, \qquad \eta_1'(-i) = \eta_2(-i) = 0.$$

**D.** A normalized quasiconformal selfmapping $w$ of $U$ is uniquely determined by its Beltrami coefficient

$$\mu = \frac{\partial w/\partial \bar{z}}{\partial w/\partial z};$$

this is a bounded measurable function or, rather, an equivalence class of such functions, that is, an element of $L_\infty(U)$, with $\|\mu\|_\infty = \operatorname{ess\,sup}|\mu(z)| < 1$. We write

$$w = w_\mu. \tag{4}$$

We denote by $z \mapsto w^\mu(z)$ a quasiconformal automorphism of the Riemann sphere $\hat{\mathbb{C}} = \mathbb{C} \cup \{\infty\}$ whose Beltrami coefficient $(\partial w^\mu/\partial \bar{z})/(\partial w^\mu/\partial z)$ equals $\mu$ in $U$ and vanishes in $L$, and we denote by $\varphi^\mu$ the Schwarzian derivative of $w^\mu|L$. It turns out that $\varphi^\mu$ and $w^\mu|L$ depend only on the equivalence class $[w_\mu]$, and that the mapping

$$[w_\mu] \mapsto \varphi^\mu \tag{5}$$

is homeomorphic bijection of $T$ onto a *bounded holomorphically convex domain* in $B$, which is contained in the ball $\|\varphi\| < \frac{3}{2}$ and contains the ball $\|\varphi\| < \frac{1}{2}$. From now on we *identify* $T$ with its image under (5). The group Mod now becomes a group of *holomorphic* mappings.

Since, for $\varphi \in T$ and $\varphi = \varphi^\mu$, the function $w^\mu|L$ depends only on $\varphi$, we may define

$$w^\mu|L = w^\varphi \qquad (\varphi = \varphi^\mu). \tag{6}$$

Since a holomorphic function is determined by its Schwarzian derivative but for a Möbius transformation (conformal selfmapping of the Riemann sphere), we have that

$$W_\varphi = \alpha_\varphi \circ w^\varphi, \tag{7}$$

where $\alpha_\varphi$ is a Möbius transformation. It is known that if $\{\varphi_j\} \subset T$ and $\lim \varphi_j = \varphi \in \partial T$, then $\lim W_{\varphi_j}(z) = W_\varphi(z)$ for all $z \in L$, and $W_\varphi$ is schlicht. But the sequences $\{\alpha_{\varphi_j}\}$, $\{w^{\varphi_j}(z)\}$ may, in general, diverge.

**E.** We recall that the group $\mathrm{Mod}_r$ acts on $T$ as the restriction of a group of linear isometries on $B$. Indeed, if $\beta$ is a real Möbius transformation and $[w_\mu] \in T$, then one computes that $\beta_*^{-1}([w_\mu]) = [\alpha \circ w_\mu \circ \beta] = [w_\nu]$ with

$$\nu(z) = \mu(\beta(z))\overline{\beta'(z)}/\beta'(z),$$

that $w^\nu = \hat{\alpha} \circ w^\mu \circ \beta$ where $\hat{\alpha}$ is a Möbius transformation, and that

$$\beta_*^{-1}(\varphi^\mu) = \{w^\nu | L, z\} = \{\hat{\alpha} \circ w^\mu \circ \beta | L, z\} = \varphi^\mu(\beta(z))\beta'(z)^2.$$

Here we used the two basic properties of the Schwarzian derivative, the *Cayley identity*

$$\{w, \zeta\}\, d\zeta^2 = \{w, z\}\, dz^2 + \{z, \zeta\}\, d\zeta^2 \tag{8}$$

for every holomorphic bijection $z \mapsto \zeta$, and the fact that $\{\sigma, z\} = 0$ for every Möbius transformation $\sigma$. The mapping

$$\varphi^\mu(z) \mapsto \varphi^\nu(z) = \varphi^\mu(\beta(z))\beta'(z)^2$$

is the desired linear isometry on $B$ onto itself.

The action of $\mathrm{Mod}_t$ on $T \subset B$ is much more complicated (cf. Gardiner [6]).

**F.** We say that a $\gamma \in \mathrm{Mod}$ *acts* on a $\varphi \in \partial T$ if there is a $\psi \in B$ such that for every sequence $\{\varphi_j\} \subset T$ with $\lim \varphi_j = \varphi$, the sequence $\{\gamma(\varphi_j)\}$ converges weakly to $\psi$. If so, we write $\psi = \gamma(\varphi)$. By Section E, every rotation acts on every $\varphi \in \partial T$, with weak convergence replaced by convergence in norm. We shall say that Mod acts on a $\varphi \in \partial T$ if every $\gamma \in \mathrm{Mod}$ does. Of course, only elements of $\mathrm{Mod}_t$ must be checked.

**G.** In order to give a sufficient condition for the action of Mod on a boundary point, we recall the definition of a *quasicircle*.

A quasicircle is the image of a circle under a quasiconformal automorphism of the Riemann sphere $\hat{\mathbb{C}}$. Every polygonal Jordan curve is a quasicircle, and so is every continuously differentiable Jordan curve. But there also are nonrectifiable quasicircles. A Jordan domain bounded by a quasicircle will be called a *quasidisc*.

***Proposition 1.*** *If $\varphi \in \partial T$ and $W_\varphi(L)$ is a union of finitely many quasidiscs,* Mod *acts on $\varphi$.*

For the proof we need two lemmas about quasidiscs.

**H.** ***Lemma 1.*** *To every quasidisc $\Delta$ there is a positive number $\varepsilon$ with the following property. Let there be given a sequence $\{f_j(z)\}$ of locally schlicht meromorphic functions defined in $\Delta$, and assume that*

$$|\{f_j, z\}\lambda_\Delta(z)^{-2}| < \varepsilon \qquad \text{for all} \quad j \quad \text{and all} \quad z \in \Delta \tag{9}$$

*where $\lambda_\Delta(z)|dz|$ is the Poincaré line element in $\Delta$. If the sequence $\{f_j(z)\}$ converges to $z$ in $\Delta$, the convergence is uniform, in the spherical metric.*

*Proof* We recall that the Poincaré line element is defined by two conditions:

$$\lambda_{h(\Delta)}(h(z))|h'(z)| = \lambda_\Delta(z) \qquad \text{for every conformal mapping } h \text{ of } \Delta, \tag{10}$$

$$\lambda_U(z) = 1/y \qquad (z = x + iy). \tag{11}$$

A meromorphic function is called locally schlicht if its poles, if any, are of order one, and at points where the function is holomorphic, its derivative does not vanish. The Schwarzian derivative of such a function is holomorphic.

According to Theorem 7 in [3], there are, for any quasidisc $\Delta$, two positive numbers, $\varepsilon$ and $\varepsilon'$, such that (9) implies that $f_j = F_j|\Delta$, where $F_j$ is a quasiconformal automorphism of $\hat{\mathbb{C}}$ with $K(F_j) < 1/\varepsilon'$. The conclusion of the lemma follows from the standard compactness property of quasiconformal mappings with uniformly bounded dilatation.

***Lemma 2.*** *Let $\{f_j(z)\}$ be a sequence of schlicht meromorphic function defined in L such that*

$$\lim f_j(z) = f(z), \qquad z \in L, \tag{12}$$

*and $f(z)$ is schlicht. If*

$$\lim \|\{f_j, z\} - \{f, z\}\| = 0, \tag{13}$$

*and if $f(L)$ is a union of finitely many quasidiscs, then the convergence (12) is uniform in the spherical metric.*

*Proof* Set $g_j = f_j \circ f^{-1}$. Then $g_j$ is schlicht in $f(L)$ and

$$\lim g_j(\zeta) = \zeta, \qquad \zeta \in f(L). \tag{14}$$

Condition (13) means [cf. (2)] that

$$\lim \sup_L |(\{f_j, z\} - \{f, z\})y^2| = \lim \sup_L |(\{f_j, z\} - \{f, z\})\lambda_U(z)^{-2}| = 0.$$

Set $\zeta = f(z)$, $z \in L$. By Cayley's identity (8) and by (10), (11), we have that

$$|\{f_j, z\} - \{f, z\}|\lambda_U(z)^{-2} = |\{g_j \circ f(z), f(z)\}f'(z)^2|\lambda_{f(L)}(f(z))^{-2}|f'(z)|^{-2}$$
$$= |\{g_j, \zeta\}\lambda_{f(L)}(\zeta)^{-2}|,$$

so that

$$\limsup_{f(L)} |\{g_j, \zeta\}\lambda_{f(L)}(\zeta)^{-2}| = 0. \tag{15}$$

Now assume that

$$f(L) = \Delta_1 \cup \Delta_2 \cup \cdots \cup \Delta_N,$$

where each $\Delta_\nu$ is a quasidisc. Since $\Delta_\nu \subset f(L)$,

$$\lambda_{\Delta_\nu}(\zeta) \geq \lambda_{f(L)}(\zeta), \qquad \zeta \in \Delta_\nu, \quad \nu = 1, \ldots, N.$$

Therefore, by (15),

$$\limsup_{\Delta_\nu} |\{g_j, \zeta\}\lambda_{\Delta_\nu}(\zeta)^{-2}| = 0, \qquad \nu = 1, \ldots, N. \tag{16}$$

By (14), (16) and Lemma 1, $g_j(\zeta)$ converges to $\zeta$ uniformly in each $\Delta_\nu$. Hence the convergence (14) is uniform, in the spherical metric, and so is (12).

**I.** Now we can prove Proposition 1. We assume that $\varphi \in \partial T$ and $W_\varphi(L)$ is a union of finitely many quasidiscs. Let $\{\varphi_j\} \subset T$ and $\lim \varphi_j = \varphi$. By Lemma 2, the convergence $W_{\varphi_j} \to W_\varphi$ is uniform in the spherical metric, and we conclude that the continuous extensions of $W_{\varphi_i}$ to $L \cup \hat{R}$ (which exists since $W_{\varphi_i}$ is a conformal mapping of $L$ onto a Jordan domain) also converge uniformly. Hence $W_\varphi$ has a continuous extension to $L \cup \hat{R}$. Choose three points $\xi_1, \xi_2, \xi_3$ on $\hat{R}$, which follow each other in this order, and such that $\zeta_1 = W_\varphi(\xi_1)$, $\zeta_2 = W_\varphi(\xi_2)$, $\zeta_3 = W_\varphi(\xi_3)$ are three distinct points on $\hat{\mathbb{C}}$, and let $\sigma$ be the real Möbius transformation which maps $(0, 1, \infty)$ onto $(\xi_1, \xi_2, \xi_3)$. Set $\hat{\varphi}(z) = \varphi(\sigma(z))\sigma'(z)^2$. Then $\hat{\varphi}$ is the image of $\varphi$ under a rotation $\gamma \in \mathrm{Mod}_r$, so that $\hat{\varphi} \in \partial T$. Also, $W_{\hat{\varphi}} = \tau \circ W_\varphi \circ \sigma$, where $\tau$ is an appropriate Möbius transformation. Hence $W_{\hat{\varphi}}(L) = \tau(W_{\hat{\varphi}}(L))$ is a finite union of quasidiscs. Noting that $W_{\hat{\varphi}}(0) = \tau(\zeta_1)$, $W_{\hat{\varphi}}(1) = \tau(\zeta_2)$, $W_{\hat{\varphi}}(\infty) = \tau(\zeta_3)$, we conclude (cf. Section E) that it suffices to prove Proposition 1 under the additional hypothesis that

$$W_\varphi(0) \neq W_\varphi(1), \qquad W_\varphi(0) \neq W_\varphi(\infty), \qquad W_\varphi(1) \neq W_\varphi(\infty). \tag{17}$$

Assuming this, recall [cf. (7)] that

$$W_{\varphi_j} = \alpha_{\varphi_j} \circ w^{\varphi_j}, \tag{18}$$

where the Möbius transformation $\alpha_{\varphi_j}$ is determined by the condition that it maps 0, 1, $\infty$ into $W_{\varphi_j}(0)$, $W_{\varphi_j}(1)$, $W_{\varphi_j}(\infty)$. Hence $\alpha_{\varphi_j}$ converges to the Möbius transformation $\alpha_\varphi$ which maps 0, 1, $\infty$ into $W_\varphi(0)$, $W_\varphi(1)$, $W_\varphi(\infty)$. We conclude that, for $z \in L$,

$$\lim w^{\varphi_j}(z) = \alpha_\varphi^{-1} \circ W_\varphi(z). \tag{19}$$

(Actually this convergence holds for $z \in L \cup \hat{R}$ and is uniform in the spherical metric.)

To establish the proposition, it suffices (cf. Section E) to show that every right translation acts on $\varphi$. We consider a fixed element $\gamma \in \mathrm{Mod}_t$; then there is a Beltrami coefficient $\theta$ such that $\gamma = (w_\theta)_*$, that is, for every Beltrami coefficient $\mu$,

$$\gamma([w_\mu]) = [w_\mu \circ w_\theta^{-1}].$$

Only $[w_\theta]$ is determined by $\gamma$, and we may choose $\theta$ as *continuous* (cf. [4, Th. 1]).

There are Beltrami coefficients $\mu_j$ with

$$\varphi_j = \varphi^{\mu_j} = \{w^{\mu_j}|L\}, \qquad w^{\varphi_j} = w^{\mu_j}|L. \tag{20}$$

(Of course, $\lim \|\mu_j\|_\infty = 1$.) We set $\psi_j = \gamma(\varphi_j)$. Then

$$\psi_j = \varphi^{\nu_j} = \{w^{\nu_j}|L\}, \qquad w^{\psi_j} = w^{\nu_j}|L, \tag{21}$$

where the Beltrami coefficient $\nu$ is determined by

$$w_{\nu_j} = w_{\mu_j} \circ w_\theta^{-1}. \tag{22}$$

Now let $q_{\mu_j}$ be the conformal mapping of the Jordan domain $w^{\mu_j}(U)$ onto $U$ which keeps 0, 1, $\infty$ fixed, and let $q_{\nu_j}$ be the similarly normalized conformal mapping of $w^{\nu_j}(U)$ onto $U$. Then

$$w_{\mu_j} = q_{\mu_j} \circ w^{\mu_j}|U, \qquad w_{\nu_j} = q_{\nu_j} \circ w^{\nu_j}|U, \tag{23}$$

since the right sides in (23) have Beltrami coefficients $\mu_j$ and $\nu_j$, respectively.

We now define mappings

$$A_j^+ : w^{\mu_j}(U) \to w^{\nu_j}(U), \qquad A_j^- : w^{\mu_j}(L) \to w^{\nu_j}(L)$$

by setting

$$A_j^+ = q_{\nu_j}^{-1} \circ q_{\mu_j}, \qquad A_j^- = w^{\nu_j} \circ w_\theta \circ (w^{\mu_j})^{-1}. \tag{24}$$

These are conformal and quasiconformal homeomorphisms between Jordan domains and are therefore restrictions of topological mappings between the closures of these domains. We claim that $A_j^+$ and $A_j^-$ coincide on the boundary curve $C_j = \partial w^{\mu_j}(U) = \partial w^{\mu_j}(L) = w^{\mu_j}(\hat{R})$. Indeed, by (20)–(24), we have

$$A_j^+ | C_j = (w_{\nu_j} \circ (w^{\nu_j})^{-1})^{-1} \circ w_{\mu_j} \circ (w^{\mu_j})^{-1} | C_j = w^{\nu_j} \circ w_{\nu_j}^{-1} \circ w_{\mu_j} \circ (w^{\mu_j})^{-1} | C_j$$
$$= w^{\nu_j} \circ w_\theta \circ (w^{\mu_j})^{-1} | C_j = A_j^- | C_j.$$

Since the quasicircle $C_j$ is a removable set for quasiconformal mappings, we conclude that $A_j^+$ and $A_j^-$ are restrictions of a quasiconformal automorphism $A_j$ of $\mathbb{C}$. We note [cf. (20) and (21)] that we can also write

$$A_j^- = A_j | w^{\varphi_j}(L) = w^{\psi_j} \circ w_\theta \circ (w^{\varphi_j})^{-1} \tag{25}$$

so that

$$w^{\psi_j} = A_j \circ w^{\varphi_j} \circ w_\theta^{-1}. \tag{26}$$

Let $\delta_j$ denote the Beltrami coefficient of $A_j$. Since the two-dimensional Lebesque measure of a quasicircle is 0, the values of $\delta_j | C_j$ are of no relevance. By (24),

$$\delta_j | w^{\mu_j}(U) = 0,$$

and by (25) $\delta_j | w^{\mu_j}(L)$ is the Beltrami coefficient of $w_\theta \circ (w^{\varphi_j})^{-1}$. Setting

$$(w^{\varphi_j})^{-1} = f_j, \tag{27}$$

we have

$$\delta_j(z) = \theta(f_j(z))\overline{f_j'(z)}/f_j'(z) \qquad \text{for} \quad z \in w^{\mu_j}(L). \tag{28}$$

We claim next that the limit

$$\delta(z) = \lim \delta_j(z) \tag{29}$$

exist a.e. in $\mathbb{C}$, and that, except on a set of measure 0,

$$\delta(z) = \begin{cases} 0 & \text{for} \quad z \notin \alpha_\varphi^{-1} \circ W_\varphi(L), \\ \theta(f(z))\overline{f'(z)}/f'(z) & \text{for} \quad z \in \alpha_\varphi^{-1} \circ W_\varphi(L), \end{cases} \tag{30}$$

where $\alpha_\varphi$ is the Möbius transformation in (19) and

$$f = (\alpha_\varphi^{-1} \circ W_\varphi)^{-1}. \tag{31}$$

Indeed, since $\{w^{\varphi_j}\}$ converges to the schlicht function $\alpha_\varphi^{-1} \circ W_\varphi$ uniformly on compact subsets of $L$, $\{f_j(z)\}$ converges to $f(z)$ uniformly on compact subsets of $\alpha_\varphi^{-1} \circ W_\varphi(L)$, and since $\theta$ is continuous, it follows from (28) that, for $z \in \alpha_\varphi^{-1} \circ W_\varphi(L)$, $\lim \delta_j(z) = \theta(f(z))\overline{f'(z)}/f'(z)$. On the other hand, since $\{w^{\varphi_j}\}$ converges to $\alpha_\varphi^{-1} \circ W_\varphi$ in $L$, uniformly with respect to the spherical metric, it follows that for every $z$ exterior to $\alpha_\varphi^{-1} \circ W_\varphi(L)$ we have that $z \notin w^{\varphi_j}(L)$ and thus $\delta_j(z) = 0$ for all sufficiently large $j$. Finally, since quasicircles have measure 0, the boundary of $\alpha_\varphi^{-1} \circ W_\varphi(L)$ has measure 0, and relations (29), (30) follow.

Since the Beltrami coefficients $\delta_j(z)$ all have the same $L_\infty$ norm, $\|\delta_j\|_\infty = \|\theta\|_\infty < 1$, and all mappings $A_j$ keep 0, 1, $\infty$ fixed, it follows (cf. [2, Th. 9]), that the sequence $\{A_j\}$ converges, uniformly with respect to the spherical metric, to a quasiconformal automorphism $A$ of $\hat{\mathbb{C}}$. This $A$ keeps 0, 1, $\infty$ fixed and has Beltrami coefficient $\delta(z)$ given by (30). Thus $A$ depends only on $\theta$ and on $\varphi$, but not on the particular sequence $\{\varphi_j\}$.

Now we can conclude from (26) that the sequence $\{w^{\psi_j}\}$ converges, uniformly with respect to the spherical metric, to the function

$$\hat{W} = A \circ \alpha_\varphi^{-1} \circ W_\varphi \circ w_\theta^{-1}. \tag{32}$$

This function is univalent and (being a normal limit of holomorphic functions) holomorphic. Set $\psi = \{\hat{W}, z\}$. By Nehari's inequality [8], $\|\psi_j\| \le \frac{3}{2}$. Since $\lim \psi_j(z) = \psi(z)$ for all $z \in L$, we conclude that $\lim \psi_j = \psi$ weakly, as asserted.

**J.** ***Lemma 3.*** *Under the hypothesis of Proposition* 1, *let* $\gamma \in \mathrm{Mod}$ *and set* $\psi = \gamma(\varphi)$. *Then* $W_\psi(L)$ *is the image of* $W_\varphi(L)$ *under a quasiconformal automorphism of* $\mathbb{C}$, *the dilatation of this automorphism depending only on* $\gamma$.

*Proof* It suffices to prove this under the hypothesis (17) and for $\gamma \in \mathrm{Mod}_t$, that is, under the circumstances of the preceding proof. Using the same notation we conclude that $W_\psi = \hat{\alpha} \circ \hat{W}$, where $\hat{W}$ is given by (32) and $\hat{\alpha}$ is a Möbius transformation. Hence $W_\psi(L) = \hat{\alpha} \circ A \circ \alpha_\varphi^{-1} \circ W_\varphi(L)$. The dilatation of $\hat{\alpha} \circ A \circ \alpha_\varphi^{-1}$ is $K(A) = (1 + \|\theta\|_\infty)/(1 - \|\theta\|_\infty)$.

**K.** We proceed to find a class of points on $\partial T$ to which Proposition 1 is applicable.

It is known that a point $\varphi \in B$ is a point of $T$ if and only if $W_\varphi(L)$ is a quasidisc. This condition can be weakened to the requirement that there be some conformal mapping $f$ of $L$ onto a quasidisc with $\{f, z\} = \varphi$. Indeed, if so, then there is a Möbius transformation $\beta$ such that $\beta \circ f = W$ satisfies $W(z) = (z+i)^{-1} + O(|z+i|)$, $z \to -i$. Hence $W = W_\varphi$ and since $W(L) = \beta \circ f(L)$ is a quasidisc, $\varphi \in T$.

A simply connected domain $\Delta \subset \hat{\mathbb{C}}$ will be called a *boundary domain* if there is a $\varphi \in \partial T$ and a Möbius transformation $\beta$ such that $W_\varphi(L) = \beta(\Delta)$. In this case the Schwarzian derivative $\psi$ of every conformal mapping $f$ of $L$ onto $\Delta$ lies on $\partial T$. Indeed, there is a real Möbius transformation $\alpha$ such that $f = W_\varphi \circ \alpha$ so that $\psi(z) = \varphi(\alpha(z))\alpha'(z)^2$. Hence $\psi$ is the image of $\varphi$ under an element of $\mathrm{Mod}_r$.

Unfortunately it is not known whether there are simply connected domains (with more than one boundary point) which are not boundary domains. Neither is it known whether all Jordan domains are boundary domains.

**L.** We shall need the fact that the property of a Jordan curve to be a quasicircle is a local property.

A *quasisegment* is the image of a (straight) segment $s$ under a quasiconformal homeomorphism of a neighborhood of $s$.

***Lemma 4*** *Let $C$ be a Jordan curve on $\hat{\mathbb{C}}$. Assume that every point $P$ on $C$ is an inner point of a quasisegment. Then $C$ is a quasicircle.*

For the proof see Lehto and Virtanen [7, pp. 99–109].

**M.** Let $\Delta_0$ be a Jordan domain in $\hat{\mathbb{C}}$, let $\Delta_1, \ldots, \Delta_n$ be Jordan domains contained in $\Delta_0$, and let $c_1, \ldots, c_n$ be continua contained in the closure $\Delta_0 \cup \partial\Delta_0$ of $\Delta_0$. Assume that (i) for each $j$, $j = 1, \ldots, n$, either the continuum $c_j$ is the unique point common to $\partial\Delta_0$ and $\partial\Delta_j$, or $\partial\Delta_j \subset \Delta_0$ and $c_j$ is a Jordan arc, with one endpoint on $\partial\Delta_0$, the other endpoint on $\partial\Delta_j$, and with all other points in $\Delta_0$, and that (ii) for $k \neq j$, $1 \leq k, j \leq n$,

$$(\Delta_k \cup \partial\Delta_k \cup c_k) \cap (\Delta_j \cup \partial\Delta_j \cup c_j) = \varnothing.$$

Then

$$\Delta = \Delta_0 - \Delta_0 \cap (\Delta_1 \cup \partial\Delta_1 \cup c_1 \cup \Delta_2 \cup \cdots \cup c_n)$$

is a simply connected domain. Assume also that (iii) if $a$ is any Jordan arc contained in the boundary $\partial\Delta$ of $\Delta$ and $P$ any inner point of $a$, there is a subarc $b$ of $a$ such that $P$ is an inner point of $b$ and $b$ is a quasisegment.

If all these conditions are satisfied, we call $\Delta$ a *domain of type $A_n$*.

***Lemma 5*** *The image of a domain of type $A_n$ under a quasiconformal automorphism of the Riemann sphere is a domain of type $A_n$.*

This follows from the definition.

***Lemma 6*** *A domain of type $A_n$ can be represented as the union of finitely many quasidiscs.*

This follows from the definition and from Lemma 4.

***Lemma 7*** *A domain of type $A_n$ is a boundary domain.*

*Proof* We assume that $z = \infty$ is exterior to $\Delta_0$ and $z = 0$ is interior to $\Delta_1$. This involves no loss of generality since it can be achieved by applying to $\Delta$ a Möbius transformation. Let $C$ be a Jordan arc joining $z = 0$ to $z = \infty$ and containing no points of $\Delta$. For $j = 2, 3, \ldots,$ let $F_j(z)$ be a single-valued branch of the function $z^{1-1/j}$ defined in the complement of $C$. This function is univalent and has continuous boundary values on both banks of $C$.

Let $f$ be a conformal mapping of $L$ onto $\Delta$. Then $f_j = F_j \circ f$ is a conformal mapping of $L$ onto $F_j(\Delta)$. Set $\varphi_j = \{f_j, z\}$, $\varphi = \{f, z\}$. By Cayley's identity (8),

$$\varphi_j(z) - \varphi(z) = \{F_j \circ f, z\} - \{f, z\} = \{F_j \circ f, f\} f'(z)^2,$$

so that

$$|\varphi_j(z) - \varphi(z)| = \frac{1}{j}\left(1 - \frac{1}{2j}\right)\left|\frac{f'(z)}{f(z)}\right|^2. \tag{33}$$

Since $\Delta$ is bounded and bounded away from 0 (both $z = 0$ and $z = \infty$ are points exterior to $\Delta$), $f(z)$ and $1/f(z)$ are bounded in $L$ and so is $yf'(z)$. It follows from (33) that

$$\lim\|\varphi_j - \varphi\| = \lim \sup_{y<0} \frac{2j-1}{2j^2}\left|\frac{f'(z)y}{f(z)}\right|^2 = 0. \tag{34}$$

Now we proceed by induction on $n$. If $n = 1$, then $F_j(\Delta)$ is, for each $j$, a Jordan domain and, by Lemma 4 and condition (iii), a quasidisc. Hence $\varphi_j \in T$, and by (34) we have that $\varphi \in T \cup \partial T$. But $\Delta$ is not a Jordan domain, so that $\varphi \notin T$ and therefore $\varphi \in \partial T$.

If $n > 1$ and the assertion is already established for domains of type $A_{n-1}$, we note that $F_j(\Delta)$ is of type $A_{n-1}$. Hence $\varphi_j \in \partial T$ and, again by (34), $\varphi \in \partial T$.

**N.** According to Lemma 7, there corresponds to every domain $\Delta$ of type $A_n$ a $\varphi \in \partial T$ with $W_\varphi(L) = \Delta$. According to Lemma 6 and Proposition 1, Mod acts on such a $\varphi$. According to Lemmas 4 and 5, $\gamma(\varphi) \in \partial T$ for all $\gamma \in$ Mod. Thus we obtain the following

***Theorem*** *There are infinitely many points on the boundary of the universal Teichmüller space which are acted on by the universal modular group, and are taken, by every element of that group, into other boundary points.*

## References

1. L. V. Ahlfors, "Lectures on Quasiconformal Mappings." Van Nostrand-Reinhold, Princeton, New Jersey, 1966.
2. L. V. Ahlfors and L. Bers, Riemann's mapping theorem for variable metrics, *Ann. of Math.* **72** (1960), 385–404.
3. L. Bers, A non-standard integral equation with applications to quasiconformal mappings, *Acta Math.* **116** (1966), 113–134.
4. L. Bers, Extremal Quasiconformal Mappings, "Advances in the Theory of Riemann Surfaces" (Ann. of Math. Studies **66**), pp. 27–52. Princeton University Press, Princeton, New Jersey, 1971.
5. L. Bers, Uniformization, moduli and Kleinian groups, *Bull. London Math. Soc.* **4** (1972), 257–300.

6. F. Gardiner, An analysis of the group operation in universal Teichmüller space, *Trans. Amer. Math. Soc.* **132** (1968), 471–486.
7. O. Lehto and K. I. Virtanen, "Quasikonforme Abbildungen." Springer-Verlag, Berlin–Heidelberg–New York, 1965.
8. Z. Nehari, Schwarzian derivatives and schlicht functions, *Bull. Amer. Math. Soc.* **55** (1949), 545–551.

Work partially supported by the National Science Foundation. In writing this note I profited from a conversation with W. Abikoff.

AMS (MOS) 1970 subject classification: 30A60

# *Differentially closed fields: a model-theoretic tour*

*Lenore Blum*

*Mills College, Oakland*
*and*
*University of California, Berkeley*

## Introduction

In his paper, "Constrained extensions of differential fields" [6], Kolchin presents a unified exposition in the language and setting of differential algebra of work done in this field by both logicians and algebraists. In the same paper he refers to a private guided tour, conducted by me, of the relevant work in model theory of A. Robinson, myself, and Shelah. The tour refers to a series of correspondence in 1972 in which I was attempting to present some of the flavor of the model theoretic approach and techniques, as specialized to the case of differential fields, in the hope that these ideas and methods would prove useful for further application. These letters were also a basis for a series of seminar talks I gave at Berkeley during that time with the same aim. Thus, the language and setting of this correspondence fell purposefully somewhere in the middle ground between logic and algebra. These letters are presented here essentially intact with some references to more recent results. It is hoped that this correspondence will further help bridge the gap between the two approaches and help make the relevant literature in logic more accessible.

What attracted me originally to the study of model theory and its applications, and what I find attractive today, is the way many very specific and concrete results follow in a natural way from logical and model theoretic properties of theories and very general principles. The work of Ax and Kochen [0] on Artin's conjecture was a beautiful example of this. In the case of differential fields (of characteristic 0) many specific results follow from the existence of a model completion for the theory (the theory of differentially closed fields) with a simple axiomatic characterization (namely that for each pair $(p, q)$ of differential polynomials in one variable there is a solution to $(p = 0,\ q \neq 0)$ if the order of $p$ is greater than the order of $q$), and the existence of differential closures.

Although somewhat unusual in format, this paper has a natural order. Section 1 discusses the *background* model theoretic work in differential algebra; Section 2 discusses the notion of *model completions*; Section 3 presents the *main results*; Section 4 lists *some properties of differential closures and some immediate consequences*; Section 5 includes extensive *notes* with proofs, and the *Appendix* presents details of the uniqueness proof.

A brief lexicon:

| *Language of logicians* | *Language of algebraists* |
|---|---|
| atomic extensions | constrained extensions |
| differentially closed fields | constrainedly closed fields |
| differential closure | constrained closure |

Berkeley
Spring 1972

Dear Professor Kolchin,

Thanks for your request for papers. Some of the results of my thesis [1] are in Sacks [11, 12]. I will try in this letter to give you some flavor of the applications of model theory to differential algebra. I'd like to stress the naturalness of some of these notions: in particular, that in a very strong sense, the class of differentially closed fields is the natural counterpart in differential algebra to the class of algebraically closed fields in ordinary algebra. Some of the applications to differential algebra do not give new results. But I believe these applications are of interest since they follow immediately and naturally from model theoretic considerations, in particular from the properties of the differential closure and the transfer principle.

I hope you will find this of interest and perhaps useful for other applications, possibly to the Galois theory.

## 1. Background

In the following, all differential fields will be ordinary and of characteristic 0. I believe much can be generalized to the partial case. Some work has been done for the characteristic $p > 0$ case by Wood [17, 18]. (Also, see Shelah [16].)

My thesis was mainly concerned with generalizing some concepts of algebra to model theory and then proving theorems in the abstract setting. It turned out that the theories of fields, ordered fields, and certain fields with valuations were examples of what I called generalized algebraic theories. Abraham Robison suggested that I investigate differential fields to see if they also fell under the general framework. He had already shown [9], using Seidenberg's elimination theory [14], that the theory of differential fields possessed a model completion (which he called the theory of differentially closed fields), and then making use of general properties (mainly due to him) of theories with model completions, he was able to give purely model theoretic proofs of certain results about differential fields, e.g. amalgamation properties, specialization of parameters, Nullstellensatz [8].

There were two features missing from Robinson's treatment that I required:

(1) A simple and explicit set of axioms for the theory of differentially closed fields. (See Theorem 1 of Section 3.) All one knew basically was that a differential field $F$ was differentially closed if and only if every finite consistent system

$$S = (f_1 = 0, \ldots, f_s = 0, J \neq 0)_{f_i,\, J \in F\{y_1, \ldots, y_n\}}$$

had a solution in $F$. (***Definition*** *$S$ is consistent* if and only if $S$ has a solution in some extension of $F$.) (See [Note 1] in Section 5.)

This would be like defining a field $F$ to be algebraically closed if and only if every finite consistent

$$S = (f_1 = 0, \ldots, f_s = 0, J \neq 0)_{f_i,\, J \in F[y_1, \ldots, y_n]}$$

has a solution in $F$ rather than equivalently, but more simply, if and only if every nonconstant polynomial in *one* variable over $F$ has a solution in $F$.

(2) The existence of differential closures (in analogy to algebraic closures but in contrast to universal extensions). (Curiously, in page 114 of his paper [9], Robinson says parenthetically that such closures do not exist. But see Theorem 2 of Section 3.)

## 2. Model completions

In the following, theories are given by first order axioms (an example will be given soon) in a fixed language, models of theories are structures that satisfy these axioms.

***Definition*** (A. Robinson) A theory $\hat{T}$ is a *model completion* of a theory $T$ if and only if

(1) $\hat{T} \supset T$, i.e. every model of $\hat{T}$ is a model of $T$.

(2) $\hat{T}$ is model consistent with respect to $T$, i.e. each model of $T$ can be extended to (embedded in) some model of $\hat{T}$.

(3) *Transfer principle* Whenever $F \subseteq F_1, F_2$ and $F$ is a model of $T$ and $F_1, F_2$ are models of $\hat{T}$, then $F_1 \equiv_F F_2$ ($F_1$ and $F_2$ are *elementarily equivalent* over $F$), i.e. any first order sentence with coefficients from $F$ is true in $F_1$ if and only if it is true in $F_2$.

*An example of a first order sentence* Suppose $F$ is a differential field and

$$S = (f_1 = 0, \ldots, f_s = 0, J \neq 0)_{f_i, J \in F\{y_1, \ldots, y_n\}}.$$

Then

$$\exists x_1 \cdots \exists x_n (f_1(x_1, \ldots, x_n) = 0 \;\&\; \cdots \;\&\; f_s(x_1, \ldots, x_n) = 0 \;\&\; J(x_1, \ldots, x_n) \neq 0)$$

is a first order sentence with coefficients from $F$ which asserts that $S$ has a solution. This sentence is to be interpreted in extensions $F' \supset F$; it is true in $F'$ just in case $S$ has a solution in $F'$. More generally, a first order sentence is a sentence in which the quantifiers $\forall$, $\exists$ range over elements (*not* subsets) of a structure.

Robinson has proved many general results about theories that possess model completions. Of particular interest, in terms of the naturalness of this notion, is his *uniqueness* result: A theory has at most *one* model completion, i.e., if $\hat{T}_1$ and $\hat{T}_2$ are both model completions of $T$, then $\hat{T}_1$ and $\hat{T}_2$ have exactly the same models (and hence, the same theorems, even though the axioms for $\hat{T}_1$ and $\hat{T}_2$ may be quite different). Hence, we can speak of *the* model completion if it exists.

*Some examples* The theory of algebraically closed fields is the model completion of the theory of fields; the theory of real closed ordered fields is the model completion of the theory of ordered fields; (with appropriate definitions) the theory of Hensel fields is the model completion of the theory of valued fields; the theory of differentially closed fields is the model completion of the theory of differential fields.

As a consequence (mainly of the transfer principle), one can give neat model-theoretic proofs of the Nullstellensatz, Hilbert's seventeenth problem [8], and the Artin conjecture (a.e.) [0].

## 3. Main results

***Theorem 1*** (Blum) *A differential field $F$ is differentially closed $\Leftrightarrow$ $F$ solves each pair* $(p = 0,\, q \neq 0)_{p,\, q \in F\{y\};\ \text{order } p > \text{order } q \geq -1}$ [Notes 2–4].

***Corollary*** (Blum) *A differential field F solves each finite consistent system* $S = (f_1 = 0, \ldots, f_s = 0, J \neq 0)_{f_i, J \in F\{y_1, \ldots, y_n\}} \Leftrightarrow F$ solves each pair $(p = 0, q \neq 0)_{p, q \in F\{y\};\ \text{order } p > \text{order } q \geq -1}$.

***Theorem 2*** (Blum, Morley, Shelah) *Each differential field has a prime differential closure, i.e. for each differential field F there is a differentially closed field* $\hat{F} \supset F$ *which can be embedded over F into any differentially closed field containing F. Moreover,* $\hat{F}$ *is unique up to F isomorphism* [Note 5].

***Corollary*** (Blum) *The theory of differential fields of characteristic* 0 *is a generalized algebraic theory (see* [1]).

## 4. Some properties of differential closures and some immediate consequences

*Notation* Suppose $F$ is a differential field. Let $\hat{F}$ be a differential closure of $F$. Let $\tilde{F}$ be the algebraic closure of $F$. Let $\mathscr{C}_F$ be the subfield of constants of $F$.

(1) $F$ and $\hat{F}$ have the same cardinality.
(2) $\hat{F}$ is *atomic* over $F$, i.e. for each $\eta_1, \ldots, \eta_n \in \hat{F}$ there is a *finite* system $S = (f_1 = 0, \ldots, f_s = 0, J \neq 0)_{f_i, J \in F\{y_1, \ldots, y_n\}}$ which completely determines the isomorphism type of $\langle \eta_1, \ldots, \eta_n \rangle$ over $F$, i.e. $\langle \mu_1, \ldots, \mu_n \rangle$ is a solution to $S$ (in some extension of $F$) $\Leftrightarrow F\langle \eta_1, \ldots, \eta_n \rangle \simeq_F F\langle \mu_1, \ldots, \mu_n \rangle$ via the map induced by sending $\eta_i$ to $\mu_i$. In particular, for each $\eta \in \hat{F}$ there is a pair $(p = 0, q \neq 0)_{p, q \in F\{y\};\ \text{order } p > \text{order } q}$ (with $p$ irreducible) which completely determines the isomorphism type of $\eta$ over $F$. Thus, $p$ is a minimal polynomial for any $\mu$ (in an extension of $F$) solving $(p = 0, q \neq 0)$ [Notes 4–6].
(3) $\hat{F}$ is differentially algebraic over $F$ (by construction or (2)).
(4) $\mathscr{C}_{\hat{F}} = \tilde{\mathscr{C}}_F$ (since if $c \in \mathscr{C}_{\hat{F}}$, then, by (2), $c \in \tilde{F}$; hence $c \in \tilde{\mathscr{C}}_F$).
(5) Suppose $S = (f_1 = 0, \ldots, f_s = 0, J \neq 0)_{f_i, J \in F\{y_1, \ldots, y_n\}}$ is consistent. Then $S$ has a solution in some differentially algebraic extension $F' \supset F$ such that $\mathscr{C}_{F'}$ is algebraic over $\mathscr{C}_F$. (*Proof* Suppose $S$ has a solution in $F_1 \supset F$. Then $S$ has a solution in $\hat{F}_1 \supset F$. By the *transfer principle*, $S$ has a solution in $\hat{F}$. Result now follows from (3) and (4).) [Note 7.]

*Recall notions of normality* (from [5]) Suppose $F \subset G$. Then $G$ is *weakly normal* over $F$ if the invariants of the group of all automorphisms of $G$ over $F$ all belong to $F$, and $G$ is *normal* over $F$ if $G$ is weakly normal over every differential field between $F$ and $G$. Suppose $\mathscr{C}_F = \mathscr{C}_G$ is algebraically closed. Then $G$ is *strongly normal* over $F$ if every isomorphism $\sigma$ of $G$ over $F$ into an extension of $G$ has the property: $G\langle \sigma G \rangle = G\langle \mathscr{C}_\sigma \rangle = \sigma G\langle \mathscr{C}_\sigma \rangle$ (where $\mathscr{C}_\sigma$ is the field of constants of $G\langle \sigma G \rangle$).

*Notation* Suppose $\eta_1, \ldots, \eta_n \in \hat{F}$. Let $G = F\langle\eta_1, \ldots, \eta_n\rangle$. Suppose $\{S_\alpha\}_{\alpha<\kappa}$ is a set of pairs, $S_\alpha = (p_\alpha = 0,\ q_\alpha \neq 0)_{p_\alpha, q_\alpha \in F\{y\};\ \text{order } p_\alpha > \text{order } q_\alpha}$. Let $X_\alpha = \{x \in \hat{F} \mid x \text{ is a solution to } S_\alpha\}$. Let $H = F\langle \bigcup X_\alpha \rangle_{\alpha<\kappa}$.

(6) Suppose $G$ is strongly normal over $F$. Then every $F$ isomorphism of $G$ into $\hat{F}$ is an automorphism (follows from (4)).

(7) $\hat{F}$ is a differential closure of $G$ [Note 5].

(8) $\hat{F}$ is a differential closure of $H$ [Note 5].

(9) $\hat{F}$ is a differential closure of $\tilde{F}$ (follows from (8) by considering the set $\{(p = 0, 1 \neq 0) \mid p \in F\{y\}, \text{order } p = 0\}$).

(10) Suppose $p \in G\{y\}$ is (absolutely) irreducible. Suppose $\{\eta_\alpha\}_{\alpha<\gamma} \subset \hat{F}$ has the property that for each $\beta < \gamma$, $p$ is a minimal polynomial for $\eta_\beta$ over $G\langle\{\eta_\alpha\}_{\alpha<\beta}\rangle$. (*Note* Any reordering of $\{\eta_\alpha\}_{\alpha<\gamma}$ also has this property. $\{\eta_\alpha\}_{\alpha<\gamma}$ is a set of *indiscernibles* over $F$.) Then $\{\eta_\alpha\}_{\alpha<\gamma}$ is countable [Note 5].

(11) Any $F$ isomorphism of $G$ into $\hat{F}$ can be extended to an automorphism of $\hat{F}$ (by (7) and uniqueness).

(12) Any $F$ automorphism of $H$ can be extended to an automorphism of $\hat{F}$ (by (8) and uniqueness).

(13) $\hat{F}$ is weakly normal over $F$, $G$, and $H$. (*Proof* For $\eta \in \hat{F} - F$ there is an $F$-isomorphism of $F\langle\eta\rangle$ into $\hat{F}$ which moves $\eta$. (Indeed, if $\eta$ is not algebraic over $F$, then there are an infinite such maps.) By (11) this isomorphism can be extended to an automorphism of $\hat{F}$. Similarly for $G$ and $H$ by (7) and (8).)

(14) Suppose every $F$-isomorphism of $G$ into $\hat{F}$ is an automorphism. Then $G$ is a normal extension of $F$ (but *not* necessarily conversely). (*Proof* Suppose $F \subset F_1 \subset G$. Then $\hat{F}$ is a differential closure of $F_1$ (by (7)). For $\eta \in G - F_1$ there is an $F_1$-isomorphism of $F_1\langle\eta\rangle$ into $\hat{F}$ which moves $\eta$. Extend to an automorphism of $\hat{F}$ (by (11)) and then restrict to $G$ to get, by hypothesis, an $F_1$-automorphism of $G$ moving $\eta$.)

(15) (Adler, Blum) Every strongly normal extension of $F$ is embeddable in $\hat{F}$ over $F$. Thus, differentially closed fields have no proper strongly normal extensions. If $\{F_i\}_{i<\omega}$ is a chain of extensions of $F$ such that $F_0 = F$ and $F_{i+1}$ is strongly normal over $F_i$, then $\{F_i\}_{i<\omega}$ is embeddable in $\hat{F}$ over $F$ [Note 8].

(16) Suppose $F\langle\mu_1, \ldots, \mu_n\rangle$ is a strongly normal extension of $F$. Then there is a finite system $S = (f_1 = 0, \ldots, f_s = 0, J \neq 0)_{f_i, J \in F\{y_1, \ldots, y_n\}}$ which completely determines the isomorphism type of $(\mu_1, \ldots, \mu_n)$ over $F$ (by (15) and (2)). Suppose $\{F_i\}_{i<\omega}$ is a chain as in (15) and $\mu_1, \ldots, \mu_n \in \bigcup_{i<\omega} F_i$. Then there is a finite system $S$ which completely determines the isomorphism type of $(\mu_1, \ldots, \mu_n)$ over $F$.

I had hoped that the following would be true: $F \subsetneqq F\langle\alpha\rangle$, $F$ differentially closed, $\alpha$ differentially algebraic over $F \Rightarrow \mathscr{C}_F \subsetneqq \mathscr{C}_{F\langle\alpha\rangle}$. (This is the case if $\alpha$ is, for example, primitive, exponential, or Weierstrassian over $F$ since there are no proper strongly normal extensions of differentially closed fields.)

For then one would have minimality of the differential closure and some simple structural characterizations of differentially closed fields (in terms of the cardinality of the constants and the differential dimension).

But, the conjecture is false, since according to some results of Rosenlicht's [10], if $\alpha$ solves, for example, $y' = y^3 - y^2$, then $\mathscr{C}_F = \mathscr{C}_{F\langle\alpha\rangle}$.

If you have any questions or comments, please write. I'll try to answer more speedily.

Sincerely,

Lenore Blum

## 5. Notes

**[1]** ***Seidenberg's Elimination Theorem*** [14] *Let*

$$S(u_1, \ldots, u_m, y_1, \ldots, y_n) = (f_1 = 0, \ldots, f_s = 0, J \neq 0)_{f_i,\, J \in I\{u_1, \ldots, u_m\}\{y_1, \ldots, y_n\}}$$

*($I$ being integer coefficients) be a finite system. Then there are a finite number of finite systems*

$$S_j(u_1, \ldots, u_m) = (f_{j_1} = 0, \ldots, f_{j_{sj}} = 0, J_j \neq 0)_{f_{ji},\, J_j \in I\{u_1, \ldots, u_m\}} \quad (j = 1, \ldots, k)$$

*such that for each differential field $F$ and each $a_1, \ldots, a_m \in F$, $S(a_1, \ldots, a_m, y_1, \ldots, y_n)$ is consistent (i.e. has a solution in some extension of $F$) $\Leftrightarrow (a_1, \ldots, a_m)$ is a solution to at least one $S_j(u_1, \ldots, u_m)$. Moreover, the procedure for producing the $S_j$'s is uniform and effective. (See [Note 6] for a proof.)*

Thus, Robinson made use of this uniform effective procedure for testing consistency in order to give first order axioms for differentially closed fields:

***Definition*** (A. Robinson) A differential field $F$ is differentially closed if and only if for each $S$ and $S_j$ $(j = 1, \ldots, k)$ as above,

$$\forall u_1 \cdots \forall u_m(\exists y_1 \cdots \exists y_n\ S(u_1, \ldots, u_m, y_1, \ldots, y_n) \leftrightarrow S_1(u_1, \ldots, u_m) \vee \cdots \vee S_k(u_1, \ldots, u_m))$$

is true in $F$.

This condition is in a sense implicitly given: For each finite system $S$ one must use Seidenberg's elimination procedure to find the corresponding $S_j$'s (to test for consistency). Also, one must consider all finite systems in $n$ variables, for each $n$.

**[2]** By convention, nonzero constant polynomials have order $-1$, while the zero polynomial has order $\infty$. Also, by convention, we say the zero polynomial is absolutely irreducible. (These conventions are not arbitrary and are consistent with proofs.) Hence, the condition ensures algebraic closedness, and more generally, solutions to each nonconstant polynomial in $F\{y\}$.

**[3]** Some other time.

**[4]** Theorem 1 and its corollary were proved in the following way:

***Definition*** $F$ is *$\kappa$-saturated* if and only if whenever $\{\phi_F(y)\}$ is a collection of less than $\kappa$-formulas in one free variable and coefficients from $F$ which is finitely satisfiable in $F$, then it is satisfiable in $F$, i.e. there is an $\eta$ in $F$ which satisfies each $\phi_F(y)$.

*Remark* It follows from the compactness theorem of first order logic that if $T$ is a consistent theory and $\kappa$ is any cardinal, then $\kappa$-saturated models of $T$ exist.

***Model completion criterion*** (Blum) *Suppose $T$ is a universal theory and (1) $\hat{T} \supset T$ and (2) $\hat{T}$ is model consistent with respect to $T$. Then $\hat{T}$ is a model completion of $T \Leftrightarrow$ every diagram of the type*

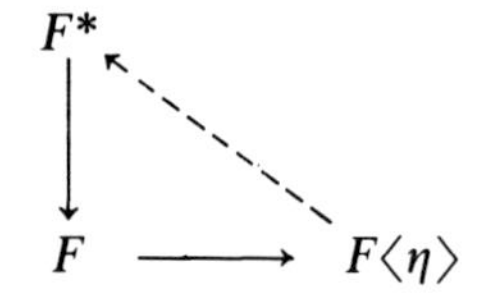


(*where* $F$, $F\langle\eta\rangle$ *are models of* $T$, $F^*$ *is a model of* $\hat{T}$ *which is* (card $F)^+$*-saturated*) *can be completed as shown.* (*For a proof see* [1] *or* [12].)

Now let $\mathrm{DF}_0$ be the theory of ordinary differential fields of characteristic 0. (If one allows symbols for additive and multiplicative inverses and identities, then $\mathrm{DF}_0$ is a universal theory, i.e., its axioms can be given using universal quantifiers ($\forall$) only.) Let $\mathrm{DCF}_0$ be the theory whose axioms include the axioms for $\mathrm{DF}_0$ plus the axioms which assert: for each pair $p$, $q$ of differential polynomials in *one* variable, there is a solution to $p$ which is a solution to $q$, provided the order of $p$ is greater than the order of $q$ ($\geq -1$). (These axioms can be written down quite formally.)

***Theorem*** $*$ (Blum) $\mathrm{DCF}_0$ *is a model completion of* $\mathrm{DF}_0$.

Theorem $*$ follows directly from the above criterion plus very elementary facts about differential fields (about characterizing and constructing simple extensions)† and *no* elimination theory.

*Proof of Theorem* $*$ Clearly, (1) $DCF_0$ extends $DF_0$; (2) follows, for example, from the construction of the differential closure [Note 5]. The diagram property follows from the fact that if $p$ is a minimal polynomial for $\eta$ over $F$, then the isomorphism type of $\eta$ over $F$ is completely determined by the set $S = \{p = 0, q \neq 0\}_{q \in F\{y\};\ \text{order } p > \text{order } q}$ whose cardinality is the same as that of $F$. For then, if $F^*$ is differentially closed, then every finite subset of $S$ (being equivalent to a pair $(p = 0, q \neq 0)_{\text{order } p > \text{order } q}$) has a solution in $F^*$; and so, if furthermore, $F^*$ is $(\text{card } F)^+$-saturated, $S$ has a solution in $F^*$.

Theorem 1 follows from Theorem $*$ by the uniqueness of model completions. The corollary follows directly from Theorem $*$ as follows:

*Proof of Corollary* ($\Rightarrow$) This follows since

$$(p = 0, q \neq 0)_{p,\, q \in F\{y\};\ \text{order } p > \text{order } q \leq -1}$$

is consistent.

($\Leftarrow$) Suppose $F$ solves each pair $(p = 0, q \neq 0)_{p,\, q \in F\{y\};\ \text{order } p > \text{order } q \geq -1}$, i.e. $F$ is a model of $DCF_0$, and suppose $S$ is a finite consistent system over $F$. Then $S$ has a solution in some extension $F' \supset F$. Let $F'' \supset F'$ be a model of $DCF_0$ (property 2 of model completions). Then $F''$ has a solution to $S$. Hence, by transfer (property 3 of model completions, $F'' \equiv_F F$), $F$ has a solution to $S$.

A general result of Robinson's says: Whenever a universal theory possesses a model completion, then it also has an elimination theorem. Hence, one gets a (purely) model theoretic proof of Seidenberg's elimination theorem, in a theoretically effective form (see [Note 6] for a simple proof of this).

**[5]** There are many different methods of constructing a prime differential closure. The following is one. (It is based on a general construction due to Morley [7] for constructing prime models over substructures for totally transcendental theories. I observed, quite easily, that the theory of differentially closed fields of characteristic 0 was totally transcendental. A complication in the characteristic $p > 0$ case, as pointed out by Wood [17, 18] is that the theory is *not* totally transcendental):

† Suppose $F\langle\eta\rangle$ is an extension of $F$. Let $p$ be a *minimal polynomial* for $\eta$ over $F$, i.e. $p$ is an irreducible polynomial in $F\{y\}$ of least order such that $p(\eta) = 0$. In case $\eta$ is differentially transcendental over $F$, then $p$ is the zero polynomial. The *order* of $\eta$ over $F$ is the order of $p$. It follows [13] that (1) if $p$ is a minimal polynomial for both $\eta$ and $\mu$ over $F$, then $F\langle\eta\rangle \simeq_F F\langle\mu\rangle$ via the map induced by sending $\eta$ to $\mu$ and (2) if $p \in F\{y\}$ is irreducible, there is an $\eta$ in some extension of $F$ such that $p$ is a minimal polynomial for $\eta$ over $F$.

We construct, by induction, a chain $\{F_i\}_{0 \le i < \omega}$ of extensions of $F$ such that $F_0 = F$ and $F_{i+1}$ has the following two properties:

(a) $F_{i+1}$ has solutions to each pair

$$(p = 0,\ q \neq 0)_{p,\, q \in F_i\{y\};\ \text{order } p > \text{order } q \ge -1}.$$

(b) Any embedding of $F_i$ into a differentially closed field $G$ can be extended to an embedding of $F_{i+1}$ into $G$.

Then we let $\hat{F} = \bigcup_{0 \le i < \omega} F_i$.

Noting that any pair $p, q \in \hat{F}\{y\}$ is already in $F_i\{y\}$ for some $i < \omega$, we see that by (a), $\hat{F}$ solves each pair $(p = 0, q \neq 0)_{p,\, q \in F\{y\};\ \text{order } p > \text{order } q \ge -1}$ and so $\hat{F}$ is differentially closed. By (b), $\hat{F}$ is prime, i.e. has the embedding property. Hence, $\hat{F}$ is a differential closure of $F$.

*Construction of* $F_{i+1}$ (given $F_i$) Let $\kappa$ be the cardinality of $F_i$ (which is seen to be the cardinality of $F$). Let $\{(p_{i_\alpha}, q_{i_\alpha})\}_{0 \le \alpha < \kappa}$ be a listing of all pairs $p_{i_\alpha}, q_{i_\alpha} \in F_i\{y\}$ such that order $p_{i_\alpha} >$ order $q_{i_\alpha} \ge -1$. Construct, by transfinite induction, a chain $\{F_{i_\alpha}\}_{0 \le \alpha < \kappa}$ of extensions of $F_i$ such that:

for $\alpha = 0$: $F_{i0} = F_i$;
for $\alpha$ a limit ordinal: $F_{i_\alpha} = \bigcup_{\beta < \alpha} F_{i_\beta}$;
for $\alpha + 1$: $F_{i_{\alpha+1}} = F_{i_\alpha}\langle \eta_{i_\alpha} \rangle$, where $\eta_{i_\alpha}$ is a solution of $(p_{i_\alpha} = 0, q_{i_\alpha} \neq 0)$ of *least order* over $F_{i_\alpha}$.

Let $F_{i+1} = \bigcup_{0 \le \alpha < \kappa} F_{i_\alpha}$. Clearly $F_{i+1}$ has property (a).

Now suppose $\phi_i$: $F_i \to G$ is an embedding of $F_i$ into a differentially closed field $G$. We construct an embedding $\phi_{i+1}$: $F_{i+1} \to G$ extending $\phi_i$. Construct, by transfinite induction, a chain $\{\phi_{i_\alpha}\}_{0 \le \alpha < \kappa}$ of extensions of $\phi_i$ such that:

for $\alpha = 0$: $\phi_{i0} = \phi_i$;
for $\alpha$ a limit ordinal: $\phi_{i_\alpha} = \bigcup_{\beta < \alpha} \phi_{i_\beta}$;
for $\alpha + 1$: $\phi_{i_{\alpha+1}}$: $F_{i_{\alpha+1}} \to G$ is defined as follows (given $\phi_{i_\alpha}$: $F_{i_\alpha} \to G$): Let $p$ be a minimal polynomial for $\eta_{i_\alpha}$ over $F_{i_\alpha}$. Let $S = (p_{i_\alpha} = 0, q_{i_\alpha} \neq 0, p = 0)$. Then $\eta_{i_\alpha}$ is a solution to $S$, and by the *least order* property of $\eta_{i_\alpha}$, $S$ completely determines the isomorphism type of $\eta_{i_\alpha}$ over $F_{i_\alpha}$. Hence (see [Note 6]), there is a $q \in F_{i_\alpha\{y\}}$, order $p >$ order $q \ge -1$ such that $S$ is equivalent to $(p = 0, q \neq 0)$, i.e. $(p = 0, q \neq 0)$ completely determines the isomorphism type of $\eta_{i_\alpha}$ over $F_{i_\alpha}$. Since $G$ is differentially closed, there is a $\mu_{i_\alpha} \in G$ which solves $(\phi_{i_\alpha}(p) = 0, \phi_{i_\alpha}(q) \neq 0)$. Clearly (by the least order property of $\eta_{i_\alpha}$),

$\phi_{i_\alpha}(p)$ is a minimal polynomial for $\mu_{i_\alpha}(F_{i_\alpha})$. Hence, mapping $\mu_{i_\alpha}$ to $\mu_{i_\alpha}$ induces an embedding $\phi_{i_{\alpha+1}}: F_{i_{\alpha+1}} \to G$ extending $\phi_{i_\alpha}$.†

Now let $\phi_{i+1} = \bigcup_{0 \leq \alpha < \kappa} \phi_{i_\alpha}$. Hence, $F_{i+1}$ has property (b).

Shelah [15] (see also Sacks [12]) shows prime closure are unique in general for totally transcendental theories. The following is an outline of his proof as specialized and modified to differential fields. The details are filled in in the Appendix.

Suppose $F$, $G$ are differential fields.

***Proposition*** *If $\hat{F}$ is a differential closure of $F$, then*

(a) *$\hat{F}$ is atomic over $F$ (see* (2) *in Section* 4 *for definition of* atomic);

(b) (Shelah) *there is no uncountable set of indiscernibles in $\hat{F}$ over $F$ (see* (10) *in Section* 4 *for definition of* set of indiscernibles).

(The proof of (a) is easy, (b) is nontrivial. See the Appendix.)

***Theorem*** (Shelah) (specialized to differential fields) *Suppose $F^*$, $G^*$ are differentially closed, $F^* \supset F$, $G^* \supset G$. Suppose*

(a) *$F^*$ is atomic over $F$ (similarly for $G^*$ over $G$);*

(b) *there is no uncountable set of indiscernibles in $F^*$ over $F$ (similarly for $G^*$ over $G$).*

*Let $S$ be a pair $(p = 0, q \neq 0)_{p,\, q \in F\{y\};\ \text{order } p > \text{order } q}$, $p$ absolutely irreducible.*

*Then any isomorphism $\phi$ of $F$ onto $G$ can be extended to an isomorphism of $\widehat{F\langle X\rangle}$ onto $\widehat{G\langle Y\rangle}$, where $X = \{x \in F^* \mid x$ is a solution $S\}$, $Y = \{y \in G^* \mid y$ is a solution to $\phi(S)\}$.*

***Corollary*** *Suppose $F^*$, $G^*$ are differentially closed, $F^* \supset F$, $G^* \supset G$. Suppose* (a) *and* (b) *above hold. Then any isomorphism of $F$ onto $G$ can be extended to an isomorphism of $F^*$ onto $G^*$.*

*Proof* Let $S = (0 = 0, 1 \neq 0)_{\text{order } 0 = \infty > \text{order } 1 = -1}$.

***Corollary*** *Suppose $\hat{F}$ is a differential closure of $F$, $\hat{G}$ a differential closure of $G$. Then any isomorphism of $F$ onto $G$ can be extended to an isomorphism of $\hat{F}$ onto $\hat{G}$.*

***Corollary*** *Any two differential closures of $F$ are isomorphic over $F$.*

***Corollary*** *A differentially closed field $F^* \supset F$ is a differential closure of $F \Leftrightarrow$* (a) *and* (b) *above hold.*

† Finding such a $\mu_{i_\alpha}$ can be done without reference to [Note 6] as follows: $S$ has a solution in $\hat{F}$ (which is differentially closed). $G$ is also differentially closed. Hence, by the transfer principle, $\phi(S)$ has a solution, say $\mu_{i_\alpha}$ in $G$. Clearly, by the least order property of $\eta_{i_\alpha}$, $\phi(p)$ must be a minimal polynomial for $\mu_{i_\alpha}$ over $\phi_{i_\alpha}(F_{i_\alpha})$.

To prove the theorem we need the following lemma:

***Lemma*** *Suppose $F^*$ is differentially closed, $F^* \supset F$, and $F^*$ atomic over $F$. Then*

(c) *$F^*$ is atomic over $F\langle \eta_1, \ldots, \eta_n \rangle$, for $\eta_1, \ldots, \eta_n \in F^*$.*
(d) (Shelah) *Suppose $\{S_\alpha\}_{\alpha<\kappa}$ is a set of pairs,*

$$S_\alpha = (p_\alpha = 0,\ q_\alpha \neq 0)_{p_\alpha,\, q_\alpha \in F\{y\};\ \text{order } p_\alpha > \text{order } q_\alpha}.$$

*Let $X_\alpha = \{x \in F^* \mid x$ is a solution to $S_\alpha\}$. Then $F^*$ is atomic over $F\langle \bigcup_{\alpha<\kappa} X_\alpha \rangle$.*

(The proof of (c) is easy, (d) nontrivial. See the Appendix.)

*Proof of Theorem* By transfinite induction on the order of $p$ (hence get result for order $p = \infty$). Suppose the result holds for all absolutely irreducible $p$ such that order $p < \delta$. Show for order $p = \delta$: Let $\{\eta_r\}_{r<r_F}$ [similarly, $\{\mu_r\}_{r<r_G}$] be a listing of a maximal subset of $F^*[G^*]$ such that for each $s < r_F$ [$s < r_G$], $p$ [$\phi(p)$] is a minimal polynomial for $\eta_s$ [$\mu_s$] over $F\langle\{\eta_r\}_{r<s}\rangle$ [$G < \{\mu_r\}_{r<s}$]. (*Note* $\{\eta_r\}_{r<r_F} \subseteq X$ [$\{\mu_r\}_{r<r_G} \subseteq Y$].) By hypothesis (b) and a reordering if necessary, we can assume $r_F, r_G \leq \omega$.

Construct, by induction, a chain $\{\phi_i\}_{0\leq i<\omega}$ of extensions of $\phi$ such that $F^*$ [$G^*$] has property (a), (b) over $F_i$, the domain of $\phi_i$ [$G_i$, the range of $\phi_i$] and such that $F_i$ [$G_i$] is algebraically closed.

(i) $i = 0$. Let $\phi_0 = \tilde{\phi}: \tilde{F} \xrightarrow{\sim} \tilde{G}$ ((a) holds by (d); (b) holds trivially).

(ii) $i + 1 = 3r + 1$. Suppose $\phi_i$ is defined. If $r \geq r_F$ let $\phi_{i+1} = \phi_i$. If $r < r_F$, then by (a), $\eta_r$ realizes an isolated type over $F_i$, i.e. there is a pair $S_i$ which completely determines the isomorphism type of $\eta_r$ over $F_i$ [Note 6]. Since $G^*$ is differentially closed, there is a $\eta_r^G \in G^*$ which is a solution to $\phi_i(S_i)$. Let

$$\phi_{i+1}: \widetilde{F_i\langle\eta_r\rangle} \xrightarrow{\sim} \widetilde{G_i\langle\eta_r^G\rangle}$$

be the extension of $\phi_i$ induced by sending $\eta_r$ to $\eta_r^G$ (*Note* $\eta_r^G \in Y$.) ((a) holds by (c) and (d); (b) holds trivially.)

(iii) $i + 1 = 3r + 2$. Suppose $\phi_i$ is defined. As in (ii) if $r < r_G$, choose $\mu_r^F \in F^*$ such that

$$\phi_{i+1}: \widetilde{F_i\langle\mu_r^F\rangle} \xrightarrow{\sim} \widetilde{G_i\langle\mu_r\rangle}$$

is the extension of $\phi_i$ induced by sending $\mu_r^F$ to $\mu_r$. (*Note* $\mu_r^F \in X$.)

(iv) $i + 1 = 3r + 3$. Suppose $\phi_i$ is defined. Let $\{S_\alpha^i\}_{\alpha<\kappa}$ be a listing of all pairs $S_\alpha^i = (p_\alpha^i = 0,\ q_\alpha^i \neq 0)_{p_\alpha^i,\, q_\alpha^i \in F_i\{y\};\ \delta > \text{order } p_\alpha^i > \text{order } q_\alpha^i \geq -1}$, $p_\alpha^i$ irreducible and $S_\alpha^i \to S$ (i.e. any solution of $S_\alpha^i$ is a solution of $S$). For $\alpha < \kappa$, let $X_\alpha^i = \{x \in F^* \mid x$ is a solution to $S_\alpha^i\}$ [$Y_\alpha^i = \{y \in G^* \mid y$ is a solution to $\phi_i(S)\}$]. (*Note* $X_\alpha^i \subseteq X$ [$Y_\alpha^i \subseteq Y$].)

Construct (by transfinite induction) a chain $\{\phi_{i_\alpha}\}_{\alpha\le\kappa}$ of extensions of $\phi_i$ such that for each $\alpha\le\kappa$, $F^*$ $[G^*]$ has properties (a) and (b) over $F_{i_\alpha}$, the domain of $\phi_{i_\alpha}$ $[G_{i_\alpha}$, the range of $\phi_{i_\alpha}]$ and such that

$$F_{i_\alpha} = \widetilde{F_i\Big\langle \bigcup_{\varepsilon<\alpha} X_\varepsilon{}^i\Big\rangle} \quad \Big[G_{i_\alpha} = \widetilde{G_i\Big\langle \bigcup_{\varepsilon<\alpha} Y_\varepsilon{}^1\Big\rangle}\Big]:$$

$\alpha = 0$: Let $\phi_{i_0} = \phi_i$.

$\alpha$ a limit ordinal: Suppose $\phi_{i_\beta}$, $\beta<\alpha$ defined. Let $\phi_{i_\alpha} = \bigcup_{\beta<\alpha}\phi_{i_\beta}$. Then

$$F_{i_\alpha} = \bigcup_{\beta<\alpha} F_{i_\beta} = \bigcup_{\beta<\alpha}\widetilde{F_i\Big\langle \bigcup_{\varepsilon<\beta} X_\varepsilon{}^i\Big\rangle} = \widetilde{F_i\Big\langle \bigcup_{\varepsilon<\alpha} X_\varepsilon{}^i\Big\rangle}$$

(since $\alpha$ is a limit ordinal). [Similarly for $G_{i_\alpha}$.] ((a) holds by (c); (b) holds trivially.)

$\alpha$ a successor (i.e. $\alpha = \beta+1$): Suppose $\phi_{i_\beta}$ is defined. Since order $p_\beta{}^i < \delta$ and $p_\beta{}^i$ is absolutely irreducible ($F_i$ is algebraically closed), by induction on $\delta$, there is a

$$\phi_{i_\alpha} = \phi_{i_{\beta+1}}: \widetilde{F_{i_\beta}\langle X_\beta{}^i\rangle} \xrightarrow{\sim} \widetilde{G_{i_\beta}\langle Y_\beta{}^i\rangle}$$

extending $\phi_{i_\beta}$. Then

$$F_{i_\alpha} = F_{i_{\beta+1}} = \widetilde{F_{i_\beta}\langle X_\beta{}^i\rangle} = \widetilde{F_i\Big\langle \bigcup_{\varepsilon<\beta} X_\varepsilon{}^i \cup X_\beta{}^i\Big\rangle} = \widetilde{F_i\Big\langle \bigcup_{\varepsilon<\alpha=\beta+1} X_\varepsilon{}^i\Big\rangle}$$

[similarly for $G_{i_\alpha} = G_{i_{\beta+1}}$]. ((a) holds by (c); (b) holds trivially.)

Now, let $\phi_{i+1} = \phi_{i_\kappa}$. So

$$F_{i+1} = \widetilde{F_i\Big\langle \bigcup_{\alpha<\kappa} X_\alpha{}^i\Big\rangle} \quad \Big[G_{i+1} = \widetilde{G_i\Big\langle \bigcup_{\alpha<\kappa} Y_\alpha{}^i\Big\rangle}\Big],$$

$\phi_{i+1}$ extends $\phi_i$, and (a) and (b) hold.

Now let $\Phi = \bigcup_{i<\omega}\phi_i$. Clearly $\Phi$ extends $\phi$. So we need only show that

$$\text{domain } \Phi = \widetilde{F\langle X\rangle} \quad [\text{similarly, range } \Phi = \widetilde{G\langle Y\rangle}]:$$

It is clear that

$$\text{domain } \Phi = \widetilde{F\Big\langle \{\eta_r\}_{r<r_F}, \{\mu_r{}^F\}_{r<r_G}, \bigcup_{\alpha<\kappa,\, r<\omega} X_\alpha^{3r+2}\Big\rangle}.$$

So it is sufficient to show that if $x\in X$, then $x\in$ domain $\Phi$: Suppose $x\in X$. So $x$ is a solution to $p$. Then, by the maximality of $\{\eta_r\}_{r<r_F}$, the order $x$ (over

$F\langle\{\eta_r\}_{r<r_F}\rangle) < \delta$. So, for some finite $r^* < r_F$, the order $x$ (over $F\langle\{\eta_r\}_{r<r^*}\rangle) < \delta$. So, order $x$ (over $F_{3r^*+2}) < \delta$. So, $x$ is a solution to $S_\alpha^{3r^*+2}$, some $\alpha < \kappa$ [Note 6]. So, $x \in X_\alpha^{3r^*+2}$, some $\alpha < \kappa$.

**[6]** A technique used in model theory is to associate with each structure $F$ and each $n < \infty$ a topological space $S_n(F)$, the *n-Stone space of F*: Suppose $F$ is a differential field:

*Points* The points of $S_n(F)$ are the isomorphism types of $n$-fold extensions of $F$ (i.e. given $\eta_1, \ldots, \eta_n$ in some extension of $F$, let $\not{p}(\eta) = \{S \mid S = (f_1 = 0, \ldots, f_s = 0, J \neq 0)_{f_i, J \in F\{y_1, \ldots, y_n\};\, s<\infty}$ and $(\eta_1, \ldots, \eta_n)$ is a solution to $S\}$. The points of $S_n(F)$ are just these $\not{p}(\eta)$'s).

*Topology* Suppose $S = (f_1 = 0, \ldots, f_s = 0, J \neq 0)_{f_i, J \in F\{y_1, \ldots, y_n\};\, s<\infty}$. Let $\mathscr{V}_S = \{\not{p} \in S_n(F) \mid S \in \not{p}\}$. The topology of $S_n(F)$ is the topology generated by the $\mathscr{V}_S$'s (which are both open and closed).

Thus, $S_n(F)$ is a totally disconnected compact Hausdorff space. (Compactness follows from the compactness theorem of first order logic plus the Elimination of Quantifiers Theorem. Refer to the additional comments that follow shortly.)

A point $\not{p} \in S_n(F)$ is isolated $\Leftrightarrow$ there is a finite system $S$ such that $\mathscr{V}_S = \{\not{p}\}$, i.e. $S$ completely determines the isomorphism type $\not{p}$. If $F' \supset F$, then $F'$ is atomic over $F$ if and only if for each $\eta_1, \ldots, \eta_n \in F'$, $\not{p}(\eta)$ is isolated.

Now *consider* $S_1(F)$: Suppose $\not{p} \in S_1(F)$. Then there is an $\eta$ in some extension of $F$ such that $\not{p} = \not{p}(\eta)$. Let $p \in F\{y\}$ be a minimal polynomial for $\eta$ over $F$. Let *order* $\not{p}$ = order $_F\, \eta$ (= order $p$). (This is independent of choice of $\eta$.) Then

$$\{\not{p}\} = \bigcap_{\substack{q \in F\{y\} \\ \text{order } \not{p} > \text{order } q \geq -1}} \mathscr{V}_{(p=0,\, q\neq 0)}.$$

Now suppose $S = (f_1 = 0, \ldots, f_s = 0, J \neq 0)_{f_i, J \in F\{y\};\, s<\infty}$ and $\not{p} \in \mathscr{V}_S$. Then, by compactness, there is a $q \in F\{y\}$, order $\not{p} >$ order $q \geq -1$ such that letting $S' = (p = 0, q \neq 0)$, $\mathscr{V}_{S'} \subset \mathscr{V}_S$ (i.e. $S' \to S$).† So, by compactness, there are a finite number of pairs

$$S_j = (p_j = 0, q_j \neq 0)_{p_j, q_j \in F\{y\};\, \text{order } p_j > \text{order } q_j \geq -1} \qquad (j = 1, \ldots, k)$$

† $\in$ This can be seen as follows: $\{\not{p}\} \cap -\mathscr{V}_S = \phi$ and so, by compactness, there are $q_1, \ldots, q_n \in F\{y\}$, order $\not{p} >$ order $q_i \geq -1$ such that $\bigcap_{i=1}^n \mathscr{V}_{(p=0,\, q_i\neq 0)} \cap -\mathscr{V}_S = \phi$. Letting $q$ be the product of the $q_i$'s we get $\mathscr{V}_{(q=0,\, q\neq 0)} \cap -\mathscr{V}_S = \phi$ or, in other words, $\mathscr{V}_{(p=0,\, q\neq 0)} \subset \mathscr{V}_S$.

such that $\mathscr{V}_S = \mathscr{V}_{S_1} \cup \cdots \cup \mathscr{V}_{S_k}$ (i.e. $S \leftrightarrow S_1 \vee \cdots \vee S_k$). (This is the reason we need only consider *pairs*.) In particular, if $\not{p}$ is isolated there is a pair

$$S = (p = 0,\ q \neq 0)_{p \text{ irreducible}, \, q \in F\{y\}; \text{ order } p > \text{ order } q \geq -1}$$

such that $\{\not{p}\} = \mathscr{V}_S$. Thus, $p$ is a minimal polynomial for any $\mu$ (in an extension of $F$) solving $(p = 0,\ q \neq 0)$.

Now suppose $\mathscr{V}$ is open. Let $\not{p} \in \mathscr{V}$ be a point of least order (in $\mathscr{V}$). Then $\{\not{p}\} = \mathscr{V}_{(p=0)} \cap \mathscr{V}$ is open and so the isolated points of $S_1(F)$ are dense. (This is the reason for the *existence* of prime differential closures.)

In general, for $S_n(F)$, the isolated points are dense. So, for each finite consistent system $S$ there is a finite consistent system $S' \supset S$ so that if $(\eta_1, \ldots, \eta_n)$ and $(\mu_1, \ldots, \mu_m)$ are both solutions to $S'$ over $F$, then $F\langle\eta_1, \ldots, \eta_n\rangle \simeq_F F\langle\mu_1, \ldots, \mu_n\rangle$.

*Additional comments* The $n$-Stone space of $F$, $S_n{}^*(F)$ is actually defined as follows: Let $\sigma(y_1, \ldots, y_n)$ be a first order formula with coefficients from $F$ and free variables among $y_1, \ldots, y_n$ (i.e. $y_i$ not bound in $\sigma$ by quantifiers $\exists y_i$ or $\forall y_i$).

*Points* For each $\eta_1, \ldots, \eta_n$ in some extension (and hence in some differentially closed extension) of $F$, let $\not{p}^*(\eta) = \{\sigma \mid \sigma(\eta_1, \ldots, \eta_n)$ true in some (and hence every) differentially closed extension of $F\langle\eta_1, \ldots, \eta_n\rangle\}$. The points of $S_n{}^*(F)$ are just these $\not{p}^*(\eta)$'s.

*Topology* The basic open sets of $S_n{}^*(F)$ are of the form $\mathscr{V}_\sigma = \{\not{p}^* \in S_n{}^*(F) \mid \sigma \in \not{p}^*\}$.

*Note* $\mathscr{V}_{\sigma_1 \& \sigma_2} = \mathscr{V}_{\sigma_1} \cap \mathscr{V}_{\sigma_2}$, $\mathscr{V}_{\sigma_1 \vee \sigma_2} = \mathscr{V}_{\sigma_1} \cup \mathscr{V}_{\sigma_2}$, $\mathscr{V}_{-\sigma} = -\mathscr{V}_\sigma$ (hence $\mathscr{V}_\sigma$ is clopen). Hence, $S_n{}^*(F)$ is a totally disconnected, compact (due to the compactness theorem for first order logic) Hausdorff space.

*Correspondence between $S_n{}^*(F)$ and $S_n(F)$* Suppose

$$S = (f_1 = 0, \ldots, f_s = 0,\ J \neq 0)_{f_i,\, J \in F\{y_1, \ldots, y_n\};\, s < \infty}.$$

Let $S^*(y_1, \ldots, y_n)$ be (the quantifier free first order formula with coefficients from $F$ and free variables among $y_1, \ldots, y_n$) $f_1 = 0 \;\&\; \cdots \;\&\; f_j = 0 \;\&\; J \neq 0$.

For each $\eta_1, \ldots, \eta_n$ in some extension of $F$, $\not{p}(\eta) = \{S \mid S^* \in \not{p}^*(\eta)\}$. On the other hand, by the transfer property, $\not{p}^*(\eta)$ is the unique point of $S_n{}^*(F)$ containing $\{S^* \mid S \in \not{p}(\eta)\}$. So the natural correspondences $\not{p}^*(\eta) \leftrightarrow \not{p}(\eta)$, $\mathscr{V}_{S^*} \leftrightarrow \mathscr{V}_S$ are well defined.

***Elimination of Quantifiers Theorem*** $(S_n{}^*(F) \leftrightarrow S_n(F))$ *For each $\sigma$ there are a finite number of finite systems $S_1, \ldots, S_k$ over $F$ such that $\mathscr{V}_\sigma = \mathscr{V}_{S_1{}^*} \cup \cdots \cup \mathscr{V}_{S_k{}^*}$ (i.e. $\vDash_{\mathrm{DEF}_0(F)} \sigma \leftrightarrow S_1{}^* \vee \cdots \vee S_k{}^*$, i.e. $\sigma(\eta_1, \ldots, \eta_n)$ is true in some (and hence every) differentially closed field containing*

*$F\langle\eta_1, \ldots, \eta_n\rangle \Leftrightarrow \langle\eta_1, \ldots, \eta_n\rangle$ is a solution to at least one $S_j$ $(j = 1, \ldots, k)$). Hence, the spaces $S_n^*(F)$ and $S_n(F)$ are the same (i.e. are naturally identified).*

*Proof* Note first that $\{\not p^*(\eta)\} = \bigcup_{S \in \not p(\eta)} \mathscr{V}_{S^*}$. So, by compactness, for each $\not p^*(\eta) \in \mathscr{V}_\sigma$ there is a finite system $S$ such that $\not p^*(\eta) \in \mathscr{V}_{S^*} \subset \mathscr{V}_\sigma$. So, again by compactness, there are a finite number of finite systems $S_1, \ldots, S_k$ such that $\mathscr{V}_\sigma = \bigcup_{j=1}^k \mathscr{V}_{S_j^*}$.

***Corollary*** (Seidenberg's Elimination Theorem) *Proof* Suppose

$$S(y_1, \ldots, y_n, u_1, \ldots, u_m) = (f_1 = 0, \ldots, f_s = 0, J \neq 0)_{f_i,\, J \in I\{y_1, \ldots, y_n\}\{u_1, \ldots, u_m\}:\, s<\infty}$$

Let $\sigma$ be $\exists u_1 \cdots \exists u_m\ S^*(y_1, \ldots, y_n, u_1, \ldots, u_m)$. Then, by above, there are a finite number of finite systems $S_1, \ldots, S_k$ over $I$ such that $\models_{\mathrm{DCF}_{0(Q)}} \sigma \leftrightarrow S_1^* \vee \cdots \vee S_k^*$. By the existential nature of $\sigma$, this is equivalent to saying: For each differential field $F$ and $\eta_1, \ldots, \eta_n$ in $F$, $S(\eta_1, \ldots, \eta_n, u_1, \ldots, u_m)$ is consistent $\Leftrightarrow (\eta_1, \ldots, \eta_n)$ is a solution to at least one $S_j$ $(j = 1, \ldots, k)$.

*Theoretical effectiveness* By the completeness theorem for first order logic, $\sigma \leftrightarrow S_1^* \vee \cdots \vee S_k^*$ must be a theorem of $\mathrm{DCF}_0$. Since $\mathrm{DCF}_0$ is effectively axiomatized, its theorems can be effectively enumerated. Hence, using a search procedure through the theorems of $\mathrm{DCF}_0$, given any $S$ one can find the corresponding $S_j$'s is a finite number of steps.

**[7]** Thus, one gets an alternate proof of Theorem 1 in "Existence Theorems Connected with the Picard–Vessiot Theory of Homogeneous Ordinary Differential Equations" [4]. (Just let $S = (f_1 = 0, \ldots, f_s = 0, J \neq 0)$ where $\{f_1, \ldots, f_s\} = \{\Sigma\}$.)

Also, to emphasize the technique of transfer, Theorem 2 of the same paper could have been gotten directly from the original theorem [3, p. 27]: The original theorem states that the following first order sentence with coefficients from $F$ is true in some extension $F' \supset F$:

$$\exists x_1 \cdots \exists x_n \left[ \mathop{\&}_{i=1}^{n} (x_i^{(n)} + p_1 x_i^{(n-1)} + \cdots + p_n x_i = 0) \right.$$
$$\left. \& \ \forall z_1 \cdots \forall z_n \left( \mathop{\&}_{i=1}^{n} (z_i' = 0) \ \& \ (z_1 x_1 + \cdots + z_n x_n = 0) \rightarrow \mathop{\&}_{i=1}^{n} z_i = 0 \right) \right].$$

This sentence is true in any extension of $F'$, and so in particular in $\hat{F}' \supset F$. By transfer, it is true in $\hat{F}$. The result now follows from (4) in Section 4.

**[8]** The proof will use ultraproducts. So some preliminaries: Let $I$ be a countable indexing set. Let $D$ be a nonprincipal ultrafilter on $I$ (thus, $D$ is a maximal filter on $I$ containing the cofinite subsets of $I$). Let $\{F_i\}_{i \in I}$ be differential fields and form the product $\prod_{i \in I} F_i$. For $(a_i)_{i \in I}$, $(b_i)_{i \in I}$ let $(a_i)_{i \in I} \sim (b_i)_{i \in I}$ if and only if $a_i = b_i$ almost everywhere with respect to $D$ (i.e. if and only if $\{i \mid a_i = b_i\} \in D$). This is an equivalence relation. Elements of the ultraproduct $\prod_{i \in I} F_i|_D$ will be the equivalence classes. Operations in the ultraproduct are defined in the obvious way. If all the $F_i$ are identical to $F$, then we have formed the ultrapower, $F^I|_D$.

*Properties of ultraproducts* (nonprincipal)

(1) ***Fundamental Theorem of Ultraproducts*** *Suppose $S(A_1, \ldots, A_m)$ is a first order sentence with coefficients $A_j \in \prod_{i \in I} F_i|_D$. Let $A_j$ be the equivalence class of $(a_{ji})_{i \in I}$. Then $S(A_1, \ldots, A_m)$ is true in $\prod_{i \in I} F_i|_D \Leftrightarrow S(a_{1_i}, \ldots, a_{m_i})$ is true a.e. (with respect to D), i.e. $\{i \mid S(a_{1_i}, \ldots, a_{m_i})$ true in $F_i\} \in D$. Hence, $\prod_{i \in I} F_i|_D$ is a differential field; it is differentially closed if each $F_i$ is. We note that $\mathscr{C}_{\prod_{i \in I} F_i|_D}$ is naturally isomorphic to $\prod_{i \in I} \mathscr{C}_{F_i}|_D$.* (Let $X \in \prod_{i \in I} F_i|_D$. Let $X$ be the equivalence class of $(X_i)_{i \in I}$. Then $X' = 0 \Leftrightarrow X_i' = 0$ a.e. (with respect to $D$).)

(2) *$\omega^+$-Saturation Let*

$$S = \{f_j = 0, J_j \neq 0\}_{j \in I;\, f_j, J_j \in \prod_{i \in I} F_i|_D\{y_1, \ldots, y_n\}}$$

*be an infinite system (fixed n). Then S has a solution in $\prod_{i \in I} F_i|_D \Leftrightarrow$ every finite subset of S has a solution in $\prod_{i \in I} F_i|_D$.*

*Proof of* (15) *in Section 4* (i) We first note that we can reduce to the case where $F$ is countable.

(ii) Let $F\langle \boldsymbol{\mu} \rangle = F\langle \mu_1, \ldots, \mu_n \rangle$ be a strongly normal extension of $F$. Then $\mathscr{C}_{F\langle \boldsymbol{\mu} \rangle} = \mathscr{C}_F$ is algebraically closed (by definition).

(iii) We next note that we can choose $F$ so that $\mathscr{C}_{\hat{F}\langle \boldsymbol{\mu} \rangle} = \mathscr{C}_F$: Let $\widehat{F\langle \boldsymbol{\mu} \rangle}$ be a differential closure of $F\langle \boldsymbol{\mu} \rangle$. Let $\hat{F}$ be a differential closure of $F$ in $\widehat{F\langle \boldsymbol{\mu} \rangle}$. Then, $\mathscr{C}_{\hat{F}\langle \boldsymbol{\mu} \rangle} \subseteq \mathscr{C}_{\widehat{F\langle \boldsymbol{\mu} \rangle}} =$ (by (4) in Section 4) $\mathscr{C}_{\widetilde{F\langle \boldsymbol{\mu} \rangle}} = \mathscr{C}_F$.

We now aim to show that $\mu_1, \ldots, \mu_n \in \hat{F}$ (and then we will be finished):

(iv) Consider the following diagram where arrows represents the natural embeddings:

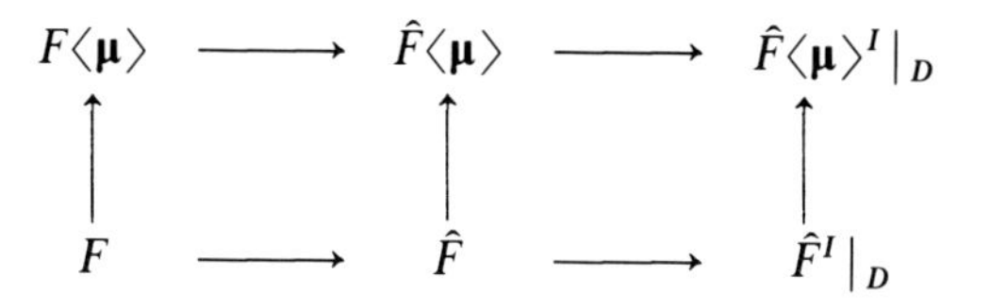

Note that both $\mathscr{C}_{\hat{F}^I}|_D$ and $\mathscr{C}_{\hat{F}\langle\boldsymbol{\mu}\rangle^I}|_D$ are naturally isomorphic to $\mathscr{C}_{F}{}^{I}|_D$. We view all the structures as substructures of the big one. Then all the subfields of constants in the right hand column get identified; we call this field $\mathscr{C}$.

(v) *Claim* There are $\eta_1, \ldots, \eta_n \in \hat{F}^I|_D$ so that mapping $\mu_i$ to $\eta_i$ induces an $\hat{F}$-isomorphism of $\hat{F}\langle\boldsymbol{\mu}\rangle$ into $\hat{F}^I|_D$ (and so induces an $F$-isomorphism of $F\langle\boldsymbol{\mu}\rangle$ into $\hat{F}^I|_D$).

*Proof* Let $S = \{f = 0 \mid f \in \hat{F}\{y_1, \ldots, y_n\} \;\&\; f(\mu_1, \ldots, \mu_n) = 0\} \cup \{J \neq 0 \mid J \in \hat{F}\{y_1, \ldots, y_n\} \;\&\; J(\mu_1, \ldots, \mu_n) \neq 0\}$. Then $S$ is countable (since $F$ and hence $\hat{F}$ is) and $S$ completely determines the isomorphism type of $(\mu_1, \ldots, \mu_n)$ over $\hat{F}$. Clearly, $S$ is consistent and so every finite subsystem of $S$ is consistent. But then, since $\hat{F}$ is differentially closed, every finite subsystem of $S$ has a solution in $\hat{F}$ and hence in $\hat{F}^I|_D$. Hence, by $\omega^+$-saturation, $S$ has a solution, say $(\eta_1, \ldots, \eta_n)$ in $\hat{F}^I|_D$.

(vi) By strong normality, $F\langle\boldsymbol{\mu}, \boldsymbol{\eta}\rangle \subseteq F\langle\boldsymbol{\eta}\rangle\langle\mathscr{C}\rangle$. But $F\langle\boldsymbol{\eta}\rangle\langle\mathscr{C}\rangle \subseteq \hat{F}^I|_D$. Hence $\mu_1, \ldots, \mu_n \in \hat{F}^I|_D$. Hence, $\mu_1, \ldots, \mu_n \in \hat{F}$.

Caracas
December 30, 1972

Dear Professor Kolchin,

The missing parts of the proofs of uniqueness (Proposition (a), (b) and Lemma (c), (d) of my last letter to you) are given on the next pages.

I had written these up a while ago after talking to you and then planned to write something about extensions $F_1$ of $F$ in $\hat{F}$ ($\hat{F}$ is a differential closure of $F$) such that each $F$ isomorphism $\sigma$: $F_1 \to \hat{F}$ has the property $\sigma F_1 \subseteq F_1$ (i.e. generalizations of strongly normal extensions) and also wanted to show: *Conjecture* Suppose $F\langle x\rangle \supset F$ and $x' = x^3 - x^2$. Then for each $\eta \in \widehat{F\langle x\rangle}$ such that $\eta' = \eta^3 - \eta^2$, either $\eta \in F$ or $\eta = x$. For then there would be an infinite number of algebraically independent solutions of $y' = y^3 - y^2$ in $\hat{\mathbb{Q}}$ and so $\hat{\mathbb{Q}}$ would not be minimal over $\mathbb{Q}$.†

But then I got into the work on a mathematical theory of Inductive Inference [2] with Manuel in Italy and so that all got put aside, so I'm sending it as is (was).

We are now on our way back to California, having stopped in Caracas to visit our families.

Best wishes and Happy New Year

Lenore Blum

† Rosenlicht [10] has now shown $\hat{\mathbb{Q}}$ not minimal over $\mathbb{Q}$ by showing $y' = y/1 + y$ has infinitely many algebraically independent solutions in $\hat{\mathbb{Q}}$ (Spring 1973).

## Appendix: Details of uniqueness proof filled in

### Definitions

Suppose $F \subseteq F_1 \subseteq F^*$.

***Definition*** *$F_1$ is normal over $F$ in $F^*$* if and only if whenever $\eta \in F_1$ and $\bar{\eta} \in F^*$ is a conjugate of $\eta$ over $F$, then $\bar{\eta} \in F_1$.

Suppose $F_1$ is atomic over $F$.

*Remark* Then (see (2) in Section 4) for each $x \in F_1$ there are $p, q \in F\{y\}$, order $p >$ order $q$ such that $p$ is a minimal polynomial for $x$ over $F$, and whenever $(p(\bar{x}) = 0,\ q(\bar{x}) \neq 0)$ for $\bar{x}$ in some extension of $F$, then $p$ is a minimal polynomial for $\bar{x}$ over $F$. So, $F_1$ is normal over $F$ in $F^* \Leftrightarrow F_1 = F\langle \bigcup_\alpha X_\alpha \rangle$ for some pairs $S_\alpha = (p_\alpha = 0,\ q_\alpha \neq 0)$, order $p_\alpha >$ order $q_\alpha$, $p_\alpha$, $q_\alpha \in F\{y\}$, and $X_\alpha = \{x \in X^* \mid (p_\alpha(x) = 0,\ q_\alpha(x) \neq 0)\}$.

Suppose $p \in F\{y\}$ and $X$ is a set of elements in some extension of $F$.

***Definition*** *$X$ is indiscernible (with respect to $p$) over $F$* if and only if whenever $X^*$ is a (finite) subset of $X$ and $x \in X - X^*$, then $p$ is a minimal polynomial for $x$ over $F\langle X^* \rangle$.

***Definition*** *$X$ is indiscernible over $F$* if and only if for some absolutely irreducible $p \in F\{y\}$, $X$ is indiscernable (with respect to $p$) over $F$.

*Remark* The above definition of indiscernability is somewhat stronger than necessary. However, with this definition the proofs become easier and neater. These (modified) results are sufficient for our purposes. Hence, we get a simplification of Shelah's proof when we specialize directly to the case of differential fields.

### Statements

pre (a) Suppose $F_1 \subset F_2 \subset F_3$, $F_2$ atomic over $F_1$, and $F_3$ atomic over $F_2$. Then $F_3$ atomic over $F_1$.

(a) Suppose $\hat{F}$ is a differential closure of $F$. Then $\hat{F}$ is atomic over $F$.

(c) Suppose $F \subset F^*$, $F^*$ atomic over $F$, and $\eta_1, \ldots, \eta_N \in F^*$. Then $F^*$ atomic over $F\langle \eta_1, \ldots, \eta_N \rangle$.

(d) (Shelah) Suppose $F \subset F_1 \subset F^*$. Suppose $F^*$ differentially closed and atomic over $F$. Suppose $F_1$ is normal over $F$ in $F^*$. Then $F^*$ atomic over $F_1$.

pre (b) (Shelah) Suppose $X$ is a set of indiscernibles over $F$ and $\eta_1, \ldots, \eta_N$ are in some extension of $F$. Then all but a finite number of elements of $X$ are indiscernible over $F\langle \eta_1, \ldots, \eta_N \rangle$.

(b) (Shelah) Suppose $\hat{F}$ is a differential closure of $F$. Then there are no uncountable sets of indiscernibles in $\hat{F}$ over $F$.

**Proofs**

pre (a) We must show that each finite sequence from $F_3$ realizes an isolated (isomorphism) type over $F_1$, i.e. we must show that for each $\eta_1, \ldots, \eta_N \in F_3$ there is a finite system $S(y_1, \ldots, y_N)$ of differential polynomial equations and inequations over $F_1$ which completely determine the isomorphism type of $\langle \eta_1, \ldots, \eta_N \rangle$ over $F_1$.

So suppose $\eta_1, \ldots, \eta_N \in F_3$. By hypothesis there is a finite system $S_1(\mu_1, \ldots, \mu_m, y_1, \ldots, y_N)$ with $\mu_i \in F_2$ which completely determine the isomorphism type of $\eta_1, \ldots, \eta_N$ over $F_2$. Again, by hypothesis there is a finite system $S_2(y_1, \ldots, y_m)$ over $F_1$ which completely determines the isomorphism type of $\langle \mu_1, \ldots, \mu_m \rangle$ over $F_1$.

Let $\sigma(y_1, \ldots, y_N)$ be the following first order formula (over $F_1$): $\exists x_1 \cdots \exists x_m(S_2^*(x_1, \ldots, x_m) \mathbin{\&} S_1^*(x_1, \ldots, x_m, y_1, \ldots, y_N))$. Then $\sigma(y_1, \ldots, y_N)$ completely determines the type of $\langle \eta_1, \ldots, \eta_N \rangle$ over $F_1$ (with respect to the theory of differentially closed fields), and so this type is isolated over $F_1$.

For let $\hat{F}_3$ be a differential closure of $F_3$. Clearly, $\hat{F}_3$ is a differentially closed field containing $F_1\langle \eta_1, \ldots, \eta_N \rangle$ and $\sigma(\eta_1, \ldots, \eta_N)$ holds in $\hat{F}_3$. Hence, by transfer, $\sigma(\eta_1, \ldots, \eta_N)$ holds in every differentially closed field containing $F_1\langle \eta_1, \ldots, \eta_N \rangle$. Conversely, if $\sigma(\bar{\eta}_1, \ldots, \bar{\eta}_N)$ holds in some differentially closed field containing $F_1\langle \bar{\eta}_1, \ldots, \bar{\eta}_N \rangle$, then by the properties of $S_1$ and $S_2$, $F_1\langle \eta_1, \ldots, \eta_N \rangle \simeq_{F_1} F_1\langle \bar{\eta}_1, \ldots, \bar{\eta}_N \rangle$ (via $\eta_i \to \bar{\eta}_i$).

Hence (by [Note 6], i.e. Elimination of Quantifiers Theorem and the fact that the type of $\langle \eta_1, \ldots, \eta_N \rangle$ is isolated over $F_1$) there is a finite system $S(y_1, \ldots, y_N)$ over $F_1$ which completely determines the isomorphism type of $\langle \eta_1, \ldots, \eta_N \rangle$ over $F_1$.

(a) We first note that (due to the embedding condition for differential closures) if $\hat{F}_1$ and $\hat{F}_2$ are differential closures of $F$ and $\hat{F}_2$ atomic over $F$, then so is $\hat{F}_1$.

So with no loss of generality we may assume $\hat{F}$ is constructed as in the existence proof. Hence, by pre (a) and transfinite induction, we need only show that $F_{i_\alpha+1}$ is atomic over $F_{i_\alpha}$ (where recall $F_{i_\alpha+1} = F_{i_\alpha}\langle \eta_{i_\alpha} \rangle$ and the isomorphism type of $\eta_{i_\alpha}$ over $F_{i_\alpha}$ is completely determined by ($p_{i_\alpha} = 0$, $q_{i_\alpha} \neq 0$, $p = 0$)).

So suppose $\eta_1, \ldots, \eta_N \in F_{i_\alpha+1}$. Then for each $j$ ($1 \leq j \leq N$) there are $f_j$, $g_j \in F_{i_\alpha}\{y\}$ such that $\eta_j = f_j(\eta_{i_\alpha})/g_j(\eta_{i_\alpha})$. Let $\sigma(y_1, \ldots, y_N)$ be the following first order formula over $F_{i_\alpha}$:

$$\exists y(p_{i_\alpha}(y) = 0 \mathbin{\&} q_{i_\alpha}(y) \neq 0 \mathbin{\&} p(y) = 0 \mathop{\&}_{j=1}^{N} g_j(y)y_j = f_j(y)).$$

Then (as in the above proof), $\sigma(y_1, \ldots, y_N)$ (interpreted in differentially closed fields) completely determines the isomorphism type of $\langle \eta_1, \ldots, \eta_N \rangle$ over $F_{i_\alpha}$. So (again by Elimination of Quantifiers Theorem) there is a finite system $S(y_1, \ldots, y_N)$ of equations and inequations over $F_{i_\alpha}$ which completely determines the isomorphism type of $\langle \eta_1, \ldots, \eta_N \rangle$ over $F_{i_\alpha}$.

(c) Suppose $\mu_1, \ldots, \mu_m \in F^*$. By hypothesis there is a finite system $S_1(y_1, \ldots, y_m, y_{m+1}, \ldots, y_{m+N})$ which completely determines the isomorphism type of $\langle \mu_1, \ldots, \mu_m, \eta_1, \ldots, \eta_N \rangle$ over $F$. Let $\sigma(y_1, \ldots, y_m)$ be the following first order formula (over $F$): $\exists y_{m+1} \cdots \exists y_{m+N}\, S_1(y_1, \ldots, y_m, y_{m+1}, \ldots, y_{m+N})$. Then (as above) $(y_1, \ldots, y_m)$ (interpreted in differentially closed fields) completely determines the isomorphism type of $\langle \mu_1, \ldots, \mu_m \rangle$ over $F$ and so (as above) there is a finite system $S(y_1, \ldots, y_m)$ over $F$ which completely determines the isomorphism type of $\langle \mu_1, \ldots, \mu_m \rangle$ over $F$.

(d) The following proof uses induction on $N$ to show that: Whenever $F \subset F_1 \subset F^*$, $F^*$ differentially closed and atomic over $F$, and $F_1$ normal in $F^*$ over $F$, then every sequence of length $N$ from $F^*$ realizes an isolated $N$-type over $F_1$. (However, a very simple argument using the fact that $F^*$ has nonconstants and the theorem on primitive elements shows that the result is proved once we have shown the case $N = 1$.)

$N = 1$ Suppose $\eta \in F^*$. Let $p$ be a minimal polynomial for $\eta$ over $F_1$. Let $\eta_1, \ldots, \eta_n \in F_1$ be such that $p \in F\langle \eta_1, \ldots, \eta_n \rangle\{y\}$, i.e. such that $p$ is a minimal polynomial for $\eta$ over $F\langle \eta_1, \ldots, \eta_n \rangle$. By (c), $F^*$ is atomic over $F\langle \eta_1, \ldots, \eta_n \rangle$ and so there is a $q \in F\langle \eta_1, \ldots, \eta_n \rangle\{y\}$ (with order $q <$ order $p$) such that $(p = 0, q \neq 0)$ completely determines the isomorphism type of $\eta$ over $F\langle \eta_1, \ldots, \eta_n \rangle$.

*Claim* $(p = 0, q \neq 0)$ also completely determines the isomorphism type of $\eta$ over $F_1$.

*Proof of claim* For suppose for contradiction there exists a $\mu$ in some extension of $F_1$ such that $(p(\mu) = 0, q(\mu) \neq 0)$, but $p$ is not a minimal polynomial for $\mu$ over $F_1$. Let $r \in F_1\{y\}$ be a minimal polynomial for $\mu$ over $F_1$. Then clearly order $r <$ order $p$. The system $(p = 0, q \neq 0, r = 0)$ is thus consistent over $F_1$ and so has a solution in $F^*$ since $F^*$ is differentially closed. Hence, there is a $\mu \in F^*$ such that $(p(\mu) = 0,\ q(\mu) \neq 0)$ and $\text{order}_{F_1} \mu < \text{order}_{F_1} \eta$.

Let $\mu_1, \ldots, \mu_m \in F_1$ be such that $\text{order}_{F_1} \mu = \text{order}_{F\langle \mu_1, \ldots, \mu_m \rangle} \mu$. We now get a contradiction by showing there are $\mu_1^*, \ldots, \mu_m^* \in F_1$ such that

$$F\langle \eta_1, \ldots, \eta_n, \mu, \mu_1, \ldots, \mu_m \rangle \underset{F\langle \eta_1, \ldots, \eta_n \rangle}{\simeq} F\langle \eta_1, \ldots, \eta_n, \eta, \mu_1^*, \ldots, \mu_m^* \rangle$$

(induced by: $\mu \to \eta$, $\mu_i \to \mu_i^*$). For then,

$$\text{order}_{F\langle \eta_1, \ldots, \eta_n, \mu_1, \ldots, \mu_m \rangle} \mu = \text{order}_{F\langle \eta_1, \ldots, \eta_n, \mu_1^*, \ldots, \mu_m^* \rangle} \eta.$$

But, by choice of $\mu_1, \ldots, \mu_m$ (and since $\eta_i \in F_1$) the LHS = $\text{order}_{F_1} \mu$ while by choice of $\eta_1, \ldots, \eta_n$ (and since $\mu_i \in F_1$) the RHS = $\text{order}_{F_1} \eta$. But this contradicts $\text{order}_{F_1} \mu < \text{order}_{F_1} \eta$.

**The existence of such** $\mu_1{}^*, \ldots, \mu_m{}^* \in F_1$

By (c), $F^*$ is atomic over $F\langle \eta_1, \ldots, \eta_n, \mu \rangle$ and so there is a finite system $S(\eta_1, \ldots, \eta_n, \mu, y_1, \ldots, y_m)$ which completely determines the isomorphism type of $\langle \mu_1, \ldots, \mu_m \rangle$ over $F\langle \eta_1, \ldots, \eta_n, \mu \rangle$.

Now,

$$F\langle \eta_1, \ldots, \eta_n, \mu \rangle \underset{F\langle \eta_1, \ldots, \eta_n \rangle}{\simeq} F\langle \eta_1, \ldots, \eta_n, \eta \rangle$$

(induced by $\mu \to \eta$) since $(p(\mu) = 0, q(\mu) \neq 0)$ and $(p = 0, q \neq 0)$ completely determines the isomorphism type of $\eta$ over $F\langle \eta_1, \ldots, \eta_n \rangle$.

So, $S(\eta_1, \ldots, \eta_n, \eta, y_1, \ldots, y_m)$ is consistent over $F\langle \eta_1, \ldots, \eta_n \rangle$ and hence has a solution $\langle \mu_1{}^*, \ldots, \mu_m{}^* \rangle$ in $F^*$ since $F^*$ is differentially closed.

Then, by the characterizing properties of $S$ we have

$$F\langle \eta_1, \ldots, \eta_n, \mu, \mu_1, \ldots, \mu_m \rangle \underset{F\langle \eta_1, \ldots, \eta_n \rangle}{\simeq} F\langle \eta_1, \ldots, \eta_n, \eta, \mu_1{}^*, \ldots, \mu_m{}^* \rangle$$

(induced by $\mu \to \eta$, $\mu_i \to \mu_i{}^*$).

It remains to show that $\mu_i{}^* \in F_1$. But this follows since, by restricting the above isomorphism, we have $F\langle \mu_i \rangle \simeq_F F\langle \mu_i{}^* \rangle$ (induced by $\mu_i \to \mu_i{}^*$). Now $\mu_i \in F_1$ and $F_1$ is normal in $F^*$ over $F$. So $\mu_i{}^* \in F_1$.

$N + 1$ Suppose $\eta_1, \ldots, \eta_N, \eta_{N+1} \in F^*$. By the inductive assumption, $\langle \eta_1, \ldots, \eta_N \rangle$ realizes an isolated $N$-type over $F_1$, i.e. there is a finite system $S_1(y_1, \ldots, y_N)$ (over $F_1$) which completely determines the isomorphism type of $\langle \eta_1, \ldots, \eta_N \rangle$ over $F_1$.

Now, $F\langle \eta_1, \ldots, \eta_N \rangle \subset F_1\langle \eta_1, \ldots, \eta_N \rangle \subset F^*$, $F^*$ is differentially closed and by (c) $F^*$ is atomic over $F\langle \eta_1, \ldots, \eta_N \rangle$. Also, it is clear that $F_1\langle \eta_1, \ldots, \eta_N \rangle$ is normal in $F^*$ over $F\langle \eta_1, \ldots, \eta_N \rangle$. Hence, by the case $N = 1$, $\eta_{N+1}$ realizes an isolated type over $F_1\langle \eta_1, \ldots, \eta_N \rangle$, i.e. there is a finite system $S_2(\eta_1, \ldots, \eta_N, y)$ (over $F_1\langle \eta_1, \ldots, \eta_N \rangle$) which completely determines the isomorphic type of $\eta_{N+1}$ over $F_1\langle \eta_1, \ldots, \eta_N \rangle$.

Thus, the finite system $S(y_1, \ldots, y_N, y_{N+1}) = S_1(y_1, \ldots, y_N) \cup S_2(y_1, \ldots, y_N, y_{N+1})$ (over $F_1$) clearly determines the isomorphism type of $\langle \eta_1, \ldots, \eta_N, \eta_{N+1} \rangle$ over $F_1$.

pre (b) We proceed by induction on $N$. Suppose $p \in F\{y\}$ is absolutely irreducible and $X$ is indiscernible (with respect to $p$) over $F$.

$N = 1$ Suppose $\eta$ is in some extension of $F$. Let $X^*$ be a finite subset of $X$ such that $\text{order}_{F\langle X \rangle} \eta = \text{order}_{F\langle X^* \rangle} \eta$ (with no loss of generality we are supposing that $\eta$, $X$ are in some universal extension of $F$).

*Claim* $X - X^*$ are indiscernible (with respect to $p$) over $F\langle\eta\rangle$.

*Proof of claim* Since $p(x) = 0$ for each $x \in X$, we need only show for $\{x_\alpha\}_{\alpha \leq \beta} \subseteq X - X^*$ (where for $\alpha \neq \alpha'$, $x_\alpha \neq x_{\alpha'}$) that

$$\text{order}_{F\langle\eta\rangle\langle\{x_\alpha\}_{\alpha<\beta}\rangle}\, x_\beta \geq \text{order}\ p.$$

So consider the diagram

$$\begin{array}{ccc} F\langle\eta\rangle\langle X^*\rangle\langle\{x_\alpha\}_{\alpha<\beta}\rangle & \longrightarrow & F\langle\eta\rangle\langle X^*\rangle\langle\{x_\alpha\}_{\alpha\leq\beta}\rangle \\ \uparrow & & \uparrow \\ F\langle X^*\rangle\langle\{x_\alpha\}_{\alpha<\beta}\rangle & \longrightarrow & F\langle X^*\rangle\langle\{x_\alpha\}_{\alpha\leq\beta}\rangle \end{array}$$

By choice of $X^*$ we have

$$\text{order}_{F\langle X^*\rangle\langle\{x_\alpha\}_{\alpha<\beta}\rangle}\, \eta = \text{order}_{F\langle X^*\rangle\langle\{x_\alpha\}_{\alpha\leq\beta}\rangle}\, \eta.$$

Therefore,

$$\text{order}_{F\langle\eta\rangle\langle X^*\rangle\langle\{x_\alpha\}_{\alpha<\beta}\rangle}\, x_\beta = \text{order}_{F\langle X^*\rangle\langle\{x_\alpha\}_{\alpha<\beta}\rangle}\, x_\beta\,.$$

Clearly, LHS $\leq \text{order}_{F\langle\eta\rangle\langle\{x_\alpha\}_{\alpha<\beta}\rangle}\, x_\beta$.

On the other hand, since $X$ is indiscernible (with respect to $p$) over $F$ and $X^* \cup \{x_\alpha\}_{\alpha\leq\beta} \subseteq X$ and $x_\beta \notin X^* \cup \{x_\alpha\}_{\alpha<\beta}$, we have that $p$ is a minimal polynomial for $x_\beta$ over $F\langle X^*\rangle\langle\{x_\alpha\}_{\alpha<\beta}\rangle$, i.e. RHS = order $p$. Thus, $\text{order}_{F\langle\eta\rangle\langle\{x_\alpha\}_{\alpha<\beta}\rangle}\, x_\beta \geq \text{order}\ p$ and we are done.

$N + 1$ Suppose $\eta_1, \ldots, \eta_N, \eta_{N+1}$ are in some extension of $F$. By the inductive assumption there is a cofinite subset $X_0$ of $X$ which is indiscernible (with respect to $p$) over $F\langle\eta_1, \ldots, \eta_N\rangle$. By the case $N = 1$ there is a cofinite subset $X_1$ of $X_0$ which is indiscernible (with respect to $p$) over $F\langle\eta_1, \ldots, \eta_N, \eta_{N+1}\rangle$. $X_1$ is clearly a cofinite subset of $X$.

(b) Suppose $p \in F\{y\}$ is absolutely irreducible.

Suppose $X \subset \hat{F}$ is an infinite set of indiscernibles (with respect to $p$) over $F$ and suppose $X_0 = \{x_\alpha\}_{\alpha<\omega}$ is a countably infinite subset of $X$ where for $\alpha \neq \alpha'$, $x_\alpha \neq x_{\alpha'}$.

Let $\widehat{F\langle X_0\rangle}$ be a differential closure of $F\langle X_0\rangle$ in $\hat{F}$. Since $\hat{F}$ is a differential closure of $F$ there is an $F$-embedding $f$: $\hat{F} \to \widehat{F\langle X_0\rangle}$.

Let $X_f = f(X)$. It is clear that $X_f$ is a set of indiscernibles (with respect to $p$) over $F$. So, by pre (b), for each $N$, $p$ is a minimal polynomial for all but a finite number of elements of $X_f$ over $F\langle\{x_\alpha\}_{\alpha<N}\rangle$.

Now, if $X_f$ is uncountable, there must be some $x \in X_f$ such that $p$ is a minimal polynomial for $x$ over $F\langle X_0\rangle$. But then, by (a), since $x \in \widehat{F\langle X_0\rangle}$, $x$

must realize an isolated type over $F\langle X_0\rangle$. Hence [Note 6] there is a $q \in F\langle X_0\rangle\{y\}$, order $q <$ order $p$ such that ($p = 0$, $q \neq 0$) completely determines the isomorphism type of $x$ over $F\langle X_0\rangle$.

Clearly for some $N$, $q \in F\langle\{x_\alpha\}_{\alpha<N}\rangle\{y\}$ and since $p$ is a minimal polynomial for $x_N$ over $F\langle\{x_\alpha\}_{\alpha<N}\rangle$ and order $q <$ order $p$, $x_N$ must solve ($p = 0$, $q \neq 0$). Hence $p$ is a minimal polynomial for $x_N$ over $F\langle X_0\rangle$. But this is impossible (since $x_N \in X_0$ and order $p > 0$).

Thus $X_f$ cannot be uncountable and so $X$ cannot be uncountable.

## References


0. Ax, J., and Kochen, S., Diophantine problems over local fields II, *Ann. of Math.* **83** (1966).
1. Blum, L., "Generalized algebraic structures: a model theoretic approach," Ph.D. dissertation, MIT, Cambridge, Massachusetts, 1968.
2. Blum, L., and Blum, M., Toward a mathematical theory of inductive inference, *Information and Control* **28**, no. 2 (June 1975).
3. Kolchin, E. R., Algebraic matrix groups and the Picard–Vessiot theory of homogeneous linear ordinary differential equations, *Ann. of Math.* **49**, No. 1 (1948).
4. Kolchin, E. R. Existence theorems connected with the Picard–Vessiot theory of homogeneous ordinary differential equations, *Bull. Amer. Math. Soc.* (1948).
5. Kolchin, E. R., Galois theory of differential fields, *Amer. J. Math.* **75** (1953).
6. Kolchin, E. R., Constrained extensions of differential fields, *Advances in Math.* **12** (1975).
7. Morley, M. D., Categoricity in power, *Trans. Amer. Math. Soc.* (Feb. 1965).
8. Robinson, A., "Introduction to Model Theory and the Metamathematics of Algebra." North-Holland Publ., Amsterdam, 1963.
9. Robinson, A., On the concept of a differentially closed field, *Bull. Res. Council Israel* Sect. F, 8.
10. Rosenlicht, M., The nonminimality of the differential closure, *Pacific J. Math.* **52**, No. 2 (1974).
11. Sacks, G. E., The differential closure of a differential field, *Bull. Amer. Math. Soc.* **78**, No. 5 (1972).
12. Sacks, G. E., "Saturated Model Theory." Benjamin, New York, 1972.
13. Seidenberg, A., Some basic theorems in differential algebra (characteristic $p$ arbitrary), *Trans. Amer. Math. Soc.* **73** (1952).
14. Seidenberg, A., An elimination theory for differential algebra, *Univ. of Calif. Publ. in Math.* New Ser. **3** (1956).
15. Shelah, S., Uniqueness and characterization of prime models over sets for totally transcendental first order theories, *J. Symbolic Logic* **37** (1972).
16. Shelah, S., Differentially closed fields (preprint).
17. Wood, C., The model theory of differential fields of characteristic $p \neq 0$, *Proc. Amer. Math. Soc.* **40**, No. 2 (1973).
18. Wood, C., Prime model extensions for differential fields of characteristic $p \neq 0$, *J. Symbolic Logic* **39**, No. 3 (1974).


Over the years I have been fortunate to have the support of and association with various mathematicians in this work. In particular, I would like to thank the following people for their enthusiastic and most welcome encouragement: M. Blum, P. Blum, D. Brown, M. Davis,

S. Kochen, E. R. Kolchin, B. Miller, M. Morley, J. Roitman, M. Rosenlicht, G. Sacks, S. Shelah, M. Singer, and S. Smale. It is with deep regret that I must posthumously thank Abraham Robinson who provided me with some of the earliest inspiration and encouragement. This research was supported in part by an AFOSR–NRC Postdoctoral Research Award and in part by the National Science Foundation, Grant MSC 75-21760.

AMS (MOS) 1970 subject classifications: 12H05, 02H15

# *On finite projective groups*

*Richard Brauer*†

*Harvard University*

## 1. Introduction

One of the oldest problems in finite group theory is that of finding the complex finite projective groups of a given degree $n$, i.e. the finite subgroups $L$ of $PGL(n, \mathbb{C})$. A celebrated theorem of Jordan [6] deals with this question. If $L$ is primitive, the order $|L|$ of $L$ lies below an upper bound $B(n)$ depending only on $n$. Explicit values of $B(n)$ have been given by Blichfeldt [1].

It is natural to replace the complex field $\mathbb{C}$ by an arbitrary field $K$ though Jordan's and Blichfeldt's theorems are not valid, if the characteristic of $K$ is different from zero [3]. For perfect $K$, it is well known what happens if $K$ is replaced by its algebraic closure. We shall therefore assume that $K$ itself is algebraically closed.

For small $n$, the finite projective groups $L$ of degree $n$ have been determined. For the most interesting of these, the factor group $L/Z(L)$ of $L$ modulo its center $Z(L)$ are simple. This leads to the following question: To what extent can the problem of finding all finite subgroups $L$ of $PGL(n, K)$ be reduced to the special case where $L/Z(L)$ is simple? This is the type of question which will be studied in the present paper.

There is a technique available by which problems of this kind can be approached. This has been best formulated by Huppert [5]. The following definition is motivated by a case which has to be studied in this work.

† Deceased.

***Definition 1.1*** An irreducible subgroup $L$ *of* $GL(n, K)$ is *strongly irreducible*, if every normal subgroup $N$ not contained in $Z(L)$ is irreducible.

In [1] and the relevant parts of [5], the groups $L$ are assumed to be irreducible and primitive. This does not fit in well here. For example, it is well known that the Valentiner group of order 1080 has a monomial representation of degree 6. It follows that there exist imprimitive, even monomial, subgroups $L$ of $GL(6, C)$ for which $L/Z(L)$ is isomorphic to the alternating group of degree 6 and hence simple. For this reason, we introduce the notion of quasi-primitivity.

***Definition 1.2*** [2] A subgroup $L$ of $GL(n, q)$ is *quasi-primitive*, if it is irreducible, and if in any normal subgroup $N$ of $L$, any two irreducible constituents are similar.

It is seen easily that if the subgroup $L$ of $GL(n, K)$ is irreducible and primitive, it is quasi-primitive. Moreover, the relevant parts of [1] and [5] remain valid, if the assumption of primitivity is replaced by that of quasi-primitivity. It is clear that strong irreducibility implies quasi-primitivity.

Finite strongly irreducible groups $L$ will be investigated in Section 3. Every such $L$ determines a finite simple group $S$. The case of a non-abelian $S$ is treated in Section 4 and the case of a cyclic $S$ in Section 5. Finally, Section 6 deals with quasi-primitive $L$ which are not strongly irreducible.

## 2. Preliminaries

If $\alpha$ is a mapping of a set $X$ and $\beta$ a mapping of the image, it will sometimes be convenient to use the notation $\alpha\beta$ rather than $\beta \circ \alpha$ for the composite mapping.

As already mentioned, $K$ will always be an algebraically closed field. We denote by $K^\times$ the multiplicative group of $K$ consisting of the nonzero elements of $K$. If $n$ is a positive integer, we denote the ring of matrices of degree $n$ with coefficients in $K$ by $K_n$. Finally, $K_n^\times$ will be the subgroup of nonsingular elements in $K_n$. Note that the center $Z(K_n^\times)$ of $K_n^\times$ consists of the *scalar matrices* $cI_n$ ($c \in K^\times$, $I_n$ the identity matrix of degree $n$). The group $GL(n, K)$ can be identified with $K_n^\times$ and $PGL(n, K)$ with $K_n^\times/Z(K_n^\times)$. By a *linear group* of degree $n$, we mean a subgroup of $K_n^\times$.

Let $G$ be an arbitrary group and let $\pi$ be a homomorphism of $G$ into $PGL(n, K) = K_n^\times/Z(K_n^\times)$. Following the classical paper [8] of Schur, we select an element $P(\sigma) \in K_n^\times$ for every $\sigma \in G$ such that $P(\sigma)$ corresponds to $\pi(\sigma) \in K_n^\times/Z(K_n^\times)$. The mapping $\sigma \mapsto P(\sigma)$ is called a *projective representation* $P$ of degree $n$. For $\sigma, \tau \in G$, we have

$$P(\sigma)P(\tau) = P(\sigma\tau)c(\sigma, \tau) \tag{2.1}$$

with $c(\sigma, \tau) \in K^\times$. The system $c(*, *)$ of these $c(\sigma, \tau)$ is called the *factor set* of $P$. Conversely, if an element $P(\sigma) \in K_n^\times$ is given for each $\sigma \in G$ and if an equation (2.1) holds, the mapping $\sigma \mapsto P(\sigma)Z(K_n^\times)$ is a homomorphism $\pi$ of $G$ into $K_n^\times/Z(K_n^\times) = PGL(n, K)$.

A projective representation will be termed a *linear representation*, if all $c(\sigma, \tau)$ are 1. In our terminology, linear representations should not be confused with representations of degree 1.

Two projective representations $P$ and $P_1$ are *associate*, if there exist elements $k(\sigma) \in K^\times$ such that $P_1(\sigma) = k(\sigma)P(\sigma)$ for $\sigma \in G$. The representations $P$ and $P_1$ are similar, if there exist $M \in K_n^\times$ such that $P_1(\sigma) = M^{-1}P(\sigma)M$ for $\sigma \in G$ ($n$ the degree of $P$, $M$ independent of $\sigma$).

It follows from (2.1) that

$$P(1) = c(1, 1)I_n, \qquad P(\sigma^{-1}) = c(\sigma, \sigma^{-1})c(1, 1)P(\sigma)^{-1}. \tag{2.2}$$

Hence, for $\sigma, \tau \in G$ we have

$$P(\sigma)^{-1}P(\tau)P(\sigma)c(\sigma, \sigma^{-1})c(1, 1) = P(\sigma^{-1}\tau\sigma)c(\sigma^{-1}, \tau)c(\sigma^{-1}\tau, \sigma). \tag{2.3}$$

The subgroup of $K_n^\times$ generated by the elements $P(\sigma)$ with $\sigma \in G$ will be denoted by $P(G)$. We shall say that $P$ is of *standard type*, if the determinant of each $P(\sigma)$ is a root of unity in $K$. In this case, (2.1) implies that if $G$ is finite, so is $P(G)$.

The projective representation $P$ will be called *irreducible*, if the linear group $L = P(G)$ is irreducible in the usual sense. On account of Schur's lemma, $Z(L)$ consists of the scalar matrices contained in $L$. If $N$ is a normal subgroup of $L$, Clifford's theorem shows that $N$ is completely reducible and that its irreducible constituents are conjugate in $L$. In particular, all irreducible constituents of $N$ have the same degree.

A projective representation $P$ of the group $G$ will be said to be *strongly irreducible* if $L = P(G)$ is strongly irreducible in the sense of Definition 1.1; $P$ will be *quasi-primitive*, if $L = P(G)$ is quasi-primitive in the sense of Definition 1.2. If $N_0$ is a normal subgroup of $L$, so is $N = N_0 Z(L)$. Since $Z(L)$ consists of scalar matrices and $N_0 \subseteq N$, it is clear that if $N_0$ is irreducible so is $N$ and vice versa. If any two irreducible constituents of $N_0$ are similar, the same is true for $N$ and vice versa. We conclude that an irreducible $P$ is strongly irreducible, if every normal subgroup $N \supset Z(L)$ of $L$ is irreducible. Similarly, $P$ is quasi-primitive, if for every normal subgroup $N \supset Z(G)$, all irreducible constituents of $N$ are similar.

We conclude this section with a lemma.

***Lemma 2A.*** *Assume that $L_1$ and $L_2$ are two finite irreducible linear groups of the same degree $n > 1$. Suppose that there exists a subset $M$ of*

*$L_1 \cap L_2$ such that, for $j = 1$ and 2, the elements of $L_j$ can be written in the form $\mu\zeta$ with $\mu \in M$ and $\zeta \in Z(L_j)$.*

(i) *If $L_1$ is strongly irreducible, so is $L_2$.*
(ii) *If $L_1$ is quasi-primitive, so is $L_2$.*

*Proof* Let $N_1 \supseteq Z(L_1)$ be a normal subgroup of $L_1$. Write $\nu_1 \in N_1$ in the form $\nu_1 = \rho\zeta_1$ with $\rho \in M$, $\zeta_1 \in Z(L_1)$. Then $\rho \in N_1 \cap M$. It follows that $N_1$ consists of the elements $\rho\zeta_1$ with $\rho \in N_1 \cap M$ and $\zeta_1 \in Z(L_1)$. Form now the set $N_2$ of elements of the form $\rho\zeta_2$ with $\rho \in N_1 \cap M$ and $\zeta_2 \in Z(L_2)$. Since $M \subseteq L_2$, then $N_2 \subseteq L_2$. It is seen without difficulty that $N_2$ is a normal subgroup of $L_2$ and that $N_1 \cap M = N_2 \cap M$.

If $N_j$ is irreducible, so is $N_j \cap M$ and vice versa $(j = 1, 2)$. Hence irreducibility of $N_1$ implies that of $N_2$. Now statement (i) is evident.

If all irreducible constituents of $N_j$ are similar, the same is true for the irreducible constituents of the set of matrices $N_j \cap M$ and vice versa $(j = 1, 2)$. We conclude that (ii) is correct too.

***Corollary 2B*** *Let P be a projective representation of a group. If P is strongly irreducible, so is every associate representation. If P is quasi-primitive, so is every associate representation.*

## 3. Finite strongly irreducible linear groups

We begin with two group theoretical lemmas.

***Lemma 3A*** *Let G be a group with a cyclic center $Z = Z(G)$. Let H and N be subgroups of G such that the following assumptions are satisfied.*

(i) *H is a finite normal subgroup of G.*
*We have $H \supset Z$ and $H/Z$ is an elementary abelian p-group for some prime number p. Moreover, we have $C_G(H) = Z$.*
(ii) *We have $[N, H] \subseteq Z$.*

*Statement* We have $N \subseteq H$.

*Proof* It follows from (i) that $Z(H) = Z$. The commutator subgroup $H'$ of $H$ is contained in $Z$ and since $H$ cannot be abelian, $H' \neq \langle 1 \rangle$. If $\tau_1, \tau_2 \in H$, we have

$$[\tau_1, \tau_2]^p = [\tau_1{}^p, \tau_2] \in [Z, H] = \langle 1 \rangle.$$

Hence $H'$ is generated by elements of order $p$. It follows that $p$ divides $|Z|$ and that $H'$ is the unique subgroup $Z_0$ of order $p$ of the cyclic group $Z$.

Set $|H/Z| = p^r$ and choose $r$ elements $\gamma_1, \gamma_2, \ldots, \gamma_r$ of $H$ such that

$$H = \langle \gamma_1, \gamma_2, \ldots, \gamma_r, Z \rangle. \tag{3.1}$$

Define a homomorphism $\varphi$ of $H$ into the direct product $Z_0^{(r)}$ of $r$ equal factors $Z_0$ by mapping

$$\tau \mapsto ([\tau, \gamma_1], [\tau, \gamma_2], \ldots, [\tau, \gamma_r])$$

($\tau \in H$). We claim that $\varphi(H) = Z_0^{(r)}$. If this was not so, there would exist $r$ integers $c_1, c_2, \ldots, c_r$ not all divisible by $p$ such that

$$[\tau, \gamma_1]^{c_1}[\tau, \gamma_2]^{c_2} \cdots [\tau, \gamma_r]^{c_r} = 1$$

for all $\tau \in H$. If we set

$$\sigma = \gamma_1^{c_1}\gamma_2^{c_2} \cdots \gamma_r^{c_r} \tag{3.2}$$

then $[\tau, \sigma] = 1$ for all $\tau \in H$. This means that $\sigma \in C_G(H) = Z$. However, (3.2) would imply that all $c_i$ are divisible by $p$, a contradiction. Thus

$$\varphi(H) = Z_0^{(r)}. \tag{3.3}$$

If $\nu \in N$, assumption (ii) implies that $[\nu, \tau] \in Z$ for all $\tau \in H$. Then

$$[\nu, \tau]^p = [\nu, \tau^p] \in [N, Z] = \langle 1 \rangle.$$

It follows that $[N, H]$ is generated by elements of $Z$ of order $p$. Thus, $[N, H] \subseteq Z_0$ and if $\nu \in N$,

$$([\nu, \gamma_1], [\nu, \gamma_2], \ldots, [\nu, \gamma_r])$$

is an element $\xi$ of $Z_0^{(r)}$. Now (3.3) shows that there exists an element $\rho$ of $H$ such that $\xi = \varphi(\rho)$. This implies that $[\nu, \gamma_i] = [\rho, \gamma_i]$ for $i = 1, 2, \ldots, r$, whence $[\nu\rho^{-1}, \gamma_i] = 1$. Thus $\nu\rho^{-1}$ centralizes all $\gamma_i$ and since it centralizes $Z$, $\nu\rho^{-1}$ centralizes $H$ (cf. (3.1)). Since $C_G(H) = Z$, then $\nu\rho^{-1} \in Z$ and this implies $N \subseteq ZH = H$ as we wished to show.

***Lemma 3B*** *Let $G$ be a group with cyclic center $Z = Z(G)$. Let $H \supset Z$ be a finite normal subgroup of $G$ such that $H/Z$ is a minimal normal subgroup of $G/Z$. Finally, assume that $C_G(H) = Z$.*

*Statement* If $N$ is a normal subgroup of $G$ and $Z \subset N$, then $H \subseteq N$.

*Proof* We have again $Z(H) = Z$. Since

$$Z \subseteq H \cap N \subseteq H$$

and since $H \cap N$ is normal in $G$, it follows from the assumptions that either $H \cap N = Z$ or $H \cap N = H$. In the latter case, $H \subseteq N$.

Assume then that we have $H \cap N = Z$. This implies that

$$[H, N] \subseteq H \cap N = Z. \tag{3.4}$$

If $\tau \in H$, if $v \in N$ and if $k$ is an integer, then

$$[\tau^k, v] = [\tau, v]^k = [\tau, v^k]. \tag{3.5}$$

It follows for $k = |H|$ that $v^k \in C(H) = Z$. Thus, each $v \in N$ has finite order.

Since $Z \subset N$, we can choose a prime $p$ for which there exists a $p$-element $v$ of $N$ which does not belong to $Z$. If we now take $k$ in (3.5) as a suitable power of $p$, we find $[\tau^k, v] = 1$. Since each element of $H$ of an order prime to $p$ can be written in the form $\tau^k$ with $\tau \in H$, $v$ centralizes all elements of $H$ of order prime to $p$ and it centralizes the group $O^{(p)}(H)$ generated by these elements. It follows that

$$v \in C_G(ZO^{(p)}(H)). \tag{3.6}$$

Since $O^{(p)}(H)$ is characteristic in $H$, it is normal in $G$ and so is $ZO^{(p)}(H)$. Also

$$Z \subseteq ZO^{(p)}(H) \subseteq H.$$

Now, $H/Z$ was a minimal normal subgroup of $G/Z$. We see that either $ZO^{(p)}(H) = Z$ or $ZO^{(p)}(H) = H$. The latter case is impossible since (3.6) would imply $v \in C_G(H) = Z$ and this is not consistent with the choice of $v$. Thus, $ZO^{(p)}(H) = Z$, i.e. $O^{(p)}(H) \subseteq Z$. Since $H/O^{(p)}(H)$ is a $p$-group, so is $H/Z$. As a minimal normal subgroup of $G/Z$, the group $H/Z$ then is an elementary abelian $p$-group.

Now, Lemma 3A can be applied. It follows that $N \subseteq H$. Then (3.4) yields $N = Z$, while we had $Z \subset N$. We have now shown that the case $H \cap N = Z$ is impossible; only the case $H \subseteq N$ remains.

***Proposition 3C*** *Let L be a finite irreducible linear group of degree $n > 1$. Then L is strongly irreducible, if and only if there exist irreducible normal subgroups $H \supset Z(L)$ such that $H/Z(L)$ is a minimal normal subgroup of $L/Z(L)$.*

*Proof* Since $L$ is irreducible and $n > 1$, we have $Z(L) \neq L$. If $L$ is strongly irreducible and if $H/Z(L)$ with $H \supset Z(L)$ is any minimal normal subgroup of $L/Z(L)$, the group $H$ is irreducible and the condition in Proposition 3C is satisfied.

Conversely, assume that there exists an irreducible normal subgroup $H \supset Z(L)$ such that $H/Z(L)$ is a minimal normal subgroup of $L/Z(L)$. Because of the irreducibility of $H$, the centralizer $C_L(H)$ consists of scalar matrices. In fact, $C_L(H)$ is the group of scalar matrices in $L$ and this is $Z(L)$. Since $Z(L)$ is isomorphic to a finite subgroup of $K^\times$, it is cyclic. Hence the assumptions of Lemma 3B are satisfied for $G = L$. It follows that if $N$ is a normal subgroup of $L$ and $N \supset Z$, then $N \supseteq H$. Since $H$ is irreducible, so is $N$. As remarked in Section 2, this suffices to show the strong irreducibility of $L$.

The proof of Proposition 3C also yields the following result.

***Corollary 3D*** *Let $L$ be a finite strongly irreducible linear group of degree $n > 1$. There exists a normal subgroup $H \supset Z(L)$ of $L$ such that $H$ is contained in every normal subgroup $N \supset Z(L)$ of $L$. Of course, $H$ is unique.*

The group $H$ in Corollary 3D has the center $Z(H) = Z(L)$. Since $H/Z(L)$ is a minimal normal subgroup of $L/Z(L)$, we obtain the result.

***Corollary 3E*** *If $L$ and $H$ are as in Corollary* 3D, *there exists a finite simple group $S$ such that*

$$H/Z(H) \cong S \times S \times \cdots \times S. \tag{3.7}$$

Here, $L$ determines $S$ uniquely apart from isomorphisms. Moreover, $L$ determines the number $r$ of factors in (3.7) uniquely.

It will be our aim to see what we can say about $L$, if $n$, $S$, and $r$ are given. We shall distinguish two cases:

*Case I* $S$ is noncyclic.

*Case II* $S$ is cyclic.

Case I will be studied in Section 4 and Case II in Section 5.

## 4. Case I

We continue the work in Section 3 using the same notation as before and assuming that we have Case I, i.e. that $S$ is noncyclic.

We set $Z(L) = Z$. Then $Z(H) = Z$. The set of normal subgroups $N \supseteq Z$ of $H$ will be denoted by $\mathcal{N}$, and $\mathcal{F}$ will denote the set of normal subgroups $F \supset Z$ of $H$ for which $F/Z$ is simple. Then $F/Z \cong S$ and $Z(F) = Z$.

It is an immediate consequence of Corollary 3E that there exist $r$ groups $F_1, F_2, \ldots, F_r \in \mathcal{F}$ such that

$$H = F_1 F_2 \cdots F_r \tag{4.1}$$

and that $F_i \cap F_j = Z$ for $i \neq j$. It is also clear that $F_i$ and $F_j$ are conjugate in $L$.

Actually, it is well known that $H$ is the central product of $F_1, F_2, \ldots, F_r$. Likewise, Propositions 4A–4C are well known.

***Proposition 4A*** *If $N \in \mathcal{N}$, if $F \in \mathcal{F}$, and if $F$ is not contained in $N$, then $F \cap N = Z$ and $F$ and $N$ centralize each other.*

*Proof* Since $Z \subseteq F \cap N \lhd F$ and since $F/Z$ is simple, we either have $F \cap N = Z$ or $F \cap N = F$. In the latter case, $F \subseteq N$ which has been excluded. Hence $F \cap N = Z$. If $\tau \in F$ and $\nu \in N$, then $[\tau, \nu] \in Z$. We may then set

$$\nu^{-1}\tau\nu = \tau\omega_\nu(\tau)$$

with $\omega_\nu(\tau) \in Z$. For fixed $\nu$, the mapping $\tau \mapsto \omega_\nu(\tau)$ is a homomorphism $\omega_\nu$ of $F$ into an abelian group; the elements of $Z(F)$ belong to the kernel of $\omega_\nu$. Since $F/Z(F)$ is simple and non-abelian, it follows that $\omega_\nu(\tau) = 1$ for all $\nu$ and $\tau$. This proves the last part of Proposition 4A.

**Proposition 4B**

(i) *Any two distinct members of $\mathscr{F}$ centralize each other.*
(ii) *$\mathscr{F}$ consists only of $F_1, F_2, \ldots, F_r$.*
(iii) *If $N_1 \in \mathscr{N}$ centralizes $N_2 \in \mathscr{N}$, then $N_1 \cap N_2 = Z$.*
(iv) *In (4.1), $H$ is the central product of the $F_i$.*

*Proof* Part (i) is a special case of Proposition 4A. If $F \in \mathscr{F}$ was different from $F_1, F_2, \ldots, F_r$, then $F$ would centralize each $F_i$ and hence $H$. But then $F \subseteq Z$ which is absurd as $F$ is not abelian. If $N_1$ and $N_2$ belong to $\mathscr{N}$ and if $N_1 \cap N_2 \neq Z$, there exists an $F \in \mathscr{F}$ which is contained in $N_1 \cap N_2$. Then $N_2$ cannot centralize $N_1$ since otherwise we would have $F \subseteq Z(N_1)$ and $F$ is non-abelian. Finally, (iv) is a consequence of (4.1), (i) and (iii).

**Proposition 4C** *If $N \in \mathscr{N}$, then $N$ is the product of the groups $F_i$ contained in $N$. (If no such $F_i$ exists, this is to mean that $N = Z$.)*

*Proof* Let $N_1$ be the product of the $F_i \subseteq N$ and let $N_2$ be the product of the remaining $F_j$. By Proposition 4A, $N$ centralizes each $F_j$ and hence it centralizes $N_2$. Clearly, $N_1 \subseteq N$. Since $N_1N_2 = H$, we have $N_1 \subseteq N \subseteq N_1N_2$. This implies that

$$N = N_1(N \cap N_2).$$

However, $N \cap N_2 = Z$ by Proposition 4B(iii) and then $N = N_1Z = N_1$ which proves Proposition 4C.

**Proposition 4D** *The group $H$ is quasi-primitive.*

*Proof* The mapping $\tau \mapsto \tau$ of $H$ can be interpreted as an irreducible linear representation $U$ of $H$. If $N \in \mathscr{N}$, Clifford's theorem implies that $U/N$ is completely reducible and that any two irreducible constituents $T_1$

and $T_2$ are conjugate in $H$. If $T_1$ has degree $m$, this means that there exists a matrix $Q \in K_m{}^\times$ and an element $\sigma \in H$ such that

$$Q^{-1}T_2(\nu)Q = T_1{}^\sigma(\nu) = T_1(\sigma\nu\sigma^{-1}) \tag{4.2}$$

for all $\nu \in N$. We can write $\sigma$ in the form $\sigma = \rho_1\rho_2 \cdots \rho_r$ with $\rho_i \in F_i$. Any two factors $\rho_i$ and $\rho_j$ commute, and if $F_i$ does not lie in $N$, $\rho_i$ commutes with $\nu$. It follows that, in (4.2), $\sigma$ can be chosen in $N$. Then

$$T_1(\sigma\nu\sigma^{-1}) = T_1(\sigma)T_1(\nu)T_1(\sigma)^{-1}$$

and $T_i(\sigma) \in K_m{}^\times$. Hence $T_1$ and $T_2$ are similar.

If $N_0 \lhd G$ and $N_0$ is not contained in $Z$, then $N = N_0Z$ belongs to $\mathcal{N}$. Since $Z$ consists of scalar matrices, it follows easily that any two irreducible constituents of $U/N_0$ are similar. This proves Proposition 4D.

If $N \in \mathcal{N}$ and if $U$ has the same significance as in the proof of Proposition 4D, we denote the irreducible constituent of $U|N$ by $T_N$, its degree by $m(n)$, its multiplicity by $e(N)$. Then $T_N$ is determined up to similarity. On comparing degrees, we find

$$n = e(N)m(N). \tag{4.3}$$

***Proposition 4E*** If $A$, $B \in \mathcal{N}$ and $A \cap B = Z$, then $m(AB) = m(A)m(B)$.

*Proof* Since $U|A$ is equal to the restriction of $U|AB$ to $A$, the irreducible constituents of $T_{AB}|A$ are similar to $T_A$. This means that there exists a matrix $Q \in K_{m(AB)}^\times$ such that

$$Q^{-1}T_{AB}(\alpha)Q = I_h \otimes T_A(\alpha) \tag{4.4}$$

for all $\alpha \in A$. Comparison of degrees shows that the integer $h$ is equal to $m(AB)/m(A)$.

If $\beta \in B$, $\beta$ and $\alpha$ commute and hence $Q^{-1}T_{AB}(\beta)Q$ commutes with $Q^{-1}T_{AB}(\alpha)Q$. It follows from Schur's lemma that $Q^{-1}T_{AB}(\beta)Q$ has the form

$$Q^{-1}T_{AB}(\beta)Q = R(\beta) \otimes I_{m(A)} \tag{4.5}$$

where $R(\beta) \in K_h{}^\times$. Clearly, the mapping $\beta \mapsto R(\beta)$ defines a linear representation $R$ of $B$ of degree $h$. By (4.4) and (4.5), we have

$$Q^{-1}T_{AB}(\alpha\beta)Q = Q^{-1}T_{AB}(\alpha)Q \cdot Q^{-1}T_{AB}(\beta)Q = R(\beta) \otimes T_A(\alpha)$$

(for $\alpha \in A$, $\beta \in B$). Since $T_{AB}$ is irreducible, this implies that $R$ is irreducible. Now (4.5) shows that $R$ is an irreducible constituent of $T_{AB}|B$. The irreducible constituents of $T_{AB}|B$ are similar to $B$. Thus, $R$ is similar to $T_B$. Since $R$ has degree $h = m(AB)/m(A)$, we have $m(AB)/m(A) = m(B)$ as we wished to prove.

***Proposition 4F*** *The degree $m = m(F_i)$ does not depend on $i$ $(i = 1, 2, \ldots, r)$.*

This follows at once from the fact that $F_1, F_2, \ldots, F_r$ are conjugate in $L$. Note that $m(H) = n$. On combining Propositions 4E, 4F, and 4B, we obtain the result.

***Proposition 4G*** *We have $n = m^r$.*

***Proposition 4H*** *If $K$ has the characteristic 0, $n$ divides $|H/Z| = |S|^r$.*

This follows from a theorem of Schur [8].

Since $F_1, F_2, \ldots, F_r$ are conjugate in $G$ and since they are the only members of $\mathcal{F}$ (cf. Propositions 4A and 4C(ii), every element $\sigma$ of $L$ induces a permutation $\pi(\sigma)$ of $\{F_1, F_2, \ldots, F_r\}$ and these $\pi(\sigma)$ define a transitive permutation representation $\pi$ of $L$. Let $L_0$ denote the kernel of $\pi$. Clearly, $H \subseteq L_0$. This yields the result.

***Proposition 4I*** *The subgroup*

$$L_0 = N_G(F_1) \sqcap N_G(F_2) \sqcap \cdots \sqcap N_G(F_r)$$

*of $L$ is normal and contains $H$. The group $L/L_0$ is isomorphic to a transitive subgroup of the symmetric group $\mathfrak{S}_r$.*

We next investigate the structure of $L_0/Z$.

***Proposition 4J*** *Let $\mathrm{Aut}(F_i)$ denote the automorphism group of $F_i$ and set*

$$E = \mathrm{Aut}(F_1) \times \mathrm{Aut}(F_2) \times \cdots \times \mathrm{Aut}(F_r). \tag{4.6}$$

*There exists a homomorphism $\varphi$ of $L_0$ into $E$ such that $\varphi$ has the kernel $Z$ and that*

$$\varphi(H) = \mathrm{Inn}(F_1) \times \mathrm{Inn}(F_2) \times \cdots \times \mathrm{Inn}(F_r) \tag{4.7}$$

*where $\mathrm{Inn}(F_i)$ is the group of inner automorphisms of $F_i$.*

*Proof* If $\sigma \in L_0$ and if $(\sigma)$ is the associated inner automorphism of $L_0$, we define $\varphi$ by the formula

$$\varphi(\sigma) = ((\sigma)|F_1, (\sigma)|F_2, \ldots, (\sigma)|F_r).$$

It is clear that $\varphi$ is a homomorphism of $L_0$ into $E$. An element $\sigma$ of $L_0$ belongs to the kernel of $\varphi$, if and only if $\sigma$ centralizes $F_1, F_2, \ldots, F_r$. This is so if and only if $\sigma$ centralizes $H$. Since $C_L(H) = Z$, $\varphi$ has the kernel $Z$.

If $\sigma \in H$, each $(\sigma)|F_i$ belongs to Inn $F_i$. Conversely, if elements $\rho_1, \rho_2, \ldots, \rho_r$ are given with $\rho_i \in F_i$ and if we set $\sigma = \rho_1\rho_2 \cdots \rho_r$, then $(\sigma)|F_i = (\rho_i)$ (cf. Proposition 4C(i)). Hence

$$\varphi(\sigma) = ((\rho_1), (\rho_2), \ldots, (\rho_r)).$$

This shows that (4.7) holds and the proof is complete.

Let $O(F_i) = \mathrm{Aut}(F_i)/\mathrm{Inn}(F_i)$ denote the outer automorphism group of $F_i$. Then Proposition 4J implies that

$$L_0/H \cong \varphi(L_0)/\varphi(H) \subseteq E/(\mathrm{Inn}(F_1) \times \cdots \times \mathrm{Inn}(F_r)),$$
$$E/(\mathrm{Inn}(F_1) \times \cdots \times \mathrm{Inn}(F_r)) \cong O(F_1) \times \cdots \times O(F_r).$$

It follows from Corollary 3E that there exists a homomorphism $\theta$ of $F_i \to S$ such that $\theta(F_i) = S$ and that the kernel is $Z$. It can be deduced that $O(F_i)$ is isomorphic to a subgroup of $O(S)$. We now have the result.

***Proposition 4K*** *The group $L_0/H$ is isomorphic to a subgroup of*

$$O(S)^{(r)} = O(S) \times O(S) \times \cdots \times O(S) \qquad (r \text{ factors}).$$

We summarize our results.

***Proposition 4L*** *Let $L$ be a finite strongly irreducible linear group of degree $n > 1$. Then $L$ determines a finite simple group $S$ up to isomorphism. If $S$ is not cyclic, $L$ also determines positive integers $m$ and $r$ such that $n = m^r$. There exist normal subgroups*

$$L \supseteq L_0 \supseteq H \supseteq Z \supseteq 1$$

*of $L$ with the following properties: The group $L/L_0$ is isomorphic to a transitive subgroup of the symmetric group $\mathfrak{S}_r$. The group $L_0/H$ is isomorphic to a subgroup of the direct product of $r$ factors $O(S)$ ($O(S)$ the outer automorphism group of $S$). The group $H/Z$ is isomorphic to the direct product of $r$ factors $S$. Finally, $Z$ is the center $Z(L)$ of $L$; it is cyclic.*

***Proposition 4M*** *Under the assumptions of Proposition* 4L, *there exists an irreducible linear group $F$ with the center $Z$ such that $F/Z \cong S$. The group $H$ is isomorphic with the central product of $r$ factors $F$.*

The following corollary is an immediate consequence of Propositions 4L and 4M.

***Corollary 4N*** *We have*

$$|L : Z| \leq r!\,|O(S)|^r |S|^r.$$

*If $O(S)$ is solvable; i.e. if the Schreier conjecture holds for $S$, the group $L_0/H$ is solvable. If $r = 1$, we have $L = L_0$. In particular, this will always be so for given $n$, if $n$ cannot be written as a power of an integer with an exponent larger than 1. If $r < 5$, $L/L_0$ is solvable.*

## 5. Case II

We continue our investigation of finite strongly irreducible linear groups $L$ of degree $n > 1$ using the same notation as before. We assume now that the simple group $S$ is cyclic, say of order $p$. The first proposition deals with some fairly obvious facts.

***Proposition 5A*** *The following results are true.*

(i) *The group $H/Z$ is elementary abelian of order $p^r$.*

(ii) *The commutator group $H'$ of $H$ is the unique subgroup $Z_0$ of order $p$ of the cyclic group $Z$. The characteristic of $K$ is different from $p$.*

(iii) *If $H_p$ is a Sylow $p$-subgroup of $H$ and if $Z_{p'}$ is the maximal subgroup of $Z$ of order prime to $p$, we have*

$$H = H_p \times Z_{p'}. \tag{5.1}$$

(iv) *The degree $n$ is a power of $p$.*

*Proof* (i) is clear (cf. Corollary 3E). It follows that $H'$ is contained in the cyclic group $Z = Z(H)$. Since $H \neq Z$, $H$ is non-abelian and $H' \neq \langle 1 \rangle$. If $\tau_1$ and $\tau_2$ are elements of $H$, we have $[\tau_1, \tau_2] \in Z$, $\tau_1{}^p \in Z$ and then

$$[\tau_1, \tau_2]^p = [\tau_1{}^p, \tau_2] = 1.$$

It follows that $H'$ is generated by elements of order $p$. Necessarily, $p$ divides $|Z|$ and $H'$ is the unique subgroup $Z_0$ of order $p$ of the cyclic group $Z$. If $Z_0$ is generated by $\varepsilon I_n$, $\varepsilon$ is a primitive $p$th root of unity in $K$. No such root of unity exists, if the characteristic of $K$ is $p$. This proves (ii).

It is clear that $|H|$ is the product of $|H_p|$ and $|Z_{p'}|$. Hence we have $H = H_p Z_{p'}$ and since necessarily $H_p \cap Z_{p'} = \langle 1 \rangle$, the product is direct. Finally, as the characteristic of $K$ is different from $p$, the degrees of all linear irreducible representations of $H_p$ are powers of $p$. Since $H$ is irreducible, $H_p$ is irreducible and its degree $n$ is a power of $p$.

The next proposition and its proof are due to Huppert [5].

***Proposition 5B*** *Consider $H/Z$ as a vector space $W$ of dimension $r$ over $GF(p)$. There exists a nondegenerate skew-symmetric bilinear form $f$ on $W$ such that if $\alpha$ is an automorphism of $H$ which fixes $Z$, the induced mapping*

$$\tau Z \mapsto \alpha(\tau) Z \qquad (\tau \in H) \tag{5.2}$$

*of $W$ leaves $f$ invariant. In particular, $r$ is even and the linear transformation $R(\alpha)$ in (5.2) is an element of the symplectic group $\mathrm{Spl}_f(r, p)$ belonging to $f$.*

*Proof* Let $\zeta$ be a generator of the cyclic group $H' = Z_0$ of order $p$. If $\tau_1$ and $\tau_2$ are elements of $H$, $[\tau_1, \tau_2]$ is a power of $\zeta$, say

$$[\tau_1, \tau_2] = \zeta^v$$

where $v = v(\tau_1, \tau_2)$ is an integer. Define a bilinear form $f$ on $W$ by taking $f(\tau_1 Z, \tau_2 Z)$ as the residue class (mod $p$) of $v(\tau_1, \tau_2)$; it follows from the rules for commutator elements that $f$ is actually bilinear. Similarly, it follows that $f$ is skew-symmetric. If $f$ was degenerate, there would exist an element $\tau_1 Z \neq 0$ in $W$ such that $f(\tau_1 Z, \tau Z) = 0$ for all $\tau \in H$. But then $[\tau_1, \tau] = 1$ for all $\tau \in H$, i.e., $\tau_1 \in Z(H) = Z$. We then have $\tau_1 Z = 0$, a contradiction.

Thus, $f$ is nondegenerate. This implies that $r$ is even and that $R(\alpha) \in \mathrm{Spl}_f(r, p)$.

Since $f$ will always be the form defined in this proof, we shall write $\mathrm{Spl}(r, p)$ for $\mathrm{Spl}_f(r, p)$.

**Proposition 5C** *The quotient groups in the normal series*

$$L \supseteq H \supseteq Z \supseteq \langle 1 \rangle$$

*of L have the following properties.*

(i) *$L/H$ is isomorphic to a subgroup $M$ of* $\mathrm{Spl}(r, p)$.
(ii) *$H/Z$ is isomorphic to the direct product of $r$ factors $S$.*
(iii) *$Z$ is cyclic.*

*Proof* The last two statements are obvious. It remains to prove (i).

If $\lambda \in L$ and if $(\lambda)$ denotes the corresponding inner automorphism of $L$, then $(\lambda)$ maps an element $\tau$ of $H$ on $\lambda^{-1}\tau\lambda = \tau[\tau, \lambda]$. Since $[\tau, \lambda] \in H$, the restriction $(\lambda)|H$ of $(\lambda)$ to $H$ is an automorphism of $H$ which fixes $Z$. We can then form $R((\lambda)|H)$ (cf. Proposition 5B).

Set

$$\varphi(\lambda) = R((\lambda)|H). \tag{5.3}$$

Then $\varphi$ is a homomorphism of $L$ into $\mathrm{Spl}(r, p)$ which maps $\lambda \in L$ on the linear transformation $\tau Z \mapsto \tau[\tau, \lambda]Z$ of $W$ $(\tau \in H)$.

The kernel $N$ of $\varphi$ is a normal subgroup of $L$ which consists of the elements $\nu \in L$ such that $[\tau, \nu] \in Z$ for all $\tau \in H$. It follows that

$$H \subseteq N, \qquad [H, N] \subseteq Z. \tag{5.4}$$

The second of these relations in conjunction with Proposition 5A(i) shows that Lemma 3A applies. This yields $N \subseteq H$. The first relation (5.4) now implies that $N = H$. Thus, $H$ is the kernel of $\varphi$ and $L/H$ is indeed isomorphic to a subgroup of $\mathrm{Spl}(r, p)$.

***Proposition 5D*** *Let $M$ be as in Proposition* 5C(i). *Then $M$ is an irreducible subgroup of the linear group* $\mathrm{Spl}(r, p) \subseteq GF(p)_r$.

*Proof* Let $\tau$ be an element of $H$ which does not lie in $Z$. Then $F = \langle \tau, Z \rangle$ is a subgroup of $H$, $F \supset Z$, and $F/Z$ has order $p$. Since $H/Z$ was a minimal normal subgroup of $L$, the conjugates of $F$ in $L$ generate $H$.

Translate this into a statement concerning $W$ in Proposition 5B. We see that if $w \neq 0$ is an element of $W$, say $w = \tau Z$ with $\tau \in H$, $\tau \notin Z$, the subspace of $W$ generated by all $w\varphi(\lambda)$ with $\lambda \in L$ and $\varphi(\lambda)$ as in (5.3) is $W$ itself. This implies the irreducibility of $W$.

A great deal of information concerning $H$ has been obtained by Rigby [7]. We quote some of the results.

***Proposition 5E*** *Let $G$ be a finite primitive linear group of degree $n > 1$ over an algebraically closed field $K$ and let $P$ be a non-abelian normal $p$-subgroup of $G$. The center $Z(P)$ of $P$ is cyclic and consists of scalar matrices. The group $P$ is isomorphic to the central product $P = Z(P)E$ of $Z(P)$ with an extraspecial group $E$. Set $|E| = p^{2e+1}$. The irreducible linear complex representations of $P$ of degree larger than* 1 *are algebraically conjugate and have degree $p^e$. Finally, if $p$ is odd, $E$ has exponent $p$.*

We refer to [7] for the proof; see also [4, §§31 and 32]. In fact, the proofs remain valid if the assumption of primitivity of $G$ is replaced by quasi-primitivity.

Hence Proposition 5E can be applied if $G$ is the given strongly irreducible group $L$ and $P$ is the group $H_p$ in (5.1). Since the characteristic of $K$ is different from $p$, $p$-groups have the same degrees for their irreducible linear representations in $K$ as they have for their irreducible linear complex representations. Since $H = H_p Z_{p'}$ and since $Z_{p'}$ consists of scalar matrices, the linear group $H_p$ of degree $n$ is irreducible. Then $n$ must also be the degree of one of its irreducible linear complex representations. It follows from Proposition 5E that $n = p^e$. If $p$ is odd, $H$ is determined up to isomorphism by $|H|$ and $|Z|$.

## 6. Quasi-primitive linear groups

The following well-known result is probably due to I. Schur.

***Proposition 6A*** *Assume that $L$ is a finite linear group of degree $n$ and that $H$ is a normal subgroup of the form*

$$H = I_a \otimes Y \tag{6.1}$$

*where $a$ is an integer with $1 < a < n$ and $Y$ an irreducible linear group of* *degree $b = n/a$. Then there exist projective representations A of degree a and B of degree b of L, both of standard type, such that*

$$\sigma = A(\sigma) \otimes B(\sigma) \qquad \text{for} \quad \sigma \in L, \tag{6.2}$$

$$A(\sigma) = I_a, \qquad B(\sigma) = Y(\sigma) \qquad \text{for} \quad \sigma \in H. \tag{6.3}$$

*The factor sets of A and B are reciprocal.*

*Proof* If $\sigma \in L$, we define a representation $Y^\sigma$ of $H$ by setting $Y^\sigma(\tau) = Y(\sigma\tau\sigma^{-1})$ for $\tau \in H$. It follows from (6.1) that

$$\sigma(I_a \otimes Y)\sigma^{-1} = I_a \otimes Y^\sigma. \tag{6.4}$$

Since then $Y$ and $Y^\sigma$ both are irreducible constituents of $H$, they are similar, i.e. there exists a matrix $B(\sigma) \in K_b^{\times}$ such that

$$Y^\sigma = B(\sigma) Y B(\sigma)^{-1}. \tag{6.5}$$

For $\sigma \in H$, we may choose $B(\sigma) = Y(\sigma)$. If $\alpha \notin H$, we take $B(\sigma)$ as a matrix of determinant 1. This is possible since (6.5) determines $B(\sigma)$ only apart from a scalar factor. It follows from (6.5) that the mapping $\sigma \mapsto B(\sigma)$ is a projective representation $B$ of $L$. Clearly, $B$ is of standard type, its degree is $b$.

Let $\tau \in H$. By (6.5),

$$Y(\sigma\tau\sigma^{-1}) = B(\sigma) Y(\tau) B(\sigma)^{-1}.$$

If we take the tensor product with $I_a$ and apply (6.1) we have

$$\sigma\tau\sigma^{-1} = (I_a \otimes B(\sigma))\tau(I_a \otimes B(\sigma)^{-1}).$$

Hence $(I_a \otimes B(\sigma)^{-1})\sigma$ centralizes $H = I_a \otimes Y$. Now, Schur's lemma shows that $(I_a \otimes B(\sigma)^{-1})\sigma$ has the form

$$(I_a \otimes B(\sigma)^{-1})\sigma = A(\sigma) \otimes I_b$$

with $A(\sigma) \in K_a^{\times}$. This implies (6.2) and the remaining statements are now obvious.

We come now to the application of Proposition 6A.

***Proposition 6B*** *Assume that L is a finite quasi-primitive linear group of degree $n > 1$ which is not strongly irreducible. Then L is similar to a group of the form $A \otimes B$ where A and B are projective representations of L of standard type which satisfy the following conditions: A of degree $a > 1$ is quasi-primitive and B of degree $b > 1$ is strongly irreducible.*

*Proof* (**α**) Choose a normal subgroup $H \supset Z(L)$ of $L$ such that $H/Z(L)$ is a minimal normal subgroup of $L/Z(L)$;

$$Z(L) \subset H \lhd L. \tag{6.6}$$

Since $L$ is not strongly irreducible, Proposition 3C shows that $H$ is reducible.

As $L$ is quasi-primitive, all irreducible constituents of $H$ are similar. Since $H$ as a normal subgroup of the irreducible group $L$ is completely reducible, $H$ is similar to a group of the form $I_a \otimes Y$ where $Y$ is irreducible. The integer $a$ is larger than 1. If we had $a = n$, $H$ would consist of scalar matrices and we would have $H = Z(L)$ contrary to (6.6). Thus, $b = n/a$ is larger than 1.

After replacing $L$ by a similar group, we may assume that (6.1) holds. Now Proposition 6A applies. It follows that there exist projective representations $A$ of degree $a$ and $B$ of degree $b$ of $L$ with the properties listed in Proposition 6A. In particular,

$$\sigma = A(\sigma) \otimes B(\sigma) \qquad \text{for} \quad \sigma \in L, \tag{6.7}$$

$$A(\tau) = I_a, \qquad B(\tau) = Y(\tau) \qquad \text{for} \quad \tau \in H. \tag{6.8}$$

We note some consequences. If $\zeta \in Z(L)$, then as $L$ is irreducible, $\zeta$ is a scalar matrix $\zeta = cI_n$ with $c \in K^\times$. By (6.6) and (6.8), $A(\zeta) = I_a$ and now (6.7) yields

$$B(\zeta) = cI_b \qquad \text{for} \quad \zeta = cI_n \in Z(L). \tag{6.9}$$

We next claim that, for $\sigma \in L$ and $\tau \in H$,

$$A(\sigma)^{-1}A(\tau)A(\sigma) = A(\sigma^{-1}\tau\sigma), \qquad B(\sigma)^{-1}B(\tau)B(\sigma) = B(\sigma^{-1}\tau\sigma). \tag{6.10}$$

The first of these relations is immediate, since $A(\tau) = A(\sigma^{-1}\tau\sigma) = I_a$. The second relation is a consequence of the first relation and of (6.7).

Finally, note that the irreducibility of $L$ implies the irreducibility of $A$ and of $B$.

**(β)** We now prove the strong irreducibility of $B$. We have to prove the strong irreducibility of

$$W = B(L), \tag{6.11}$$

a finite irreducible linear group of degree $b > 1$. The group $Y(H) = B(H)$ is an irreducible subgroup of $W$; it follows from (6.10) that $B(H)$ is normal in $W$. Equation (6.9) shows that $B(Z(L)) \subseteq Z(W)$. On the other hand, if we had $B(H) \subseteq Z(W)$, $B(H) = Y(H)$ would be abelian. This is impossible since $Y$ is an irreducible linear representation of $H$ of degree $b > 1$. It follows that

$$Z(W) \subset B(H)Z(W) \lhd W.$$

Since $B(H)$ is irreducible, so is $B(H)Z(W)$.

If $W$ is not strongly irreducible, it follows from Proposition 3C that there exists a normal subgroup $N$ of $W$ with

$$Z(W) \subset N \subset B(H)Z(W), \qquad N \lhd W. \tag{6.12}$$

Then

$$N = (N \cap B(H))Z(W). \tag{6.13}$$

Define a subset $M$ of $H$ by the formula

$$M = \{\mu \in H \,|\, B(\mu) \in N\}. \tag{6.14}$$

Then $B(M) = N \cap B(H)$ and (6.13) reads

$$N = B(M)Z(W). \tag{6.15}$$

Since $Z(L) \subset H$ and, by (6.9), $B(Z(L)) \subseteq Z(W)$, it follows from (6.12) and (6.14) that $Z(L) \subseteq M$. As $B|H = Y$ is a linear representation of $H$, $M$ is a subgroup of $H$. If $\tau \in M$ in (6.10), then $B(\tau) \in N \lhd W$. We see that $B(\sigma^{-1}\tau\sigma) \in N$. Here, $\sigma^{-1}\tau\sigma \in H$ and then, by (6.14), $\sigma^{-1}\tau\sigma \in M$. It follows that $M \lhd L$. Thus

$$Z(L) \subseteq M \subseteq H, \qquad M \lhd L.$$

However, $H$ has been chosen such that $H/Z(L)$ was a minimal normal subgroup of $L/Z(L)$. We therefore have $M = Z(L)$ or $M = H$. In the former case, (6.15) shows that $N = Z(W)$ and, in the latter case, that $N = B(H)Z(W)$. In either case, we have a contradiction with (6.12).

It follows that $W$ is strongly irreducible as we wished to show.

($\gamma$) It remains to prove that $A$ is quasi-primitive, i.e. we have to prove the quasi-primitivity of

$$X = A(L). \tag{6.16}$$

Since $A$ is irreducible, so is $X$.

Suppose that $\lambda_1$ and $\lambda_2$ belong to $L$ and that $\xi_1$ and $\xi_2$ belong to $Z(X)$. Then

$$A(\lambda_1)\xi_1 A(\lambda_2)\xi_2 = A(\lambda_1, \lambda_2) \cdot c(\lambda_1, \lambda_2)I_a \cdot \xi_1, \xi_2$$

where $c(*, *)$ is the factor set of $A$. All terms here except possibly $c(\lambda_1, \lambda_2)I_a$ belong to $X$. Then $c(\lambda_1, \lambda_2)I_a$ also belongs to $X$ and hence to $Z(X)$. Thus, the set of elements $A(\lambda)\xi$ with $\lambda \in L$ and $\xi \in Z(X)$ is closed under multiplication. It follows that

$$X = \{A(\lambda)\xi \,|\, \lambda \in L, \xi \in Z(X)\}.$$

If $X$ is not quasi-primitive, there exists a normal subgroup $P$ of $X$ such that $P$ has two irreducible constituents $T_1$ and $T_2$ which are not similar.

Since we may replace $P$ by $PZ(X)$, we may assume that $P \supseteq Z(X)$. It follows from Clifford's theorem that $P$ is completely reducible and that all irreducible constituents of $P$ have the same degree $t$.

Define a subset $R$ of $L$ by the formula

$$R = \{\rho \in L \mid A(\rho) \in P\}.$$

It is seen without difficulty that

$$P = A(R)Z(X) \tag{6.17}$$

and that $R$ is a subgroup of $L$ such that

$$H \subseteq R \lhd L. \tag{6.18}$$

Define now a projective representation of $R$ by the formula $\rho \mapsto T_i(A(\rho))$. The factor set is the factor set $c(*, *)$ of $A$ with the arguments restricted to $R$. Since $B$ has the factor set $c(*, *)^{-1}$ a linear representation $U_i$ of $R$ is defined by the formula

$$U_i(\rho) = T_i(A(\rho)) \otimes B(\rho).$$

Here, $U_i$ is a constituent of $R$, because $T_i$ was a constituent of $P$. The degree of $U_i$ is $tb$.

Since $L$ was irreducible and $R \lhd L$, $R$ is completely reducible and all its irreducible constituents are similar. It follows that $U_1$ and $U_2$ are similar. There exists a matrix $Q \in K_{tb}^{\times}$ such that

$$Q^{-1}(T_1(A(\rho)) \otimes B(\rho))Q = T_2(A(\rho)) \otimes B(\rho) \tag{6.19}$$

for each $\rho \in R$.

In particular, for $\rho \in H \subseteq R$, we have $A(\rho) = I_a$ and hence $T_i(A(\rho)) = I_t$. Thus,

$$Q^{-1}(I_t \otimes B(\rho))Q = I_t \otimes B(\rho)$$

for all $\rho \in H$. Since $B|H = Y$ is irreducible, Schur's lemma implies that $Q$ has the form $Q = Q_0 \otimes I_b$ with $Q_0 \in K_t^{\times}$. On substituting this in (6.19), we see easily that

$$Q_0^{-1}T_1(A(\rho))Q_0 = T_2(A(\rho))$$

for $\rho \in R$. Of course, for $\xi \in Z(X)$,

$$Q_0^{-1}T_1(\xi)Q_0 = T_2(\xi),$$

since $\xi$ is a scalar matrix $cI_a$ with $c \in K^{\times}$ and then $T_i(\xi) = cI_t$.

The last two formulas in conjunction with (6.17) show that the representations $T_1$ and $T_2$ of $P$ are similar. This is a contradiction. Hence $X$ is quasi-primitive as we had to show.

***Remark 6C*** If $L$ in Proposition 6B contains a scalar matrix $\xi = cI_n$ with $c \in K^\times$, then $A(\xi)$ and $B(\xi)$ are scalar matrices.

This follows from (6.7) and the fact that similarity transformations preserve scalar matrices.

The following proposition is a corollary of Proposition 6B.

***Proposition 6D*** *Let $G$ be a finite group. Let $P$ be a projective representation of $G$ of degree $n > 1$. Assume that $P$ is of standard type and that it is quasi-primitive but not strongly irreducible. Then there exist projective representations $P_1$ and $P_2$ of $G$ of standard type with the following properties: $P$ is similar to $P_1 \otimes P_2$. The representation $P_1$ is quasi-primitive of degree $a > 1$. The representation $P_2$ is strongly irreducible of degree $b > 1$, $ab = n$.*

*Proof* Set $L = P(G)$. Then $L$ is a finite linear group of degree $n$ which is quasi-primitive but not strongly irreducible. Hence Proposition 6B can be applied to $L$. Let $A$ and $B$ have the significance given in Proposition 6B and set $P_1 = A \cdot P$, $P_2 = B \cdot P$. As is seen easily, $P_1$ and $P_2$ are projective representations of $G$ of standard type; $P_1$ has degree $a > 1$ and $P_2$ degree $b > 1$. It is also clear that $P_1 \otimes P_2$ is similar to $P$.

The group $L$ consists of the elements $\lambda = P(\sigma)\xi$ with $\sigma \in G$, $\xi \in Z(L)$. If $c(*, *)$ is the factor set of $A$, then

$$A(\lambda) = A(P(\sigma))A(\xi) \cdot uI_a \tag{6.20}$$

with $\mu = c(P(\sigma), \xi)^{-1}$. As shown by (6.20) $uI_a$ belongs to $A(L)$ and hence to $Z(A(L))$. By (6C), $A(\xi) \in Z(A(L))$ and (6.20) takes the form

$$A(\lambda) = A(P(\sigma))\eta_0 = P_1(\sigma)\eta_0$$

with $\sigma \in G$, $\eta_0 \in Z(A(L))$. Now, $A(L)$ consists of the elements $\alpha = A(\lambda)\eta_1$ with $\lambda \in L$ and $\eta_1 \in Z(A(L))$. We find

$$\alpha = P_1(\sigma)\eta$$

where $\sigma \in G$ and $\eta \in Z(A(L))$.

On the other hand, $P_1(G)$ consists of the elements $P_1(\sigma)\zeta$ with $\sigma \in G$, $\zeta \in Z(P_1(G))$. It is now clear that Lemma 2A can be applied to $L_1 = A(L)$ and $L_2 = P_1(G)$. The set $M$ is to be taken as the set of all $P_1(\sigma)$ with $\sigma \in G$.

Since the projective representation $A$ of $L$ is quasi-primitive, so is the group $L_1 = A(L)$. By Lemma 2A, $L_2 = P_1(G)$ is quasi-primitive and so then is $P_1$.

An analogous argument can be applied with $A$ replaced by $B$ and $P_1$ by $P_2$. Since $B$ was strongly irreducible, we find that $P_2$ is strongly irreducible. This concludes the proof.

If in Proposition 6D, the representation $P_1$ is not strongly irreducible, we can apply Proposition 6D again to $P_1$. Continuing in this manner, we obtain the result.

***Proposition 6E*** *Let $G$ be a finite group and let $P$ of degree $n > 1$ be a quasi-primitive projective representation of $G$ of standard type. Then there exist strongly irreducible projective representations $P_i$ of $G$ of degree $n_i > 1$ $(i = 1, 2, \ldots, k)$ such that $P$ is similar to*

$$Q^{-1}PQ = P_1 \otimes P_2 \otimes \cdots \otimes P_k \tag{6.21}$$

*$(Q \in K_n^{\times})$. The $P_i$ may be chosen of standard type.*

There is no uniqueness theorem connected with Remark 6C. Not even the number $k$ of factors in (6.21) is determined uniquely by $P$. Of course, we have

$$n = n_1, n_2, \ldots, n_k. \tag{6.22}$$

In particular, Proposition 6E can be applied when $L$ is a finite quasi-primitive linear group of degree $n > 1$ and $P$ the projective representation of $L$ with $P(\lambda) = \lambda \in K_n^{\times}$ for $\lambda \in L$. Then $L$ itself is similar to $P_1 \otimes P_2 \otimes \cdots \otimes P_k$.

If $L$ is not strongly irreducible, necessarily $k > 1$. If the finite strongly irreducible linear groups are known of degrees which are proper divisors of $n$, our result can be used to obtain information concerning $L$.

## References


1. H. F. Blichfeldt, "Finite Collineation Groups." Univ. of Chicago Press, Chicago, Illinois, 1917.
2. R. Brauer, On the order of finite projective groups in a given dimension, *Nachr. Akad. Wiss. Göttingen Math.-Phys. Kl.* No. 11 (1969), 103–106.
3. R. Brauer and W. Feit, An analogue of Jordan's theorem in characteristic *p*, *Ann. of Math.* (2), **84** (1966), 119–131.
4. L. Dornhoff, "Group Representation Theory," Part A. Dekker, New York, 1971.
5. B. Huppert, Lineare auflösbare Gruppen, *Math. Z.* **68** (1957), 126–150.
6. C. Jordan, Mémoire sur les équations differentielles linéaires à intégrale algébrique, *J. Reine Angew. Math.* **84** (1878), 89–215.
7. J. F. Rigby, Primitive linear groups containing a normal nilpotent subgroup larger than the centre of the group, *J. London Math. Soc.* **35** (1960), 389–400.
8. I. Schur, Über die Darstellung der endlichen Gruppen durch gebrochene lineare Substitutionen, *J. Reine Angew. Math.* **127** (1904), 20–50; "Gesammelte Abhandlungen"(A. Brauer and H. Rohrbach, eds.), Vol. 1, pp. 86–116. Springer-Verlag, Berlin–Heidelberg–New York, 1973.


This research has been supported by Grant MPS 74–24740 of the National Science Foundation.

AMS (MOS) 1970 subject classification: 20C99

# *Unipotent differential algebraic groups*

*Phyllis J. Cassidy*
*Smith College*

## Introduction of subject matter and notation

A differential algebraic group is, roughly speaking, a group object in the category of differential algebraic sets. The most concrete example, studied in [1] (which we shall call DAG), has as its underlying set a differential variety in the sense of Kolchin and Ritt. Thus, it is the solution set in affine space of finitely many polynomial differential equations. The coordinate functions of the group laws are everywhere defined differential rational functions. In particular, every algebraic group is differential algebraic. In a work now in preparation [8], Kolchin is developing a general theory of differential algebraic groups. The method is axiomatic, along the lines used in his book [7] to develop the theory of algebraic groups.

Throughout, $\mathscr{U}$ will be a fixed differential field of characteristic 0, that is, a field, of which $\mathbf{Q}$ is a subfield, equipped with a finite set $\Delta$ of derivation operators that commute with each other. $\mathscr{K}$ will denote the field of constants of $\mathscr{U}$. We suppose that $\mathscr{U}$ is universal over $\mathbf{Q}$, i.e., that for every finitely differentially generated extension $\mathscr{F}$ of $\mathbf{Q}$ in $\mathscr{U}$ and every finitely differentially generated extension $\mathscr{G}$ of $\mathscr{F}$ there is a differential embedding of $\mathscr{G}$ in $\mathscr{U}$ that leaves the elements of $\mathscr{F}$ fixed. All differential fields, except those for which the contrary is obvious or explicitly stated, are differential subfields of $\mathscr{U}$ over which $\mathscr{U}$ is universal.

We shall use the prefix "$\boldsymbol{\Delta}$-" instead of "differential algebraic" and "differential rational," and "$\boldsymbol{\Delta}$-$\mathscr{F}$-group" in place of "differential algebraic group defined over $\mathscr{F}$."

A $\boldsymbol{\Delta}$-$\mathscr{F}$-group $G$ is *$\mathscr{F}$-unipotent* if $G$ has a normal sequence of $\boldsymbol{\Delta}$-$\mathscr{F}$-subgroups whose successive quotients are isomorphic over $\mathscr{F}$ to subgroups of the additive group $\mathbf{G}_a$ of $\mathscr{U}$. Every $\boldsymbol{\Delta}$-$\mathscr{F}$-subgroup of $\mathbf{G}_a{}^n$ is defined by homogeneous linear differential equations with coefficients in $\mathscr{F}$. Thus, if $\mathscr{U}$ is ordinary, with derivation operator $\delta$, i.e. with $\delta$-groups, a $\delta$-$\mathscr{F}$-group $G$ is $\mathscr{F}$-unipotent if and only if there is a finitely generated Picard–Vessiot extension $\mathscr{G}$ of $\mathscr{F}$ such that $G$ has a normal sequence of $\delta$-$\mathscr{G}$-subgroups whose successive quotients are isomorphic over $\mathscr{G}$ to $\mathbf{G}_a$ or to the additive group $(\mathbf{G}_a)_{\mathscr{K}}$ of the field of constants of $\mathscr{U}$.

Unipotent differential algebraic groups bear a striking resemblance to unipotent algebraic groups defined over a field $k$ of characteristic $p > 0$. This resemblance is most marked when $\mathscr{U}$ is an ordinary differential field. For any differential field $\mathscr{F}$, the ring $\mathscr{F}[\delta]$ of linear differential operators is a left and right principal ideal domain. If $k$ is perfect, this is also true of the ring $k[F]$ of noncommuting polynomials in the Frobenius operator $F$, where $F(x) = x^p$. Since every unipotent $\boldsymbol{\Delta}$-group is connected, the analog of connected unipotent algebraic group seems to be a unipotent $\boldsymbol{\Delta}$-group whose underlying differential algebraic set is isomorphic to an affine space. If $G$ is commutative, following Rosenlicht [14], we call $G$ a *vector group* if $G$ is a $\mathbf{G}_{\mathrm{m}}$-module, where $\mathbf{G}_{\mathrm{m}}$ is the multiplicative group of $\mathscr{U}$, and if $G$ is isomorphic as a $\mathbf{G}_m$-module to $\mathbf{G}_{\mathrm{a}}{}^n$, where $\mathbf{G}_{\mathrm{m}}$ acts in the usual way on $\mathbf{G}_{\mathrm{a}}{}^n$. Since the characteristic of $\mathscr{U}$ is 0, the *algebraic* group $\mathbf{G}_{\mathrm{a}}{}^n$ has defined on it a unique structure of vector group. However, just as in the case of algebraic groups defined over fields of characteristic $p > 0$, this is not the case when $\mathbf{G}_{\mathrm{a}}{}^n$ is viewed as a *differential algebraic* group. Suppose, for example, $\mathscr{U}$ is an ordinary differential field. Then every invertible matrix in $GL_{\mathscr{U}[\delta]}(n)$ defines a vector group structure on $\mathbf{G}_{\mathrm{a}}{}^n$. (For a discussion of this phenomenon in characteristic $p > 0$, see Rosenlicht [14, p. 687].) Following Tits [16] we call a unipotent $\boldsymbol{\Delta}$-group $G$ *wound over* $\mathbf{G}_{\mathrm{a}}$ if there is no nontrivial $\boldsymbol{\Delta}$-homomorphism from $\mathbf{G}_{\mathrm{a}}$ into $G$. An equivalent characterization is that $G$ have differential dimension 0. In Theorem 4, which uses the fact that $\mathscr{F}[\delta]$ is a left and right principal ideal domain, we show that if $\mathscr{U}$ is ordinary and $G$ is a commutative unipotent linear $\delta$-$\mathscr{F}$-group, then $G$ is isomorphic over $\mathscr{F}$ to a direct product of a vector group defined over $\mathscr{F}$ and a group wound over $\mathbf{G}_{\mathrm{a}}$. Furthermore, there is a finitely generated extension $\mathscr{G}$ of $\mathscr{F}$ such that $G$ is isomorphic over $\mathscr{G}$ to $\mathbf{G}_{\mathrm{a}}{}^d \times (\mathbf{G}_{\mathrm{a}})_{\mathscr{K}}{}^e$, where $d$ is the differential dimension of $G$. If the field of constants of $\mathscr{F}$ is algebraically closed, $\mathscr{G}$ may be taken to be a Picard–Vessiot extension of $\mathscr{F}$.

If we want to study unipotent $\boldsymbol{\Delta}$-groups whose underlying sets are isomorphic to affine space, we must first describe all extensions of $\mathbf{G}_{\mathrm{a}}$ by $\mathbf{G}_{\mathrm{a}}$.

Kolchin's work on constrained cohomology in [8] (for a discussion of constrained cohomology see also Kovacic's paper [9] in this volume) shows that every extension of $\mathbf{G}_a$ by $\mathbf{G}_a$ admits a differential rational cross section. It follows quite easily that every extension of $\mathbf{G}_a$ by $\mathbf{G}_a$ is central, since it is not hard to see that $\mathbf{G}_a$ does not act nontrivially on $\mathbf{G}_a$ (Corollary 2 of Proposition 19). Thus, we are led to investigate Cent Ext($\mathbf{G}_a$, $\mathbf{G}_a$). In Section 4 we develop along the lines of Mac Lane's "Homology" the basic notions of central extensions of a $\mathbf{\Delta}$-group $A$ by a commutative $\mathbf{\Delta}$-group $B$. For a discussion of group extensions in categories of groups with structure, which includes the category of $\mathbf{\Delta}$-groups, see Hochschild's paper [4] in this volume.

The existence of cross sections leads us to the second cohomology group. We adapt rational cohomology to our category in Section 3. In particular, we show that, for every $n$, $H^n(\mathbf{G}_a, \mathbf{G}_a)$ is in a natural way a left module over the ring $\mathscr{U}[\mathbf{\Delta}]$ of linear differential operators.

Cent Ext($\mathbf{G}_a$, $\mathbf{G}_a$) is an abelian group with respect to the Baer sum of extensions and is, in addition, a left module over $\mathrm{Hom}(\mathbf{G}_a, \mathbf{G}_a) = \mathscr{U}[\mathbf{\Delta}]$. Since we do not have "group chunks" in the sense of Weil, and since $\mathbf{\Delta}$-2-cocycles and $\mathbf{\Delta}$-cross sections are not necessarily everywhere defined, we are forced in Theorem 5 to show by explicit computation that every $\mathbf{\Delta}$-2-cocycle is cohomologous to a differential polynomial 2-cocycle. In the course of the proof we show that every symmetric $\mathbf{\Delta}$-2-cocycle is a coboundary. In Theorem 6 we prove that $H^2(\mathbf{G}_a, \mathbf{G}_a)$ is isomorphic as a left module over $\mathscr{U}[\mathbf{\Delta}]$ to Cent Ext($\mathbf{G}_a$, $\mathbf{G}_a$). In particular, therefore, every abelian extension of $\mathbf{G}_a$ by $\mathbf{G}_a$ splits.

So, to determine the extensions of $\mathbf{G}_a$ by $\mathbf{G}_a$ up to isomorphism, it suffices to describe the second cohomology group. In contrast to the situation for algebraic groups over fields of characteristic 0, where the second cohomology group is trivial, if $\mathscr{U}$ is an ordinary differential field, we show in Theorem 8 that $H^2(\mathbf{G}_a, \mathbf{G}_a)$ is a free left module over $\mathscr{U}[\delta]$, with basis given by the cohomology classes of the 2-forms $x \cdot \delta^j y = xy^{(j)}$, where $j$ is an odd positive integer. In [15] Serre refers to [11], in which Lazard computes the symmetric 2-cocycles from $\mathbf{G}_a \times \mathbf{G}_a$ into $\mathbf{G}_a$. The analogous paper in differential algebra is Ritt's "Associative differential operations" [13], in which he exhibits up to a naturally defined equivalence the two formal groups in one differential parameter. In this paper Ritt computes all differential polynomials $f$ in two variables that satisfy the functional equation $f(x, y) + f(x + y, z) = f(y, z) + f(x, y + z)$. He shows that $f$ is cohomologous to the 2-form $\sum_{i<j} a_{ij} x^{(i)} y^{(j)}$, $a_{ij} \in \mathscr{U}$. So, every $\delta$-group extension of $\mathbf{G}_a$ by $\mathbf{G}_a$ is isomorphic to a group whose underlying set is the plane and whose law of composition is given by the formula

$$(u_1, u_2)(v_1, v_2) = (u_1 + v_1, u_2 + v_2 + \sum_{i<j} a_{ij} u_1^{(i)} v_1^{(j)}).$$

In a paper in preparation we show that every $\delta$-group whose underlying set is the plane $\mathcal{U}^2$ is unipotent and is isomorphic to a group of this form.

## 1. The structure of unipotent differential algebraic groups

In [8] Kolchin proves the existence in the category of $\mathbf{\Delta}$-$\mathcal{F}$-groups of quotients. If $G$ is a $\mathbf{\Delta}$-$\mathcal{F}$-group and $H$ is a normal $\mathbf{\Delta}$-$\mathcal{F}$-subgroup of $G$, then $G/H$ can be given a structure of $\mathbf{\Delta}$-$\mathcal{F}$-group in such a way that the quotient homomorphism $\pi$ is a $\mathbf{\Delta}$-$\mathcal{F}$-homomorphism. Given a $\mathbf{\Delta}$-$\mathcal{F}$-homomorphism $\varphi\colon G \to G'$, with $G'$ a $\mathbf{\Delta}$-$\mathcal{F}$-group and $H \subset \ker \varphi$, there is a unique $\mathbf{\Delta}$-$\mathcal{F}$-homomorphism $\psi\colon G/H \to G'$ such that $\psi \circ \pi = \varphi$. $\psi$ is a $\mathbf{\Delta}$-$\mathcal{F}$-isomorphism if $\varphi$ is surjective and $\ker \varphi = H$. Kolchin [8] also shows that if $H_1$ and $H_2$ are $\mathbf{\Delta}$-$\mathcal{F}$-subgroups of a $\mathbf{\Delta}$-$\mathcal{F}$-group $G$, with $H_1$ normal in $G$, then $H_1 H_2$ is a $\mathbf{\Delta}$-$\mathcal{F}$-subgroup of $G$. These results enable us to state, and mimic the proofs of, the Noether isomorphism theorems and the Jordan–Hölder–Schreier theorem. In adapting Zassenhaus' proof of the latter theorem, we must construct groups $G^{i+1}(G^i \cap H^j)$, where $G = G^0 \supset G^1 \supset \cdots \supset G^r = \{1\}$ and $G = H^0 \supset H^1 \supset \cdots \supset H^s = \{1\}$ are normal sequences of $\mathbf{\Delta}$-$\mathcal{F}$-subgroups of a $\mathbf{\Delta}$-$\mathcal{F}$-group $G$.

***Proposition 1*** *Let $G$ be a $\mathbf{\Delta}$-$\mathcal{F}$-group and let $H_1$ and $H_2$ be $\mathbf{\Delta}$-$\mathcal{F}$-subgroups of $G$, with $H_1$ normal in $G$. Then $H_1 H_2/H_1$ is $\mathbf{\Delta}$-$\mathcal{F}$-isomorphic to $H_2/H_1 \cap H_2$.*

***Proposition 2*** *Let $G$ be a $\mathbf{\Delta}$-$\mathcal{F}$-group and let $H_1$ and $H_2$ be normal $\mathbf{\Delta}$-$\mathcal{F}$-subgroups of $G$, with $H_1 \subset H_2$. Then $G/H_2$ is $\mathbf{\Delta}$-$\mathcal{F}$-isomorphic to $(G/H_1)/(H_2/H_1)$.*

***Proposition 3*** *Let $G$ be a $\mathbf{\Delta}$-$\mathcal{F}$-group. Two normal sequences of $\mathbf{\Delta}$-$\mathcal{F}$-subgroups of $G$, ending with the trivial subgroup, have $\mathbf{\Delta}$-$\mathcal{F}$-isomorphic refinements.*

***Definition*** A $\mathbf{\Delta}$-$\mathcal{F}$-group $G$ is *$\mathcal{F}$-unipotent* if $G$ has a normal sequence of $\mathbf{\Delta}$-$\mathcal{F}$-subgroups whose successive quotients are $\mathbf{\Delta}$-$\mathcal{F}$-isomorphic to $\mathbf{\Delta}$-$\mathcal{F}$-subgroups of $\mathbf{G}_a$. We call $G$ *unipotent* if there is an extension field $\mathcal{G}$ of $\mathcal{F}$ such that $G$ is $\mathcal{G}$-unipotent.

The definition given in [3, p. 485] implies, since the characteristic of $\mathcal{F}$ is zero, that an $\mathcal{F}$-group $G$ is $\mathcal{F}$-unipotent if $G$ has a normal sequence of $\mathcal{F}$-subgroups whose successive quotients are $\mathcal{F}$-isomorphic to $\mathbf{G}_a$. Thus, every $\mathcal{F}$-unipotent $\mathcal{F}$-group is an $\mathcal{F}$-unipotent $\mathbf{\Delta}$-$\mathcal{F}$-group.

***Proposition 4*** *Let $H$ be a $\mathbf{\Delta}$-$\mathcal{F}$-subgroup of $\mathbf{G}_a{}^n$. There exists a natural number $r$ and a $\mathbf{\Delta}$-$\mathcal{F}$-homomorphism $\varphi\colon \mathbf{G}_a{}^n \to \mathbf{G}_a{}^r$ such that $H = \ker \varphi$.*

*Proof* There exist homogeneous linear differential polynomials $L_1, \ldots, L_r$ in $\mathscr{F}\{y_1, \ldots, y_n\}$ such that $H$ is the set of zeros in $\mathbf{G}_a{}^n$ of $L_1, \ldots, L_r$. Define $\varphi(x) = (L_1(x), \ldots, L_r(x))$.

**Corollary** *Let $G$ be a $\mathbf{\Delta}$-$\mathscr{F}$-subgroup of $\mathbf{G}_a{}^n$ and let $\alpha\colon G \to G'$ be a surjective $\mathbf{\Delta}$-$\mathscr{F}$-homomorphism from $G$ onto a $\mathbf{\Delta}$-$\mathscr{F}$-group $G'$. Then $G'$ is $\mathbf{\Delta}$-$\mathscr{F}$-isomorphic to a $\mathbf{\Delta}$-$\mathscr{F}$-subgroup of $\mathbf{G}_a{}^r$ for some $r$.*

*Proof* We may assume that $G'$ is a $\mathbf{\Delta}$-$\mathscr{F}$-subgroup of $GL(q)$ for some $q$ [DAG, Props. 10 and 15]. $\alpha$ extends to a $\mathbf{\Delta}$-$\mathscr{F}$-homomorphism $\tilde{\alpha}$ from the smallest algebraic subgroup $A(G)$ of $\mathbf{G}_a{}^n$ containing $G$ into $GL(q)$ [2, Prop. 3]. $A(G)$ is $\mathscr{F}$-isomorphic to $\mathbf{G}_a{}^k$ for some $k$. ker $\tilde{\alpha}$ is a $\mathbf{\Delta}$-$\mathscr{F}$-subgroup of $A(G)$. Therefore, by the proposition, there is a natural number $r$ and a $\mathbf{\Delta}$-$\mathscr{F}$-homomorphism $\varphi\colon A(G) \to \mathbf{G}_a{}^r$ such that ker $\varphi$ = ker $\tilde{\alpha}$. Therefore, $\tilde{\alpha}(A(G))$ is $\mathbf{\Delta}$-$\mathscr{F}$-isomorphic to $\varphi(A(G))$, whence $G'$ is $\mathbf{\Delta}$-$\mathscr{F}$-isomorphic to a $\mathbf{\Delta}$-$\mathscr{F}$-subgroup of $\mathbf{G}_a{}^r$.

If $\mathscr{U}$ is an ordinary differential field with derivation operator $\delta$, i.e., with $\delta$-groups, and $H$ is a $\delta$-$\mathscr{F}$-subgroup of $\mathbf{G}_a$, we can strengthen Proposition 4.

**Proposition 5** *Let $H$ be a $\delta$-$\mathscr{F}$-subgroup of $\mathbf{G}_a$. There exists a $\delta$-$\mathscr{F}$-endomorphism $\varphi$ of $\mathbf{G}_a$ such that the sequence*

$$0 \to H \to \mathbf{G}_a \xrightarrow{\varphi} \mathbf{G}_a \to 0$$

*is exact.*

*Proof* There exists a single homogeneous linear differential polynomial $L$ in $\mathscr{F}\{y\}$ such that $H$ is the set of zeros in $\mathbf{G}_a$ of $L$. We define $\varphi(x) = L(x)$. $\varphi$ is the desired $\delta$-$\mathscr{F}$-endomorphism of $\mathbf{G}_a$.

**Corollary 1** *Let $G'$ be a $\delta$-$\mathscr{F}$-group and let $\varphi\colon \mathbf{G}_a \to G'$ be a surjective $\delta$-$\mathscr{F}$-homomorphism. Then $G'$ is $\delta$-$\mathscr{F}$-isomorphic to $\mathbf{G}_a$.*

**Corollary 2** *Let $G$ be a $\delta$-$\mathscr{F}$-subgroup of $\mathbf{G}_a$ and let $\varphi$ be a surjective $\delta$-$\mathscr{F}$-homomorphism from $G$ onto a $\delta$-$\mathscr{F}$-group $G'$. Then $G'$ is $\delta$-$\mathscr{F}$-isomorphic to a $\delta$-$\mathscr{F}$-subgroup of $\mathbf{G}_a$.*

***Definition*** A $\mathbf{\Delta}$-group $G$ is *elementary* if every nontrivial $\mathbf{\Delta}$-homomorphic image of $G$ is $\mathbf{\Delta}$-isomorphic to $G$.

**Proposition 6** *The $\delta$-groups $\mathbf{G}_a$ and $(\mathbf{G}_a)_{\mathscr{K}}$ are elementary.*

*Proof* Since $(\mathbf{G}_a)_{\mathscr{K}}$ has no nontrivial proper $\delta$-subgroups, it is clearly elementary. The fact that $\mathbf{G}_a$ is elementary follows immediately from Corollary 1 of Proposition 5.

We shall see in Section 2 that $\mathbf{G}_a$ and $(\mathbf{G}_a)_{\mathscr{K}}$ are the only commutative elementary unipotent $\delta$-groups up to $\delta$-isomorphism.

***Theorem 1*** *Let $G$ be a $\mathbf{\Delta}$-$\mathscr{F}$-group.*

1. *If $G$ is $\mathscr{F}$-unipotent, then every $\mathbf{\Delta}$-$\mathscr{F}$-subgroup of $G$ is $\mathscr{F}$-unipotent.*
2. *If $H$ and $G'$ are $\mathscr{F}$-unipotent $\mathbf{\Delta}$-$\mathscr{F}$-groups and $\varphi\colon G \to G'$ is a surjective $\mathbf{\Delta}$-$\mathscr{F}$-homomorphism with kernel $H$, then $G$ is $\mathscr{F}$-unipotent.*
3. *If $G$ is $\mathscr{F}$-unipotent, $G'$ is a $\mathbf{\Delta}$-$\mathscr{F}$-group, and $\varphi\colon G \to G'$ is a surjective $\mathbf{\Delta}$-$\mathscr{F}$-homomorphism, then $G'$ is unipotent.*

*Proof* Let $G = G^0 \supset G^1 \supset \cdots \supset G^r = \{1\}$ be a normal sequence of $\mathbf{\Delta}$-$\mathscr{F}$-subgroups of $G$ such that $G^i/G^{i+1}$ is $\mathbf{\Delta}$-$\mathscr{F}$-isomorphic to a $\mathbf{\Delta}$-$\mathscr{F}$-subgroup of $\mathbf{G}_a$. Let $H = H^0 \supset H^1 \supset \cdots \supset H^r = \{1\}$ be the normal sequence defined by the formula $H^i = H \cap G^i$. By Proposition 1, $H^i/H^{i+1}$ is $\mathbf{\Delta}$-$\mathscr{F}$-isomorphic to $H^iG^{i+1}/G^{i+1}$, which is a $\mathbf{\Delta}$-$\mathscr{F}$-subgroup of $G^i/G^{i+1}$.

To prove 2, let $H = H^0 \supset H^1 \supset \cdots \supset H^r = \{1\}$ be a normal sequence of $\mathbf{\Delta}$-$\mathscr{F}$-subgroups of $H$ such that $H^i/H^{i+1}$ is $\mathbf{\Delta}$-$\mathscr{F}$-isomorphic to a $\mathbf{\Delta}$-$\mathscr{F}$-subgroup of $\mathbf{G}_a$ and let $G' = G'^0 \supset G'^1 \supset \cdots \supset G'^s = \{1\}$ be a parallel sequence for $G'$. Consider the normal sequence $G = \varphi^{-1}G'^0 \supset \varphi^{-1}G'^1 \supset \cdots \supset \varphi^{-1}G'^s = H = H^0 \supset H^1 \supset \cdots \supset H^r = \{1\}$. Proposition 2 implies that $\varphi^{-1}G'^i/\varphi^{-1}G'^{i+1}$ is isomorphic to $G'^i/G'^{i+1}$, hence is $\mathbf{\Delta}$-$\mathscr{F}$-isomorphic to a $\mathbf{\Delta}$-$\mathscr{F}$-subgroup of $\mathbf{G}_a$. This sequence is thus the desired normal sequence for $G$.

Let $G = G^0 \supset G^1 \supset \cdots \supset G^r = \{1\}$ be a normal sequence of $\mathbf{\Delta}$-$\mathscr{F}$-subgroups of $G$ whose successive quotients are $\mathbf{\Delta}$-$\mathscr{F}$-isomorphic to $\mathbf{\Delta}$-$\mathscr{F}$-subgroups of $\mathbf{G}_a$. Proposition 2 implies that $\varphi G^i/\varphi G^{i+1}$ is $\mathbf{\Delta}$-$\mathscr{F}$-isomorphic to $G^i \cdot \ker \varphi/G^{i+1} \cdot \ker \varphi$. Proposition 1 implies that the latter group is $\mathbf{\Delta}$-$\mathscr{F}$-isomorphic to $G^i/G^i \cap (G^{i+1} \cdot \ker \varphi)$, which is an image of $G^i/G^{i+1}$. By hypothesis, $G^i/G^{i+1}$ is $\mathbf{\Delta}$-$\mathscr{F}$-isomorphic to a $\mathbf{\Delta}$-$\mathscr{F}$-subgroup of $\mathbf{G}_a$. So, $\varphi G^i/\varphi G^{i+1}$ is the image under a $\mathbf{\Delta}$-$\mathscr{F}$-homomorphism of a $\mathbf{\Delta}$-$\mathscr{F}$-subgroup of $\mathbf{G}_a$. The corollary of Proposition 4 implies that it is then $\mathbf{\Delta}$-$\mathscr{F}$-isomorphic to a $\mathbf{\Delta}$-$\mathscr{F}$-subgroup of $\mathbf{G}_a{}^r$ for some $r$. Part 1 of this theorem then implies that $\varphi G^i/\varphi G^{i+1}$ is $\mathscr{F}$-unipotent, $0 \le i \le r-1$, hence is a multiple extension of $\mathbf{\Delta}$-$\mathscr{F}$-subgroups of $\mathbf{G}_a$. As in part 2, this implies that $\varphi G$ is a multiple extension of $\mathbf{\Delta}$-$\mathscr{F}$-subgroups of $\mathbf{G}_a$.

***Corollary 1*** *Let $G$ be a unipotent $\mathbf{\Delta}$-$\mathscr{F}$-group. Then every nontrivial element of $G$ has infinite order.*

*Proof* Let $x \neq 1$ be in $G$. Let $G(x)$ be the smallest $\mathbf{\Delta}$-subgroup of $G$ containing $x$. $G(x)$ is unipotent, by the theorem. If $x$ has finite order, then $G(x)$ is a nontrivial finite unipotent $\mathbf{\Delta}$-group, which is impossible.

***Corollary 2*** *Every unipotent $\mathbf{\Delta}$-group $G$ is connected.*

*Proof* If $G$ does not equal its identity component $G_0$, the theorem implies that the finite nontrivial $\mathbf{\Delta}$-group $G/G_0$ is unipotent, which is not possible.

***Proposition 7*** *Let $G$ be a $\delta$-$\mathscr{F}$-group. If $G$ is $\mathscr{F}$-unipotent, there is a finitely generated extension $\mathscr{G}$ of $\mathscr{F}$ such that $G$ has a normal sequence of $\delta$-$\mathscr{G}$-subgroups of $G$ whose successive quotients are $\delta$-$\mathscr{G}$-isomorphic to $\mathbf{G}_{\mathrm{a}}$ or to $(\mathbf{G}_{\mathrm{a}})_{\mathscr{K}}$. If the field of constants of $\mathscr{F}$ is algebraically closed, $\mathscr{G}$ may be taken to be Picard–Vessiot over $\mathscr{F}$.*

*Proof* Since $G$ is $\mathscr{F}$-unipotent, $G$ has a normal sequence $G = G^0 \supset G^1 \supset \cdots \supset G^r = \{1\}$ of $\delta$-$\mathscr{F}$-subgroups such that $G^i/G^{i+1}$ is $\delta$-$\mathscr{F}$-isomorphic to a $\delta$-$\mathscr{F}$-subgroup $H^i$ of $\mathbf{G}_{\mathrm{a}}$. When $H_i$ is a proper $\delta$-subgroup of $\mathbf{G}_{\mathrm{a}}$, let $\eta_i = (\eta_{i1}, \ldots, \eta_{in_i})$ be a fundamental system of zeros of the defining linear homogeneous differential polynomial $L_i$ of $H_i$. Let $\mathscr{G} = \mathscr{F}\langle(\eta_i)\rangle_{0 \le i \le r-1}$. Then $H^i$ is $\delta$-$\mathscr{G}$-isomorphic to $(\mathbf{G}_{\mathrm{a}})^{n_i}_{\mathscr{K}}$. So, a suitable refinement will give us the desired sequence.

We denote by $T(n)$ the subgroup of $GL(n)$ consisting of all upper triangular matrices, i.e., all those matrices whose entries below the principal diagonal are zero. $T(n)$ is a connected $\mathbf{Q}$-group. We denote by $T(n, 1)$ the $\mathbf{Q}$-subgroup of $T(n)$ consisting of all matrices in $T(n)$ for which $a_{ii} = 1$, $1 \le i \le n$. $T(n, 1)$ is a $\mathbf{Q}$-unipotent $\mathbf{Q}$-group.

A matrix $x$ in $GL(n)$ is *unipotent* if the matrix $x - 1$ is nilpotent, i.e., if all the characteristic values of $x$ equal 1. In particular, every element of $T(n, 1)$ is unipotent.

Let $G$ be a $\mathbf{\Delta}$-$\mathscr{F}$-subgroup of $GL(n)$ every element of which is unipotent. Then the smallest algebraic subgroup $A(G)$ containing $G$ shares this property. Kolchin proved (see [7, p. 369]) that $A(G)$ is conjugate over any field $\mathscr{F}$ of definition to an algebraic subgroup of $T(n, 1)$. It follows, since $A(G)$ is defined over any field of definition for $G$, that $G$ is conjugate over $\mathscr{F}$ to a differential algebraic subgroup of $T(n, 1)$. In particular, every $\mathbf{\Delta}$-$\mathscr{F}$-subgroup of $GL(n)$, every element of which is unipotent, is $\mathscr{F}$-unipotent.

We saw in Corollary 1 of the structure theorem that every nontrivial element of a unipotent differential algebraic group has infinite order. We shall show now that this characterizes linear unipotent $\mathbf{\Delta}$-$\mathscr{F}$-groups.

***Proposition 8*** *Let $G$ be a linear $\mathbf{\Delta}$-$\mathscr{F}$-group containing no nontrivial element of finite order. Then $G$ is $\mathscr{F}$-unipotent.*

*Proof* We may assume that $G$ is a $\mathbf{\Delta}$-$\mathscr{F}$-subgroup of $GL(n)$ for some $n$. We shall show that every element of $G$ is a unipotent matrix. Let $x \ne 1$ be in $G$. Let $H$ be the identity component of $G(x)$. Then, since $H$ is commutative, so is $A(H)$. Therefore, $A(H)$ can be written as a direct product $T \times U$, where

$T$ is a torus and $U$ consists of unipotent matrices. Let $K$ be the identity component of $H \cap T$. $K$ contains no nontrivial element of finite order. Therefore, $K = \{1\}$ [DAG, p. 939]. It follows that $H = \{1\}$ since $H \cap T$ is finite. Consider the projection homomorphism $\varphi$ from $T \times U$ onto $U$. The restriction of $\varphi$ to $H$ is an isomorphism. Therefore, $H$ consists of unipotent matrices [DAG, p. 942]. Let $y$ be in $G(x)$. $y = s \cdot u$, where $s$ is semisimple and $u$ is unipotent and $s \cdot u = u \cdot s$. For some positive integer $m$, $y^m = s^m \cdot u^m$ is in $H$. $y^m$ is, therefore, unipotent. Now, $s^m = y^m u^{-m}$, whence, since $y^m$ and $u^{-m}$ commute and are unipotent, it follows that $s^m$ is both semisimple and unipotent. Therefore, $s^m = 1$, whence $y^m = u^m$. It follows that $u^m$ is in $H$. Now, since $u$ is unipotent, the smallest $\mathbf{\Delta}$-subgroup $G(u)$ of $GL(n)$ containing $u$ is isomorphic to $(\mathbf{G}_a)_{\mathscr{K}}$. Therefore, $G(u) \cap H = \{1\}$ or $G(u)$. Since $u^m$ is in $H$, $G(u) \cap H = G(u)$. Thus, $u$ is an element of $H$. Since $y = s \cdot u$, it now follows that $s$ is in $G(x)$. But, $s^m = 1$, whence $s = 1$, since $G(x)$ contains no nontrivial element of finite order. Therefore, $y = u$.

The following theorem follows from the preceding discussion.

***Theorem 2*** *Let $G$ be a linear $\mathbf{\Delta}$-$\mathscr{F}$-group. The following conditions are equivalent:*

1. *$G$ is $\mathscr{F}$-unipotent.*
2. *$G$ contains no nontrivial element of finite order.*
3. *$G$ is $\mathbf{\Delta}$-$\mathscr{F}$-isomorphic to a $\mathbf{\Delta}$-$\mathscr{F}$-subgroup of $T(n, 1)$.*

***Corollary 1*** *Every unipotent linear $\mathbf{\Delta}$-$\mathscr{F}$-group is $\mathscr{F}$-unipotent.*

***Corollary 2*** *Let $G$ be a unipotent $\mathbf{\Delta}$-$\mathscr{F}$-subgroup of $GL(n)$ and let $x \neq 1$ be in $G$. Then $x$ is a unipotent matrix and $G(x)$ is $\mathbf{\Delta}$-isomorphic to $(\mathbf{G}_a)_{\mathscr{K}}$. The smallest algebraic subgroup of $GL(n)$ containing $x$ is isomorphic to $\mathbf{G}_a$.*

Let $G$ be an $\mathscr{F}$-unipotent $\delta$-$\mathscr{F}$-group. In rough analogy to a definition used by Tits [16, p. 128] we say that $G$ is *$\mathscr{F}$-wound* over $\mathbf{G}_a$ (resp. over $(\mathbf{G}_a)_{\mathscr{K}}$) if every $\delta$-$\mathscr{F}$-homomorphism $\varphi$: $\mathbf{G}_a \to G$ (resp. $\varphi$: $(\mathbf{G}_a)_{\mathscr{K}} \to G$) is trivial.

***Proposition 9*** *Let $G$ be an $\mathscr{F}$-unipotent $\delta$-$\mathscr{F}$-group.*

1. *If $G$ is $\mathscr{F}$-wound over $\mathbf{G}_a$ (resp. over $(\mathbf{G}_a)_{\mathscr{K}}$), then every $\delta$-$\mathscr{F}$-subgroup of $G$ is $\mathscr{F}$-wound over $\mathbf{G}_a$ (resp. over $(\mathbf{G}_a)_{\mathscr{K}}$).*
2. *$G$ is $\mathscr{F}$-wound over $\mathbf{G}_a$ (resp. over $(\mathbf{G}_a)_{\mathscr{K}}$) if and only if $G$ contains no $\delta$-$\mathscr{F}$-subgroup $\delta$-$\mathscr{F}$-isomorphic to $\mathbf{G}_a$ (resp. to $(\mathbf{G}_a)_{\mathscr{K}}$).*
3. *If $G$ is $\mathscr{F}$-wound over $(\mathbf{G}_a)_{\mathscr{K}}$, then $G$ is $\mathscr{F}$-wound over $\mathbf{G}_a$.*

*Proof* 1 is clear. Now, if $G$ contains a $\delta$-$\mathscr{F}$-subgroup $\mathscr{F}$-isomorphic to $\mathbf{G}_a$ (resp. to $(\mathbf{G}_a)_{\mathscr{K}}$), then $G$ cannot, of course, be $\mathscr{F}$-wound over $\mathbf{G}_a$ (resp. over $(\mathbf{G}_a)_{\mathscr{K}}$). Suppose the condition of 2 is satisfied, and there is a nontrivial

$\delta$-$\mathscr{F}$-homomorphism $\varphi\colon \mathbf{G}_a \to G$ (resp. $\varphi\colon (\mathbf{G}_a)_{\mathscr{K}} \to G$). Then, since $\mathbf{G}_a$ (resp. $(\mathbf{G}_a)_{\mathscr{K}}$) is elementary, $\varphi\mathbf{G}_a$ (resp. $\varphi(\mathbf{G}_a)_{\mathscr{K}}$) is a $\delta$-$\mathscr{F}$-subgroup of $G$ that is $\delta$-$\mathscr{F}$-isomorphic to $\mathbf{G}_a$ (resp. to $(\mathbf{G}_a)_{\mathscr{K}}$). Suppose $G$ is $\mathscr{F}$-wound over $(\mathbf{G}_a)_{\mathscr{K}}$, and not over $\mathbf{G}_a$. Then there is a $\delta$-$\mathscr{F}$-subgroup of $G$ that is $\delta$-$\mathscr{F}$-isomorphic to $\mathbf{G}_a$. This isomorphism produces, by restriction, a differential algebraic subgroup of $G$ that is $\delta$-$\mathscr{F}$-isomorphic to $(\mathbf{G}_a)_{\mathscr{K}}$.

## 2. Commutative linear unipotent differential algebraic groups

Let $G$ be a $\mathbf{\Delta}$-$\mathscr{F}$-subgroup of $GL(n)$. The Lie algebra $\ell(G)$ of matrices of $G$ is a $\mathbf{\Delta}$-$\mathscr{F}$-subalgebra of $\mathfrak{gl}(n)$ (a $\mathscr{K}$-subalgebra that is also $\mathbf{\Delta}$-$\mathscr{F}$-closed). If $G$ is an algebraic subgroup, then $\ell(G)$ is isomorphic as an additive group to $\mathbf{G}_a{}^r$, where $r$ is the dimension of $G$. Suppose $G$ is an $\mathscr{F}$-unipotent $\mathbf{\Delta}$-$\mathscr{F}$-subgroup of $GL(n)$. Then $A(G)$ is an $\mathscr{F}$-unipotent $\mathscr{F}$-subgroup of $GL(n)$. Let $x$ be in $A(G)$. $x = 1 + n$, where $n$ is nilpotent. $\operatorname{Log} x = n - n^2/2 + n^3/3 - \cdots$. $\operatorname{Log} x$ is a polynomial in $n$, and its entries are polynomial functions in the entries of $x$. The map $x \mapsto \log x$ is an $\mathscr{F}$-isomorphism of algebraic sets from $A(G)$ onto $\ell(A(G)) = A(\ell(G))$, which maps $G$ isomorphically onto $\ell(G)$. If $G$, hence $A(G)$, is commutative, this map is an $\mathscr{F}$-isomorphism of algebraic groups. Therefore, we have proved the following proposition.

***Proposition 10*** *Let $G$ be a unipotent $\mathbf{\Delta}$-$\mathscr{F}$-subgroup of $GL(n)$. Then $A(G)$ is isomorphic over $\mathscr{F}$ as an algebraic set to $\mathbf{G}_a{}^r$ for some $r$. This isomorphism maps $G$ onto a differential algebraic subgroup of $\mathbf{G}_a{}^r$. If $G$, hence $A(G)$, is commutative, the isomorphism is an $\mathscr{F}$-isomorphism of algebraic groups.*

Since the defining differential ideal of a $\mathbf{\Delta}$-$\mathscr{F}$-subgroup $G$ of $\mathbf{G}_a{}^n$ is homogeneous linear, $G$ is a $\mathbf{\Delta}$-$\mathscr{F}$-closed subspace of the $\mathscr{K}$-vector space $\mathscr{U}^n$, i.e., $G$ is both a $\mathbf{\Delta}$-$\mathscr{F}$-set and a vector space over the field $\mathscr{K}$ of constants. It is natural to ask whether $G$ decomposes into a direct product of simpler groups. If $G$ is an algebraic subgroup of $\mathbf{G}_a{}^n$, defined over $\mathscr{F}$, then $G$ is isomorphic over $\mathscr{F}$ to $\mathbf{G}_a{}^r$, where $r$ is the dimension of $G$; in fact, there is an $\mathscr{F}$-automorphism of the algebraic group $\mathbf{G}_a{}^n$ such that the image of $G$ under the automorphism is $0 \times \cdots \times 0 \times \mathbf{G}_a{}^r$. If $\mathscr{U}$ is an ordinary differential field, then we have a decomposition of an arbitrary $\delta$-$\mathscr{F}$-subgroup $G$ of $\mathbf{G}_a{}^n$ that is similar to the decomposition of a commutative unipotent algebraic group of exponent $p$ (Tits [16, p. 128]).

*We shall assume until further notice that $\mathscr{U}$ is an ordinary differential field, with derivation operator $\delta$.*

Let $G$ be a $\delta$-$\mathscr{F}$-subgroup of $\mathbf{G}_a{}^n$, let $V$ be the differential vector space over $\mathscr{F}$ of homogeneous linear differential polynomials in $y_1, \ldots, y_n$, and let $\mathfrak{i}$ be the defining differential ideal of $G$. The differential vector subspace $W = \mathfrak{i} \cap V$ generates $\mathfrak{i}$ as a differential ideal in $\mathscr{U}\{y_1, \ldots, y_n\}$.

The ring $\mathscr{F}[\delta]$ of linear differential operators in $\delta$ with coefficients in $\mathscr{F}$ is a noncommutative polynomial ring whose elements are denoted by

$$f(\delta) = a_0\delta + a_1\delta + \cdots + a_m\delta^m, \qquad a_i \in \mathscr{U}, \quad 0 \leq i \leq m.$$

The product $\delta a = a\delta + a'$, where $a' = \delta(a)$. This ring bears a great deal of similarity to the ring $k[F]$ of noncommutative polynomials in the Frobenius operator $F$, where $k$ is a field of prime characteristic $p$ and $F(x) = x^p$. The product in the latter ring is given by the formula $Fa = a^pF$. There is one striking difference, however. $\mathscr{F}[\delta]$ is a left and right principal ideal domain, independent of the coefficient field $\mathscr{F}$, whereas $k[F]$ is a left and right principal ideal domain only if $F$ is an automorphism of $k$, that is, $k$ is perfect.

$V$ is a free left $\mathscr{F}[\delta]$-module with basis $y_1, \ldots, y_n$, since a typical homogeneous differential polynomial $L$ in $\mathscr{F}\{y_1, \ldots, y_n\}$ can be written

$$\begin{aligned} L &= \sum_{1 \leq i \leq n} \sum_{0 \leq j \leq m_i} a_{ji} y_i^{(j)} \\ &= \sum_{1 \leq i \leq n} f_i(\delta) y_i, \qquad \text{where } f_i(\delta) = \sum_{0 \leq j \leq m_i} a_{ji}\,\delta^j. \end{aligned}$$

$W$ is a submodule of $V$. A "theorem on elementary divisors" holds for free finitely generated left modules over $\mathscr{F}[\delta]$.

***Theorem 3*** (paraphrased from Van der Waerden [17], "Moderne Algebra," 1st ed: §106, pp. 120–125) *Let $\mathscr{F}[\delta]$ be the ring of linear differential operators in $\delta$, let $V$ be the free left $\mathscr{F}[\delta]$-module with basis $y_1, \ldots, y_n$, and let $W \subset V$ be a submodule. Then $W$ is free and the rank $s$ of $W$ is $\leq n$. Moreover, there is a basis $z_1, \ldots, z_n$ of $V$ and $f_1(\delta), \ldots, f_s(\delta)$ in $V$, with $f_i \mid f_{i+1}$ (left and right) such that $f_1(\delta)z_1, \ldots, f_s(\delta)z_s$ is a basis of $W$.*

Since $z_1, \ldots, z_n$ is a basis of $V$, there is an invertible matrix $D = (g_{ij}(\delta))$ in $GL_{\mathscr{F}[\delta]}(n)$ such that

$$z_j = \sum_{1 \leq i \leq n} g_{ij}(\delta) y_i \qquad (1 \leq i \leq n).$$

The matrix $D$ defines a $\delta$-$\mathscr{F}$-automorphism $T$ of the $\delta$-$\mathscr{F}$-group $\mathbf{G}_a{}^n$, by the formula

$$v_j = \sum_{1 \leq i \leq n} g_{ij}(\delta) u_i \qquad (1 \leq j \leq n).$$

The $\delta$-$\mathscr{F}$-subgroup $T(G)$ of $\mathbf{G}_a{}^n$ is the set of all $(v_1, \ldots, v_n)$ in $\mathbf{G}_a{}^n$ such that $f_1(\delta)v_1 = \cdots = f_s(\delta)v_s = 0$. Let $G_i$ be the $\delta$-$\mathscr{F}$-subgroup of $\mathbf{G}_a$ consisting of the set of zeros of $f_i(\delta)y$, $1 \leq i \leq s$. $G_i$ is a finite-dimensional vector space over $\mathscr{K}$ of dimension $m_i$ equal to the order of $f_i(\delta)y$. Let $\mathscr{G}$ be the extension of $\mathscr{F}$ generated by a fundamental system of zeros of $f_i(\delta)y$, $1 \leq i \leq s$. Then $T(G)$

is $\delta$-$\mathscr{G}$-isomorphic to $(\mathbf{G}_a)_{\mathscr{K}}{}^m \times \mathbf{G}_a{}^r$, where $m = m_1 + \cdots + m_s$ and $r = n - s$ is the differential dimension of $G$. Thus we have proved the following theorem.

***Theorem 4*** *Let $\mathscr{U}$ be an ordinary differential field, with derivation operator $\delta$, and let $G$ be a $\delta$-$\mathscr{F}$-subgroup of $\mathbf{G}_a{}^n$. There exist homogeneous linear differential polynomials $f_1(\delta)y, \ldots, f_s(\delta)y$ in $\mathscr{F}\{y\}$ such that $G$ is $\delta$-$\mathscr{F}$-isomorphic to $G_1 \times \cdots \times G_s$, where $r = n - s$ is the differential dimension of $G$, and $G_i$ is the set of zeros in $\mathbf{G}_a$ of $f_i(\delta)y$, $1 \leq i \leq s$. Let $\mathscr{G}$ be the smallest extension of $\mathscr{F}$ containing a fundamental system of zeros of $f_i(\delta)y$, $1 \leq i \leq s$. Then $G$ is $\delta$-$\mathscr{G}$-isomorphic to $(\mathbf{G}_a)_{\mathscr{K}}{}^m \times \mathbf{G}_a{}^r$, where $m = m_1 + \cdots + m_s$.*

Theorem 4 says that if $G$ is a commutative unipotent linear $\delta$-$\mathscr{F}$-group, then there exists a finitely generated extension $\mathscr{G}$ of $\mathscr{F}$ such that $G$ is the direct product of $\mathscr{G}$-elementary unipotent differential algebraic groups.

We now relinquish the hypothesis that the universal differential field $\mathscr{U}$ be ordinary.

Let $A$ and $B$ be $\mathbf{\Delta}$-$\mathscr{F}$-groups. We say that *$A$ acts on $B$* if there is an everywhere defined $\mathbf{\Delta}$-$\mathscr{F}$-map $\alpha$: $A \times B \to B$, sending $(a, b) \to {}^a b$, such that

1. ${}^{a_1 a_2}b = {}^{a_1}({}^{a_2}b)$,
2. ${}^1 b = b$,
3. ${}^a(b_1 b_2) = {}^a b_1 \, {}^a b_2$.

We call $B$ an *$A$-module.* If $B$ and $B'$ are $A$-modules, a $\mathbf{\Delta}$-$\mathscr{F}$-homomorphism $\varphi$: $B \to B'$ is called an $A$-homomorphism if $\varphi({}^a b) = {}^a\varphi(b)$, $a \in A$, $b \in B$. The multiplicative group $\mathbf{G}_m$ of $\mathscr{U}$ acts on $\mathbf{G}_a{}^n$ in a natural way, defining the usual vector space structure. We call this the *natural action.* A $\mathbf{\Delta}$-$\mathscr{F}$-group $G$ is called a *vector group* defined over $\mathscr{F}$ if there is an action of the $\mathbf{\Delta}$-$\mathscr{F}$-group $\mathbf{G}_m$ on $G$ and a $\mathbf{\Delta}$-$\mathscr{F}$-isomorphism $\varphi$ from the $\mathbf{G}_m$-module G onto the $\mathbf{G}_m$-module $\mathbf{G}_a{}^r$ relative to the natural action (see Rosenlicht [14]). We often call $G$ simply a vector group. It follows immediately from the definition that a vector group defined over $\mathscr{F}$ is a commutative linear $\mathscr{F}$-unipotent $\mathbf{\Delta}$-$\mathscr{F}$-group. Moreover, since the characteristic of $\mathscr{U}$ is 0, every commutative unipotent *algebraic* group defined over $\mathscr{F}$ is a vector group defined over $\mathscr{F}$. However, vector groups are certainly not restricted to algebraic groups. For example, let $G \subset \mathbf{G}_a{}^2$ be defined by the differential polynomial $y_1{}' - y_2$ (where, of course, $\mathscr{U}$ is an ordinary differential field). The automorphism $T(u_1, u_2) = (u_1, u_1{}' - u_2)$ maps $G$ onto $\mathbf{G}_a \times 0$, whence $G$ is a vector group that is not an algebraic group.

We saw above that a commutative linear unipotent $\delta$-group can be written as the direct product of a vector group and a group that is isomorphic to a direct product of copies of $(\mathbf{G}_a)_{\mathscr{K}}$. *We shall assume again that $\mathscr{U}$ is ordinary.*

***Proposition 11*** *Let $G$ be a commutative linear unipotent $\delta$-$\mathscr{F}$-group. The following conditions are equivalent:*

1. *$G$ is $\mathscr{F}$-wound over $\mathbf{G}_a$.*
2. *There exists a finitely generated extension $\mathscr{G}$ of $\mathscr{F}$ such that $G$ is $\delta$-$\mathscr{G}$-isomorphic to $(\mathbf{G}_a)_{\mathscr{K}}{}^r$ for some $r$.*
3. *The differential dimension of $G$ is 0.*

*Proof* Suppose $G$ is $\mathscr{F}$-wound over $\mathbf{G}_a$. Theorem 4 states that $G = W \times V$, where $W$ is a $\delta$-$\mathscr{F}$-subgroup of $G$ and satisfies condition 2, and $V$ is a vector group defined over $\mathscr{F}$. If $V \neq 0$, then $V$ is the direct product of $\delta$-$\mathscr{F}$-subgroups isomorphic to $\mathbf{G}_a$, which contradicts condition 1. Thus, condition 1 implies 2. That condition 2 implies 3 is obvious. Now, suppose $\varphi$: $\mathbf{G}_a \to G$ is a $\delta$-$\mathscr{F}$-homomorphism. diff dim(ker $\varphi$) + diff dim($\varphi(\mathbf{G}_a)$ = diff dim($\mathbf{G}_a$). Therefore, if we assume condition 3, diff dim(ker $\varphi$) + 0 = 1. Thus, the differential dimension of the kernel of $\varphi$ is 1. But, every proper $\delta$-$\mathscr{F}$-subgroup of $\mathbf{G}_a$ has differential dimension 0. Thus, ker $\varphi = \mathbf{G}_a$, whence condition 3 implies 1.

***Corollary*** *Let $G$ be a commutative unipotent linear $\delta$-$\mathscr{F}$-group. Let $\mathscr{G}$ be an extension field of $\mathscr{F}$. If $G$ is $\mathscr{F}$-wound over $\mathbf{G}_a$, then $G$ is $\mathscr{G}$-wound over $\mathbf{G}_a$.*

*Proof* Suppose $G$ is $\mathscr{F}$-wound over $\mathbf{G}_a$. Then the differential dimension of $G$ is 0. Therefore, $G$ is $\mathscr{G}$-wound over $\mathbf{G}_a$.

We shall say simply that $G$ is *wound* over $\mathbf{G}_a$.

We can restate Theorem 4. *Every commutative linear unipotent $\delta$-$\mathscr{F}$-group $G$ is the direct product $W \times V$, where $W$ is a $\delta$-$\mathscr{F}$-group wound over $\mathbf{G}_a$ and $V$ is a vector group defined over $\mathscr{F}$.*

Suppose $G$ is a linear unipotent $\delta$-$\mathscr{F}$-group (not necessarily commutative). If $G$ is $\mathscr{F}$-wound over $\mathbf{G}_a$ (resp. over $(\mathbf{G}_a)_{\mathscr{K}}$), then so is $\ell(G)$. For, suppose $\varphi$: $\mathbf{G}_a \to \ell(G)$ (resp. $\varphi$: $(\mathbf{G}_a)_{\mathscr{K}} \to \ell(G)$) is a $\delta$-$\mathscr{F}$-homomorphism. Then $\psi = \exp \circ \varphi$ is a $\delta$-$\mathscr{F}$-homomorphism from $\mathbf{G}_a \to G$ (resp. $(\mathbf{G}_a)_{\mathscr{K}} \to G$). Therefore, for all $a \in \mathscr{U}$, $1 = \psi(a) = \exp(\varphi(a))$. Thus, $\varphi(a) = 0$, and $\ell(G)$ is $\mathscr{F}$-wound over $\mathbf{G}_a$ (resp. over $(\mathbf{G}_a)_{\mathscr{K}}$). A parallel proof (replacing exp by log) shows that if $\ell(G)$ is $\mathscr{F}$-wound over $\mathbf{G}_a$ or $(\mathbf{G}_a)_{\mathscr{K}}$, so is $G$. In particular, if $G$ is $\mathscr{F}$-wound over $\mathbf{G}_a$, then the differential dimension of $\ell(G)$, hence of $G$ [DAG, p. 926] is 0, and $\ell(G)$ is $\delta$-$\mathscr{G}$-isomorphic to $(\mathbf{G}_a)_{\mathscr{K}}{}^r$, where $\mathscr{G}$ is a finitely generated extension of $\mathscr{F}$, which can be taken to be Picard–Vessiot over $\mathscr{F}$ if the field of constants of $\mathscr{F}$ is algebraically closed. We can restate these facts in the form of a proposition.

***Proposition 12*** *Let $G$ be a linear unipotent $\delta$-$\mathscr{F}$-group. Then $G$ is $\mathscr{F}$-wound over $\mathbf{G}_a$ (resp. $\mathscr{F}$-wound over $(\mathbf{G}_a)_{\mathscr{K}}$) if and only if $\ell(G)$ is $\mathscr{F}$-wound over $\mathbf{G}_a$ (resp. over $(\mathbf{G}_a)_{\mathscr{K}}$). Moreover, the following conditions are equivalent:*

1. *$G$ is $\mathscr{F}$-wound over $\mathbf{G}_a$.*
2. *There is a finitely generated extension $\mathscr{G}$ of $\mathscr{F}$ such that the underlying differential algebraic set of $G$ is $\delta$-$\mathscr{F}$-isomorphic to $(\mathbf{G}_a)_{\mathscr{K}}{}^r$ for some $r$. If the field of constants of $\mathscr{F}$ is algebraically closed, $\mathscr{G}$ can be taken to be Picard–Vessiot over $\mathscr{F}$.*
3. *The differential dimension of $G$ is 0.*

***Corollary*** *Let $G$ be a linear unipotent $\delta$-$\mathscr{F}$-group of differential dimension $> 0$. Then $G$ contains an element $x \neq 1$ such that $x$ is rational over $\mathscr{F}$.*

If $G$ is a $\delta$-$\mathscr{F}$-subgroup of $\mathbf{G}_a{}^n$ we can characterize $G$ as a group wound over $\mathbf{G}_a$ or as a vector group by the nature of the module of homogeneous linear differential polynomial functions on $G$.

***Proposition 13*** *Let $G$ be a $\delta$-$\mathscr{F}$-subgroup of $\mathbf{G}_a{}^n$ of differential dimension $r$, and let $\bar{V}$ be the left module of homogeneous linear differential polynomial functions on $G$. Then $G$ is a vector group defined over $\mathscr{F}$ if and only if $\bar{V}$ is free of rank $n - r$. $G$ is wound over $\mathbf{G}_a$ if and only if $\bar{V}$ is a torsion module.*

*Proof* As above, let $V$ be the left module over $\mathscr{F}[\delta]$ of homogeneous linear differential polynomials in $y_1, \ldots, y_n$ with coefficients in $\mathscr{F}$, and let $W = \mathfrak{i} \cap V$, where $\mathfrak{i}$ is the defining differential ideal of $G$. Then $V/W$ can be identified with $\bar{V}$. Let $z_1, \ldots, z_n, f_1(\delta), \ldots, f_{n-r}(\delta)$ be as in Theorem 4, and let $\bar{z}_1, \ldots, \bar{z}_n$ be the images in $\bar{V}$ of the differential indeterminates. Then $\bar{z}_1, \ldots, \bar{z}_n$ generate $\bar{V}$ as a left $\mathscr{F}[\delta]$-module. Let $\bar{V}_T$ (resp. $\bar{V}_F$) be the submodule generated by $\bar{z}_1, \ldots, \bar{z}_{n-r}$ (resp. $\bar{z}_{n-r+1}, \ldots, \bar{z}_n$). Then $\bar{V}_T$is the torsion part of $\bar{V}$, $\bar{V}_F$ is free, and $\bar{V} = \bar{V}_T \oplus \bar{V}_F$ Furthermore, $G$ is $\delta$-$\mathscr{F}$-isomorphic to $G_1 \times \cdots \times G_{n-r} \times \mathbf{G}_a{}^r$, where $G_i$ is the zero-set of $f_i(\delta)y$ in $\mathbf{G}_a$. By Proposition 11, $G$ is wound over $\mathbf{G}_a$ if and only if $r = 0$, i.e., if and only if $\bar{V} = \bar{V}_T$. Suppose $\bar{V} = \bar{V}_F$. Then $\bar{z}_1 = \cdots = \bar{z}_{n-r} = 0$, $\bar{z}_{n-r+1}, \ldots, \bar{z}_n$ are linearly independent over $\mathscr{F}[\delta]$, and therefore $G$ is $\delta$-$\mathscr{F}$-isomorphic to $\mathbf{G}_a{}^r$. Suppose $G$ is a vector group defined over $\mathscr{F}$. Let $\sigma$ be a $\delta$-$\mathscr{F}$-isomorphism from $G$ onto $\mathbf{G}_a{}^r$. Let $V'$ be the left $\mathscr{F}[\delta]$-module of homogeneous linear differential polynomial functions on $\mathbf{G}_a{}^r$ defined over $\mathscr{F}$. We define an isomorphism $\sigma^*$ from $V'$ onto $\bar{V}$ by the formula $\sigma^*(L(x)) = L(\sigma(x))$, $x \in G$. But, $V'$ is clearly free of rank $r$, whence so is $\bar{V}$.

***Proposition 14*** *Let $G$ be a vector group defined over $\mathscr{F}$, and let $\varphi\colon G \to G'$ be a $\delta$-$\mathscr{F}$-homomorphism. Then $\varphi G$ is a vector group defined over $\mathscr{F}$.*

*Proof* $\varphi G$ is a commutative linear unipotent $\delta$-$\mathscr{F}$-group. Therefore, Theorem 4 implies that $\varphi G = W \times V$, where $W$ and $V$ are defined over $\mathscr{F}$, $W$ is wound over $\mathbf{G}_a$, and $V$ is a vector group defined over $\mathscr{F}$. Let $\mathrm{pr}_1$ be the projection of $\varphi G$ on $W$. Let $x \in G$. Since $G$ is a vector group, there is a

$\delta$-subgroup $H$ of $G$ containing $x$ and isomorphic to $\mathbf{G}_a$. Since $W$ is wound over $\mathbf{G}_a$, $(\mathrm{pr}_1 \circ \varphi)(H) = \{1\}$. Therefore, $\mathrm{pr}_1(\varphi x) = 1$. Since $x$ was arbitrarily chosen, $\mathrm{pr}_1(\varphi G) = W = \{1\}$, whence $\varphi G = V$ is a vector group defined over $\mathscr{F}$.

***Proposition 15*** *Let $G$ be a commutative linear unipotent $\delta$-$\mathscr{F}$-group. Let $V$ be a $\delta$-subgroup of $G$ such that $V$ is a vector group defined over $\mathscr{F}$ and $G = W \times V$, with $W$ a $\delta$-$\mathscr{F}$-subgroup wound over $\mathbf{G}_a$. Then $V$ contains every $\delta$-subgroup $V'$ of $G$ such that $V'$ is a vector group.*

*Proof* Let $\mathrm{pr}_1$ be the projection of $G$ on $W$. Then by Proposition 14, $\mathrm{pr}_1(V')$ is a vector group. Since $W$ is wound, $\mathrm{pr}_1(V') = \{1\}$. Therefore, $V' \subset \ker(\mathrm{pr}_1) = V$.

***Corollary 1*** *Let $G$ be a commutative linear $\mathscr{F}$-unipotent $\delta$-$\mathscr{F}$-group. There is a unique maximal vector subgroup $V$ of $G$, and $V$ is defined over $\mathscr{F}$.*

***Corollary 2*** *Let $G$ be a commutative linear unipotent $\delta$-$\mathscr{F}$-group, and let $V'$ be a $\delta$-$\mathscr{F}$-subgroup of $G$ such that $V'$ is a vector group. Then $V'$ is an $\mathscr{F}$-direct factor of $G$.*

*Proof* By Corollary 1, $V' \subset V$, where $V$ is a vector group defined over $\mathscr{F}$ and $G = W \times V$, $W$ wound over $\mathbf{G}_a$ and defined over $\mathscr{F}$. So it suffices to show that $V'$ is an $\mathscr{F}$-direct factor of $V$. Without loss of generality we may assume that $V = \mathbf{G}_a{}^n$. Then there is an automorphism $T$ of $V$, with $T$ defined over $\mathscr{F}$, such that $T(V')$ is the set of $(u_1, \ldots, u_n) \in \mathbf{G}_a{}^n$ such that $f_1(\delta)u_1 = \cdots = f_s(\delta)u_s = 0$, with $f_1(\delta), \ldots, f_s(\delta)$ in $\mathscr{F}[\delta]$. Thus $T(V') = V'' \times W''$, where $V'' = 0 \times \cdots \times 0 \times \mathbf{G}_a^{n-s}$, and $W'' = G_1 \times \cdots \times G_s \times 0 \times \cdots \times 0$, where $G_i$ is the zero-set in $\mathbf{G}_a$ of $f_i(\delta)y$. By Proposition 14, $T(V')$ is a vector group. Therefore, $W = \{0\}$, and $T(V') = V''$, which clearly has a supplement in $V$ defined over $\mathscr{F}$, whence so does $V'$.

***Proposition 16*** *Let $1 \to V \to G \to V' \to 1$ be an exact sequence of commutative linear $\delta$-$\mathscr{F}$-groups, with $V$ and $V'$ vector groups defined over $\mathscr{F}$. Then $G$ is a vector group defined over $\mathscr{F}$.*

*Proof* By Theorem 1, $G$ is $\mathscr{F}$-unipotent. Therefore, $G = W \times \tilde{V}$, with $W$ a $\delta$-$\mathscr{F}$-subgroup of $G$ wound over $\mathbf{G}_a$ and $\tilde{V}$ a vector group defined over $\mathscr{F}$. By Proposition 25, $V \subset \tilde{V}$. Now, $W$ is $\delta$-$\mathscr{F}$-isomorphic to $G/\tilde{V}$, which is $\delta$-$\mathscr{F}$-isomorphic to $(G/V)/(\tilde{V}/V)$. $G/V$ is $\delta$-$\mathscr{F}$-isomorphic to $V'$. Therefore, $W$ is a $\delta$-$\mathscr{F}$-homomorphic image of a vector group defined over $\mathscr{F}$, whence, by Proposition 14, $W = \{1\}$, and $G$ is a vector group defined over $\mathscr{F}$.

***Corollary 1*** *Let $0 \to \mathbf{G}_a \to G \to \mathbf{G}_a \to 0$ be an exact sequence of linear commutative $\delta$-$\mathscr{F}$-groups. Then $G$ is $\delta$-$\mathscr{F}$-isomorphic to $\mathbf{G}_a \times \mathbf{G}_a$.*

***Corollary 2*** *Let $1 \to H \to G \to G' \to 1$ be an exact sequence of linear $\delta$-$\mathscr{F}$-groups such that $G'$ and $H$ are vector groups defined over $\mathscr{F}$. Then the underlying differential algebraic set of $G$ is $\delta$-$\mathscr{F}$-isomorphic to affine space.*

*Proof* The given exact sequence gives rise to an exact sequence of $\delta$-$\mathscr{F}$-Lie algebras

$$0 \to \ell(H) \to \ell(G) \to \ell(G') \to 0.$$

Since $H$ and $G'$ are vector groups and $\ell(H)$ and $\ell(G')$ are isomorphic to $H$ and $G'$ respectively as $\delta$-$\mathscr{F}$-groups, it follows from Corollary 1 that $\ell(G)$ is a vector group. Therefore, the underlying differential algebraic set of $G$ is $\delta$-$\mathscr{F}$-isomorphic to affine space (see Proposition 10).

***Corollary 3*** *Let $0 \to \mathbf{G}_a \to G \to \mathbf{G}_a \to 0$ be an exact sequence of linear $\delta$-$\mathscr{F}$-groups. Then the underlying differential algebraic set of $G$ is $\delta$-$\mathscr{F}$-isomorphic to $\mathbf{G}_a \times \mathbf{G}_a$.*

## 3. Differential rational cohomology

In this section we do not require $\mathscr{U}$ to be an ordinary differential field. The cohomology groups constructed here are adaptations to the category of $\mathbf{\Delta}$-groups of the Hochschild cohomology groups (Hochschild and Serre [5]; see also Demazure and Gabriel [3, II, §3]).

Let $A$ and $B$ be $\mathbf{\Delta}$-groups, with $B$ commutative and written additively. Suppose $A$ acts on $B$. We define $C_{\mathbf{\Delta}}{}^0(A, B) = C^0_{\text{mor}}(A, B) = B$. If $n$ is a positive integer, we define $C_{\mathbf{\Delta}}{}^n(A, B)$ to be the set of $\mathbf{\Delta}$-maps $A^n \to B$. $C^n_{\text{mor}}(A, B)$ is the subset of $C_{\mathbf{\Delta}}{}^n(A, B)$ consisting of everywhere defined $\mathbf{\Delta}$-maps. $C_{\mathbf{\Delta}}{}^n(A, B)$ is an additive group in the usual way, and $C^n_{\text{mor}}(A, B)$ is a subgroup. We call $C_{\mathbf{\Delta}}{}^n(A, B)$ the *group of $\mathbf{\Delta}$-n cochains of A into B*, and we call $C^n_{\text{mor}}(A, B)$ the *group of everywhere defined $\mathbf{\Delta}$-n-cochains of A into B*. We define the coboundary homomorphism $\partial^n$: $C_{\mathbf{\Delta}}{}^n(A, B) \to C_{\mathbf{\Delta}}^{n+1}(A, B)$ as follows: First, $\partial^0(b)(a) = {}^a b - b$, for $a \in A, b \in B$. Suppose, now, that $n > 0$. Let $\partial^n = \sum_{0 \le i \le n-1} (-1)^i \partial_i{}^n$, where

$$\partial_0{}^n f(a_1, \ldots, a_{n+1}) = {}^{a_1} f(a_2, \ldots, a_n),$$

$$\partial_i{}^n f(a_1, \ldots, a_{n+1}) = f(a_1, \ldots, a_i a_{i+1}, \ldots, a_{n+1}), \qquad \text{if} \quad 1 \le i \le n,$$

$$\partial^n_{n+1} f(a_1, \ldots, a_{n+1}) = f(a_1, \ldots, a_n).$$

In these formulas, $a_1, \ldots, a_{n+1}$ are independent generic points for components of $A$ over a differential field of definition for the components and for $f$. One verifies that $\partial^n \circ \partial^{n-1} = 0$, $n > 0$, and that $\partial^n(C^n_{\text{mor}}(A, B)) \subset C^{n+1}_{\text{mor}}(A, B)$. Define the group $Z_{\mathbf{\Delta}}{}^n(A, B)$ of *$\mathbf{\Delta}$-n-cocycles* to be ker $\partial^n$ and the group

$B_{\Delta}{}^{n}(A, B)$ to be im $\partial^{n-1}$ if $n > 0$ and $\{0\}$ if $n = 0$. The subgroup $Z^{n}_{\mathrm{mor}}(A, B)$ is defined to be ker $\partial^{n} | C^{n}_{\mathrm{mor}}(A, B)$, and $B^{n}_{\mathrm{mor}}(A, B)$ is im $\partial^{n-1} | C^{n}_{\mathrm{mor}}(A, B)$ if $n > 0$ and $= 0$ if $n = 0$. $B_{\Delta}{}^{n}(A, B)$ is called the *group of* $\Delta$-*n*-*coboundaries* and $B^{n}_{\mathrm{mor}}(A, B)$ is called the *group of everywhere defined* $\Delta$-*n*-*coboundaries*. We define the *n*th $\Delta$-*cohomology group* $H_{\Delta}{}^{n}(A, B) = Z_{\Delta}{}^{n}(A, B)/B_{\Delta}{}^{n}(A, B)$ and $H^{n}_{\mathrm{mor}}(A, B) = Z^{n}_{\mathrm{mor}}(A, B)/B^{n}_{\mathrm{mor}}(A, B)$. If $f$ and $g$ are in the same coset we call them *cohomologous*.

Clearly, $H_{\Delta}{}^{0}(A, B) = H^{0}_{\mathrm{mor}}(A, B) = {}^{A}B$, the subgroup of $B$ consisting of the fixed points under the action of $A$ on $B$. Evidently, $f$ is in $Z_{\Delta}{}^{1}(A, B)$ if ${}^{a_1}f(a_2) - f(a_1 a_2) + f(a_1) = 0$ whenever $a_1, a_2$ and $a_1 a_2$ are in the domain of $f$. So, $Z_{\Delta}{}^{1}(A, B)$ is the group of $\Delta$-*crossed homomorphisms* from $A$ into $B$. $f$ is in $B_{\Delta}{}^{1}(A, B)$ if there is an element $b \in B$ such that for all $a \in A$ $f(a) = {}^{a}b - b$.

Let $A$ be a connected $\Delta$-subgroup of $GL(n)$. $A$ acts on the additive group of $\ell(A)$ by conjugation. It is well known that $H^{1}_{\mathrm{rat}}(SL(n), \mathfrak{sl}(n)) = \{0\}$, where $H^{1}_{\mathrm{rat}}(SL(n), \mathfrak{sl}(n))$ is the first rational cohomology group (with respect to the action by conjugation). However, if we consider $SL(n)$ and $\mathfrak{sl}(n)$ as $\Delta$-groups, the first cohomology group is not trivial.

***Proposition 17*** *Let* $A \neq \{1\}$ *be a connected* $\mathscr{F}$-*subgroup of* $GL(n)$. *Then* $H_{\Delta}{}^{1}(A, \ell(A)) \neq \{0\}$, *where the action of* $A$ on $\ell(A)$ *is by conjugation.*

*Proof* Let $\delta_i \in \Delta$. Define the $\Delta$-crossed homomorphism $\ell\,\delta_i\colon A \to \ell(A)$ by the formula $\ell\,\delta_i(a) = \delta_i(a) \cdot a^{-1}$, where, if $a = (a_{kl})$, then we have $\delta_i(a) = (\delta_i(a_{kl}))$. $\ell\,\delta_i$ is surjective (Kovacic [10, p. 272]). Suppose there is an element $x \in \ell(A)$ such that for all $a \in A$, $\ell\,\delta_i(a) = axa^{-1} - x$. Then $-x = \ell\,\delta_i(a) - axa^{-1} = \ell\,\delta_i(a) + a(-x)a^{-1}$. Let $b \in A$ be such that $\ell\,\delta_i(b) = -x$. Then $\ell\,\delta_i(b) = \ell\,\delta_i(ab)$, whence $b^{-1}ab \in A_{\mathscr{K}_i}$, where $\mathscr{K}_i$ is the constant field of $\delta_i$. Thus, $b^{-1}Ab \subset A_{\mathscr{K}_i}$, which is absurd. So, the logarithmic derivative map does not split.

In this paper we are interested chiefly in $H_{\Delta}{}^{2}(A, B)$. $f$ is in $Z_{\Delta}{}^{2}(A, B)$ if generically ${}^{a_1}f(a_2, a_3) - f(a_1 a_2, a_3) + f(a_1, a_2 a_3) - f(a_1, a_2) = 0$.

***Proposition 18*** *The inclusion homomorphism* $Z^{2}_{\mathrm{mor}}(A, B) \to Z_{\Delta}{}^{2}(A, B)$ *induces an injective homomorphism* $H^{2}_{\mathrm{mor}}(A, B) \to H_{\Delta}{}^{2}(A, B)$.

*Proof* Let $f \in B^{2}_{\mathrm{mor}}(A, B)$. Then $f$ is everywhere defined on $A \times A$. Since $f$ is a coboundary, there is a $\Delta$-map $h$ from $A$ into $B$ such that generically ${}^{a_1}h(a_2) - h(a_1 a_2) + h(a_1) = f(a_1, a_2)$. Let $U$ be the domain of definition of $h$. Then, whenever $a_1$, $a_2$ are in $U$ the above relation holds. Therefore, $h$ is defined on $U \cdot U$. But, $U \cdot U = A$ since $U$ is a dense open subset of $A$ and $A$ is a $\Delta$-group. Therefore, $h \in C^{1}_{\mathrm{mor}}(A, B)$. The proposition follows immediately.

Let $A$ and $B$ be commutative $\mathbf{\Delta}$-groups and suppose the action of $A$ on $B$ is the trivial action. A $\mathbf{\Delta}$-2-cocycle $f$ in $Z_{\mathbf{\Delta}}{}^2(A, B)$ is *symmetric* if $f(a_1, a_2) = f(a_2, a_1)$ whenever $a_1$, $a_2$ are independent generic points for components over some field of definition for the components of $A$ and for $f$. We denote the subgroup of symmetric $\mathbf{\Delta}$-2-cocycles by $Z_{\mathbf{\Delta}}{}^2(A, B)_s$. Clearly, $B_{\mathbf{\Delta}}{}^2(A, B) \subset Z_{\mathbf{\Delta}}{}^2(A, B)_s$. We define the *second symmetric* $\mathbf{\Delta}$*-cohomology group* $H_{\mathbf{\Delta}}{}^2(A, B)_s = Z_{\mathbf{\Delta}}{}^2(A, B)_s / B_{\mathbf{\Delta}}{}^2(A, B)$. Evidently, the inclusion homomorphism $Z_{\mathbf{\Delta}}{}^2(A, B)_s \to Z_{\mathbf{\Delta}}{}^2(A, B)$ induces an injection $H_{\mathbf{\Delta}}{}^2(A, B)_s \to H_{\mathbf{\Delta}}{}^2(A, B)$. In a similar manner we define $H^2_{\mathrm{mor}}(A, B)_s$.

We now suppose that $A = B = \mathbf{G}_a$ and write everything additively. Let $\mathrm{Hom}_{\mathbf{\Delta}}(\mathbf{G}_a, \mathbf{G}_a)$ be the group of $\mathbf{\Delta}$-endomorphisms of $\mathbf{G}_a$.

***Proposition 19*** *Let $\lambda$ be in $\mathrm{Hom}_{\mathbf{\Delta}}(\mathbf{G}_a, \mathbf{G}_a)$. Then there is a unique $f \in \mathscr{U}[\mathbf{\Delta}]$ such that for all $x \in \mathbf{G}_a$, $\lambda(x) = f(\delta)x$, where $\delta = (\delta_1, \ldots, \delta_m)$.*

*Proof* Let $P(y) \in \mathscr{U}\{y\}$ be such that for all $x \in \mathbf{G}_a\, \lambda(x) = P(x)$. Since $P(0) = 0$, $P$ has zero constant term. Let $n$ be a positive integer and let $x$ be in $\mathbf{G}_a$. $P(nx) = nP(x)$. Therefore, for every $c \in \mathscr{K}$, $P(cx) = cP(x)$, whence $P$ is homogeneous of degree 1. It follows that there is a unique $f(\delta) \in \mathscr{U}[\mathbf{\Delta}]$ such that $P(y) = f(\delta)y$.

Thus, we may identify $\mathrm{Hom}_{\mathbf{\Delta}}(\mathbf{G}_a, \mathbf{G}_a)$ with the noncommutative polynomial ring $\mathscr{U}[\mathbf{\Delta}]$.

***Corollary 1*** *$\lambda$ is a $\mathbf{\Delta}$-automorphism of $\mathbf{G}_a$ if and only if there is an element $a \in \mathbf{G}_m$ such that for all $x$ in $\mathbf{G}_a\, \lambda(x) = ax$.*

***Corollary 2*** *The only action of $\mathbf{G}_a$ on $\mathbf{G}_a$ is the trivial action.*

*Proof* The map $(a, x) \to {}^a x$ is an everywhere defined $\mathbf{\Delta}$-map from $\mathbf{G}_a \times \mathbf{G}_a \to \mathbf{G}_a$. For fixed $a \in \mathbf{G}_a$ the map $x \to {}^a x$ is a $\mathbf{\Delta}$-automorphism of $\mathbf{G}_a$. Therefore, by Corollary 1, there is an element $\tilde{a} \in \mathbf{G}_m$ such that for all $x \in \mathbf{G}_a$, ${}^a x = \tilde{a}x$. In particular, ${}^a 1 = \tilde{a}$ and ${}^{ab}1 = \tilde{a}\tilde{b}$. Therefore, the map $a \to {}^a 1$ is a $\mathbf{\Delta}$-homomorphism $\mathbf{G}_a \to \mathbf{G}_m$. Since the only such homomorphism is the trivial homomorphism [DAG, p. 942], it follows that for all $a \in \mathbf{G}_a$, ${}^a 1 = \tilde{a} = 1$. Thus, the action is trivial.

It follows from Corollary 2 that we may speak of $H_{\mathbf{\Delta}}{}^n(\mathbf{G}_a, \mathbf{G}_a)$ without ambiguity. $H_{\mathbf{\Delta}}{}^0(\mathbf{G}_a, \mathbf{G}_a) = \mathbf{G}_a$ and $H_{\mathbf{\Delta}}{}^1(\mathbf{G}_a, \mathbf{G}_a) = \mathrm{Hom}_{\mathbf{\Delta}}(\mathbf{G}_a, \mathbf{G}_a) = \mathscr{U}[\mathbf{\Delta}]$. This parallels the case of algebraic groups with respect to a coefficient field $k$ of prime characteristic, where $H^1_{\mathrm{rat}}(\mathbf{G}_a, \mathbf{G}_a) = k[F]$.

The functional equation for $\mathbf{\Delta}$-2-cocycles in $Z_{\mathbf{\Delta}}{}^2(\mathbf{G}_a, \mathbf{G}_a)$ is given by the formula (where $x$, $y$, $z$ are differential indeterminates)

$$f(y, z) - f(x + y, z) + f(x, y + z) - f(x, y) = 0.$$

$f$ is a coboundary if there is a function $h \in \mathcal{U}\langle x \rangle$ such that

$$f(x, y) = h(x + y) - h(x) - h(y).$$

Let $n$ be a natural number. $C_{\mathbf{\Delta}}{}^n(\mathbf{G}_a, \mathbf{G}_a) = \mathcal{U}\langle \mathbf{G}_a{}^n \rangle$, the field of differential rational functions on $\mathbf{G}_a{}^n$, hence is a left $\mathcal{U}[\mathbf{\Delta}]$-module. Let $y_1, \ldots, y_{n+1}$ be differential indeterminates. Then for $\delta \in \mathbf{\Delta}$

$$\delta(\partial^n f)(y_1, \ldots, y_{n+1}) = \sum_{0 \le i \le n} (-1)^i\, \delta(\partial_i{}^n f)(y_1, \ldots, y_{n+1}),$$

$$\begin{aligned}\delta(\partial_0{}^n f)(y_1, \ldots, y_{n+1}) &= \delta(f(y_2, \ldots, y_{n+1})) \\ &= \delta f(y_2, \ldots, y_{n+1}) \\ &= \partial_0{}^n(\delta f)(y_1, \ldots, y_{n+1}).\end{aligned}$$

For $1 \le i \le n$,

$$\begin{aligned}\delta(\partial_i{}^n f)(y_1, \ldots, y_{n+1}) &= \delta(f(y_1, \ldots, y_i y_{i+1}, \ldots, y_{n+1})) \\ &= \delta f(y_1, \ldots, y_i y_{i+1}, \ldots, y_{n+1}) \\ &= \partial_i{}^n(\delta f)(y_1, \ldots, y_i y_{i+1}, \ldots, y_{n+1}), \\ \delta(\partial_{n+1}^n f)(y_1, \ldots, y_{n+1}) &= \delta(f(y_1, \ldots, y_n)) \\ &= \delta f(y_1, \ldots, y_n) \\ &= \partial_{n+1}^n(\delta f)(y_1, \ldots, y_n).\end{aligned}$$

Thus, $\delta \circ \partial^n = \partial^n$. Clearly, for all $a \in \mathcal{U}$, $\partial^n(af) = a\, \partial^n f$. Therefore, $\partial^n$ is a homomorphism of left $\mathcal{U}[\mathbf{\Delta}]$-modules. It follows that $\ker(\partial^n) = Z_{\mathbf{\Delta}}{}^n(\mathbf{G}_a, \mathbf{G}_a)$ and $\operatorname{im} \partial^{n-1} = B_{\mathbf{\Delta}}{}^n(\mathbf{G}_a, \mathbf{G}_a)$ are submodules of $C_{\mathbf{\Delta}}{}^n(\mathbf{G}_a, \mathbf{G}_a)$. Therefore, $H_{\mathbf{\Delta}}{}^n(\mathbf{G}_a, \mathbf{G}_a)$ is a left module over $\mathcal{U}[\mathbf{\Delta}] = H_{\mathbf{\Delta}}{}^1(\mathbf{G}_a, \mathbf{G}_a)$. Clearly, $H^n_{\mathrm{mor}}(\mathbf{G}_a, \mathbf{G}_a)$ is a submodule.

## 4. Extensions of differential algebraic groups

Let $A$, $B$, $C$ be $\mathbf{\Delta}$-groups, with $B$ commutative and written additively. A short exact sequence

$$E\colon\quad 0 \to B \xrightarrow{\iota} C \xrightarrow{\pi} A \to 1,$$

with $B$ central in $C$, and $\iota$, $\pi$ $\mathbf{\Delta}$-homomorphisms, is called a *central extension* of $A$ by $B$.

If

$$E'\colon\quad 0 \to B' \xrightarrow{\iota'} C' \xrightarrow{\pi'} A' \to 1$$

is a central extension of $A'$ by $B'$, a morphism $\Gamma: E \to E'$ is a triple $(\beta, \gamma, \alpha)$ of $\mathbf{\Delta}$-homomorphisms $\beta: B \to B'$, $\gamma: C \to C'$, $\alpha: A \to A'$ such that the following diagram is commutative:

$$\begin{array}{ccccccccc} E: & 0 & \longrightarrow & B & \xrightarrow{\iota} & C & \xrightarrow{\pi} & A & \longrightarrow & 1 \\ \Gamma\downarrow & & & \beta\downarrow & & \gamma\downarrow & & \alpha\downarrow & & \\ E' & 0 & \longrightarrow & B' & \xrightarrow{\iota'} & C' & \xrightarrow{\pi'} & A' & \longrightarrow & 1 \end{array}$$

If $E''$ is a central extension of $A''$ by $B''$ and $\Gamma' = (\beta', \gamma', \alpha')$ is a morphism from $E'$ to $E''$, then $\Gamma'\Gamma = (\beta'\beta, \gamma'\gamma, \alpha'\alpha)$ is a morphism from $E$ to $E''$.

Two central extensions $E$ and $E'$ of $A$ by $B$ are *equivalent* if there is a morphism $\Gamma = (1_B, \gamma, 1_A)$ from $E$ to $E'$. Since $\gamma$ is a bijective $\mathbf{\Delta}$-homomorphism, $\gamma$ is a $\mathbf{\Delta}$-isomorphism. Equivalence of central extensions of $A$ by $B$ is an equivalence relation. If $E$ is equivalent to $E'$, we write $E \equiv E'$. We denote by $\mathrm{Cent\ Ext}_{\mathbf{\Delta}}(A, B)$ the set of equivalence classes of central extensions of $A$ by $B$. One central extension of $A$ by $B$ is the direct product $B \times A$. A central extension $E$ of $A$ is said to be *split* if it is equivalent to the direct product extension.

*Let $E$ be a central extension of $A$ by $B$, and let $\alpha: A' \to A$ be a $\mathbf{\Delta}$-homomorphism. There exists a central extension, called $E\alpha$, of $A'$ by $B$ and a morphism $\Gamma = (1_B, \gamma, \alpha)$ from $E\alpha$ to $E$, that is, the following diagram is commutative:*

$$\begin{array}{ccccccccc} E: & 0 & \longrightarrow & B & \xrightarrow{\iota} & C & \xrightarrow{\pi} & A & \longrightarrow & 1 \\ \Gamma\uparrow & & & 1_B\uparrow & & \gamma\uparrow & & \alpha\uparrow & & \\ E\alpha: & 0 & \longrightarrow & B & \xrightarrow{\iota'} & C' & \xrightarrow{\pi'} & A' & \longrightarrow & 1 \end{array}$$

*The pair $(\Gamma, E\alpha)$ is unique up to equivalence. Moreover, $(\Gamma, E\alpha)$ is couniversal. Whenever $E_1$ is a central extension of $B_1$ by $A'$ such that there is a morphism $\Gamma_1: E_1 \to E$, with $\Gamma_1 = (\beta_1, \gamma_1, \alpha)$, then $\Gamma_1$ factors through $E\alpha$, that is, there is a morphism $\Gamma_2 = (\beta_1, \gamma_2, 1)$ such that $\Gamma_1 = \Gamma\Gamma_2$. Symbolically,*

$$E_1 \xrightarrow{\Gamma_2} E\alpha \xrightarrow{\Gamma} E.$$

Following Mac Lane [12, p. 65], we define $C'$ to be the $\mathbf{\Delta}$-subgroup of $C \times A'$ consisting of all pairs $(c, a')$ such that $\pi(c) = \alpha(a')$. Define $\iota'(b) = (\iota(b), 1)$ and $\pi'(c, a') = a'$. Then $\iota'$ and $\pi'$ are $\mathbf{\Delta}$-homomorphisms and the sequence

$$E: \quad 0 \to B \xrightarrow{\iota'} C' \xrightarrow{\pi'} A' \to 1$$

is exact. We now define $\gamma(c, a') = c$. Then $\Gamma = (1_B, \gamma, \alpha)$ is a morphism from $E$ to $E\alpha$. Suppose $E_1$ is a central extension of $A'$ by $B_1$ and $\Gamma_1 = (\beta_1, \gamma_1, \alpha)$ is a morphism from $E_1 \to E$. Define $\gamma_2(c_1) = (\gamma_1(c_1), \pi_1(c_1))$. $\Gamma_2 = (\beta_1, \gamma_2, 1)$ is the desired morphism.

It follows immediately from the couniversality of $(\Gamma, E\alpha)$ that the pair is unique up to equivalence, and that $E1_A \equiv E$ and $E(\alpha\alpha') = (E\alpha)\alpha'$. Thus, Cent $\text{Ext}_\Delta(?, B)$ is a contravariant functor from the category of differential algebraic groups to the category of sets.

*Let $E$ be a central extension of $A$ by $B$ and let $B'$ be a commutative $\Delta$-group. Let $\beta\colon B \to B'$ be a $\Delta$-homomorphism. There exists a central extension, called $\beta E$, of $A$ by $B'$, and a morphism $\Gamma = (\beta, \gamma, 1_A)$, with $\Gamma\colon E \to \beta E$, that is, the following diagram is commutative:*

$$\begin{array}{ccccccccc}
E\colon & 0 & \longrightarrow & B & \xrightarrow{\iota} & C & \xrightarrow{\pi} & A & \longrightarrow & 1 \\
\Gamma\downarrow & & & \downarrow\beta & & \downarrow\gamma & & \downarrow 1_A & & \\
E\colon & 0 & \longrightarrow & B' & \xrightarrow{\iota'} & C' & \xrightarrow{\pi'} & A & \longrightarrow & 1
\end{array}$$

*The pair $(\Gamma, \beta E)$ is unique up to equivalence. Moreover, $(\Gamma, \beta E)$ is universal, that is, whenever $E_1$ is an extension of $B'$ by $A_1$ and $\Gamma_1 = (\beta, \gamma_1, \alpha_1)$ is a morphism from $E$ to $E_1$, then $\Gamma_1$ factors through $\beta E$. There is a morphism $\Gamma_2 = (1_B, \gamma_2, \alpha_1)$ from $\beta E$ to $E_1$ such that $\Gamma_2\Gamma = \Gamma_1$.* Symbolically,

$$E \xrightarrow{\Gamma} \beta E \xrightarrow{\Gamma_2} E_1.$$

Following Mac Lane, as above, we define $C'$ to be the quotient of $B' \times C$ by the normal $\Delta$-subgroup $N$ consisting of all pairs $(-\beta b, \iota b)$, with $b \in B$. Let $\text{cl}(b', c)$ be the coset belonging to $(b', c)$. We define $\iota'(b') = \text{cl}(b', 1)$ and $\pi'(\text{cl}(b', c)) = \pi(c)$. $\iota'$ and $\pi'$ are $\Delta$-homomorphisms and the sequence

$$\beta E\colon \quad 0 \to B' \xrightarrow{\iota'} C' \xrightarrow{\pi'} A \to 1$$

is exact. We now define $\gamma(c) = \text{cl}(0, c)$. $\gamma$ is a $\Delta$-homomorphism and $\Gamma = (\beta, \gamma, 1_A)$ is a morphism from $E$ to $\beta E$. Now, suppose $E_1$ is a central extension of $A_1$ by $B'$ such that there is a morphism $\Gamma_1 = (\beta, \gamma_1, \alpha_1)$ from $E \to E_1$. Define $\gamma_2(\text{cl}(b', c)) = \iota_1(b')\gamma_1(c)$. Then $\Gamma_2 = (1, \gamma_2, \alpha_1)$ is the desired morphism.

It follows immediately from the universality of $(\Gamma, \beta E)$ that the pair is unique up to equivalence, and that $1_B E \equiv E$ and $(\beta'\beta)E = \beta'(\beta E)$. Thus, Cent $\text{Ext}_\Delta(A, ?)$ is a covariant functor from the category of differential algebraic groups to the category of sets.

Now, let $E$ be a central extension of $A$ by $B$, let $A_1$ be a $\Delta$-group and $B'$ be a commutative $\Delta$-group. Let $\beta\colon B \to B'$ and $\alpha\colon A' \to A$ be $\Delta$-homomorphisms.

Then $\beta(E\alpha) \equiv (\beta E)\alpha$. Thus, Cent $\text{Ext}_{\Delta}(?, ?)$ is a bifunctor from the category of differential algebraic groups to the category of sets, contravariant in the first variable and covariant in the second variable.

We now define the "Baer sum" of two elements of Cent $\text{Ext}_{\Delta}(A, B)$. For details see Mac Lane [12, Chap. III, §2 and Chap. IV, §3]. For ease, we shall identify the extension $E$ with its class.

Let $G$ be a $\Delta$-group. The diagonal homomorphism $\Delta_G\colon G \to G \times G$ is given by the formula $\Delta_G(g) = (g, g)$. Suppose $G$ is commutative and is written additively. Then the codiagonal map $\nabla_G\colon G \times G \to G$, defined by the formula $\nabla_G(g_1, g_2) = g_1 + g_2$ is a $\Delta$-homomorphism. Now, let

$$E_1\colon\ 0 \longrightarrow B \xrightarrow{\iota_1} C_1 \xrightarrow{\pi_1} A \longrightarrow 1$$

and

$$E_2\colon\ 0 \longrightarrow B \xrightarrow{\iota_2} C_2 \xrightarrow{\pi_2} A \longrightarrow 1$$

be central extensions of $A$ by $B$. Then

$$E_1 \times E_2\colon\ 0 \longrightarrow B \times B \xrightarrow{\iota_1 \times \iota_2} C_1 \times C_2 \xrightarrow{\pi_1 \times \pi_2} A \times A \longrightarrow 1$$

is a central extension of $A \times A$ by $B \times B$. We define the sum $E_1 + E_2 = \nabla_B E_1 \times E_2\, \Delta_A$. With respect to this binary operation, Cent $\text{Ext}_{\Delta}(A, B)$ is a commutative group. The class of the split extension $0 \to B \to B \times A \to A \to 1$ is the zero element of the group. The class of $(-1_B)E$ is the additive inverse of the class of $E$. Suppose $B'$ is a commutative $\Delta$-group (resp. $A'$ is a $\Delta$-group) and $\beta\colon B \to B'$ (resp. $\alpha\colon A' \to A$) is a $\Delta$-homomorphism. Then

$$\beta(E_1 + E_2) = \beta E_1 + \beta E_2, \qquad (E_1 + E_2)\alpha = E_1\alpha + E_2\alpha.$$

Therefore, the maps $\beta_*$: Cent $\text{Ext}_{\Delta}(A, B) \to$ Cent $\text{Ext}_{\Delta}(A, B')$ and $\alpha^*$: Cent $\text{Ext}_{\Delta}(A, B) \to$ Cent $\text{Ext}_{\Delta}(A', B)$, defined by the formulas $\beta_*(E) = \beta E$ and $\alpha^*(E) = E\alpha$, are group homomorphisms. Furthermore, if $\beta_1$ and $\beta_2$ are in $\text{Hom}_{\Delta}(B, B')$, then $(\beta_1 + \beta_2)E = \beta_1 E + \beta_2 E$. Thus, Cent $\text{Ext}_{\Delta}(A, ?)$ is a covariant *additive* functor. This gives us an isomorphism

$$\text{Cent Ext}_{\Delta}(A, B_1 \times B_2) \approx \text{Cent Ext}_{\Delta}(A, B_1) \times \text{Cent Ext}_{\Delta}(A, B_2).$$

Moreover, Cent $\text{Ext}_{\Delta}(A, B)$ is a left module over $\text{Hom}_{\Delta}(B, B)$. In particular, Cent $\text{Ext}_{\Delta}(\mathbf{G}_a, \mathbf{G}_a)$ is a left module over $\mathscr{U}[\Delta]$.

Suppose, in addition, $A$ is commutative, and let Ab $\text{Ext}_{\Delta}(A, B)$ be the subset of Cent $\text{Ext}_{\Delta}(A, B)$ consisting of the equivalence classes of commutative extensions of $A$ by $B$. Ab $\text{Ext}_{\Delta}(A, B)$ is evidently a subgroup of Cent $\text{Ext}_{\Delta}(A, B)$. Ab $\text{Ext}_{\Delta}(?, ?)$ is a bifunctor contravariant in the first variable and covariant in the second variable. Furthermore, Ab $\text{Ext}_{\Delta}(A, ?)$ is, of course, additive. As in the category of algebraic groups (see [14, p. 165]) it follows from the commutativity of the extensions considered that Ab $\mathrm{Ext}^{\Delta}(?, ?)$ is an additive bifunctor from the category of commutative $\Delta$-groups to the category of commutative groups. In particular,

$$\text{Ab Ext}_{\Delta}(A_1 \times A_2, B) \approx \text{Ab Ext}_{\Delta}(A_1, B) \times \text{Ab Ext}_{\Delta}(A_2, B),$$

$$\text{Ab Ext}_{\Delta}(A, B_1 \times B_2) \approx \text{Ab Ext}_{\Delta}(A, B_1) \times \text{Ab Ext}_{\Delta}(A, B_2).$$

Moreover, Ab $\mathrm{Ext}_{\Delta}(A, B)$ is a submodule of the left module Cent $\mathrm{Ext}_{\Delta}(A, B)$ over $\mathrm{Hom}_{\Delta}(B, B)$ and is, in addition, a right module over $\mathrm{Hom}_{\Delta}(A, A)$. In fact, Ab $\mathrm{Ext}_{\Delta}(A, B)$ is a left-$\mathrm{Hom}_{\Delta}(B, B)$ and a right-$\mathrm{Hom}_{\Delta}(A, A)$ bimodule.

Let

$$E\colon \quad 0 \to B \overset{\iota}{\to} C \overset{\pi}{\to} A \to 1$$

be a central extension of $A$ by $B$. We say that $E$ admits a $\mathbf{\Delta}$-cross section if there is a $\mathbf{\Delta}$-map $\sigma\colon A \to C$ such that $\pi \circ \sigma = 1_A$ generically.

***Proposition 20*** *Every central extension of* $\mathbf{G}_a$ by $\mathbf{G}_a$ *admits a* $\mathbf{\Delta}$*-cross section.*

*Proof* In [8] Kolchin develops a cohomology theory for constrained extensions of a differential field (see also Kovacic [9]). If the target group is an $\mathcal{F}$-group (algebraic group defined over $\mathcal{F}$), Kolchin proves (Theorem 13) that the constrained cohomology set $H_{\Delta}{}^1(\mathcal{F}, G)$ is equal to the Galois cohomology set $H^1(\mathcal{F}, G)$. Kolchin then shows that if $G$ is any $\mathbf{\Delta}$-$\mathcal{F}$-group there is a bijection from the set of $\mathcal{F}$-isomorphism classes of principal homogeneous $\mathbf{\Delta}$-$\mathcal{F}$-spaces for $G$ onto $H_{\Delta}{}^1(\mathcal{F}, G)$. It is well known that the Galois cohomology set $H^1(\mathcal{F}, \mathbf{G}_a)$ is trivial. It follows, therefore, that every principal homogeneous $\mathbf{\Delta}$-$\mathcal{F}$-space for $\mathbf{G}_a$ contains an $\mathcal{F}$-rational point.

Suppose

$$E\colon \quad 0 \to \mathbf{G}_a \overset{\iota}{\to} C \overset{\pi}{\to} \mathbf{G}_a \to 0$$

is a central extension of $\mathbf{G}_a$ by $\mathbf{G}_a$. Let $\mathcal{F}$ be a differential field of definition for $C$, $\iota$, $\pi$. Let $x$ be generic for $\mathbf{G}_a$ over $\mathcal{F}$ and let $X = \pi^{-1}(x)$. Then $X$ is a principal homogeneous $\mathbf{\Delta}$-$\mathcal{G}$-space for $\mathbf{G}_a$, where $\mathcal{G} = \mathcal{F}\langle x\rangle$ (Kolchin [8]). To say that $E$ has a $\mathbf{\Delta}$-$\mathcal{F}$-cross section $\sigma$ is equivalent to saying that $X$ has a point $\sigma(x)$ rational over $\mathcal{F}\langle x\rangle$.

We state without proof two propositions whose proofs follow *mutatatis mutandi* the proofs of Ths. A8 and A9 of [6, p. 151, 154].

Let

$$E\colon \quad 0 \to A' \overset{\iota}{\to} A \overset{\pi}{\to} A'' \to 1$$

be a central extension of $A''$ by $A'$, let $B$ be a commutative $\mathbf{\Delta}$-group, and let $\alpha \in \mathrm{Hom}_{\Delta}(A', B)$. Then $\alpha E \in \mathrm{Cent\ Ext}_{\Delta}(A'', B)$. Define $\partial\alpha = \alpha E$. Then

$$\partial(\alpha_1 + \alpha_2) = (\alpha_1 + \alpha_2)E = a_1 E + \alpha_2 E = \partial\alpha_1 + \partial\alpha_2 .$$

Therefore, $\partial\colon \mathrm{Hom}_{\Delta}(A', B) \to \mathrm{Cent\ Ext}_{\Delta}(A'', B)$ is a homomorphism.

***Proposition 21*** *Let*

$$E\colon\ 0 \to A' \xrightarrow{\iota} A \xrightarrow{\pi} A'' \to 1$$

*be a central extension of $A''$ by $A'$, and let $B$ be a commutative $\mathbf{\Delta}$-group. Consider the following sequence of additive groups:*

$$0 \longrightarrow \mathrm{Hom}_{\Delta}(A'', B) \xrightarrow{\circ\pi} \mathrm{Hom}_{\Delta}(A, B) \xrightarrow{\circ\iota} \mathrm{Hom}_{\Delta}(A', B)$$

$$\xrightarrow{\partial} \mathrm{Cent\ Ext}_{\Delta}(A'', B) \xrightarrow{\pi^*} \mathrm{Cent\ Ext}_{\Delta}(A, B) \xrightarrow{\iota^*} \mathrm{Cent\ Ext}_{\Delta}(A', B).$$

1. *The sequence is exact, except possibly at* $\mathrm{Cent\ Ext}_{\Delta}(A, B)$, *where only* $\iota^*\pi^* = 0$ *holds in general.*
2. *If* $\mathrm{Hom}_{\Delta}(A', B) = 0$, *then the sequence is exact.*

*Note* Exactness at $\mathrm{Hom}_{\Delta}(A', B)$ is equivalent to saying that a $\mathbf{\Delta}$-homomorphism $\alpha'\colon A' \to B$ can be extended to a $\mathbf{\Delta}$-homomorphism $A \to B$ if and only if $\alpha' E$ splits.

Let

$$E\colon\ 0 \to B' \xrightarrow{\iota} B \xrightarrow{\pi} B'' \to 0$$

be a commutative extension of commutative $\mathbf{\Delta}$-groups, and let $A$ be a commutative $\mathbf{\Delta}$-group. Let $\alpha \in \mathrm{Hom}_{\Delta}(A, B'')$. Then $\alpha^*(E) = E\alpha \in \mathrm{Ab\ Ext}_{\Delta}(A, B')$. We define $\partial(\alpha) = E\alpha$. Since the groups are commutative,

$$\partial\colon \mathrm{Hom}_{\Delta}(A, B'') \to \mathrm{Ab\ Ext}_{\Delta}(A, B')$$

is a homomorphism.

***Proposition 22*** *Let $A$, $B'$, $B$, $B''$ be commutative $\mathbf{\Delta}$-groups. Let*

$$E\colon\ 0 \to B' \xrightarrow{\iota} B \xrightarrow{\pi} B'' \to 0$$

*be in* $\mathrm{Ab\ Ext}_{\Delta}(B'', B')$. *Then the following sequence of additive groups is exact:*

$$0 \longrightarrow \mathrm{Hom}_{\Delta}(A, B') \xrightarrow{\iota\circ} \mathrm{Hom}_{\Delta}(A, B) \xrightarrow{\pi\circ} \mathrm{Hom}_{\Delta}(A, B'')$$

$$\xrightarrow{\partial} \mathrm{Ab\ Ext}_{\Delta}(A, B') \xrightarrow{\iota_*} \mathrm{Ab\ Ext}_{\Delta}(A, B) \xrightarrow{\pi_*} \mathrm{Ab\ Ext}_{\Delta}(A, B'').$$

*Note* Exactness at $\mathrm{Hom}_\Delta(A, B'')$ is equivalent to saying that if $\alpha: A \to B''$ is a $\Delta$-homomorphism, there exists a $\Delta$-homomorphism $\tilde{\alpha}: A \to B$ such that $\alpha = \pi \circ \tilde{\alpha}$ if and only if $E\alpha$ splits.

As an application of Proposition 22, consider the exact sequence

$$E: \quad 1 \to B \xrightarrow{\iota} \mathbf{G}_{\mathrm{m}} \xrightarrow{\pi} \mathbf{G}_{\mathrm{a}} \to 0,$$

where $B$ is an infinite $\delta$-subgroup of $\mathbf{G}_{\mathrm{m}}$ and $\pi = L \circ \ell\delta$, where $L \in \mathscr{U}[\delta]$ vanishes on $\ell\,\delta B$. The short exact sequence gives rise to a long exact sequence

$$0 \longrightarrow \mathrm{Hom}_\delta(\mathbf{G}_{\mathrm{a}}, B) \xrightarrow{\iota\,\circ} \mathrm{Hom}_\delta(\mathbf{G}_{\mathrm{a}}, \mathbf{G}_{\mathrm{m}}) \xrightarrow{\pi\,\circ} \mathrm{Hom}_\delta(\mathbf{G}_{\mathrm{a}}, \mathbf{G}_{\mathrm{a}})$$

$$\xrightarrow{\partial} \mathrm{Ab\ Ext}_\delta(\mathbf{G}_{\mathrm{a}}, B) \xrightarrow{\iota_*} \mathrm{Ab\ Ext}_\delta(\mathbf{G}_{\mathrm{a}}, \mathbf{G}_{\mathrm{m}}) \xrightarrow{\pi_*} \mathrm{Ab\ Ext}_\delta(\mathbf{G}_{\mathrm{a}}, \mathbf{G}_{\mathrm{a}}).$$

Now, as we remarked on p. 99, $\mathrm{Hom}_\delta(\mathbf{G}_{\mathrm{a}}, B)$ and $\mathrm{Hom}_\delta(\mathbf{G}_{\mathrm{a}}, \mathbf{G}_{\mathrm{m}})$ are both equal to 0. Furthermore, by Corollary 1 of Proposition 16,

$$\mathrm{Ab\ Ext}_\delta(\mathbf{G}_{\mathrm{a}}, \mathbf{G}_{\mathrm{a}}) = 0.$$

Thus, the following sequence is exact:

$$0 \to \mathscr{U}[\delta] \xrightarrow{\partial} \mathrm{Ab\ Ext}_\delta(\mathbf{G}_{\mathrm{a}}, B) \xrightarrow{\iota_*} \mathrm{Ab\ Ext}_\delta(\mathbf{G}_{\mathrm{a}}, \mathbf{G}_{\mathrm{m}}) \to 0$$

So, $\mathrm{Ab\ Ext}_\delta(\mathbf{G}_{\mathrm{a}}, \mathbf{G}_{\mathrm{m}}) \approx \mathrm{Ab\ Ext}_\delta(\mathbf{G}_{\mathrm{a}}, B)/\partial u[\delta]$. Now, if $P \in \mathscr{U}[\delta]$, $\partial P$ is $EP$: $1 \to B \to C' \to \mathbf{G}_{\mathrm{a}} \to 0$, where $C'$ is the $\delta$-subgroup of $C'$ defined by the differential equation $L(\ell\,\delta u_1) = P(u_2)$. The injectivity of $\partial$ implies that if $P \neq 0$, $C'$ cannot be written as a direct product $B \times \mathbf{G}_{\mathrm{a}}$.

***Proposition 23*** *Let $B$ be a commutative $\Delta$-group of differential dimension 0. Let $A$ be a commutative $\Delta$-group such that every proper $\Delta$-subgroup has smaller differential dimension. Then* $\mathrm{Cent\ Ext}_\Delta(A, B) = \mathrm{Ab\ Ext}_\Delta(A, B)$.

*Proof* Let

$$E: \quad 0 \to B \xrightarrow{\iota} C \xrightarrow{\pi} A \to 0$$

be a central extension. Since $A$ is commutative, $[C, C] \subset \iota(B)$, hence is central in $C$. Let $x \in C$ and let $\varphi_x(a) = axa^{-1}x^{-1}$. Then the centrality of the commutator subgroup implies that $\varphi_x$ is a $\Delta$-homomorphism from $C$ into $\iota(B)$. We now compare differential dimensions. First, diff dim $B$ + diff dim $A$ = diff dim $C$. Therefore, diff dim $C$ = diff dim $A$. Second, diff dim ker $\varphi_x$ + diff dim $\varphi_x(C)$ = diff dim $C$ = diff dim $A$. Therefore, diff dim ker $\varphi_x$ = diff dim $A$. Now, diff dim ker $\pi$ = diff dim $B = 0$. Therefore, diff dim $\pi(\ker \varphi_x)$ = diff dim ker $\varphi_x$ = diff dim $A$. It follows that

$\pi(\ker \varphi_x) = A$. But, since $\iota(B)$ is central in $C$, $\iota(B) = \ker \pi \subset \ker \varphi_x$. Therefore, $\ker \varphi_x = C$, and $C$ is commutative.

***Corollary*** *Let $B$ be a commutative $\boldsymbol{\Delta}$-group of $\boldsymbol{\Delta}$-dimension* 0 *and let $A$ be a commutative $\boldsymbol{\Delta}$-group whose Lie algebra is a vector group. Then*

$$\text{Cent Ext}_{\boldsymbol{\Delta}}(A, B) = \text{Ab Ext}_{\boldsymbol{\Delta}}(A, B).$$

*Proof* If $A'$ is a proper $\boldsymbol{\Delta}$-subgroup of $A$ such that diff dim $A'$ = diff dim $A$, then the Lie algebra of $A'$ is a proper $\boldsymbol{\Delta}$-subalgebra of the Lie algebra of $A$ whose differential dimension equals that of the Lie algebra of $A$ (Kolchin [8]). Since the Lie algebra of $A$ is a vector group, this is not possible.

In particular, if $B$ is a commutative $\boldsymbol{\Delta}$-group of differential dimension 0,

$$\text{Cent Ext}_{\boldsymbol{\Delta}}(\mathbf{G}_{\mathrm{a}}{}^n, B) = \text{Ab Ext}_{\boldsymbol{\Delta}}(\mathbf{G}_{\mathrm{a}}{}^n, B)$$

and

$$\text{Cent Ext}_{\boldsymbol{\Delta}}(\mathbf{G}_{\mathrm{M}}{}^n, B) = \text{Ab Ext}_{\boldsymbol{\Delta}}(\mathbf{G}_{\mathrm{M}}{}^n, B).$$

***Proposition 24*** *Let $B$ be a proper $\delta$-subgroup of $\mathbf{G}_{\mathrm{a}}$. Then* Cent $\text{Ext}_\delta(\mathbf{G}_{\mathrm{a}}, B)$ *is a right module over $\mathscr{U}[\delta]$ of rank* 1. *It is generated by the class of the extension*

$$E\colon\ 0 \longrightarrow B \xrightarrow{\text{incl}} \mathbf{G}_{\mathrm{a}} \xrightarrow{L} \mathbf{G}_{\mathrm{a}} \longrightarrow 0,$$

*where $L$ is the defining homogeneous linear differential polynomial of $B$. If $P \in \mathscr{U}[\delta]$, then* $\text{cl}(EP) = 0$ *if and only if $P \in L\mathscr{U}[\delta]$.*

*Proof* By Proposition 23, Cent $\text{Ext}_\delta(\mathbf{G}_{\mathrm{a}}, B)$ = Ab $\text{Ext}_\delta(\mathbf{G}_{\mathrm{a}}, B)$. In particular, then, as we observed on p. 104, Cent $\text{Ext}_\delta(\mathbf{G}_{\mathrm{a}}, B)$ is a right module over $\text{Hom}_\delta(\mathbf{G}_{\mathrm{a}}, \mathbf{G}_{\mathrm{a}}) = \mathscr{U}[\delta]$. By Proposition 22, the short exact sequence gives rise to a long exact sequence

$$0 \to \text{Hom}_\delta(\mathbf{G}_{\mathrm{a}}, B) \to \text{Hom}_\delta(\mathbf{G}_{\mathrm{a}}, \mathbf{G}_{\mathrm{a}}) \xrightarrow{L\circ} \text{Hom}_\delta(\mathbf{G}_{\mathrm{a}}, \mathbf{G}_{\mathrm{a}})$$

$$\xrightarrow{\partial} \text{Ab Ext}_\delta(\mathbf{G}_{\mathrm{a}}, B) \to \text{Ab Ext}_\delta(\mathbf{G}_{\mathrm{a}}, \mathbf{G}_{\mathrm{a}}).$$

Let $\varphi \in \text{Hom}_\delta(\mathbf{G}_{\mathrm{a}}, B)$. Then diff dim $\ker \varphi$ + diff dim $\varphi(\mathbf{G}_{\mathrm{a}})$ = diff dim $\mathbf{G}_{\mathrm{a}}$. Therefore, $\ker \varphi = \mathbf{G}_{\mathrm{a}}$. Thus, $\text{Hom}_\delta(\mathbf{G}_{\mathrm{a}}, B) = 0$. By Corollary 1 of Proposition 16, Ab $\text{Ext}_\delta(\mathbf{G}_{\mathrm{a}}, \mathbf{G}_{\mathrm{a}}) = 0$. We may rewrite the exact sequence

$$0 \to \mathscr{U}[\delta] \xrightarrow{L\circ} \mathscr{U}[\delta] \xrightarrow{\gamma} \text{Ab Ext}_\delta(\mathbf{G}_{\mathrm{a}}, B) \to 0.$$

Let $P \in \mathscr{U}[\delta]$. Then $\partial P = EP$. Since $\partial$ is surjective, this proves the assertion.

***Corollary*** *Let* $\tilde{E}: 0 \to (\mathbf{G}_a)_{\mathscr{K}} \to C \to \mathbf{G}_a \to 0$ *be a central extension of the* $\delta$*-group* $\mathbf{G}_a$ *by the* $\delta$*-group* $(\mathbf{G}_a)_{\mathscr{K}}$. *Either* $C$ *splits or* $C$ *is* $\delta$*-isomorphic to* $\mathbf{G}_a$.

*Proof* By Proposition 24, we may assume that $\tilde{E} = EP$, where

$$E = 0 \to (\mathbf{G}_a)_{\mathscr{K}} \to \mathbf{G}_a \overset{\delta}{\to} \mathbf{G}_a \to 0,$$

and $P \in \mathscr{U}[\delta]$. Therefore, $C$ is the $\delta$-subgroup of $\mathbf{G}_a \times \mathbf{G}_a$ consisting of all pairs $(u_1, u_2)$ such that $\delta u_1 = Pu_2$. $P = a + \delta M$, $M \in \mathscr{U}[\delta]$, $a \in \mathscr{U}$. Moreover, $C$ splits if and only if $a = 0$. We rewrite the defining equation of $C$: $\delta(u_1 - Mu_2) = au_2$. Let $\varphi$: $C \to \mathbf{G}_a$ be defined by the formula $\varphi(u_1, u_2) = u_1 - Mu_2$. $\varphi$ is a surjective $\delta$-homomorphism. Suppose $\varphi(u_1, u_2) = 0$. Then $u_2 = 0$, whence $u_1 = 0$. Thus, $\varphi$ is the desired $\delta$-isomorphism.

## 5. The groups $H_\Delta{}^2(\mathbf{G}_a, \mathbf{G}_a)$ and Cent $\mathrm{Ext}_\Delta(\mathbf{G}_a, \mathbf{G}_a)$

***Theorem 5*** $H_\Delta{}^2(\mathbf{G}_a, \mathbf{G}_a) = H^2_{\mathrm{mor}}(\mathbf{G}_a, \mathbf{G}_a)$.

*Proof* The proof consists of a sequence of lemmas.

***Lemma 1*** *Let* $f(x, y) \in Z_\Delta{}^2(\mathbf{G}_a, \mathbf{G}_a)$. *There exist differential polynomials* $A(x, y)$ *and* $Q(x)$ *such that* $f(x, y) = A(x, y)/Q(x)Q(y)Q(x + y)$.

*Proof* We may suppose that $f(x, y) \notin \mathscr{U}\{x, y\}$. We write $f(x, y) = A(x, y)/B(x, y)$, where $A$, $B \in \mathscr{U}\{x, y\}$ and are relatively prime. Now, write $B(x, y) = P(x)Q(y)R(x, y)$, where no irreducible factor of $R(x, y)$ is in $\mathscr{U}\{x\}$ or in $\mathscr{U}\{y\}$. Let $z$ be a differential indeterminate. The functional equation satisfied by 2-cocycles and the relative primeness of $A$ and $B$ imply the existence of differential polynomials $C_i(x, y, z)$, $1 \leq i \leq 3$, such that

$$B(x, y)C_1(x, y, z) = B(y, z)B(x + y, z)B(x, y + z),$$

$$B(y, z)C_2(x, y, z) = B(x, y)B(x + y, z)B(x, y + z),$$

$$B(x + y, z)C_3(x, y, z) = B(x, y)B(y, z)B(x, y + z).$$

Suppose $B(x, y) = aR(x, y)$, $a \in \mathscr{U}^*$. The first equation then reads

$$aR(x, y)C_1(x, y, z) = a^3R(y, z)R(x + y, z)R(x, y + z).$$

Let $D$ be an irreducible factor of $R$. By definition of $R$, $D(x, y)$ cannot be an irreducible factor of $R(y, z)$ or of $R(x + y, z)$ or of $R(x, y + z)$. This contradiction forces us to assume that $P(x) \notin \mathscr{U}$ or $Q(y) \notin \mathscr{U}$. Suppose $Q(y) \notin \mathscr{U}$ and let $D$ be an irreducible factor of $Q(y)$. The first equation reads (after cancellation)

$$\begin{aligned} &Q(y)R(x, y)C_1(x, y, z) \\ &\quad = P(y)Q^2(z)R(y, z)P(x + y)R(x + y, z)Q(y + z)R(x, y + z). \end{aligned}$$

$D(y)$ cannot divide $Q^2(z)P(x+y)Q(y+z)$, and, by definition of $R$, $D(y)$ cannot be an irreducible factor of $R(y, z)R(x+y, z)R(x, y+z)$. Therefore, $D$ divides $P$. Thus, $Q(y)$ divides $P(y)$ and, in particular, $P \notin \mathscr{U}$. A parallel argument, which makes use of the second equation, shows that $P(y)$ divides $Q(y)$. Therefore, we may assume that $P(y) = Q(y)$. Thus, $B(x, y) = Q(x)Q(y)R(x, y)$. After cancellation, the third equation reads

$$Q(x+y)R(x+y, z)C_3(x, y, z)$$
$$= Q^2(x)Q^2(y)Q(y+z)R(x, y)R(y, z)R(x, y+z).$$

We conclude that $Q(x+y)$ divides $R(x, y)$. The first equation now reads

$$Q(y)R(x, y)C_1(x, y, z)$$
$$= Q(y)Q^2(z)R(y, z)Q(x+y)R(x+y, z)Q(y+z)R(x, y+z).$$

Let $D$ be an irreducible factor of $R$. Since $D(x, y)$ contains derivatives of both $x$ and $y$, $D(x, y)$ cannot divide $Q(y)Q^2(z)Q(y+z)$. $D(x, y)$ cannot be an irreducible factor of $R(y, z)R(x+y, z)R(x, y+z)$. Therefore, $D(x, y)$ divides $Q(x+y)$. Thus, $R(x, y)$ divides $Q(x+y)$, and we may assume that $R(x, y) = Q(x+y)$.

***Lemma 2*** *Let* $f(x, y) \in Z_{\mathbf{\Delta}}{}^2(\mathbf{G}_a, \mathbf{G}_a)$. *Then there is a symmetric* $\mathbf{\Delta}$-2-*cocycle* $s(x, y)$ *such that* $f(x, y) - s(x, y) \in \mathscr{U}\{x, y\}$.

*Proof* Write $f(x, y) = A(x, y)/Q(x)Q(y)Q(x+y)$ (by Lemma 1). The functional equation then implies that

$$A(y, z)Q(x)Q(x+y)Q(x+y+z) - A(x+y, z)Q(x)Q(y)Q(y+z)$$
$$+ A(x, y+z)Q(y)Q(z)Q(x+y) - A(x, y)Q(z)Q(y+z)$$
$$+ Q(x+y+z) = 0.$$

Set $z = x$. After cancellation, the equation reads

$$Q(2x+y)(A(y, x) - A(x, y)) + Q(y)(A(x, x+y) - A(x+y, x)) = 0.$$

Thus, $Q(y)$ divides $Q(2x+y)(A(y, x) - A(x, y))$. Since $Q(y)$ and $Q(2x+y)$ are relatively prime, $Q(y)$ divides $A(y, x) - A(x, y)$. Therefore, $Q(x)$ divides $A(x, y) - A(y, x)$, and thus also divides $A(y, x) - A(x, y)$. Similarly, $Q(2x+y) = Q(x+(x+y))$ divides $A(x, x+y) - A(x+y, x)$. Therefore, $Q(x+y)$ divides $A(x, y) - A(y, x)$. Thus, $f(x, y) - f(y, x) \in \mathscr{U}\{x, y\}$. But, $f(x, y) = \frac{1}{2}(f(x, y) + f(y, x)) + \frac{1}{2}(f(x, y) - f(y, x))$. Therefore, $f(x, y) = s(x, y) + P(x, y)$, where $s(x, y)$ is symmetric and $P(x, y) \in \mathscr{U}\{x, y\}$.

***Lemma 3*** $H_{\mathbf{\Delta}}{}^2(\mathbf{G}_a, \mathbf{G}_a)_s = 0$.

*Proof* Let $s(x, y) \in Z_{\Delta}{}^2(\mathbf{G}_a, \mathbf{G}_a)_s$. There is a natural number $r$ such that $s(x, y) \in \mathcal{G} = \mathcal{U}(\theta x, \theta' y)_{\theta, \theta' \in \Theta(r)}$, where $\Theta(r)$ is the set of all $\theta \in \Theta$ with ord $\theta \leq r$. $\mathcal{G}$ is the field of rational functions on $\mathbf{G}_a{}^N$, where

$$N = 2\binom{r+m}{m}.$$

So, $s$ is a symmetric rational 2-cocycle from $\mathbf{G}_a{}^N$ into $\mathbf{G}_a$. By Prop. 4b of [15], $H^2_{\text{rat}}(\mathbf{G}_a{}^N, \mathbf{G}_a)_s$ is isomorphic to the subgroup of Ab Ext$(\mathbf{G}_a{}^N, \mathbf{G}_a)$ composed of those extensions that admit a rational cross section. By Prop. 6 of [15], $H^2_{\text{rat}}(\mathbf{G}_a{}^N, \mathbf{G}_a)_s \approx$ Ab Ext$(\mathbf{G}_a{}^N, \mathbf{G}_a)$ since all extensions of $\mathbf{G}_a{}^N$ by $\mathbf{G}_a$ admit a rational cross section. By the corollary of Prop. 8 of [15], Ab Ext$(\mathbf{G}_a{}^N, \mathbf{G}_a) = 0$. Thus, $H^2_{\text{rat}}(\mathbf{G}_a{}^N, \mathbf{G}_a)_s = 0$. Therefore, $s$ is the coboundary of a rational function $h$ in $\mathcal{U}(\theta x)_{\theta \in \Theta(r)}$. But, then, there is a differential rational function $h \in \mathcal{U}\langle x \rangle$ such that $s(x, y) = h(x + y) - h(x) - h(y)$.

The proof of the theorem now follows immediately.

*Note* Let $f \in Z_{\Delta}{}^2(A, B)$ and suppose $f$ is defined at $(1, 1)$. Let $g = f - f(1, 1)$. Then $g$ and $f$ are cohomologous and $g(1, 1) = 0$. We call $g$ *normalized*. We will assume in the proof of the following theorem that if $f$ is defined at $(1, 1)$, then $f$ is normalized.

***Theorem 6*** *$H_{\Delta}{}^2(\mathbf{G}_a, \mathbf{G}_a)$ is isomorphic as a left module over $\mathcal{U}[\Delta]$ to* Cent Ext$_{\Delta}(\mathbf{G}_a, \mathbf{G}_a)$.

*Proof* Let $F \in H_{\Delta}{}^2(\mathbf{G}_a, \mathbf{G}_a)$. By Theorem 5, $F = \text{cl}(f)$, where $f \in Z^2_{\text{mor}}(\mathbf{G}_a, \mathbf{G}_a)$. Let $C = \mathbf{G}_a \times \mathbf{G}_a$, together with the binary operation defined by the formula $(u_1, u_2)(v_1, v_2) = (u_1 + v_1, u_2 + v_2 + f(u_1, v_1))$. The functional equation for 2-cocycles implies associativity and the fact that $f(u, 0) = f(0, v) = f(0, 0) = 0$ (since $f$ is assumed to be normalized). Thus, the identity element is $(0, 0)$ and $(u_1, u_2)^{-1} = (-u_1, -u_2 - f(u_1, -u_1))$. $0 \times \mathbf{G}_a$ is a central $\Delta$-subgroup of $C$ and $\iota$: $\mathbf{G}_a \to 0 \times \mathbf{G}_a$, defined by the formula $\iota(u) = (0, u)$ is a $\Delta$-isomorphism. We define $\pi(u_1, u_2)$ to be $u_1$. We call

$$E: \quad 0 \to \mathbf{G}_a \xrightarrow{\iota} C \xrightarrow{\pi} \mathbf{G}_a \to 0$$

the extension associated with $f$, and we define $\varepsilon(F) = \text{cl}(E)$. If $g \in Z^2_{\text{mor}}(\mathbf{G}_a, \mathbf{G}_a)$ is cohomologous to $f$, there is an element $h(x) \in \mathcal{U}\{x\}$ such that $g(x, y) = f(x, y) + h(x + y) - h(x) - h(y)$. Let

$$E': \quad 0 \to \mathbf{G}_a \xrightarrow{\iota'} C' \xrightarrow{\pi'} \mathbf{G}_a \to 0$$

be the extension associated with $g$. We define a $\mathbf{\Delta}$-isomorphism $\gamma: C \to C'$ by the formula $\gamma(u_1, u_2) = (u_1, u_2 + h(u_1))$. Let $\Gamma = (1_{\mathbf{G}_a}, \gamma, 1_{\mathbf{G}_a})$. $\Gamma$ is an isomorphism of extensions.

Let

$$E: \quad 0 \to \mathbf{G}_a \xrightarrow{\iota} C \xrightarrow{\pi} \mathbf{G}_a \to 0$$

be a central extension of $\mathbf{G}_a$ by $\mathbf{G}_a$. By Proposition 20, $E$ admits a $\mathbf{\Delta}$-cross section $\sigma: \mathbf{G}_a \to C$ (which, of course, need not be everywhere defined). Let $f(x, y)$ be the unique element of $\mathscr{U}\langle x, y\rangle$ such that whenever $u$ and $v$ are independent generic points for $\mathbf{G}_a$ over some field of definition for $C$, $\iota$, $\pi$, $\sigma$, then $\iota f(u, v) = \sigma(u)\sigma(v)\sigma(u + v)^{-1}$. Then $f$ is in $Z_{\mathbf{\Delta}}{}^2(\mathbf{G}_a, \mathbf{G}_a)$. We define $\kappa(\mathrm{cl}(E)) = \mathrm{cl}(f)$. Suppose $\sigma'$: $\mathbf{G}_a \to C$ is another cross section. Let $\mathscr{F}$ be a differential field of definition for everything in sight, and let $u$ be generic for $\mathbf{G}_a$ over $\mathscr{F}$. $\pi\sigma u = \pi\sigma' u = u$. Thus, the formula $\iota h(u) = \sigma'(u)\sigma(u)^{-1}$ defines a $\mathbf{\Delta}$-map from $\mathbf{G}_a$ to $\mathbf{G}_a$. Let $v$ be a generic point for $\mathbf{G}_a$ over $\mathscr{F}$ independent from $u$. Let $g(u, v) = \sigma'(u)\sigma'(v)\sigma'(u + v)^{-1}$. Let $g(x, y)$ be the associated 2-cocycle. Then $\iota f(u, v) = \iota(g(u, v) + h(u + v) - h(u) - h(v))$. Thus $f$ and $g$ are cohomologous.

It is instructive to see precisely how $\varepsilon$ transforms addition of 2-cocycles into the Baer sum of extensions, and derivatives of 2-cocycles into homomorphisms of extensions.

Let $f_i$ be in $Z^2_{\mathrm{mor}}(\mathbf{G}_a, \mathbf{G}_a)$ and let $E_i$ be the extension associated with $f_i$, $i = 1, 2$.

$$E_i = 0 \to \mathbf{G}_a \xrightarrow{\iota_i} C_i \xrightarrow{\pi_i} \mathbf{G}_a \to 0.$$

Let $C' = \mathbf{G}_a \times \mathbf{G}_a \times \mathbf{G}_a$, together with the binary operation $(u_1, u_2, u_3)(v_1, v_2, v_3) = (u_1 + v_1, u_2 + v_2 + f_1(u_1, v_1), u_3 + v_3 + f_2(u_1, v_1))$. $C'$ is a $\mathbf{\Delta}$-group. Let $N$ be the central $\mathbf{\Delta}$-subgroup of $\mathbf{G}_a \times C'$ consisting of all triples $(-(u_2 + u_3), 0, u_2, u_3)$. Let $C = \mathbf{G}_a \times C'/N$. Let $\iota: \mathbf{G}_a \to C$ be defined by the formula $\iota(u) = \mathrm{cl}(u, 0, 0)$, and let $\pi(\mathrm{cl}(u, u_1, u_2, u_3)) = u_1$. Since $\mathrm{cl}(u + u_2 + u_3, 0, 0, 0) = \mathrm{cl}(u, 0, u_2, u_3)$, it follows that $\iota(\mathbf{G}_a) = \ker \pi$. The extension

$$0 \to \mathbf{G}_a \xrightarrow{\iota} C \xrightarrow{\pi} \mathbf{G}_a \to 0$$

is equivalent to the Baer sum $E_1 + E_2$. Let $\varphi(u, u_1, u_2, u_3) = (u_1, u + u_2 + u_3)$. $\varphi$ is a $\mathbf{\Delta}$-homomorphism from $\mathbf{G}_a \times C'$ onto the group $C''$ associated with $f_1 + f_2$, with kernel $N$. The induced isomorphism from $C$ onto $C''$ gives rise to an isomorphism of extensions.

Let $L$ be in $\mathscr{U}[\mathbf{\Delta}]$ and let $f$ be in $Z^2_{\mathrm{mor}}(\mathbf{G}_a, \mathbf{G}_a)$. Let

$$E: \quad 0 \to \mathbf{G}_a \xrightarrow{\iota} C \xrightarrow{\pi} \mathbf{G}_a \to 0$$

be the extension associated with $f$. Let

$$E'\colon\quad 0 \to \mathbf{G}_a \xrightarrow{\iota'} C' \xrightarrow{\pi'} \mathbf{G}_a \to 0$$

be $LE$. $C'$ is the quotient of $\mathbf{G}_a \times C$ by the normal $\mathbf{\Delta}$-subgroup $N$ consisting of all triples $(-Lu, 0, u)$, $u \in \mathscr{U}$. Let

$$E''\colon\quad 0 \to \mathbf{G}_a \xrightarrow{\iota''} C'' \xrightarrow{\pi''} \mathbf{G}_a \to 0$$

be the extension associated with $Lf$. We define a $\mathbf{\Delta}$-homomorphism $\varphi$: $\mathbf{G}_a \times C \to C''$, with kernel $N$, by the formula $\varphi(u, u_1, u_2) = (u_1, Lu_2 + u)$. The induced isomorphism $\gamma\colon C' \to C''$ gives rise to an isomorphism of extensions. Thus, $\varepsilon$ is a homomorphism of left $\mathscr{U}[\mathbf{\Delta}]$-modules.

We must show that $\varepsilon$ and $\kappa$ are mutually inverse. Let $f$ be in $Z^2_{\text{mor}}(\mathbf{G}_a, \mathbf{G}_a)$ and let

$$E\colon\quad 0 \to \mathbf{G}_a \xrightarrow{\iota} C \xrightarrow{\pi} \mathbf{G}_a \to 0$$

be the extension associated with $f$. Let $\sigma\colon \mathbf{G}_a \to C$ be defined by the formula $\sigma(u) = (u, 0)$. Then $\sigma$ is an everywhere defined $\mathbf{\Delta}$-cross section and $\iota f(u, v) = \sigma(u)\sigma(v)\sigma(u + v)^{-1}$. We showed above that if $\sigma'\colon \mathbf{G}_a \to C$ is any $\mathbf{\Delta}$-cross section, then the $\mathbf{\Delta}$-2-cocycle defined by $\sigma'$ is cohomologous to $f$. Therefore, $\kappa\varepsilon(\text{cl}(f)) = \text{cl}(f)$.

We must now show that $\varepsilon\kappa(\text{cl}(E)) = \text{cl}(E)$.

$$E = 0 \to \mathbf{G}_a \xrightarrow{\iota} C \xrightarrow{\pi} \mathbf{G}_a \to 0.$$

Let $\sigma$ be a $\mathbf{\Delta}$-cross section from $\mathbf{G}_a$ to $C$. Let $\mathscr{F}$ be a differential field of definition for everything in sight and let $u$, $v$ be independent generic points for $\mathbf{G}_a$ over $\mathscr{F}$. Let $\iota f(u, v) = \sigma(u)\sigma(v)\sigma(u + v)^{-1}$. Let $g \in Z^2_{\text{mor}}(\mathbf{G}_a, \mathbf{G}_a)$ be cohomologous to $f$ ($f$ itself need not be everywhere defined), and let

$$E'\colon\quad 0 \to \mathbf{G}_a \xrightarrow{\iota'} C' \xrightarrow{\pi'} \mathbf{G}_a \to 0$$

be the extension associated with $g$. We must show that $E$ and $E'$ are isomorphic. There is an element $h \in \mathscr{U}\langle x \rangle$ such that $f(x, y) = g(x, y) + h(x) + h(y) - h(x + y)$. Let $u$ be generic for $\mathbf{G}_a$ over $\mathscr{F}$. We define $\gamma(\sigma u) = (u, h(u))$. Let $v \in \mathbf{G}_a$. We define $\gamma(\iota v) = (0, v)$. For every $w \in C$ generic over $\mathscr{F}$ there exists a unique pair $(u_1, u_2) \in \mathbf{G}_a \times \mathbf{G}_a$ such that $w = \sigma u_1 \cdot \iota u_2$. In fact, $u_1 = \pi(w)$ and $u_2 = (\sigma u_1)^{-1} w$. In particular, $u_1$ is generic for $\mathbf{G}_a$ over $\mathscr{F}$, whence $\gamma$ is defined at $\sigma u_1$ and we may define $\gamma(w) = \gamma(\sigma u_1) \cdot \gamma(\iota u_2) = (u_1, u_2 + h(u_1))$, since $g(u_1, 0) = 0$. Thus, $\gamma(w) = (\pi(w), j((\sigma(\pi(w))^{-1} w)$, where $j$ is the inverse of the isomorphism $\iota$. It follows that $\gamma$ defines a unique $\mathbf{\Delta}$-$\mathscr{F}$-map from $C$ to $C'$. Let $z = \sigma v_1 \cdot \iota v_2$ be a generic point for $C$ over $\mathscr{F}$

independent from $w$. An easy computation shows that $\gamma(wz) = \gamma(w)\gamma(z)$. So, $\gamma$ is a $\Delta$-homomorphism [DAG, Prop. 6, p. 908]. Since $\gamma$ has the generic inverse $\gamma^{-}(u_1, u_2) = \sigma u_1\, \iota(u_2 - h(u_1))$, $\gamma$ is an isomorphism. By definition, $\gamma(\iota u) = (0, u) = \iota'(u)$, for $u \in \mathbf{G}_a$, and if $w = \sigma u_1 \cdot \iota u_2$ is generic for $C$ over $\mathcal{F}$, $\pi'(\gamma(w)) = \pi'(u_1, u_2 + h(u_1)) = u_1 = \pi(w)$. So, $\Gamma = (1_{\mathbf{G}_a}, \gamma, 1_{\mathbf{G}_a})$ is an isomorphism of extensions. Therefore, $\varepsilon(\kappa(\mathrm{cl}(E)) = \mathrm{cl}(E)$. It follows, in particular, that every central extension of $\mathbf{G}_a$ by $\mathbf{G}_a$ admits an everywhere defined $\Delta$-cross section.

***Corollary*** Ab $\mathrm{Ext}_{\Delta}(\mathbf{G}_a, \mathbf{G}_a) = 0$.

*Proof* Clearly, $\varepsilon(H_{\Delta}{}^2(\mathbf{G}_a, \mathbf{G}_a)_s = \mathrm{Ab\ Ext}_{\Delta}(\mathbf{G}_a, \mathbf{G}_a)$. But, by Lemma 3, $H_{\Delta}{}^2(\mathbf{G}_a, \mathbf{G}_a)_s = 0$.

*Note* If $\mathcal{U}$ is an ordinary differential field, the corollary follows from Corollary 1 of Prop. 16.

*We will assume in the remainder of this section that $\mathcal{U}$ is an ordinary differential field with derivation operator $\delta$. We denote $\delta^i u$ by $u^{(i)}$.*

We proved in Theorem 5 that $H_{\delta}{}^2(\mathbf{G}_a, \mathbf{G}_a) = H^2_{\mathrm{mor}}(\mathbf{G}_a, \mathbf{G}_a)$. In fact, we proved directly that every cohomology class is represented by a differential polynomial 2-cocycle. Ritt proves [13, pp. 759–761] that every differential polynomial 2-cocycle is cohomologous to a 2-form.

***Theorem 7*** (Ritt) *Let $f(x, y)$ be such that $f(y, z) - f(x + y, z) + f(x, y + z) - f(x, y) = 0$, where $x, y, z$ are differential indeterminates. Then there is a differential polynomial $h(x)$ and elements $a_{ij} \in \mathcal{U}$ $(0 \le i \le g - 1, 1 \le j \le g, i < j)$ such that $f(x, y) = \sum a_{ij} x^{(i)} y^{(j)} + h(x + y) - h(x) - h(y)$.*

If $f$ is a differential polynomial 2-cocycle and $0 \to \mathbf{G}_a \to C \to \mathbf{G}_a \to 0$ is the extension associated with $f$, we denote $C$ by the symbol $\mathbf{G}_a \times_f \mathbf{G}_a$.

***Corollary*** *Every central extension of $\mathbf{G}_a$ by $\mathbf{G}_a$ is isomorphic to a central extension of the form*

$$0 \to \mathbf{G}_a \to \mathbf{G}_a \times_f \mathbf{G}_a \to \mathbf{G}_a \to 0,$$

*where $f(x, y) = \sum_{i<j} a_{ij} x^{(i)} y^{(j)}$.*

The second cohomology group is a vector space over $\mathcal{U}$ as well as a left module over the ring $\mathcal{U}[\delta]$ of linear differential operators. We shall distinguish the two structures on $H_{\delta}{}^2(\mathbf{G}_a, \mathbf{G}_a)$ by using Latin letters to denote the $\mathcal{U}$-vector space structure and script letters to denote the $\mathcal{U}[\delta]$-module structure.

Theorems 5 and 7 show that the family $\mathrm{cl}(x^{(i)}y^{(j)})$, $i < j$, generates $V = H_{\delta}{}^2(\mathbf{G}_a, \mathbf{G}_a)$ over $\mathcal{U}$. Since $\sum_{i<j} a_{ij} x^{(i)} y^{(j)}$ is symmetric if and only if

$a_{ij} = 0$ for all pairs $(i, j)$, it follows that the family is a basis of $V$. If we use induction on $i$ we can easily show that the cohomology classes of the 2-forms $xy^{(j)}$, $j < 1$, generate $\mathscr{V} = H_\delta{}^2(\mathbf{G}_a, \mathbf{G}_a)$ over $\mathscr{U}[\delta]$. For,

$$x^{(i)}y^{(j)} = \delta(x^{(i-1)}y^{(j)}) - x^{(i-1)}y^{(j+1)},$$

for $i \geq 1$. They are not, however, linearly independent. For, $\delta(xy') = xy'' + x'y'$; since $x'y'$ is symmetric, $\delta(\mathrm{cl}(xy')) = \mathrm{cl}(xy'')$.

***Theorem 8*** *The cohomology classes of the differential polynomials $x \cdot \delta^j y = xy^{(j)}$, $j$ an odd positive integer, form a basis of $H_\delta{}^2(\mathbf{G}_a, \mathbf{G}_a)$ as a left module over the ring $\mathscr{U}[\delta]$ of linear differential operators.*

*Proof* We define a grading of $V = H_\delta{}^2(\mathbf{G}_a, \mathbf{G}_a)$ as follows. The zero cohomology class is isobaric of weight $r$ for every positive integer $r$. If $F$ is nonzero, there is a unique 2-form $f(x, y) = \sum_{i<j} a_{ij}x^{(i)}y^{(j)}$ such that $F = \mathrm{cl}(f)$. We define the weight of $F$ to be the weight of the differential polynomial $f$, i.e. $\max\{i + j \mid a_{ij} \neq 0\}$. If $r$ is a positive integer, we let $V_r$ be the vector space over $\mathscr{U}$ consisting of all cohomology classes that are isobaric of weight $r$. Then $V$ is the direct sum of the subspaces $V_r$, $r \geq 1$. The cohomology classes of the differential polynomials $x^{(i)}y^{(r-i)}$, $0 \leq i < r/2$, form a basis of $V_r$. Thus, if $[r/2]$ is the integer part of $r/2$, the dimension of $V_r$ is $r/2 = [r/2]$ if $r$ is even and is $[r/2] + 1$ if $r$ is odd. It is easy to see that the cardinality of the set of cohomology classes of the differential polynomials $(xy^{(j)})^{(r-j)}$, $1 \leq j \leq r$, $j$ odd, is the number of odd integers $j$ such that $1 \leq j \leq r$, and thus is $r/2$ if $r$ is even and $[r/2] + 1$ if $r$ is odd. Therefore, it suffices to show that these cohomology classes generate $V_r$. We use induction on $r$. $V_1$ is 1-dimensional and contains $\mathrm{cl}(xy')$. Assume the family generates $V_{r-1}$, $r \geq 2$, and let $\mathscr{W}$ be the submodule generated by the family $\mathrm{cl}(xy^{(j)})$, $j$ odd. By the discussion preceding the theorem, if $0 \leq i < r/2$, $x^{(i)}y^{(r-i)}$ is cohomologous to $f(x, y) = \sum_{i \leq j \leq k} D_j xy^{(j)}$, where $D_j \in \mathscr{U}[\delta]$, and the weight of $D_j xy^{(j)} = r$ (since $\mathrm{cl}(f(x, y)$ is isobaric of weight $r$). If the order of $D_j$ is $> 0$, then $j < r$, whence the induction assumption tells us that $\mathrm{cl}(xy^{(j)}) \in \mathscr{W}$, and therefore $D_j(\mathrm{cl}(xy^{(j)}) \in \mathscr{W}$. Thus, it clearly suffices to show that $xy^{(r)} \in \mathscr{W}$. Since this is so if $r$ is odd we may suppose $r = 2k$, $k$ a positive integer. But,

$$\delta(x^{(i)}y^{(2k-i-1)}) = x^{(i)}y^{(2k-i)} + x^{(i+1)}y^{(2k-i-1)}, \qquad 0 \leq i \leq k - 1.$$

Hence, by the induction assumption,

$$\mathrm{cl}\left(\sum_{0 \leq i \leq k-1} (-1)^{(i)}\, \delta(x^{(i)}y^{(2k-i-1)})\ \mathrm{cl}(xy^{(2k)} \pm x^{(k)}y^{(k)}) = \mathrm{cl}(xy^{(2k)})\right)$$

is in $\mathscr{W}$, since $x^{(i)}y^{(2k-i-1)}$ is isobaric of weight $2k - 1$. This ends the proof of the theorem.

*Note* If $k$ has characteristic $p > 0$, $H^2_{\text{rat}}(\mathbf{G}_a, \mathbf{G}_a) = H^2_{\text{mor}}(\mathbf{G}_a, \mathbf{G}_a)$ is a free left module over $k[F]$. The family $\text{cl}(x \cdot F^j y) = \text{cl}(xy^{p^j})$, $j$ a positive integer, forms a basis of a submodule supplementary to $H^2_{\text{mor}}(\mathbf{G}_a, \mathbf{G}_a)_s$ (see Demazure and Gabriel [3, II, §3, No. 4]). So, the analogies are again striking. However, zest is provided by the need to consider parity in the differential algebraic case.

## References


1. P. J. Cassidy, Differential algebraic groups, *Amer. J. Math.* **94** (1972), 891–954.
2. P. J. Cassidy, The differential rational representation algebra on a linear differential algebraic group, *J. Algebra* **37** (1975), 223–238.
3. M. Demazure and P. Gabriel, "Groupes algebriques," Vol. 1. Masson, Paris, 1970.
4. G. Hochschild, Basic constructions in group extension theory, this volume.
5. G. Hochschild and J.-P. Serre, Cohomology of group extensions, *Trans. Amer. Math. Soc.* **74** (1953), 110–134.
6. T. Kambayashi, M. Miyanishi, and M. Takeuchi, "Unipotent Algebraic Groups" (Lect. Notes in Math. **414**), Springer-Verlag, New York, 1974.
7. E. R. Kolchin, "Differential Algebra and Algebraic Groups." Academic Press, New York, 1973.
8. E. R. Kolchin, "Differential Algebraic Groups" (in preparation).
9. J. Kovacic, Constrained cohomology, this volume.
10. J. Kovacic, The inverse problem in the Galois theory of differential fields, *Ann. of Math.* **89** (1969), 583–608.
11. M. Lazard, Sur les groupes de Lie formels à un paramètre, *Bull. Soc. Math. France* **83** (1955), 251–274.
12. S. Mac Lane, "Homology." Academic Press, New York, 1963.
13. J. F. Ritt, Associative differential operations, *Ann. of Math.* **52** (1950), 756–765.
14. M. Rosenlicht, Extensions of vector groups by Abelian varieties, *Amer. J. Math.* **80** (1958), 685–714.
15. J.-P. Serre, "Groupes algebriques et corps de classes." Hermann, Paris, 1959.
16. J. Tits, "Lectures on Algebraic Groups," mimeographed Lecture Notes, Yale Univ. Math. Dept., New Haven, 1967.
17. B. L. van der Waerden, "Moderne Algebra," Vol. II. Springer-Verlag, Berlin, 1931.


AMS (MOS) 1970 subject classification: 12H05

# *Solutions in the general solution*

*Richard M. Cohn*

*Rutgers University*

## 1. Introduction

**1.** In the same remarkable paper [8] in which he presented the low power theorem, Ritt gave criteria for solving the following problem: determine whether a given solution of a second order differential polynomial is in its general solution. Important parts of his work use analytic methods. The work of Morrison [7] opens up the possibility of approaching that problem and its generalizations algebraically. In this paper I initiate this study. A related matter was treated in [2].

**2.** Ritt's treatment of the problem made use of what he called $y$-solutions for a differential polynomial $A$. These are power series in $y$ for its derivative which annul $A$ formally. (Ritt required the coefficients to be analytic with a common point of analyticity, but this is not relevant to the considerations which follow.) The existence of a $y$-solution is equivalent to the existence of a differential field containing a solution and admitting a discrete rank 1 valuation whose valuation domain and valuation ideal are closed under the derivation. Such a valuation is also a differential valuation as defined by Blum [1]. It is an immediate consequence of Morrison's results that a necessary and sufficient condition for a solution of an algebraically irreducible differential polynomial of any order to be in its general solution can be stated in terms of the existence of a chain of rank 1 valuations which are differential valuations. For the second order case this condition is easily recast in the form stated by Ritt [8, Sec. 67], except that the discrete valuation of rank 1 (equivalently, the $y$-solution) of Ritt's condition is replaced by an arbitrary valuation of rank 1. (See the criteria of Section 8.) This is a significant change and must be regarded as a weakening since necessity is the important feature of these criteria, sufficiency being obvious. In particular, it seems unlikely that one can replace the algorithm which Ritt found by an algorithm based directly on the existence of an arbitrary rank 1 valuation. The advantage claimed for the present approach is its purely algebraic character and consequent simplicity and generality.

The main result of this paper is the proposition of Section 11. It provides a test for determining whether a rank 1 valuation is a differential valuation.

This proposition is also related to a theorem of Morrison as explained in Section 11. By its use the criteria of Section 8 are replaced by weaker criteria in Section 12.

**3.** Throughout this paper all rings are assumed to contain the identity. Rings and fields are denoted by italic letters, differential rings and fields by script letters. The derivation in an ordinary ring or field is denoted by $\partial$; however, we frequently use subscripts to indicate derivatives.

## 2. Differentially closed places

**4.** Let $\mathscr{K}$ be a differential field of characteristic 0. A *differentially closed place* of $\mathscr{K}$ is an equicharacteristic place $\phi$ of $\mathscr{K}$ such that the place ring $R$ and valuation ideal $I$ are closed under the derivations of $\mathscr{K}$. It is worth noting that $R$ must be closed if $I$ is closed. For suppose there exist $x \in R$, $\partial \in \Delta$ such that $\partial x \notin R$. Let $y = 1/\partial x$, $z = x/\partial x$. Then $y$, $z \in I$, and therefore $\partial y$, $\partial z \in I$, since $I$ is closed. This leads to a contradiction, since $1 = \partial z - x\,\partial y$.

Blum [1] defines a *differential place* of a differential field $\mathscr{K}$ of characteristic 0 to be a differential homomorphism $\phi$ of a differential subring $\mathscr{R}$ of $\mathscr{K}$ such that $\mathscr{K}$ is the quotient field of $\mathscr{R}$ and $\phi$ cannot be extended to a differential subring of $\mathscr{K}$ properly containing $\mathscr{R}$. A differential place is not necessarily a place, as shown by Morrison [7]. In connection with a differential place Blum defines a mapping of $\mathscr{K} - \{0\}$ onto a partially ordered group, which he calls a differential valuation. It is a valuation if and only if the differential place is a place. These definitions provide the alternate terminology "differential place which is a place" for "differentially closed place" and "differential valuation which is a valuation" for "valuation at a differentially closed place."

Another related but distinct concept is that of *valued differential field* introduced by Kolchin [3] to study approximation problems. A valued differential field is a differential field $\mathscr{K}$ and a valuation of $\mathscr{K}$ related to the derivations of $\mathscr{K}$ by requirements similar to but not equivalent to the property of valuations at differentially closed places stated in Theorem 1. The place determined by the valuation of a valued differential field may fail to be differentially closed.

We shall assume hereafter that all differential fields considered are ordinary and have characteristic 0. Several results extend in an obvious way to partial differential fields.

**5.** In this section a basic property of valuations at differentially closed places is demonstrated. The essential ideas are to be found in Morrison [7], but we shall need the specific formulation given here.

***Definition*** Let $F$ be a field, $T$ a subfield of $F$, and $\partial: T \to F$ a derivation. Let $v$ be a valuation of $F$. Let $S \subseteq T$. Then $\partial$ is *nearly monotone* on $S$ with respect to $v$ if for each positive integer $n$ and each $x \in S$, $(n+1)v(\partial x) \geqq nv(x)$. $\partial$ is *monotone* on $S$ with respect to $v$ if for each $x \in S$, $v(\partial x) \geqq v(x)$.

*Remark* It is easily verified that if $\partial$ is nearly monotone on $S$, $x \neq 0$, and $x \in I \cap S$, where $I$ denotes the valuation ideal of $v$, then $(n+1)v(\partial x) > nv(x)$, since equality would imply $(n+2)v(\partial x) < (n+1)v(x)$.

*Remark* If $\partial$ is nearly monotone on $S$ with respect to $v$ and $v$ is of rank 1, then $\partial$ is monotone on $S$ with respect to $v$.

*Remark* If $\partial$ is monotone on $S$, then $\partial$ is monotone on quotients of elements of $S$, since $\partial(a/b)/(a/b) = \partial a/a - \partial b/b$.

***Theorem 1*** *Let $v$ be a valuation of a differential field $\mathscr{K}$ with derivation $\partial$. Let $\mathscr{K}$ and its residue field have characteristic* 0. *Then $v$ is a valuation at a differentially closed place if and only if $\partial$ is nearly monotone on the valuation domain $R$ of $v$.*

*Proof* Sufficiency is evident. To prove necessity, suppose the condition fails, and let $n$ be the minimal positive integer such that for some $y \in R$, $(n+1)v(y_1) < nv(y)$. Letting $I$ denote the valuation ideal of $v$, this relation may be rewritten

$$y^n = zy_1^{n+1}, \qquad z \in I. \tag{1}$$

Differentiation yields

$$ny^{n-1}y_1 = z_1 y_1^{n+1} + (n+1)zy_1{}^n y_2. \tag{2}$$

Now by minimality of $n$ if $n > 1$, and trivially if $n = 1$, $v(y_1{}^n) \geqq v(y^{n-1})$.

Since $z \in I$, $z_1 \in I$ and therefore $v(z_1) > 0$. This shows that the valuation of the left hand side of (2) is less than that of the first term on the right hand side. Therefore

$$v(y^{n-1}y_1) = v(zy_1{}^n y_2). \tag{3}$$

Taking $n$th powers of the parenthetical expressions in (3), substituting from (1), and reducing, one obtains

$$v(y_1^{n-1}) = v(zy_2{}^n). \tag{4}$$

This contradicts the minimality of $n$ if $n > 1$, and contradicts $zy_2 \in I$ if $n = 1$.

*Remark* Using the Domination Lemma [4, p. 181] one can prove Theorem 1 at once from (1); but Theorem 1 is a much easier result than the Domination Lemma.

An immediate corollary is the following result due to Morrison [7].

***Corollary*** *Let $v$ be a valuation at a differentially closed place of the differential field $\mathscr{K}$, and let $\mathscr{R}$ be the valuation domain of $v$. Then every prime ideal of $\mathscr{R}$ is a differential ideal.*

**6.** As observed by Morrison [7, Sec. 4], not every differential place is a place. It is equivalent to state that not every differential homomorphism of a differential integral domain $\mathscr{D}$ onto a differential integral domain extends to a differential homomorphism of a place ring of the quotient field of $\mathscr{D}$. It is nevertheless an important consequence of her work that in a large class of cases differential specializations are closely related to differential places which are places. For our purpose we reformulate part of this result [7, Prop. 3] as follows.

***Theorem 2*** *Let $k$ be a differential field of characteristic* 0, *and let $\mathscr{D} = k\{x\}$ be a differential integral domain, where $x$ denotes a finite family of elements each differentially algebraic over $k$. Let $\bar{x}$ denote a differential $k$-specialization of $x$. There exist families $x^{(0)} = x$, $x^{(1)}, \ldots, x^{(h)} = \bar{x}$ in extensions of $k$ and differentially closed places $\phi^{(i)}$ of $k\langle x^{(i)}\rangle/k$, $i = 0, \ldots, h-1$, such that $\phi^{(i)}x^{(i)} = x^{(i+1)}$ and the valuation $v_i$ associated with $\phi^{(i)}$ has* rank 1.

*Proof* Let $d = \text{t.d. } k\langle x\rangle/k - \text{t.d. } k\langle \bar{x}\rangle/k$, where t.d. denotes transcendence degree. Then $d$ is finite. If $d = 0$, the conclusion is trivial. Let $d \geqq 1$. By [7, Prop. 3], there exists a differential $k$-specialization $x^*$ of $x$ and a differential place $\phi^*$ of $k\langle x^*\rangle/k$ such that $\phi^*$ is a place with rank 1 valuation and $\phi^* x^* = \bar{x}$.

Of course, t.d. $k\langle x^*\rangle/k \geqq 1$. Therefore, if $d = 1$, the specialization from $x$ to $x^*$ is generic, and we may put $x = x^*$, $h = 1$, $\phi^{(0)} = \phi^*$. This completes the proof in that case.

Let $d > 1$. If the specialization from $x$ to $x^*$ is generic, the proof is completed as above. If not, then since $1 \leqq$ t.d. $\mathscr{k}\langle x\rangle\mathscr{k}$ − t.d. $\mathscr{k}\langle x^*\rangle/\mathscr{k} < d$, we may assume by induction that there is a chain of differential places as in the theorem connecting $x$ and $x^*$. These differential places and $\phi^*$ form the chain connecting $x$ and $\bar{x}$.

**7.** An examination of the work of Blum [1] and Morrison [7] shows that Theorem 2 depends ultimately on Levi's study [6] of the differential ideal $[y^p]$ which led to his algebraic proof of the sufficiency of the low power condition. It is therefore of interest to note that sufficiency of the low power condition can easily be recovered from Theorem 2 and the properties of valuations at differentially closed places. Considering the simplest case, let $A = Y^d + H$ be a differential polynomial with $d > 0$ and all terms of $H$ of degree exceeding $d$. Let $y \neq 0$ be a solution of $A$. Let $\phi$ be a place of $\mathscr{k}\langle y\rangle$ such that $\phi$ is the identity on $\mathscr{k}$ and $\phi y_i = 0$, $i = 0, 1, \ldots$ . It follows at once from the form of $A$ that the derivation of $\mathscr{k}\langle y\rangle$ is not nearly monotone on $\mathscr{k}\{y\}$ with respect to the valuation associated with $\phi$. By Theorem 1, $\phi$ is not a differentially closed place, and therefore by Theorem 2 (or directly by [7, Prop. 3]) there is no solution of $A$ different from 0 which specializes differentially to 0. Therefore the manifold of $Y$ is a singular component of the manifold of $A$.

## 3. Preliminary criteria

**8.** Let $\mathscr{k}$ be an ordinary differential field of characteristic 0 and $Y$ a differential indeterminate over $\mathscr{k}$. Let $A$ be an algebraically irreducible element of $\mathscr{k}\{Y\} - \mathscr{k}$ and $y$ a generic zero of the general solution of $A$. Let $\bar{y}$ be an element of an extension of $\mathscr{k}$. The following criterion is an immediate consequence of Theorem 2.

*The necessary and sufficient condition that $\bar{y}$ be in the general solution of $A$ is that there exist elements $y^{(0)} = y, y^{(1)}, \ldots, y^{(h)} = \bar{y}$ lying in extensions of $\mathscr{k}$ and differential places $\phi^{(i)}$ of $\mathscr{k}\langle y^{(i)}\rangle$, $i = 0, 1, \ldots, h-1$, such that each $\phi^{(i)}$ is a place with associated valuation of* rank 1 *and $\phi^{(i)}y^{(i)} = y^{(i+1)}$, and $\phi^{(i)}$ is the identity on $\mathscr{k}$.*

Let t.d. $\mathscr{k}\langle\bar{y}\rangle/\mathscr{k} = n$, and suppose $A$ is of order $n + 2$. The preceding criterion may be reformulated so as to resemble more closely the result due to Ritt mentioned in the Introduction. Let $B^{(1)}, \ldots, B^{(t)}$ be the algebraically irreducible differential polynomials of order $n + 1$ whose general solutions are the singular components of order $n + 1$ of $A$ which contain $\bar{y}$. Let $B = B^{(1)} \cdots B^{(t)}$.

*In order that $\bar{y}$ be in the general solution of $A$, it is necessary and sufficient that there exist a solution $y^*$ of $A$ not annuling $B$ and a differential place $\phi$ of $\mathscr{k}\langle y^*\rangle$ such that $\phi$ is a place with valuation of* rank 1, *$\phi$ is the identity on $\mathscr{k}$, and $\phi y^* = \bar{y}$. (Equivalently: It is necessary and sufficient that there exist a* rank 1 *valuation $v$ of $\mathscr{k}\langle y^*\rangle$ such that the derivation of $\mathscr{k}\langle y^*\rangle$ is monotone with respect to $v$, $\mathscr{k}$ and $y^*$ are in the valuation domain of $v$, and for an appropriate choice $\phi$ of the place map associated with $v$, $\phi y^* = \bar{y}$.)*

*Proof Sufficiency* Suppose the solution $y^*$ exists. Since t.d. $\mathscr{k}\langle y^*\rangle/\mathscr{k} > n$ and $y^*$ annuls no $B^{(i)}$, $y^*$ is not in any singular component of the manifold of $A$. Hence $y^*$ is in the general solution, and so is its specialization $\bar{y}$.

*Necessity* Using Morrison [7, Prop. 3], choose $y^*$ exactly as $x^*$ was chosen in the proof of Theorem 2. Then $\phi$ exists as described. Also t.d. $\mathscr{k}\langle y^*\rangle/\mathscr{k} > n$. A solution common to the general solution of $A$ and a singular component of the manifold of $A$ is of order at most $n$. Hence, $y^*$ is not in any singular component of $A$ and so is not in the general solution of any $B^{(i)}$. A singular solution of any $B^{(i)}$ is of order at most $n$, so that $y^*$ is not a singular solution of any $B^{(i)}$. Hence, $y^*$ is not a solution of $B$.

## 4. Algebraic extensions

**9.** We now work toward a proof of the proposition of Section 11 whose purpose is to facilitate determining whether a valuation of finite rank is a differential valuation. This proposition is closely related to the result obtained by Morrison on the extension of a differential valuation from a differential field $\mathscr{k}$ to a differential field algebraic over $\mathscr{k}$ [7, Th. 3 and Prop. 10]. In fact, these two results are different generalizations of the special case of the proposition of Section 11 in which $K$ is itself a differential field. It is important for our purpose, however, that the proposition of Section 11 does not require $K$ to be a differential field.

**10.** ***Lemma*** *Let $H$ be a field, $\phi$ a place of $H$ with* rank 1 *valuation $v$ and residue field $F$. Let $H$ be complete under $v$. Let $F$ (therefore also $H$) be of characteristic* 0. *Then $H$ contains a subfield $G$ which is mapped isomorphically onto $F$ by $\phi$ (field of representatives). $G$ is algebraically closed in $H$.*

*Proof* The place ring $R$ of $v$ contains fields, e.g., the rationals. By Zorn's lemma there is a maximal field $G$ in $R$. It will be shown that $G$ has the properties required by the lemma.

Of course $\phi$ maps $G$ isomorphically onto a subfield $F'$ of $F$. Let $y \in F$. It must be shown that $y \in F'$.

Observe first that $y$ is algebraic over $F'$. For suppose it is not. Let $x \in R$ be such that $y = \phi x$. Then $x$ is transcendental over $G$, and $\phi$ is finite on $G(x)$. Therefore $G \neq G(x) \subseteq R$, which contradicts the maximality of $G$.

Let the minimal polynomial for $y$ over $F'$ be $Q(Y) = Y^n + \phi(a_1)Y^{n-1} + \cdots + \phi(a_n)$, the $a_i$ in $G$, $Y$ an indeterminate over $F$. Then $Q(Y) = (Y - y)B$, where $B$ has coefficients in $F$.

Let $P(X) = X^n + a_1X^{n-1} + \cdots + a_n$, $X$ an indeterminate over $G$. Since $Y - y$ and $B$ have no common factors, it follows from Hensel's reducibility criterion that $P(X) = (X - t)A$, where $t$ and the coefficients of $A$ are in $R$, and $\phi t = y$. Since $t$ is algebraic over $G$, $G(t) = G[t]$. Now $t \in R$, since $\phi t \in F$; and therefore $G(t) \subseteq R$. By maximality, $t \in G$. Therefore $y \in F'$.

It remains only to prove that $G$ is algebraically closed in $H$. Let $x \in H$ be algebraic over $G$. Replacing $x$ by $1/x$ if necessary we may assume $x \in R$. Let $A(X) = X^m + b_1X^{m-1} + \cdots + b_m$ be the minimal polynomial for $x$ over $G$. Then $y = \phi x$ is a solution of $B(Y) = Y^m + \phi(b_1)Y^{m-1} + \cdots + \phi(b_m)$, which has coefficients in $F' = F$. Since $B(Y)$ has the linear factor $Y - y$ and $\phi$ is an isomorphism of $G$ onto $F$, $A(X)$ has a linear factor. Since $A(X)$ is irreducible, it is linear and $x \in G$.

Note that if $H_0$ is a subfield of $H$ and $G_0$ a subfield of $H_0$ mapped isomorphically by $\phi$ onto the residue field of $H_0$, then one may choose $G$ so that $G_0 \subseteq G$. This is so since a maximal field in $R \cap G_0$ extends to a maximal field in $R$.

**11.** ***Proposition*** *Let $L/K$ be a field extension with $L : K < \infty$. Let $\phi$ be a place of $L$ with valuation $v$ of finite rank and such that the residue field (and therefore also $L$) is of characteristic 0. Let $R$ be the place ring of $\phi$. Let $\partial : L \to L$ be a derivation which is nearly monotone on $R \cap K$ with respect to $v$. Then $\partial$ is nearly monotone on $R$.*

*Proof* We consider first the case that $v$ has rank 1. In this case $\partial$ is monotone on $K$ by the second and third remarks preceding Theorem 1.

Without loss of generality choose $L$ normal over $K$. It will be shown that we may assume $K$ and $L$ complete under $v$.

Let $\bar{L}$ be the completion of $L$. We use $\bar{v}$ to denote the extension of $v$ to $\bar{L}$. Limits of those Cauchy sequences which consist only of elements of $K$ constitute a subfield $\bar{K}$ of $\bar{L}$, and $\bar{K}$ is the completion of $K$ under the contraction of $v$ to $K$. Adjoining to $\bar{K}$ a set of generators of $L$ over $K$ generates a field $M$ complete under $\bar{v}$ and containing $L$. Then $M = \bar{L}$; and so $\bar{L} : \bar{K} < \infty$, and $\bar{L}$ is normal over $\bar{K}$. It will be shown that $\partial$ extends to a derivation $\bar{\partial} : \bar{L} \to \bar{L}$, which is monotone on $\bar{K}$ with respect to $\bar{v}$.

Let $x \in \bar{K}$. Let $\{x_i'\}$, $\{x_i''\}$ be Cauchy sequences in $K$ which converge to $x$. Let $y_i = x_i'$ if $i$ is odd, $y_i = x_i''$ if $i$ is even. Then $\{y_i\}$ is also a Cauchy sequence converging to $x$. The monotonicity of $\partial$ implies that $\{\partial x_i'\}$, $\{\partial x_i''\}$, and $\{\partial y_i\}$ are Cauchy sequences of elements of $L$. Evidently these sequences have the same limits. It has thus been shown that if $\{x_i\}$ is any Cauchy sequence of elements of $K$ which converges to $x$, then $\{\partial x_i\}$ is a Cauchy sequence whose limit depends only on $x$, not on the choice of the $x_i$. Let $\partial^* x$ denote this limit. Consideration of constant sequences shows that $\partial^* x = \partial x$, $x \in K$. It is routine to show that $\partial^*$ is a derivation of $\bar{K}$ into $\bar{L}$ and is monotone on $\bar{K}$ with respect to $\bar{v}$.

Because $\bar{L}$ is algebraic over $\bar{K}$ there is a unique derivation $\bar{\partial}: \bar{L} \to \bar{L}$ which extends $\partial^*$. Since the restriction of $\partial^*$ to $K$ has no extension except $\partial$ to a derivation of $L$, it must be that $\bar{\partial}$ is an extension of $\partial$. To prove the proposition when $v$ is of rank 1, it suffices to show that $\bar{\partial}$ is monotone on $\bar{L}$. This completes the proof of the claim that $K$ and $L$ may be assumed to be complete under $v$, and we make this assumption henceforth.

It follows from the assumption of completeness that there is just one extension to $L$ of the contraction of $v$ to $K$. This implies that the decomposition field of $v$ is $K$ itself. Let $K_T$ denote the inertia field of $v$ and let $F^*$, $F_T{}^*$ denote the residue field of $K$, $K_T$, respectively. By the lemma of Section 10 there exist fields $F \subseteq K$, $F_T \subseteq K_T$, such that $\phi$ is an isomorphism on $F$ and $F_T$, and $\phi F = F^*$, $\phi F_T = F_T{}^*$. By the final observation of Section 10 we may choose $F \subseteq F_T$. The following are then true:

(a) $K_T = K(F_T)$, and $F_T$ and $K$ are linearly disjoint over $F$. Indeed, it follows from [10, p. 71, Eq. 5] and from the fact that $K$ is the decomposition field of $v$ that $K_T : K = F_T : F$. Let $u_1, \ldots, u_r$ be a linear basis for $F_T$ over $F$. To prove (a) it is only necessary to show that the $u_i$ are linearly independent over $K$. If not, let $\sum k_i u_i = 0$ be an equation of linear dependence over $K$. We may assume $v(k_i) \geqq 0$ for all $i$ and $k_j = 1$ for some $j$. Applying $\phi$ yields $\sum \phi(k_i)\phi(u_i) = 0$, which is a contradiction since the $\phi(k_i)$ are in $F^*$ and are not all 0.

(b) The Galois group of $L/K_T$ is abelian. This follows from [10, p. 78, Th. 25] and the fact that the large ramification group $G_v$ is (1) when the residue field has characteristic 0 [10, p. 77, Th. 24].

(c) $F_T{}^*$ is the residue field of $L$. This is so since the residue field of $L$ is always purely inseparable over the residue field of $K_T$ [10, p. 70, Th. 22].

It will first be shown that $\partial$ is monotone on $K_T$. Let $K_T = K(x)$, $x \in F_T$. This is possible by (a) and characteristic 0. Let $X^n + a_1 X^{n-1} + \cdots + a_n$ be a minimal polynomial for $x$ over $K$. It follows from (a) that the $a_i$ are in $F$, so that $v(a_i) = 0$ for each nonzero $a_i$. In particular, $v(a_n) = 0$. Let $x_1 = x, x_2, \ldots, x_n$ be the conjugates of $x$ over $F$. Since $F_T$ is normal over

$F$ [10, p. 69, Th. 21], the $x_i$ are in $K_T$. Because $K$ is complete, $nv(x_i) = v(a_n) = 0$, so $v(x_i) = 0$, $1 \leqq i \leqq n$. We suppose that there is some $i$ for which $v(\partial x_i) < 0$ and obtain a contradiction.

Since the $x_i$ are the solutions of a polynomial with coefficients in $F$ there are equations

$$\sum_{i=1}^{n} x_i^{\,j} = b_j, \qquad b_j \in F; \qquad 1 \leqq j \leqq n. \tag{1}$$

Differentiation yields

$$\sum_{i=1}^{n} x_i^{j-1}\, \partial x_i = \partial b_j/j; \qquad 1 \leqq j \leqq n. \tag{2}$$

Since $F$ is contained in the valuation domain $R$ of $\phi$ and $\partial$ is monotone on $K$, each $\partial b_j \in R$. Choose $i$ so that $v(\partial x_i)$ is minimal. Say $i = 1$. By assumption $v(\partial x_1) < 0$. Now $\phi$ is finite on the $x_i$ and on each $\partial x_i/\partial x_1$, and $\phi(\partial b_j/\partial x_1)$ is 0, $1 \leqq j \leqq n$. Dividing each equation of (2) by $\partial x_1$ and applying $\phi$, we therefore obtain

$$\sum_{i=1}^{n} \phi(x_i^{j-1})\phi(\partial x_i/\partial x_1) = 0, \qquad 1 \leqq j \leqq n. \tag{3}$$

The $\phi x_i$ are all distinct since $\phi$ acts isomorphically on $F_T$. Therefore (3) is a set of homogeneous linear relations satisfied by the $\phi(\partial x_i/\partial x_1)$ with determinant a Vandermonde determinant and therefore different from 0. Therefore, $\phi(\partial x_i/\partial x_1) = 0$, $1 \leqq i \leqq n$. For $i = 1$, this yields a contradiction. Therefore $v(\partial x) \geqq 0$.

Let $y \in K_T$. Then $y = \sum_{i=0}^{n-1} h_i x^i$, the $h_i \in K$. It must be shown that $v(\partial y) \geqq v(y)$. Let $j$ be such that $v(h_j) \leqq v(h_i)$, $0 \leqq i \leqq n-1$. Since $v(\partial s) \geqq v(s)$, $v(\partial t) \geqq v(t)$ implies $v(\partial(st)) \geqq v(st)$, we may replace $y$ by $y/h_j$, or equivalently assume $h_j = 1$, $v(h_i) \geqq 0$, $0 \leqq i \leqq n-1$. Computing $\partial y$ and using $v(x) = 0$, $v(\partial x) \geqq 0$, it is evident that $v(\partial y) \geqq 0$. It remains to prove $v(y) = 0$.

Now $\phi y = \sum_{i=0}^{n-1} \phi h_i(\phi x)^i$, and $\phi h_j = 1$. Since $\phi x$ satisfies no equation of degree less than $n$ over $F^*$, it follows that $\phi y$ is finite and not 0. Hence, $v(y) = 0$. This completes the proof that $\partial$ is monotone on $K_T$.

It follows from (b) that $L$ may be obtained from $K_T$ by a succession of cyclic extensions. Hence, to complete the proof of the proposition it suffices to consider fields $K_1$, $K_2$ with $K_T \subseteq K_1 \subsetneqq K_2 \subseteq L$, such that there is no extension intermediate to $K_1$ and $K_2$, and $K_2 = K_1(x)$, $x^p \in K_1$, $p$ a prime. It must be shown that if $\partial$ is monotone on $K_1$, then $\partial$ is monotone on $K_2$.

We may assume $v(x) \geqq 0$. It will be shown that there exists no $a \in K_1$ with $v(x) = v(a)$. Supposing the contrary and replacing $x$ by $x/a$ if necessary, we may assume $v(x) = 0$. Let $x^p = b \in K_1$. Then $(\phi x)^p - \phi b = 0$. By (c), $\phi x \in F_T{}^*$, $\phi b \in F_T{}^*$. Therefore the polynomial $P(Y) = Y^p - \phi b$ of $F_T{}^*[Y]$ has the factor $Y - \phi x$. Of course $P(Y)$ has no repeated factor. Since $K_1$ is complete under $v$ because it is algebraic over $K$, Hensel's reducibility criterion is applicable and shows that $X^p - b$ is reducible in $K_1[X]$ contradicting the fact that it is the minimal polynomial for $x$ over $K_1$. It follows at once that for each $i$, $1 \leqq i \leqq p$, there exists no $a \in K_1$ such that $v(x^i) = v(a)$; for $K_1(x^i) = K_2$.

Let $y \in K_2$, $y \neq 0$. Then $y = \sum_{i=0}^{p-1} a_i x^i$, $a_i \in K_1$, $0 \leqq i < p$, and some $a_i \neq 0$, By the result of the preceding paragraph, no two nonzero terms of this sum have equal valuations. Let $a_j x^j$ be the term of the sum with minimum valuation. Then $v(y) = v(a_j x^j)$. Of course, $v(\partial a_k) \geqq v(a_k)$, $1 \leqq k < p$; and it follows at once from $x^p = b$, $v(\partial b) \geqq v(b)$, that $v(\partial x) \geqq v(x)$. For any $k$, $1 \leqq k \leqq p$,

$$v(\partial(a_k x^k)) = v(ka_k x^{k-1}\,\partial x + x^k\,\partial a_k) \geqq v(a_k x^k) \geqq v(a_j x^j).$$

This implies $v(\partial y) \geqq v(y)$. This completes the proof for $v$ of rank 1.

In the general case we proceed by induction on the rank of $v$. We may write $\phi = \psi\theta$, where $\theta$ is a place of $L$ with valuation $v_\theta$ of rank 1 and $\psi$ is a place of the residue field $\bar{L}$ of $\theta$. Let $v_\alpha$ be the valuation, $R_\alpha$ the place ring, and $I_\alpha$ the valuation ideal of $\alpha$, $\alpha = \psi, \theta, \phi$.

Let $x \in I_\theta \cap K$, $x \neq 0$. Since $I_\theta \subseteq I_\phi$, it follows from the hypothesis and the first remark preceding Theorem 1 that if $p$ is any positive integer, then $(\partial x)^p/x^{p-1} \in I_\phi$. Since also $I_\phi \subseteq R_\theta$ this shows that $\partial$ is nearly monotone with respect to $v_\theta$ on $I_\theta \cap K$. Since $v_\theta$ has rank 1 we know by the second remark preceding Theorem 1 that $\partial$ is monotone with respect to $v_\theta$ on $I_\theta \cap K$.

There exists b $\in I_\theta \cap K, b \neq 0$, for otherwise $\theta$ would act as an isomorphism on $K$, hence on $L$. Let $c \in R_\theta \cap K$. Then $c = bc/b$ is a quotient of elements of $I_\theta \cap K$. By the third remark preceding Theorem 1, $\partial$ is monotone on $R_\theta \cap K$. From the case of rank 1, $\theta$ is a differential place of $L$, and $\bar{L}$ is a differential field with derivation $\bar{\partial}$ induced by $\partial$.

Let $\bar{K} = \theta(R_\theta \cap K)$. Then $\bar{K} \subseteq \bar{L}$, and $\bar{K}$ is a field. It will be shown that $\bar{L} : \bar{K}$ is finite. Let $t > L : K$. Let $\bar{u}_1, \ldots, \bar{u}_t$ be elements of $\bar{L}$. Choose $u_1, \ldots, u_t$ in $L$ such that $\theta u_i = \bar{u}_i$. There exist $c_1, \ldots, c_t$ in $K$ and not all 0 such that $\sum c_i u_i = 0$. The $c_i$ may be chosen so that $v_\theta c_i \geqq 0$, $1 \leqq i \leqq t$, and $v_\theta c_j = 0$ for some $j$. Then applying $\theta$ to the preceding sum one obtains an equation of linear dependence of the $\bar{u}_i$ over $\bar{K}$.

Let $\bar{u} \in R_\psi \cap \bar{K}$, $\bar{u} \neq 0$. Then $\bar{u} = \theta u$, $u \in R_\theta \cap K$. Since $\psi\bar{u} = \phi u$ is finite, $u \in R_\phi \cap K$. From the hypothesis it follows that if $p$ is any positive integer, then $(\partial u)^p/u^{p-1} \in R_\phi$. Therefore

$$\psi(\bar{\partial}\bar{u})^p/\bar{u}^{p-1}) = \psi\theta((\partial u)^p/u^{p-1}) = \phi((\partial u)^p/u^{p-1})$$

is finite, and so $(\bar{\partial}\bar{u})^p/\bar{u}^{p-1} \in R_\psi$. This shows that $\psi$ is nearly monotone on $R_\psi \cap \bar{K}$. Since rank $\psi <$ rank $\phi$, it follows from the induction hypothesis that $\psi$ is nearly monotone on $R_\psi$. Hence, $\psi$ is a differential place of $\bar{L}$.

Let $x \in I_\phi$. Then $\psi\theta x = 0$, so that $\theta x \in I_\psi$. Since $\psi$ is a differential place, $\partial\theta x = \theta\, \partial x \in I_\psi$. It follows that $\phi\, \partial x = \psi\theta\, \partial x = 0$, so that $\partial x \in I_\phi$. This shows that $I_\phi$ is closed under $\partial$. As noted in the first paragraph of Section 4, this implies that $\phi$ is a differential place. (Equivalently, $\partial$ is nearly monotone on $R_\phi$ with respect to $v_\phi = v$.)

## 5. Final criteria

**12.** Let $\mathscr{k}$ be an ordinary differential field of characteristic 0, and let $Y$ be a differential indeterminate. Let $A$ be an algebraically irreducible differential polynomial of order $n > 0$ in $\mathscr{k}\{Y\}$. Let $y$ be a generic zero of the general solution of $A$ and $\bar{y}$ an element of a differential extension field of $k$. We shall use $a$ to denote $(y, y_1, \ldots, y_n)$, $\bar{a}$ to denote $(\bar{y}, \bar{y}_1, \ldots, \bar{y}_n)$, and $a_0$ to denote $(y, y_1, \ldots, y_{n-1})$.

***Theorem 3*** (a) *In order that $\bar{y}$ be in the general solution of $A$, it is necessary and sufficient that there exist specializations (not assumed to be differential specializations) $a = a^{(0)} \to a^{(1)} \to \cdots \to a^{(h)} = \bar{a}$ over $\mathscr{k}$ and places $\phi^{(i)}$ of $\mathscr{k}(a^{(i)})$, $0 \leqq i < h$, such that $\phi^{(i)}$ is the identity on $\mathscr{k}$, $\phi^{(i)}a^{(i)} = a^{(i+1)}$, the valuation $v^{(i)}$ associated with $\phi^{(i)}$ has* rank 1, *and for each $x \in \mathscr{k}[a_0]$, $v^{(i)}(\phi^{(i)} \cdots \phi^{(0)}\, \partial x) \geqq v^{(i)}(\phi^{(i)} \cdots \phi^{(0)} x)$.*

(b) *In order that $\bar{y}$ be in the general solution of $A$, it is sufficient that there exist a place $\phi$ of $\mathscr{k}(a)$ with valuation $v$ and valuation domain $R$ such that $\phi$ is the identity on $\mathscr{k}$, $\phi a = \bar{a}$, and $\partial$ is nearly monotone with respect to $v$ on $R \cap \mathscr{k}(a_0)$.*

*Proof* The necessity of (a) is evident from the first criterion of Section 8. To prove sufficiency of (a), note that it follows at once from the proposition of Section 11 that $\phi^{(0)}$ is a differential place of $\mathscr{k}\langle a \rangle$. Therefore $\partial$ induces a derivation on $\mathscr{k}(a^{(1)})$ and under the induced derivation $a^{(1)}$ is a differential specialization of $a$ over $\mathscr{k}$. This completes the proof if $h = 1$. If $h > 1$, induction on $h$ shows that $a^{(h)}$ with the appropriate derivation is a differential specialization of $a^{(1)}$ over $\mathscr{k}$, and therefore of $a$ over $\mathscr{k}$.

The sufficiency of (b) is an immediate consequence of the proposition of Section 11.

## 6. Several indeterminates

**13.** The methods discussed in this paper do not extend in any obvious way to the problem of determining which solutions of a differential polynomial in several indeterminates are in its general solution. Nevertheless,

Ritt [9] solved this problem for the case of a polynomial in two indeterminates and of order at most 1 in each by methods resembling those of [8].

A theorem of Lando [5, Th. 1] on realizations of differential kernels immediately yields as a special case the following results for differential polynomials in several indeterminates:

*Let $\mathscr{k}$ be an ordinary differential field of characteristic* 0, $Y = (Y^{(1)}, \ldots, Y^{(n)})$ *differential indeterminates over $\mathscr{k}$. Let $A \in \mathscr{k}\{Y\}$ be algebraically irreducible and of order $r_i \geqq 1$ in $Y^{(i)}$, $1 \leqq i \leqq n$. Let $y = (y^{(1)}, \ldots, y^{(n)})$ be a solution of $A$ such that the set $\{y_j^{(i)} | 1 \leqq i \leqq n,\ 0 \leqq j < r_i\}$ is algebraically independent over $\mathscr{k}$. Then $y$ is in the general solution of $A$.*

We conclude with the remark that the question of determining which realizations of a differential kernel are specializations of the principal realization is a natural generalization of the problem discussed in this paper.

## References


1. P. Blum, Complete models of differential fields, *Trans. Amer. Math. Soc.* **137** (1969), 309–325.
2. R. M. Cohn, The general solution of a first order differential polynomial, *Proc. Amer. Math. Soc.* **54** (1976).
3. E. R. Kolchin, Rational approximation to solutions of algebraic differential equations, *Proc. Amer. Math. Soc.* **10** (1959), 238–244.
4. E. R. Kolchin, "Differential Algebra and Algebraic Groups." Academic Press, New York, 1973.
5. B. A. Lando, Jacobi's bound for the order of systems of first order differential equations, *Trans. Amer. Math. Soc.* **152** (1970), 119–135.
6. H. Levi, On the structure of differential polynomials and on their theory of ideals, *Trans. Amer. Math. Soc.* **51** (1942), 532–568.
7. S. D. Morrison, Extensions of differential places, *Amer. J. Math.* (to appear).
8. J. F. Ritt, On the singular solutions of algebraic differential equations, *Ann. of Math.* **37** (1936), 552–617.
9. J. F. Ritt, On certain points in the theory of algebraic differential equations, *Amer. J. Math.* **60** (1938), 1–43.
10. O. Zariski and P. Samuel, "Commutative Algebra," Vol. 2. Van Nostrand-Reinhold, Princeton, New Jersey, 1960.


The author is grateful to the Rutgers University Faculty Academic Study Plan for leave which facilitated the preparation of this paper.

AMS (MOS) 1970 subject classification: 12H05

# *Folk theorems on elliptic equations*

*Jean Dieudonné*

*Université de Nice*

To E. Kolchin on his 60th birthday.

**1.** I do not believe that any of the results in this note is unknown to specialists. They belong to the category of "folk theorems" which are almost impossible to locate in the literature; but when you mention them to specialists, they will always realize that they have known them all the time. This attitude should not be judged too critically; these theorems are not deep, and it is very natural that specialists should concentrate their efforts on more important problems. Nevertheless, I think it may interest some nonspecialists to see how the recent progress in the theory of partial differential equations yields very easily properties of elliptic operators which, in classical books, are invariably restricted to operators of the second order.

**2.** My starting point is the Gårding–Višik theorem on self-adjoint elliptic operators on a bounded open set in $\mathbf{R}^n$. Let X be a *bounded* open set in $\mathbf{R}^n$, $P = \sum_\alpha a_\alpha \mathrm{D}^\alpha$ a differential operator of even order $2p$, having the following properties:

(1) $P$ is the restriction to X of an operator $P_1$, written in the same way, where the $a_\alpha$ are real $\mathrm{C}^\infty$ functions in a bounded neighborhood $\mathrm{X}_1$ of $\bar{\mathrm{X}}$.

(2) $P_1$ (hence also $P$) is equal to its formal adjoint (in the sense of differential operators).

(3) $P_1$ is elliptic, meaning that its symbol $\sum_{|\alpha|=2p} a_\alpha(x)\xi^\alpha$ is a polynomial in $\xi$ which does not vanish for $\xi \neq 0$, and we assume in addition that its values are always $\geq 0$.

One may consider $P$ as a hermitian (unbounded) operator in Hilbert space $L^2(X)$, whose domain is the space $\mathscr{D}(X)$ of $C^\infty$ functions with compact support. The Gårding–Višik theorem then asserts the existence of a particular self-adjoint extension $A_P$ of $P$. Its domain $\operatorname{dom}(A_P)$ is the set of functions $f$ in the Sobolev space $H_0{}^p(X)$ such that the distribution $P \cdot f$ is a function of $L^2(X)$, and is the value of $A_P \cdot f$. The spectrum S of $A_P$ is a sequence $(\lambda_k)$ of real numbers tending to $+\infty$, and for any complex number $\zeta \notin S$, $(A_P - \zeta I)^{-1} = G_\zeta$ is a *compact* operator in $L^2(X)$. We are interested in what follows in properties of $G_\zeta$ (usually called the *Green operator* of $P - \zeta I$). Note that the self-adjointness of $A_P$ implies that $G_\zeta{}^* = G_{\bar{\zeta}}$.

**3.** For $\zeta \notin S$, it follows from the definitions and the "interior regularity" of elliptic operators that $G_\zeta$ is a bijection of

$$\mathscr{E}(X) \cap L^2(X) \text{ onto } \mathscr{E}(X) \cap \operatorname{dom}(A_P)$$

($\mathscr{E}(X)$ being the space of $C^\infty$ functions in X). The first remark is that, restricted to $\mathscr{D}(X)$, $G_\zeta$ is a *pseudo-differential operator* which is a *parametrix* of $P - \zeta I$. The continuity of $G_\zeta$ as a mapping of the Hilbert space $L^2(X)$ into itself implies, by Schwartz's kernel theorem, that $G_\zeta$, as a mapping of $\mathscr{D}(X)$ into $\mathscr{E}(X)$, is defined by a kernel distribution $G_\zeta \in \mathscr{D}'(X \times X)$; due to the relation $G_\zeta{}^* = G_{\bar{\zeta}}$, it is easily seen that the distribution $G_\zeta$ is *regular* in the sense of Schwartz. Now, as $P - \zeta I$ is a properly supported elliptic operator, it has a properly supported parametrix $Q_\zeta$; in other words, $(P - \zeta I)Q_\zeta = I - R_\zeta$, where $R_\zeta$ is a smoothing operator, which is properly supported. This implies that $G_\zeta$ considered as a mapping of $\mathscr{D}(X)$ into $\mathscr{E}(X)$ satisfies the relation $G_\zeta = Q_\zeta + G_\zeta R_\zeta$; but by Schwartz's kernel theorem, the composite $G_\zeta R_\zeta$ of a regular map and a smoothing operator, properly supported, is again a smoothing operator, and this shows that $G_\zeta$ is also a parametrix of $P - \zeta I$, but in general it will *not* be properly supported.

Now, an integration by parts in the manner of Hörmander [1] easily proves that any pseudo-differential operator of order $< -1$ is in fact an *integral* operator, with a kernel which is locally integrable and of class $C^\infty$ outside of the diagonal of $X \times X$, but will in general be unbounded in the neighborhood of the diagonal if its order is $\geq -n$. We may therefore write for $u \in \mathscr{D}(X)$

$$(G_\zeta \cdot u)(x) = \int_X G(\zeta, x, y)u(y)\, dy,$$

where, for each $\zeta \notin S$, $(x, y) \mapsto G(\zeta, x, y)$ is a function of the preceding type; G is called the *Green function* of $P$ and one has

$$G(\bar{\zeta}, x, y) = \overline{G(\zeta, y, x)}.$$

**4.** Simple examples show that the iterates $G_\zeta^q$ of the Green operator have no simple relation to the Green operators of the iterates $P^q$ of $P$. Nevertheless, we shall see that we may obtain useful information on $G_\zeta$ by the iteration process. We may assume that 0 is not in the spectrum of $A_P$, by replacing $P$ by $P + bI$ with a convenient $b > 0$; we will then take $\zeta = 0$ and write simply $G$ instead of $G_0$; $G$ is therefore the inverse of $A_P$.

As $G$ applies $\mathrm{L}^2(\mathrm{X})$ into $\mathrm{dom}(A_P)$, the same is true for any $G^q$, hence by induction it follows that $P^q G^q$ is defined in $\mathrm{L}^2(\mathrm{X})$ and is the identity in that space. Take $q$ such that $2pq > n$; then $P_1{}^q$ has a parametrix $Q_1$ which is a pseudo-differential operator, properly supported, of order $-2pq$, hence an integral operator with *continuous* kernel $\mathrm{Q}_1$; one has $Q_1 P_1{}^q = I + K_1$, where $K_1$ is a smoothing operator, properly supported, hence has a $\mathrm{C}^\infty$ kernel $\mathrm{K}_1$. Let Q and K be the restrictions of $\mathrm{Q}_1$ and $\mathrm{K}_1$ to $\mathrm{X} \times \mathrm{X}$, which are therefore *bounded* uniformly continuous functions in $\mathrm{X} \times \mathrm{X}$, and let $Q$ and $K$ be the corresponding integral operators; for any function $f \in \mathscr{D}(\mathrm{X})$, we still have $Q \cdot (P^q \cdot f) = f + K \cdot f$ since $P^q \cdot f \in \mathscr{D}(\mathrm{X})$.

We shall *assume* that this relation still holds for $f \in \mathrm{dom}(A_P)$.

**5.** We may now apply the preceding relation for $f = G^q \cdot u$, where $u$ is any function in $\mathrm{L}^2(\mathrm{X})$, and we get

$$Q \cdot u = G^q \cdot u + K \cdot (G^q \cdot u). \tag{1}$$

Now, as the kernels of $Q$ and $K$ are bounded functions in $\mathrm{X} \times \mathrm{X}$, it is clear that they are continuous maps of $\mathrm{L}^1(\mathrm{X})$ into the Banach space $\mathscr{B}(\mathrm{X})$ of bounded functions, and a fortiori they are continuous maps of $\mathrm{L}^2(\mathrm{X})$ into $\mathscr{B}(\mathrm{X})$. As $G^q$ is a continuous map of $\mathrm{L}^2(\mathrm{X})$ into itself, it follows from (1) that $G^q$ also maps continuously $\mathrm{L}^2(\mathrm{X})$ into $\mathscr{B}(\mathrm{X})$. Relation (1) also shows, as in Section 2, that $G^q$ is a pseudo-differential operator of order $-2pq$, hence is an integral operator with continuous kernel

$$(G^q \cdot f)(x) = \int_{\mathrm{X}} \mathrm{H}(x, y) f(y)\, dy.$$

Now the fact that there is a constant C such that for every $x \in \mathrm{X}$

$$|(G \cdot f)(x)| \leq \mathrm{C}\mathrm{N}_2(f)$$

where $\mathrm{N}_2$ is the Hilbert norm in $\mathrm{L}^2(\mathrm{X})$, implies that for every $x \in \mathrm{X}$, every $x \in \mathrm{X}$, $y \mapsto \mathrm{H}(x, y)$ belongs to $\mathrm{L}^2(\mathrm{X})$ and

$$\int_{\mathrm{X}} |\mathrm{H}(x, y)|^2\, dy \leq \mathrm{C}^2. \tag{2}$$

But $G^q$ being self-adjoint, $H(y, x) = \overline{H(x, y)}$, hence we also have $\int_X |H(x, y)|^2 \, dx \leq C^2$. The operator $G^{2q}$ is then an integral operator with kernel

$$H_1(x, y) = \int_X H(x, z)H(z, y) \, dz,$$

and it follows immediately from the Cauchy–Schwarz inequality that $|H_1(x, y)| \leq C^2$; $G^{2q}$ is thus an integral operator with *bounded continuous kernel.* This also shows of course that $G^q$ is a Hilbert–Schmidt operator, which is well known.

**6.** Let $(\mu_k)$ be the sequence of eigenvalues of $A_P$, each repeated a number of times equal to its multiplicity. For each $\mu_k$, let $u_k$ be an eigenfunction, normalized by $N_2(u_k) = 1$, and such that the $u_k$ form a Hilbert basis of $L^2(X)$. It is clear that $G^q \cdot u_k = \mu_k^{-q} u_k$, hence equation (1) gives the relation

$$u_k = \mu_k{}^q Q \cdot u_k - K \cdot u_k$$

which already proves that the $u_k$ are *bounded* functions in X. But, for any $r$, if we take $q$ such that $2pq > n + r$, the kernels of $Q$ and $K$ are $C^r$ functions which are bounded together with their derivatives of order $\leq r$ in $X \times X$. Therefore the $u_k$ are $C^\infty$ functions such that each of their derivatives is *bounded* in X.

Furthermore, the components of the function $y \mapsto H(y, x)$ with respect to the basis $(u_k)$ of $L^2(X)$ are equal to $\mu_k^{-q} \cdot \overline{u_k(x)}$; we have therefore the Parseval relation

$$\sum_k \frac{1}{\mu_k^{2q}} |u_k(x)|^2 = \int_X |H(y, x)|^2 \, dy \leq C^2$$

and by integration

$$\sum_k \frac{1}{\mu_k^{2q}} = \iint_{X \times X} |H(y, x)|^2 \, dy \, dx \leq C^2 m(X).$$

**7.** Let $f$ be a function of $\mathscr{D}(X)$; then $P^q \cdot f$ belongs to $\mathscr{D}(X)$, and by induction on $q$ we have $G^q \cdot (P^q \cdot f) = f$. If $P^q \cdot f = \sum_k a_k u_k$ in $L^2(X)$, with $\sum_k |a_k|^2 < +\infty$, we have therefore for the development $f = \sum_k c_k u_k$ the relation $c_k = \mu_k^{-q} a_k$, in other words, the series $\sum_k \mu_k^{2q} |c_k|^2$ converges for any integer $q$. Furthermore, as, for $q$ large enough, $G^q$ is a continuous map of $L^2(X)$ into the Banach space $\mathscr{C}_b^{(r)}(X)$ of functions of class $C^r$ which are bounded with their derivatives of order $\leq r$, we see that each of the series $\sum_k c_k D^\nu u_k$ converges uniformly to $D^\nu f$ in X.

**8.** When $2p > n$, we may take $q = 1$. There is then for the Green function a result quite similar to Mercer's theorem. As there is only a finite number of eigenvalues $\mu_k$ which are $< 0$, we may suppose, by considering $P + bI$ instead of $P$, that they are all $> 0$; for any function $f = \sum_k c_k u_k$ we then have

$$(G \cdot f \mid f) = \sum_k \frac{1}{\mu_k} |c_k|^2 \geq 0,$$

equality being only possible for $f = 0$ almost everywhere.

The first step consists in proving that the continuous function $\mathrm{G}(x, x)$ is everywhere $\geq 0$. This is done by contradiction; if $\mathrm{G}(x_0, x_0) < 0$, one may assume that $\mathscr{R}\mathrm{G}(x, y) \leq -\delta < 0$ for $x$ and $y$ in a neighborhood V of $x_0$; taking $f \in \mathscr{D}(\mathrm{X})$ real valued with compact support in V and such that $\int_\mathrm{X} f(x)\, dx \neq 0$, one gets a contradiction between $(G \cdot f \mid f) \geq 0$ and

$$\mathscr{R} \iint_{\mathrm{X}\times\mathrm{X}} \mathrm{G}(x, y) f(x) f(y)\, dx\, dy \leq -\delta\left(\int_\mathrm{X} f(x)\, dx\right)^2.$$

For each $r > 0$, let

$$\mathrm{G}_r(x, y) = \mathrm{G}(x, y) - \sum_{k=1}^{r} \frac{1}{\mu_k} u_k(x)\overline{u_k(y)};$$

then

$$(G_r \cdot f \mid f) = (G \cdot f \mid f) - \sum_{k=1}^{r} \frac{1}{\mu_k} |(f \mid u_k)|^2.$$

But one checks that if $h = f - \sum_{k=1}^{r} (f \mid u_k) u_k$, one has $(G_r \cdot f \mid f) = (G \cdot h \mid h) \geq 0$; the same argument as above proves that $\mathrm{G}_r(x, x) \geq 0$, hence

$$\sum_{k=1}^{r} \frac{1}{\mu_k} |u_k(x)|^2 \leq \mathrm{G}(x, x)$$

and the series

$$\sum_{k=1}^{\infty} \frac{1}{\mu_k} |u_k(x)|^2$$

is therefore convergent. The Cauchy–Schwarz inequality then shows that

$$\sum_{k=r}^{s} \frac{1}{\mu_k} |u_k(x)\overline{u_k(y)}| \leq \left(\sum_{k=r}^{s} \frac{1}{\mu_k} |u_k(x)|^2\right)^{1/2} \left(\sum_{k=r}^{s} \frac{1}{\mu_k} |u_k(y)|^2\right)^{1/2}$$

therefore the series

$$\sum_{k=1}^{\infty} \frac{1}{\mu_k} u_k(x)\overline{u_k(y)}$$

is convergent, and for each $x$ (resp. $y$) the convergence is uniform in $y$ (resp. $x$); the sum

$$\sum_{k=1}^{\infty} \frac{1}{\mu_k} u_k(x)\overline{u_k(y)}$$

is thus separately continuous. However, that series converges to $G(x, y)$ *in the space* $L^2(X \times X)$, hence

$$\sum_{k=1}^{\infty} \frac{1}{\mu_k} u_k(x)\overline{u_k(y)} = G(x, y)$$

except perhaps in a null set; from which one easily concludes that in fact the relation holds everywhere.

## References

The properties of pseudo-differential operators which we have used can be found in

1. L. Hörmander, Pseudo-differential operators and hypoelliptic equations, *in* "Singular Integrals" (*Proc. Symp. Pure Math.* **10**). Amer. Math. Soc., Providence, Rhode Island, 1967.
2. M. Taylor, "Pseudo-Differential Operators" (Lect. Notes in Math. **416**). Springer-Verlag, Berlin, 1974.

They will also be discussed at length in Chap. XXIII of my "Treatise on Analysis."

# *Limit properties of stochastic matrices*

*Samuel Eilenberg*
*Columbia University*

*Alex Heller*
*The Graduate Center of CUNY*

We propose to prove

***Theorem A*** *For any $n \times n$ stochastic matrix $M$ of real numbers, there exists an integer $p > 1$ (called the period of $M$) such that the powers $M^{pr}$ $(r \geq 1)$ converge to a stochastic matrix $L$ and such that the $p$ matrices*

$$L, \quad ML, \quad M^2L, \ldots, M^{p-1}L$$

*are distinct.*

Later in the paper we shall be able to give more detailed information as to how the integers $p$ and the matrix $L$ can be effectively computed. We shall also know more precisely how fast $M^{pr}$ converges to $L$.

In principle the theorem is supposed to be "known." However, an explicit statement in full generality could not be found.

Since the theorem was needed in the theory of automata, it was essential to produce a proof which is effective and provides finitistic methods for finding $p$ and $L$.

Experimenting with examples indicated that the integer $p$ and the position of zeros in the matrix $L$ depend only on the position of zeros in the matrix

$M$ and not on actual numerical values of its entries. This observation is indeed correct and paves the way to a finitistic treatment of a considerable part of the proof.

## 1. Preliminaries

We consider the Euclidean $n$-dimensional vector space $R^n$. A vector

$$x = (x_1, \ldots, x_n)$$

is said to be *nonnegative* (resp. *positive*) if $x_i \geq 0$ (resp. $x_i > 0$) for all $1 \leq i \leq n$. The vector $x$ is said to be *stochastic* if it is nonnegative and if

$$\sum_{i=1}^{n} x_i = 1$$

The following norm in $R^n$ will be used:

$$\|x\| = \sum_{i=1}^{n} |x_i|$$

Thus stochastic vectors have norm 1. For any two stochastic vectors $x$ and $y$ we have

$$\|x - y\| \leq 2$$

with equality holding iff

$$x \cdot y = \sum_{i=1}^{n} x_i y_i = 0$$

Let $M$ be a $p \times q$-matrix of real numbers. The $i$th row of $M$ will be denoted by $M_{i*}$, while the $j$th column will be denoted by $M_{*j}$. The matrix is *stochastic* if each of its rows is stochastic. The norm of $M$ is defined by

$$\|M\| = \sup_i \|M_{i*}\|$$

Thus each stochastic matrix has norm 1.

For any vector $x$

$$\|xM\| \leq \sum_{i,j} |x_i M_{ij}| \leq \|x\| \ \|M\|$$

By taking for $x$ the vectors $(0, \ldots, 0, 1, 0, \ldots, 0)$ we see that

$$\|M\| = \sup \frac{\|xM\|}{\|x\|}$$

for all vectors $x \neq 0$.

Since $(ML)_{i*} = M_{i*} L$ we also have

$$\|ML\| \leq \|M\| \ \|L\|$$

A matrix $M$ is said to be *constant* if all its rows are equal. For any vector $x$ we then have

$$xM = (\sum x_i)m$$

where $m$ is the row of $M$.

## 2. Shrinking matrices

Given any $n \times n$ stochastic matrix $M$ we define

$$\delta_M = \tfrac{1}{2} \sup_{i,j} \|M_{i*} - M_{j*}\|$$

***Proposition 2.1*** *For any pair $x$, $y$ of stochastic vectors*

$$\|xM - yM\| \leq \|x - y\| \, \delta_M$$

*Proof* Define

$$I = \{i \,|\, x_i > y_i\}, \qquad J = \{j \,|\, y_j > x_j\}$$

$$u_i = x_i - y_i \qquad \text{for} \quad i \in I$$

$$v_j = y_j - x_j \qquad \text{for} \quad j \in J$$

Since

$$\sum_{i=1}^{n} x_i - y_i = 0$$

we have

$$w = \sum_{i \in I} u_i = \sum_{j \in J} v_j = \tfrac{1}{2}\|x - y\|$$

Without loss we may assume that $x \neq y$ so that $w \neq 0$. We have

$$\|xM - yM\| = \| \sum (x_i - y_i) M_{i*} \| = \left\| \sum_{i \in I} u_i M_{i*} - \sum_{j \in J} v_j M_{j*} \right\|$$

$$= \frac{1}{w} \left\| \sum_{i \in I, j \in J} u_i v_j M_{i*} - \sum_{i \in I, j \in J} u_i v_j M_{j*} \right\|$$

$$= \frac{1}{w} \left\| \sum_{i \in I, j \in J} u_i v_j (M_{i*} - M_{j*}) \right\|$$

$$\leq \frac{1}{w} \sum_{i \in I, j \in J} u_i v_j \|M_{i*} - M_{j*}\|$$

$$\leq \frac{1}{w} 2 \, \delta_M \sum_{i \in I, j \in J} u_i v_j = \frac{1}{w} 2 \, \delta_M \, w^2$$

$$= 2w \, \delta_M = \|x - y\| \, \delta_M \quad \blacksquare$$

The stochastic $n \times n$ matrix $M$ is said to be *shrinking* if $\delta_M < 1$. Proposition 2.1 justifies this terminology. Note that $M$ is shrinking iff

$$\|M_{i*} - M_{j*}\| < 2$$

for all indices $1 \leq i, j \leq n$. This is equivalent to

$$M_{i*} \cdot M_{j*} \neq 0$$

Thus we obtain

***Proposition 2.2*** *The stochastic $n \times n$ matrix $M$ is shrinking if and only if for any indices $1 \leq i, j \leq n$ there exists an index $1 \leq k \leq n$ such that*

$$M_{ik} \neq 0 \neq M_{jk} \quad \blacksquare$$

***Theorem 2.3*** *For any shrinking stochastic $n \times n$ matrix $M$, the limit*

$$\lim_{r \to s} M^r = M^\omega$$

*exists and is a constant stochastic matrix. Further*

$$\|M^r - M^\omega\| \leq 2\, \delta_M{}^r \tag{2.1}$$

*Proof* From Proposition 2.1 we deduce that

$$\|xM^r - xM^{r+q}\| \leq \|x - xM^q\|\, \delta_M{}^r \leq 2\, \delta_M{}^r$$

Thus

$$\|M^r - M^{r+q}\| \leq 2\, \delta_M{}^r$$

for all $r \geq 0, q \geq 0$. This implies that the limit $M^\omega$ exists and that (2.1) holds. Since also

$$\|xM^r - yM^r\| \leq \|x - y\|\, \delta_M{}^r \leq 2\, \delta_M{}^r$$

it follows that $xM^\omega = yM^\omega$ so that $M^\omega$ is a constant $\blacksquare$

***Proposition 2.4*** *Let $M$ be a $n \times n$ stochastic matrix with convergent powers*

$$\lim_{r \to \infty} M^r = M^\omega$$

*and let $v$ be a stochastic $n$-vector. The following conditions are then equivalent:*

(i) *$M^\omega$ is constant with $v$ as a row*
(ii) *The equations*

$$xM = x, \qquad \sum_{i=1}^{n} x_i = 1$$

*have a unique solution $x = v$.*

*Proof* (i) ⇒ (ii) Since $M^\omega M = M^\omega$ it follows that $vM = v$. Thus $v$ is a solution of the above system. Suppose $x$ is another solution. Since $xM = x$ it follows that $xM^r = x$ and thus $xM^\omega = x$. Since $M^\omega$ is a constant with row $v$, it follows that $xM^\omega = (\sum_{i=1}^n x_i)v$. Thus $x = v$ since $\sum x_i = 1$.

(ii) ⇒ (i) Since $vM = v$ we have $vM^\omega = v$. Since $M^\omega$ is a constant, $v$ must be a row of $M^\omega$ ∎

***Corollary 2.5*** *If the sequence $(M^r)$ converges to a constant with row $v$, then the components of $v$ depend rationally upon the components of $M$.*

*Proof* The equations

$$xM = x, \qquad \sum_{i=1}^{n} x_i = 1$$

form a system of $n + 1$ linear equations with $n$ unknowns, about which we know that the solution is unique. Thus the equations may be solved by elimination ∎

The above argument provides an effective procedure for finding $v$.

***Example 2.1*** Consider the most general $2 \times 2$ stochastic matrix

$$M = \begin{vmatrix} 1 - r & r \\ s & 1 - s \end{vmatrix}$$

with $0 \le r \le 1$ and $0 \le s \le 1$. If $0 < s + r < 2$ the matrix has a positive column and therefore by Proposition 2.2 it is shrinking. Thus $M^r$ converges to a constant matrix $M^\omega$ with row $(u, v)$. To find this vector we apply Proposition 2.4 and consider the equations

$$(u, v)\begin{vmatrix} 1 - r & r \\ s & 1 - s \end{vmatrix} = (u, v), \qquad u + v = 1 \tag{2.2}$$

or more explicitly

$$\begin{aligned} u(1 - r) + vs &= u \\ ur + v(1 - s) &= v \\ u + v \qquad &= 1 \end{aligned}$$

The first two equations both reduce to

$$ur - vs = 0$$

and combined with the third equation yield

$$u = s/(r + s), \qquad v = r/(r + s)$$

If $r + s = 0$, then

$$M = \begin{vmatrix} 1 & 0 \\ 0 & 1 \end{vmatrix}$$

and $M^\omega = M$. If $r + s = 2$, then

$$M = \begin{vmatrix} 0 & 1 \\ 1 & 0 \end{vmatrix}$$

The powers of $M$ are alternatively

$$\begin{vmatrix} 0 & 1 \\ 1 & 0 \end{vmatrix} \quad \text{and} \quad \begin{vmatrix} 1 & 0 \\ 0 & 1 \end{vmatrix}$$

and thus the limit $M^\omega$ does not exist. Note however that $u = v = \frac{1}{2}$ is still the unique solution of the system of equations (2.2).

## 3. Analysis of supports

We denote by $\mathscr{B}_{n,n}$ the set of all $n \times n$ matrices with entries 0 or 1. Such matrices are multiplied using the ordinary rules of matrix multiplication but with the rule $1 + 1 = 2$ replaced by $1 + 1 = 1$. This ensures that $\mathscr{B}_{n,n}$ is closed under multiplication and forms a finite monoid with $2^{(n2)}$ elements.

With each stochastic matrix $M$ we associate the matrix $M^\dagger$, called the *support* of $M$, which is obtained by replacing each nonzero entry in $M$ by 1. Clearly $(MN)^\dagger = M^\dagger N^\dagger$ and thus a morphism is obtained from the monoid of stochastic matrices to the monoid $\mathscr{B}_{n,n}$. The powers $(M^\dagger)^r = (M^r)^\dagger$, $r > 1$, form a cyclic subsemigroup of $\mathscr{B}_{n,n}$. It is a known fact that each finite cyclic semigroup contains a unique idempotent element $e = ee$. Let, then, $N$ be the idempotent in the cyclic semigroup $\{(M^\dagger)^r | r \geq 1\}$. The least integer $d$ such that $(M^d)^\dagger = N$ will be called the *idempotency exponent* of $M$. Thus $d$ is the least integer $\geq 1$ such that the matrices $M^d$ and $M^{2d}$ have zeros in the same positions.

To further analyze the situation it will be convenient to regard the matrix $M^\dagger$ as a relation

$$f_M : Q \to Q$$

where $Q = \{1, \ldots, n\}$. For each $i \in Q$ the relation $f_M$ assigns the subset

$$if_M = \{j | M_{ij} \neq 0\}$$

Since the matrix $M$ is stochastic no row of $M$ is zero and therefore

$$if \neq \varnothing \qquad \text{for all} \quad i \in Q \tag{3.1}$$

Composition of relations matches the multiplication of matrices in $\mathscr{B}_{n,n}$. Therefore the relation

$$g = f_M{}^d$$

where $d$ is the idempotency exponent of $M$ satisfies

$$gg = g \tag{3.2}$$

$$ig \neq \varnothing \qquad \text{for all} \quad i \in Q \tag{3.3}$$

A subset $X$ of $Q$ is said to be *closed* if $Xg = X$. For any nonempty subset $X$ of $Q$, the set $Xg$ is closed since $Xgg = Xg$. We shall be interested in the nonempty, minimal closed subsets

$$E_1, \ldots, E_k$$

of $Q$. Note that since $Qg$ is closed and nonempty, there exists at least one such set; thus $k \geq 1$.

We note that

$$q \in E_i \qquad \text{implies} \quad qg = E_i \tag{3.4}$$

and this implies

$$E_i \cap E_j = \varnothing \qquad \text{for} \quad i \neq j \tag{3.5}$$

The sets $E_1, \ldots, E_k$ are called the *essential* sets of $M$. The set

$$D = Q - (E_1 \cup \cdots \cup E_k)$$

is called the *inessential* set. We prove

$$i \in D \qquad \text{implies} \quad ig - D \neq \varnothing \tag{3.6}$$

Indeed suppose $ig \subset D$. Since $ig$ is closed and nonempty, it would have to contain an essential set. However $D$ does not contain any such set.

Next we consider the effect of the relation $f = f_M$ on the essential sets (recall that the essential sets were defined using the relation $g = f^d$). We claim:

$$\text{If } E \text{ is an essential set then so is } Ef. \tag{3.7}$$

We first note that $Ef$ is closed; indeed $Efg = Egf = Ef$. Since $Ef \neq \varnothing$ it must contain an essential set

$$E' \subset Ef$$

This implies

$$E'f^{d-1} \subset Eff^{d-1} = Eg = E$$

Since $E'f^{d-1}$ also is closed, it contains an essential set

$$E'' \subset E'f^{d-1}$$

Since $E'' \subset E$ we must have $E'' = E$. Thus $E \subset E'f^{d-1}$ and therefore

$$Ef \subset E'f^{d-1}f = E'g = E'$$

Thus $Ef = E'$ as required

Property (3.7) yields a function

$$\varphi: \{1, \ldots, k\} \to \{1, \ldots, k\}$$

such that

$$E_i f = E_{i\varphi}$$

Since $E_i f^d = E_i g = E_i$ it follows that

$$\varphi^d = \text{identity}$$

Thus $\varphi$ is a permutation and its order $p$ is a divisor of $d$. The integer $p \geq 1$ is called the *period* of $f = f_M$ (or the *period* of the stochastic matrix $M$).

One can go a step further. If the permutation $\varphi$ is decomposed into cycles, there results a break up of the class of all essential sets $E_1, \ldots, E_k$ into cycles $C_1, \ldots, C_l$ $(l \leq k)$ with *local* periods $p_1, \ldots, p_l$. The period $p$ is then the least common multiple of $p_1, \ldots, p_l$. Within each cycle the essential sets are permuted cyclically by $f$.

*Remark* The period $p$ just defined should not be confused with the period of the cyclic semigroup consisting of the relation $f$ and its positive powers. This period is defined as the least integer $q$ such that $f^k = f^{k+q}$ holds for some $k \geq 1$. This integer $q$ also is a divisor of $d$ but may differ from $p$. This may be seen in the example of the matrix

$$M = \begin{vmatrix} 1 & 0 & 0 \\ \frac{1}{2} & 0 & \frac{1}{2} \\ \frac{1}{2} & \frac{1}{2} & 0 \end{vmatrix}$$

The powers $M^r$ converge to the constant matrix with row (1, 0, 0) and thus by our theorem $p = 1$. However the matrix $B = M^\dagger$ satisfies

$$B \neq B^2, \qquad B = B^3$$

Thus $q = 2$. Also $d = 2$ since $B^2 = B^4$.

## 4. The case $d = 1$

We shall consider here the case of a stochastic matrix $L$ for which the idempotency index is 1. This means that its support relation $f: Q \to Q$ is *idempotent* i.e. satisfies $ff = f$. Using the decomposition

$$Q = E_1 \cup \cdots \cup E_k \cup D$$

we may write the matrix $L$ in the form

$$L = \begin{array}{c|c|c|c|c|c|}
 & E_1 & E_2 & & E_k & D \\
\hline
E_1 & P_1 & & \cdots & & \\
\hline
E_2 & & P_2 & \cdots & & \\
\hline
 & \vdots & \vdots & & \vdots & \\
\hline
E_k & & & \cdots & P_k & \\
\hline
D & R_1 & R_2 & \cdots & R_k & S \\
\hline
\end{array}$$

The condition

$$qf = E_i \qquad \text{for all} \quad q \in E_i$$

shows that the unmarked spaces carry zeros and that the square matrices $P_1, \ldots, P_k$ are stochastic and positive. The condition (3.6),

$$qf - D \neq \varnothing \qquad \text{for all} \quad q \in D$$

implies that each row of $S$ has sum $<1$. Thus

$$\|S\| < 1$$

***Theorem 4.1*** *If L is a stochastic matrix with idempotency index* 1, *then the limit*

$$\lim_{r \to \infty} L^r = L^\omega$$

*exists and*

$$\|L^r - L^\omega\| \leq ab^r$$

*for suitable real numbers* $0 < a$, $0 < b < 1$. *Further, with the notations above, the stochastic matrix* $L^\omega$ *is*

$$L^\omega = \begin{array}{|c|c|c|c|c|} \hline P_1^\omega & & \cdots & & \\ \hline & P_2^\omega & \cdots & & \\ \hline & \vdots & \vdots & \vdots & \\ \hline & & \cdots & P_k^\omega & \\ \hline R_1^\omega & R_2^\omega & & R_k^\omega & \\ \hline \end{array}$$

*with zeros in unmarked rectangles and with*

$$P_i^\omega = \lim_{r \to \infty} P_i^r, \qquad R_i^\omega = S^* R_i P_i^\omega, \qquad S^* = \sum_{r=0}^{\infty} S^r = (I - S)^{-1}$$

*Proof* Each power $L^r$ of $L$ will have the same form as $L$ but with $P_i$ and $S$ replaced by their powers $P_i^r$ and $S^r$ and with each $R_i$ replaced by $R_i^{(r)}$ defined inductively by the formulas

$$R_i^{(0)} = 0, \qquad R_i^{(r+1)} = R_i P_i^r + S R_i^{(r)}$$

Since the matrices $P_i$ are positive it follows from (2.1) that

$$\|P_i^r - P_i^\omega\| \leq ab^r \tag{4.1}$$

(with suitable $0 < a$, $0 < b < 1$). Since $\|S\| < 1$ it follows that

$$\|S^r\| \leq \|S\|^r$$

so that

$$\|S^r - 0\| \leq ab^r$$

(with $a$ and $b$ suitably chosen). Further, since the series $\sum_{r=0}^{\infty} \|S^r\|$ converges the series

$$S^* = \sum_{r=0}^{\infty} S^r$$

converges and

$$S^* = (I - S)^{-1}$$

(Note: the matrix $S^*$ is nonnegative but need not be a stochastic.)

There remains to be proved that

$$\|R_i^{(r)} - R_i^{\omega}\| < ab^r$$

for suitable $0 < a$, $0 < b < 1$, as this will imply $\lim R_i^{(r)} = R_i^{\omega}$ and with it also $\lim L^r = L^{\omega}$.

We note that

$$R_i P_i^{\omega} + SR_i^{\omega} = R_i P_i^{\omega} + SS^*R_i P_i^{\omega} = S^*R_i P_i^{\omega} = R_i^{\omega}$$

and hence that

$$R_i^{(r+1)} - R_i^{\omega} = R_i(P_i^r - P_i^{\omega}) + S(R_i^{(r)} - R_i^{\omega})$$

Thus denoting

$$e_r = \|R_i^{(r)} - R_i^{\omega}\|$$

we have

$$e_{r+1} \leq \|R_i\| ab^r + \|S\| e_r$$

with $a$ and $b$ taken from (4.1). Consequently

$$e_{r+1} \leq cd^r + de_r$$

with

$$e = \|R_i\| a, \qquad d = \sup(b, \|S\|) < 1$$

An easy induction then shows

$$e_r \leq \left((r-1)c + \frac{e_1}{d}\right) d^r$$

Replacing $d$ by $v$ such that $t < v < 1$ we obtain

$$e_r \leq uv^r$$

where $u$ is chosen so that

$$\left((r-1)c + \frac{e_1}{t}\right)\left(\frac{t}{v}\right)^r + e_0 \leq u$$

for all $r \geq 1$ ▮

***Proposition 4.2*** *Each row of $R_i^\omega$ is a multiple of the row of the constant matrix $P_i^\omega$.*

*Proof* From the formulas

$$R_i^\omega = S^* R_i P_i^\omega, \qquad P_i^\omega P^\omega = P_i^\omega$$

we deduce that

$$R_i^\omega P_i^\omega = P_i^\omega$$

Let $r_i$ be any row of $R_i^\omega$ and let $p_i$ be a row of $P_i^\omega$. Since $P_i^\omega$ is a constant we obtain

$$r_i = |r_i| p_i \qquad ▮$$

***Corollary 4.3*** *Each row $L_{q,*}^\omega$ with $q \in D$ is a linear combination with non-negative coefficients of rows $L_{e,*}^\omega$ with $e \in E_1 \cup \cdots \cup E_k$* ▮

***Proposition 4.5*** *The matrix $L^\omega$ depends rationally upon $L$.*

*Proof* In view of Proposition 2.5, it suffices to prove this for the matrix $S^* = (I - S)^{-1}$. This follows from the fact that the equation

$$S^*(I - S) = I$$

yields a system of $n^2$ linear equations with $n^2$ unknowns, having a unique solution ▮

## 5. The general case

We shall now consider an arbitrary stochastic matrix $M$ with idempotency index $d$ and period $p$. The matrix

$$L = M^d$$

has an idempotent support and thus the limit

$$\lim_{r \to \infty} L^r = L^\omega$$

exists and has the form given in Theorem 4.1. We further have the decomposition

$$Q = E_1 \cup \cdots \cup E_k \cup D$$

and the permutation

$$\varphi: \{1, \ldots, i\} \to \{1, \ldots, k\}$$

such that

$$E_i f = E_{i\varphi}$$

where $f: Q \to Q$ is the support of $M$ and $g = f^d$ is the support of $L$. The permutation $\varphi$ has order $p$. Since $ML^r = L^r M$ we have

$$ML^\omega = L^\omega M$$

***Proposition 5.1*** *The equality*

$$L^\omega M^u = L^\omega M^v$$

holds if and only if

$$u \equiv v \bmod p.$$

In particular

$$L^\omega M^p = L^\omega$$

*Proof* For each $1 \leq i \leq k$ consider the stochastic vector

$$x_i = L^\omega_{q,*} \qquad \text{with} \quad q \in E_i$$

The choice of $q$ within $E_i$ is immaterial, since the rows $L^\omega_{q,*}$ are the same for all $q \in E_i$. We assert that

$$x_i M = x_{i\varphi} \tag{5.1}$$

Let $y = x_i M$. Since $E_i f = E_{i\varphi}$ it follows that $y$ and $x_{i\varphi}$ have zero in exactly the same positions (i.e. the positions $j \in Q - E_{i\varphi}$). Since $L^\omega = L^\omega L^\omega$ we have $x_i = x_i L^\omega$. Therefore

$$y = x_i M = x_i L^\omega M = x_i M L^\omega = y L^\omega$$

Since $y$ and $x_{i\varphi}$ have zeros in the same positions, it follows from the explicit form of $L^\omega$ that $yL^\omega = x_{i\varphi}$. Thus $x_{i\varphi} = yL^\omega = y = x_i M$. This proves (5.1).

Now assume $L^\omega M^u = L^\omega M^v$. This implies $x_i M^u = x_i M^v$ and by (5.1)

$$x_{i\varphi^u} = x_{i\varphi^v} \qquad \text{for all} \quad 1 \leq i \leq k$$

This implies $i\varphi^u = i\varphi^v$ and thus $\varphi^u = \varphi^v$. Consequently $u \equiv v \bmod p$.

Conversely if $u \equiv v \bmod p$, then $\varphi^u = \varphi^v$. From (5.1) we then deduce

$$x_i M^u = x_i M^v$$

for all $1 \leq i \leq k$. Since by Proposition 4.4, any row of $L^\omega$ either is one of the $x_i$ or is a linear combination of them, it follows that $L^\omega M^u = L^\omega M^v$ ∎

***Theorem 5.2*** *Let M be a stochastic matrix with period p. Then*

$$\lim_{r\to\infty} M^{rp+j} = L^{\omega}M^{j}$$

*for* $0 \le j < p$ *and these p limits are distinct from each other. Further*

$$\|M^{rp+j} - L^{\omega}M^{j}\| < ab^{r} \tag{5.2}$$

*for suitable real numbers* $0 < a$, $0 < b < 1$.

*Proof* Let $d = ep$ be the idempotency index of $M$ and let $L = M^d$. Then $L$ has idempotency index 1 and thus by Theorem 4.1

$$\lim_{r\to\infty} M^{rd} = \lim_{r\to\infty} L^{r} = L^{\omega}$$

Consequently

$$\lim_{r\to\infty} M^{rd+u} = L^{\omega}M^{u}$$

for all $0 \le u < d$. However by Proposition 5.1, these limits are equal only for $u$'s that are congruent mod $p$. This proves the first part of the theorem.

To prove (5.2) set

$$r = qe + t, \qquad 0 \le t < e$$

Then

$$q \ge r/e$$

Further by Theorem 4.1 we have

$$\begin{aligned}\|M^{rp+d} - L^{\omega}M^{j}\| &= \|M^{j+tp}M^{qd} - M^{j+tp}L^{\omega}\| \\ &\le \|M^{d+tp}\|\,\|M^{qd} - L^{\omega}\| \le \|M^{qd} - L^{\omega}\| \le ab^{q} \le a(b^{1/e})^{r}\end{aligned}$$

as required ▮

We have seen that if $M$ is a stochastic matrix with period $p$, then the powers $\{M^r \mid r \ge 0\}$ break into $p$ sequences with distinct limits. It is natural to consider the average of these limits

$$C = (1/p)L^{\omega}(I + M + M^2 + \cdots + M^{p-1})$$

***Proposition 5.3*** *For each stochastic matrix, the averages*

$$C_r = (1/r)(I + M + \cdots + M^{r-1})$$

*converge and*

$$\lim C_r = C$$

*Proof* Define

$$D_q = (1/q)(I + M^p + M^{2p} + \cdots + M^{(q-1)p})$$

Since $\lim_{q\to\infty} M^{qp} = L^\omega$ it follows from a standard theorem that

$$\lim_{q\to\infty} D_q = L^\omega$$

Given $r \geq 0$ write

$$r = qp + k, \qquad 0 \leq k < p$$

The identity

$$C_r = (qp/r)D_q C_p + (k/r)M^{pq}C_k$$

then holds. As $r$ tends to $\infty$ the last term approaches zero while the first one approaches $L^\omega C_p = C$ as required ▌

The matrix $C$ is called the Cesaro limit of the matrices $\{M^r\}$ and the notation

$$C = c\lim_{r\to\infty} M^r$$

is used.

***Corollary 5.4*** *For any stochastic matrix $M$ the matrices $L^\omega M^i$ and $C$ depend rationally upon $M$* ▌

AMS (MOS) 1970 subject classification: 15A51

# *A fixed-point characterization of linearly reductive groups*

*J. Fogarty and P. Norman*

*University of Massachusetts*
*Amherst*

## 1. Introduction

In [2] the first author gave a fixed-point characterization of unipotent groups, viz., if $G$ is a connected linear algebraic group over the field $k$, then $G$ is unipotent if and only if, for all proper geometrically connected $G$-schemes $X$ over $k$, the fixed-point scheme $X^G$ is connected. Our objective here is to give an analogous characterization of linearly reductive groups. We prove the following:

***Theorem*** *Let $G$ be a linear algebraic group over the field $k$. Then $G$ is linearly reductive if and only if, for all smooth algebraic $G$-schemes $X$ over $k$, the fixed-point scheme $X^G$ is smooth.*

By a linear algebraic group over $k$, we mean a reduced affine group scheme of finite type over $k$. If $G$ is a linear algebraic group over $k$, by a $G$-scheme $X$ over $k$ we mean a scheme $X$ over $k$, plus a map of schemes $\alpha: G \times X \to X$ satisfying the usual axioms for an action of $G$ on $X$ (cf. [4]). If X is a $G$-scheme, the fixed-point scheme $X^G$ is the largest closed subscheme of $X$ on which $G$ acts trivially (cf. [2]).

Let $E$ be a vector space over $k$, let $S^{\cdot}E$ be the symmetric algebra of $E$, and let $V = \mathrm{Spec}(S^{\cdot}E)$. An action $\alpha: G \times V \to V$ is linear if the induced map $\tilde{\alpha}: S^{\cdot}E \to H^0(G, \mathfrak{O}_G) \otimes S^{\cdot}E$ takes $E$ to $k \otimes E$ (identifying $E$ with $S^1E$). If this is the case, we say that $E$ is a rational $G$-module. We recall that a linear algebraic group $G$ is linearly reductive if every rational $G$-module is completely reducible, i.e., if $E$ is a rational $G$-module, and $E_0$ is a $G$-submodule of $E$, then there is a $G$-submodule $E_1$ of $E$ such that $E = E_0 \oplus E_1$. $E_1$ is called a $G$-complement for $E_0$ in $E$.

## 2. Proof of the theorem

We show first that if $G$ is a linearly reductive linear algebraic group over $k$ and $X$ is a smooth $G$-scheme of finite type over $k$, then $X^G$ is smooth. We may assume that $k$ is algebraically closed, for if $\bar{k}$ is an algebraic closure of $k$, then $(X \times \bar{k})^{G \times \bar{k}} = (X^G) \times \bar{k}$. If $x$ is a closed point of $X^G$ let $R = \mathfrak{O}_{X,x}$ be the local ring of $x$ on $X$. If $\mathfrak{J}$ is the ideal in $\mathfrak{O}_X$ defining $X^G$, then the stalk $\mathfrak{J}_x$ of $\mathfrak{J}$ at $x$ is the ideal in $R$ generated by all $f^g - f$, $f \in R, g \in G(k)$. We want to show that $J = \mathfrak{J}_x$ is generated by part of a regular system of parameters of $R$. Let $\hat{R}$ be the completion of $R$. Then $G(k)$ operates on $\hat{R}$ and $J\hat{R}$ is the ideal in $\hat{R}$ generated by all $f^g - f$, $f \in \hat{R}$, $g \in G(k)$. If $\mathfrak{m}$ is the maximal ideal of $\hat{R}$, we show that $\mathfrak{m}^2$ has a $G$-complement in $\mathfrak{m}$. $\hat{R}/\mathfrak{m}^n$ is a rational $G$-module for all $n$, so having chosen $G$-complements $V_j$ for $\mathfrak{m}^2/\mathfrak{m}^j$ in $\mathfrak{m}/\mathfrak{m}^j$ such that reduction $\mathfrak{m}^{j-1}$ maps $V_j$ onto $V_{j-1}$, we let $V = \varprojlim V_j$. Then $V$ is a $G$-submodule of $\hat{R}$ and is a $G$-complement for $\mathfrak{m}^2$ in $\mathfrak{m}$. Let $W = V \cap J\hat{R}$, and let $I = W\hat{R}$. Then $G(k)$ operates trivially on $\mathfrak{m}/I + \mathfrak{m}^2$ and hence trivially on $gr(\hat{R}/I)$. Therefore $G(k)$ operates trivially on $\hat{R}/I + \mathfrak{m}^j$ for all $j$, since $G$ is linearly reductive. Hence $G(k)$ operates trivially on $\hat{R}/I$. Therefore $J\hat{R} \subset I$, and hence $J\hat{R} = I$. But $I$ is clearly generated by part of a regular system of parameters of $\hat{R}$, and so $J$ is generated by part of a regular system of parameters of $R$.

Now suppose that $G$ has the smooth fixed-point property, i.e., for all smooth algebraic $G$-schemes $X$, $X^G$ is smooth. Let $G_0$ be the component of the identity in $G$. We show that $G_0$ also has the smooth fixed-point property. Let $X_0$ be a smooth algebraic $G_0$-scheme. Let $\Gamma = G/G_0$. Then $\Gamma$ is finite. Let $X = \prod_{u \in \Gamma} X_{0,u}$ be the product of copies of $X_0$ indexed by the cosets of $G_0$ and let $G$ act on $X$ via $(x_{u_1}, \ldots, x_{u_r})^{gh} = (x^g_{u_1 h}, \ldots, x^g_{u_r h})$, $g \in G_0(k)$, $h \in G(k)$. If $\Delta: X_0 \to X$ is the diagonal map, then $X^G = \Delta(X_0^{G_0})$, so $X^G$ smooth implies $X_0^{G_0}$ smooth.

If $G$ has the smooth fixed-point property, then so does $\Gamma$. Now $\Gamma$ is linearly reductive except when the characteristic divides the order of $\Gamma$. Assume char $k = p > 0$ and that $\Gamma$ has a $p$-Sylow subgroup $P$. $P$ acts on the projective line $\mathbf{P}^1$ as follows. Choose a normal subgroup $N$ in $P$ with $P/N \approx \mathbf{Z}/p\mathbf{Z}$, and let $P$ act on $\mathbf{P}^1$ via the representation $t \mapsto \begin{pmatrix} 1 & t \\ 0 & 1 \end{pmatrix}$ of $P/N$. Let $E$ be the space of this representation.

$P$ has a unique fixed point $x_0$ on $\mathbf{P}^1$ and acts trivially on the tangent space $T_{x_0}(\mathbf{P}^1)$ to $\mathbf{P}^1$ at $x_0$. If $m$ is the index of $P$ in $\Gamma$, let $X = (\mathbf{P}^1)^m$ be the $m$-fold product. $H^0(X, \mathfrak{O}_X(1)) \approx E^m$ ($m$-fold direct sum) and $\Gamma$ operates here via the representation induced from the representation $E$ of $P$. Also $X$ is embedded in $\mathbf{P}(H^0(X, \mathfrak{O}_X(1))^*)$ and is stable under the induced action of $\Gamma$. If $1 = g_1, \ldots, g_m$ is a set of right coset representatives for $P$ in $\Gamma$, then $X^\Gamma$ has $x = (x_0^{g_1}, \ldots, x_0^{g_m})$ as its only fixed closed point. Also, the tangent space $T_x(X)$ to $X$ at $x$, as $\Gamma$-module, is simply the representation of $\Gamma$ induced by the trivial 1-dimensional representation of $P$. Thus $T_X(X)^\Gamma \neq (0)$, i.e., $X^\Gamma$ is not reduced.

We are thus reduced to the case where $G$ is connected.

We deal first with the case of characteristic $p > 0$, and $G$ not solvable. Let $H$ be the radical of $G$, i.e., the maximal connected normal solvable closed subgroup. Then $G/H$ is semisimple. We show that $G/H$ does not have the smooth fixed-point property. Clearly we may assume that $H = (1)$. Suppose first that $G$ is simply connected. Then there is an irreducible rational $G$-module $E$ of dimension $p^r$($r$ is the number of positive roots of $G$), viz., the Steinberg module corresponding to $(p-1)\rho$, where $\rho$ is half the sum of the positive roots (cf. [3]). Then $G$ acts on $SL(E)$ by inner automorphisms and $SL(E)^G$ is the centralizer of $G$ in $SL(E)$. By Schur's lemma, the centralizer of $G$ in $GL(E)$ is the group $D \approx GL(1)$ of invertible homotheties of $E$, and $D \cap SL(E) \approx \mu_{p^r}$ which is not reduced. Since the center of the image of $G$ in $SL(E)$ lies in $D$, the action of $G$ on $SL(E)$ factors through the adjoint group $G'$ of $G$, i.e., the center of $G$ acts trivially on $SL(E)$. On the other hand, if $G$ is any connected semisimple group and $\tilde{G}$ is its simply connected covering group, the map $\tilde{G} \to G$ is a central isogeny. This shows that in positive characteristic, a connected nonsolvable group cannot have the smooth fixed-point property.

*Remark* It was known already (cf. [5]) that a connected nonsolvable group in positive characteristic is not linearly reductive. Now let $G$ be a connected solvable group. Since the additive group $G_a$ does not have the smooth fixed-point property (cf. [2, §6]), it follows that if $G$ has the smooth fixed-point property, then there are no nonzero rational homomorphisms $G \to G_a$. Since $G$ is solvable, it suffices, by the Lie–Kolchin theorem, in order to prove $G$ linearly reductive, to show that every exact sequence

$$0 \to E \to V \to k \to 0 \tag{1}$$

of rational $G$-modules, with $\dim E = 1$, splits. ($k$ is the trivial 1-dimensional $G$-module.) If $E$ is trivial, then the image of $G$ in $GL(V)$ is unipotent. By the above remark, $V$ must then be trivial. If $E$ is not trivial, we may choose a basis for $V$ so that the image $H$ of $G$ in $GL(V)$ lies in the Borel subgroup given by

$$B(k) = \left\{\begin{pmatrix} a & 0 \\ c & b \end{pmatrix} : ab \neq 0\right\}.$$

Then $b = 1$ for all elements of $H(k)$ and $a \neq 1$ for some element of $H(k)$. If (1) does not split, then we may assume that $c \neq 0$ for some element of $H(k)$. In other words, $H$ is isomorphic to the affine group $AF(1)$ of the line. Let $k[A, B]$ be the coordinate ring of $H$ and let $H$ act on $\mathbf{A}^2$ via the map $k[X, Y] \to k[A, B] \otimes k[X, Y]$ given by

$$X \mapsto 1 \otimes X,$$

$$Y \mapsto A \otimes Y + C \otimes X^2.$$

It is not hard to show this gives an action. Now $(\mathbf{A}^2)^H$ is defined by the ideal $I$ in $k[X, Y]$ generated by all $f^h - f$, $f \in k[X, Y]$, $h \in H(k)$. If $h_0 = \left(\begin{smallmatrix} 1 & 0 \\ -1 & 1 \end{smallmatrix}\right)$, then $Y^{h_0} - Y = X^2$, so $X^2 \in I$. But $X \notin I$, so that $(\mathbf{A}^2)^H$ is not reduced, and $G$ cannot have the smooth fixed-point property.

Now assume that the characteristic is zero. Then the connected linear algebraic group $G$ is the semidirect product $H \cdot U$ where $H$ is a reductive group (i.e., the radical of $H$ is a torus) and $U$ is the unipotent radical of $G$. Since a reductive group is linearly reductive in characteristic zero (cf. [1]), it suffices to show that $U = (1)$. If $U'$ is the commutator subgroup of $U$, we may replace $G$ with $G/U'$ and assume that $U$ is commutative. Then $U$ is a vector group whose composition we write additively.

$G(k)$ is the set of pairs $(h, u)$, $h \in H(k)$, $u \in U(k)$, and $(h, u)(h', u') = (hh', u + huh^{-1})$. Let $X = U \times \mathbf{A}1$, $U = \operatorname{Spec}(k[Y_1, \ldots, Y_n])$, $\mathbf{A}^1 = \operatorname{Spec}(k[T])$, and let $G$ act on $X$ via

$$(h, u_1, \ldots, u_n)\colon \begin{cases} T \mapsto T \\ Y_i \mapsto Y_i^h + u_i T^2. \end{cases}$$

If $I$ is the ideal of $X^G$ and $u = (1, u_2, \ldots, u_n)$, we have

$$Y_1^{(1,u)} - Y_1 = Y_1 + T^2 - Y_1 = T^2.$$

Thus $T^2 \in I$, but evidently $T \notin I$ and $X^G$ is not reduced. (cf. [6]).

*Remarks* It would be of some interest, of course, to have a proof that the smooth fixed-point property implies linear reductivity which does not use the structure of linear algebraic groups. It would also be of considerable interest to give examples of semisimple groups $G$ in characteristic $p > 0$, and smooth $G$-schemes $X$ of finite type, with $(X^G)_{\mathrm{red}}$ not smooth.

## References

1. Borel, A. Linear algebraic groups, *AMS Proc. Symp. Pure Math.* **9**, *Providence* (1966).
2. Fogarty, J., Fixed point schemes, *Amer. J. Math.* **95** (1973), 1.
3. Haboush, W., Reductive groups are linearly reductive, *Ann. of Math.* **102** (1975), 1.
4. Mumford, D., "Geometric Invariant Theory." Springer-Verlag, Berlin, 1965.
5. Nagata, M., Complete reducibility of rational representations of a matrix group, *J. Math. Kyoto Univ.* **1** (1961), 87.
6. Norman, P., A fixed point criterion for linear reductivity, *Proc. Amer. Math. Soc.* **50** (1975).

AMS (MOS) 1970 subject classification 20G99

# *Orthogonal and unitary invariants of families of subspaces*

*P. X. Gallagher and R. J. Proulx*

*Columbia University*

Let $V$ be an $n$-dimensional real (or complex) Hilbert space, and let $W_1, \ldots, W_k$ be subspaces of $V$, of dimensions $p_1, \ldots, p_k$. Denote by $P_j$ the orthogonal projection of $V$ onto $W_j$. In this paper, we show that the numbers

$$\operatorname{tr}(P_{j_1} \cdots P_{j_m}) \qquad (1 \leqslant j_\mu \leqslant k, \quad 1 \leqslant m \leqslant 4n^2) \tag{1}$$

determine the subspaces $W_1, \ldots, W_k$ up to a linear isometry of $V$.

For a pair of subspaces $W_1$ and $W_2$, a second system of invariants is the sequence of Jordan angles $\theta_1, \ldots, \theta_p$ with $p = \min\{p_1, p_2\}$. The angles are related to the trace invariants by

$$\operatorname{tr}(P_1 P_2)^r = \sum_{i=1}^{p} \cos^{2r} \theta_i. \tag{2}$$

The dimensions $p_1, p_2$ and the angles $\theta_1, \ldots, \theta_p$ also determine the pair $W_1, W_2$ up to a linear isometry.

If $W_1, \ldots, W_k$ are 1-dimensional subspaces, and $k \geqslant 3$, then the angles $\theta_{ij}$ between $W_i$ and $W_j$ for $1 \leqslant i < j \leqslant k$ are not in general sufficient to determine the subspaces up to a linear isometry. The additional invariants which are necessary can be described in terms of a cohomology class of a graph associated to the system of subspaces. Alternatively, the trace invariants (1) with $m \leqslant n$ suffice in this case, if $n \geqslant 4$.

Our proof that a $k$-tuple of subspaces is determined up to a linear isometry by the traces (1) is similar to the argument sketched by Kaplansky [3, pp. 72–73] for a result of Specht [6] and Pearcy [4] on the orthogonal (or unitary) invariants of a single endomorphism. A recent result of Procesi [5] implies that each trace invariant is a polynomial in the trace invariants with $m \leqslant 2^n - 1$.

## 1. Pairs of subspaces

In this section we define the Jordan angles [2] between a pair of subspaces, and relate the angles to the trace invariants.

***Lemma 1*** *Let $W$ and $X$ be subspaces of dimensions $p$ and $q$ in an $n$-dimensional real (or complex)* Hilbert *space $V$. There is an orthonormal basis $w_1, \ldots, w_p$ of $W$ and an orthonormal basis $x_1, \ldots, x_q$ of $X$ such that $(w_i, x_j) = 0$ for $i \neq j$.*

*Proof* We may suppose that $W$ and $X$ are not orthogonal. Let $w_1$ and $x_1$ be unit vectors in $W$ and $X$ which maximize $|(w_1, x_1)|$. Then $x_1$ is orthogonal to the orthogonal complement of $w_1$ in $W$. In fact, if $w_2$ is a unit vector in $W$ orthogonal to $w_1$, then

$$w = \frac{(x_1, w_1)w_1 + (x_1, w_2)w_2}{(|(w_1, x_1)|^2 + |(w_2, x_1)|^2)^{1/2}}$$

is a unit vector in $W$. The maximality condition $|(w, x_1)| \leqslant |(w_1, x_1)|$ reduces to

$$|(w_1, x_1)|^2 + |(w_2, x_1)|^2 \leqslant |(w_1, x_1)|^2,$$

from which $(w_2, x_1) = 0$. Similarly, $w_1$ is orthogonal to the orthogonal complement of $x_1$ in $X$. The result now follows by induction on $\min\{p, q\}$.

Let $P$ and $Q$ be the orthogonal projections of $V$ onto the subspaces $W$ and $X$ of Lemma 1. Supposing $p \leqslant q$, we have, for $i \leqslant p$,

$$Px_i = (x_i, w_i)w_i, \qquad \text{and} \qquad Qw_i = (w_i, x_i)x_i,$$

so $PQPw_i = |(w_i, x_i)|^2 w_i$. Since $PQP$ annihilates the orthogonal complement of $W$, it follows that the numbers $|(w_i, x_i)|^2 = \cos^2 \theta_i$, together with $n - p$ zeros, are the eigenvalues of $PQP$. In particular, the system of *Jordan angles* $\theta_1, \ldots, \theta_p$ (taken in $[0, \pi/2]$) depend only on $V$, $W$, and $X$. Relation (2) follows from the fact that $(PQ)^r = P(PQ)^r$ has the same trace as $(PQ)^r P = (PQP)^r$.

**Lemma 2** *Let $v_1, \ldots, v_N$ and $v_1', \ldots, v_N'$ be elements of a finite-dimensional* Hilbert *space $V$. If $(v_i, v_j) = (v_i', v_j')$ for all $i$, $j$, then there is a linear isometry of $V$ sending $v_i$ to $v_i'$ for all $i$.*

*Proof* It follows from the assumption that $\sum c_i v_i$ has the same norm as $\sum c_i v_i'$, for all choices of the coefficients $c_i$. In particular, $\sum c_i v_i = 0$ only if $\sum c_i v_i' = 0$, so $\sum c_i v_i \mapsto \sum c_i v_i'$ is a well-defined linear isometry from the span of the $v_i$ to the span of the $v_i'$. This, together with an arbitrary linear isometry connecting the orthogonal complements of these two subspaces, gives a linear isometry of $V$ sending $v_i$ to $v_i'$.

**Theorem 1** *Let $W$, $X$ and $W'$, $X'$ be two pairs of subspaces of dimensions $p$, $q$ in a finite-dimensional* Hilbert *space $V$. If the angles between $W$ and $X$ are the same as the angles between $W'$ and $X'$, then there is a linear isometry of $V$ which sends $W$ to $W'$ and $X$ to $X'$.*

*Proof* By the assumption, there are orthonormal bases $w_1, \ldots, w_p$ and $x_1, \ldots, x_q$ of $W$ and $X$, and orthonormal bases $w_1', \ldots, w_p'$ and $x_1', \ldots, x_q'$ of $W'$ and $X'$, so that $(w_i, x_j) = (w_i', x_j') = 0$ for $i \neq j$, and $|(w_i, x_i)| = |(w_i', x_i')|$ for $i \leqslant \min\{p, q\}$. After modifying the basis vectors by unimodular scalar factors, we may suppose that $(w_i, x_i) = (w_i', x_i')$ for $i \leqslant \min\{p, q\}$. It follows from Lemma 2 that there is a linear isometry of $V$ sending $w_i$ to $w_i'$ and $x_j$ to $x_j'$ for all $i$, $j$.

## 2. Systems of lines

To each $k$-tuple of one-dimensional subspaces $W_1, \ldots, W_k$ of the $n$-dimensional real (or complex) Hilbert space $V$, we associate a *graph*, with *vertices* $1, \ldots, k$ and *edges* consisting of ordered pairs $(i, j)$ with $i \neq j$ for which $W_i$ and $W_j$ are not orthogonal. The free multiplicative abelian group on the edges, modulo the relations $(i, j)(j, i) = 1$, is the group $C$ of *chains*. Chains of the form

$$\gamma = (j_1 j_2)(j_2 j_3) \cdots (j_m j_1) \tag{3}$$

are *circuits*. The subgroup $\Gamma$ generated by the circuits is the *homology group* of the graph.

Let $w_1, \ldots, w_k$ be unit vectors, with $w_i \in W_i$ for each $i$. For each edge $(ij)$, we have $(w_i, w_j) = z_{ij} \cos \theta_{ij}$, where $\theta_{ij}$ is the angle between $W_i$ and $W_j$ and $z_{ij}$ is a well-defined element of $U$, the group of unimodular scalars. Since $z_{ij} z_{ji} = 1$, the map $(ij) \mapsto z_{ij}$ extends uniquely to a homomorphism $f\colon C \to U$. Although $f$ depends on the choice of the unit vectors, the restriction $\varphi = f\,|\,\Gamma$ depends only on the system of lines, since replacing each $w_i$ by

$\zeta_i w_i$, with $\zeta_i \in U$, replaces each $z_{ij}$ by $\zeta_i \zeta_j^{-1} z_{ij}$, and hence leaves $f(\gamma)$ fixed, for each circuit $\gamma$. In this way we define the *cohomology class* $\varphi \in \mathrm{Hom}(\Gamma, U)$ of the system of lines.

***Theorem 2*** *Let $W_1, \ldots, W_k$ and $W_1', \ldots, W_k'$ be two k-tuples of one-dimensional subspaces of the n-dimensional real or complex Hilbert space V. Assume that the angle between $W_i$ and $W_j$ is the same as the angle between $W_i'$ and $W_j'$, for each i, j. Assume also that the corresponding cohomology classes are the same. Then there is a linear isometry of V sending $W_i$ to $W_i'$ for each i.*

*Proof* Let $w_1, \ldots, w_k$ and $w_1', \ldots, w_k'$ be unit vectors, with $w_i \in W_i$ and $w_i' \in W_i'$ for each $i$. The first assumption gives $|(w_i, w_j)| = |(w_i', w_j')|$ for each $i, j$. For each edge $(ij)$, let

$$(w_i, w_j) = z_{ij} \cos \theta_{ij}, \qquad \text{and} \qquad (w_i', w_j') = z_{ij}' \cos \theta_{ij}. \tag{4}$$

Expressed in terms of $\zeta_{ij} = z_{ij}^{-1} z_{ij}'$, the second assumption is that

$$\zeta_{j_1 j_2} \zeta_{j_2 j_3} \cdots \zeta_{j_m j_1} = 1, \tag{5}$$

for each circuit (3). It follows from this that there are numbers $\zeta_i \in U$ such that $\zeta_{ij} = \zeta_i^{-1} \zeta_j$ for each edge $(ij)$. In proving this, we may suppose the graph is connected. Choose $\zeta_1$ arbitrarily, and put

$$\zeta_i = \zeta_1 \zeta_{1 j_1} \zeta_{j_1 j_2} \cdots \zeta_{j_\mu i}, \tag{6}$$

where $(1j_1), (j_1 j_2), \ldots, (j_\mu i)$ are the edges of any *path* from 1 to $i$. It follows from (5) that $\zeta_i$ is well defined, and it follows from (6) that $\zeta_i \zeta_{ij} = \zeta_j$, for each edge $(ij)$, as asserted. For each edge $(ij)$,

$$(w_i, w_j) = z_{ij} \cos \theta_{ij} = \zeta_i \zeta_j^{-1} z_{ij}' \cos \theta_{ij} = (\zeta_i w_i', \zeta_j w_j').$$

By Lemma 2, there is a linear isometry of $V$ sending $w_i$ to $\zeta_i w_i'$ and hence $W_i$ to $W_i'$ for each $i$.

Next we relate the cohomology class $\varphi$ of a $k$-tuple of lines $W_1, \ldots, W_k$ to the trace invariants. Since $P_j w_i = (w_i, w_j) w_j$, we have

$$P_{j_1} P_{j_2} \cdots P_{j_m} w_i = (w_{j_2}, w_{j_1}) \cdots (w_i, w_{j_m}) w_{j_1},$$

and hence

$$\mathrm{tr}(P_{j_1} P_{j_2} \cdots P_{j_m}) = (w_{j_2}, w_{j_1}) \cdots (w_{j_1}, w_{j_m}). \tag{7}$$

Thus the trace invariant (7) vanishes unless there are edges from $j_1$ to $j_2, \ldots, j_m$ to $j_1$, in which case

$$\mathrm{tr}(P_{j_1} P_{j_2} \cdots P_{j_m}) = \cos \theta_{j_1 j_2} \cdots \cos \theta_{j_m j_1} \varphi(\gamma^{-1}),$$

where $\gamma = (j_1 j_2) \cdots (j_m j_1)$. It follows that the angles $\theta_{ij}$ and the class $\varphi$ determine and are determined by the system of all traces (7).

***Lemma 3*** *The homology group of the graph associated to a system of one-dimensional subspaces* $W_1, \ldots, W_k$ *in a* Hilbert *space of dimension* $n \geqslant 4$ *is generated by the circuits* $\gamma = (j_1 j_2) \cdots (j_m j_1)$ *with* $m \leqslant n$.

*Proof* Each circuit is a product of *minimal* circuits, i.e. circuits $\gamma$ as above for which there is no edge from $j_\mu$ to $j_\nu$ for $1 \leqslant \mu < \nu \leqslant m$, except for $\mu = \nu - 1$ and $\mu = 1$, $\nu = m$. It suffices to show that if $\gamma$ is a minimal circuit with $m \geqslant 5$, then the subspaces $W_{j_1}, \ldots, W_{j_m}$ are linearly independent, i.e.

$$W_{j_\nu} \subset \sum_{\mu \neq \nu} W_{j_\mu} \tag{8}$$

is impossible, for each $\nu$. The argument for $\nu = 3$ is typical. Let $v_1, \ldots, v_m$ be unit vectors, with $v_\mu \in W_{j_\mu}$. Assuming (8) with $\nu = 3$, we would have $v_3 = av_2 + bv_4$, since $v_3$ is orthogonal to $v_1, v_5, \ldots, v_m$. Since $v_1$ is orthogonal to $v_3$ and $v_4$, but not to $v_2$, we get $a = 0$. Similarly, since $v_5$ is orthogonal to $v_2$ and $v_3$, but not to $v_4$, we get $b = 0$. Hence $v_3 = 0$, a contradiction.

***Corollary*** *A system of one-dimensional subspaces* $W_1, \ldots, W_k$ *in a* Hilbert *space* $V$ *of dimension* $n \geqslant 4$ *is determined up to a linear isometry of* $V$ *by the numbers*

$$\operatorname{tr}(P_{j_1} \cdots P_{j_m}) \qquad (1 \leqslant j_\mu \leqslant k, \quad m \leqslant n),$$

*where* $P_j$ *is the orthogonal projection of* $V$ *onto* $W_j$.

## 3. Representations of $H^*$-algebras

An $H^*$-*algebra* is a finite-dimensional† real (or complex) associative algebra $\mathscr{A}$, together with a positive definite quadratic (or Hermitian) form $(x, y)$ on $\mathscr{A}$, and a map $x \mapsto x^*$ on $\mathscr{A}$, satisfying

$$(xy, z) = (y, x^*z) = (x, zy^*).$$

It follows from these axioms that the map * is a linear (or conjugate-linear) involutory antiautomorphism of $\mathscr{A}$.

For example, if $V$ is a finite-dimensional Hilbert space, then the full endomorphism algebra $\mathscr{E}(V)$, together with the usual adjoint map and the inner product $(S, T) = \operatorname{tr}(ST^*)$, is an $H^*$-algebra.

A *representation* of a real (or complex) $H^*$-algebra $\mathscr{A}$ on a finite-dimensional real (or complex) Hilbert space $V$ is an algebra homomorphism $R\colon \mathscr{A} \to \mathscr{E}(V)$ satisfying $R(x^*) = R(x)^*$. If $W$ is an $\mathscr{A}$-invariant subspace of $V$, then the orthogonal complement of $W$ in $V$ is also $\mathscr{A}$-invariant. It follows that each representation of an $H^*$-algebra is a direct sum of

† Infinite-dimensional $H^*$-algebras were first defined and studied by Ambrose [1].

irreducible representations. The *character* $\chi$ of a representation $R$ is defined by $\chi(x) = \mathrm{tr}(R(x))$; we have $\chi(x^*) = \chi(x)$ (or $\bar{\chi}(x)$), and $\chi_{R \oplus S} = \chi_R + \chi_S$. Two representations $R$ and $R'$ of an $H^*$-algebra $\mathscr{A}$ on spaces $V$ and $V'$ are *equivalent* if there is a linear isometry $U\colon V \to V'$ such that $R'(x) = UR(x)U^{-1}$.

***Theorem 3*** *Two representations $R$ and $R'$ of an $H^*$-algebra $\mathscr{A}$ are equivalent if and only if their characters are equal.*

*Proof* Assume first that $R$ and $R'$ are irreducible. Let $b_1, \ldots, b_n$ be an orthonormal basis for $\mathscr{A}$. For each linear map $T\colon V \to V'$, put

$$T_{\mathscr{A}} = \sum_{i=1}^{n} R'(b_i)TR(b_i^*).$$

For each $x \in \mathscr{A}$, we have

$$R'(x)T_{\mathscr{A}} = \sum_{i=1}^{n} R'(xb_i)TR(b_i^*).$$

Since $xb_i = \sum_j (xb_i, b_j)b_j$, this is

$$\sum_i \sum_j (xb_i, b_j)R'(b_j)TR(b_i^*) = \sum_j \sum_i (b_j^*x, b_i^*)R'(b_j)TR(b_i^*).$$

Since $b_j^*x = \sum_i (b_j^*x, b_i^*)b_i^*$ (because $b_1^*, \ldots, b_n^*$ is also an orthonormal basis for $\mathscr{A}$), this is

$$\sum_{j=1}^{n} R'(b_j)TR(b_j^*x) = T_{\mathscr{A}} R(x).$$

Thus

$$R'(x)T_{\mathscr{A}} = T_{\mathscr{A}} R(x). \tag{9}$$

Taking the adjoint of (9), we get

$$T_{\mathscr{A}}^*R'(x^*) = R(x^*)T_{\mathscr{A}}^*. \tag{10}$$

Replacing $x^*$ by $x$ in (10), and combining the result with (9), we conclude that $T_{\mathscr{A}}^*T_{\mathscr{A}}$ commutes with $R(x)$. Since $T_{\mathscr{A}}^*T_{\mathscr{A}}$ is self-adjoint, it has a real eigenvalue $\lambda$. Hence $T_{\mathscr{A}}^*T_{\mathscr{A}} - \lambda I$ is singular and commutes with $R(x)$, so $T_{\mathscr{A}}^*T_{\mathscr{A}} = \lambda I$, by Schur's lemma. Taking the trace, we see that $\lambda \geqslant 0$, with equality only if $T_{\mathscr{A}} = 0$. If $\lambda > 0$, put $U = \lambda^{-1/2}T_{\mathscr{A}}$. Then $U^*U = I$, so $U$ is a linear isometry, and $R'(x) = UR(x)U^{-1}$, so $R$ and $R'$ are equivalent. It follows that if $R$ and $R'$ are not equivalent, then $T_{\mathscr{A}} = 0$, for each $T$.

Let $E$ and $E'$ be endomorphisms of $V$ and $V'$. Then the map $T \mapsto E'TE$ is an endomorphism of the space of linear maps from $V$ to $V'$; its trace is $\mathrm{tr}(E') \cdot \mathrm{tr}(E)$. It follows from the result above that if $R$ and $R'$ are inequivalent irreducible representations of $\mathscr{A}$, with characters $\chi$ and $\chi'$, then

$$\sum_{i=1}^{n} \chi'(b_i)\chi(b_i{}^*) = 0.$$

From this one deduces in the usual way that the characters of the (equivalence classes of) irreducible representations of $\mathscr{A}$ are linearly independent. Hence the character of a representation determines the number of occurrences of any given irreducible representation (up to equivalence) in any direct sum decomposition into irreducible representations. It follows that any two representations with the same character are equivalent.

## 4. Systems of subspaces

Using the algebra generated by the orthogonal projections $P_j$, we prove the result stated at the beginning.

***Theorem 4*** *Let $V$ be an $n$-dimensional real (or complex)* Hilbert *space. Let $W_1, \ldots, W_k$ and $W_1', \ldots, W_k'$ be subspaces of $V$. Let $P_i$ and $P_i'$ be the orthogonal projections of $V$ onto $W_i$ and $W_i'$ respectively. Assume that*

$$\mathrm{tr}(P_{j_1} \cdots P_{j_m}) = \mathrm{tr}(P'_{j_1} \cdots P'_{j_m}) \tag{11}$$

*for all $m$-tuples of indices $j_1, \ldots, j_m$, for all $m \leqslant 4n^2$. Then there is a linear isometry $U$ of $V$ such that $UW_i = W_i'$ for each $i$.*

*Proof* Let $\mathscr{A}$ be the subalgebra of $\mathscr{E}(V)$ generated by the $P_i$. Since orthogonal projections are self-adjoint, $\mathscr{A}$ is an $H^*$-algebra, with adjoint map and inner product inherited from $\mathscr{E}(V)$. As a vector space, $\mathscr{A}$ is spanned by the set of all *monomials* $M = P_{j_1} \cdots P_{j_m}$; this includes $I$, the monomial of degree 0. In fact, $\mathscr{A}$ is spanned by the system $\mathscr{S}_{n^2}$ of monomials of degree $m \leqslant n^2$, since the subspace $\mathscr{A}_l$ spanned by the monomials of degree $\leqslant l$ strictly increases until $\mathscr{A}_l = \mathscr{A}$, and $\dim \mathscr{A} \leqslant n^2$. Similarly, the $H^*$-algebra $\mathscr{A}'$ generated by the $P_i'$ is spanned by the system $\mathscr{S}'_{n^2}$ of monomials $M' = P'_{j_1} \cdots P'_{j_m}$ of degree $\leqslant n^2$.

It follows from the hypothesis that $(M, N) = (M', N')$ for all $M, N \in \mathscr{S}_{2n^2}$. By Lemma 2, there is therefore a vector space isomorphism $\mathscr{A} \to \mathscr{A}'$ which sends $M$ to $M'$ for each $M \in \mathscr{S}_{2n^2}$. For $M, N \in \mathscr{S}_{n^2}$, we have $MN \in \mathscr{S}_{2n^2}$, so $MN \mapsto (MN)' = M'N'$. It follows that our map is an algebra homomorphism. Evidently it commutes with *.

The inclusion $\mathscr{A} \subset \mathscr{E}(V)$ and the composite $\mathscr{A} \to \mathscr{A}' \subset \mathscr{E}(V)$ give two representations of $\mathscr{A}$ on $V$. These representations have equal characters, by our assumption (11). It follows by Theorem 3 that they are equivalent: there is a linear isometry $U$ of $V$ such that $P_i' = UP_iU^{-1}$ and hence $W_i' = UW_i$ for all $i$.

A slightly more elaborate argument, as in [4], would replace $4n^2$ by $2n^2$ in this result.

## References

1. Ambrose, W., Structure theorems for a special class of Banach algebras, *Trans. Amer. Math. Soc.* **57** (1945), 364–386.
2. Jordan, C., Essai sur la géométrie à *n* dimensions, *Bull. Soc. Math. France* **3** (1875), 103–174.
3. Kaplansky, I., "Linear Algebra and Geometry," 2nd ed. Chelsea, New York, 1974.
4. Pearcy, C., A complete set of unitary invariants for operators generating finite $W^*$-algebras of type I, *Pacific J. Math.* **12** (1962), 1405–1414.
5. Procesi, C., The invariant theory of $n \times n$ matrices, *Advances in Math.* **19** (1976), 306–381.
6. Specht, W., Zur Theorie der Matrixen II, *Jahresber. Math. Ver.* **50** (1940), 19–23.

AMS (MOS) 1970 subject classifications: 15A72, 46K15

# *The Macdonald–Kac formulas as a consequence of the Euler–Poincaré principle*

*Howard Garland and James Lepowsky*
*Yale University*

**Introduction**

The present note is a summary of the paper [2], where we computed the homology of certain infinite-dimensional Lie algebras, and used these computations, together with the Euler–Poincaré principle, to obtain the identities of Macdonald [6] and Kac [4]. Our results include the results in Kostant's paper [5], and also include infinite-dimensional analogues of Kostant's results.

This paper is organized as follows: In Section 1 we introduce certain (possibly infinite-dimensional) Lie algebras associated with symmetrizable Cartan matrices, and we introduce the related notions of the Weyl groups and root systems of these Lie algebras. These Lie algebras and related notions were first defined and studied by Kac [3] and Moody [7], and Section 1 is essentially a brief introduction to their ideas. We take one departure, however, which is to introduce certain derivations $D_i$ of a Kac–Moody algebra $\mathfrak{g}$, and then define an extended algebra $\mathfrak{g}^e$, obtained by adjoining such derivations to $\mathfrak{g}$.

In Section 2 we follow Kac [4] and introduce the notion of a quasisimple module. In Section 3 we define a certain subalgebra $\mathfrak{u}^-$ of $\mathfrak{g}$, associated with a *certain* (finite-type) subset of "simple roots." The subalgebra $\mathfrak{u}^-$ is the analogue of the nilpotent radical of a parabolic subalgebra for the case when $\mathfrak{g}$ is finite-dimensional (indeed, when $\mathfrak{g}$ is finite-dimensional, $\mathfrak{u}^-$ is just that—the nilpotent radical of a parabolic subalgebra). We then introduce certain homology groups $H_i(\mathfrak{u}^-, X^t)$ associated with the quasisimple module $X$, and observe that $H_i(\mathfrak{u}^-, X^t)$ has the structure of an $\mathfrak{r}^e$-module, for a certain finite-dimensional, reductive Lie algebra $\mathfrak{r}^e \subseteq \mathfrak{g}^e$. Indeed, $\mathfrak{r}^e$ normalizes $\mathfrak{u}^-$, $\mathfrak{r}^e \cap \mathfrak{u}^- = 0$, and $\mathfrak{r}^e \oplus \mathfrak{u}^-$ is the analogue of the parabolic subalgebra, which we mentioned earlier. In Theorem 3.2, we then determine the structure of $H_i(\mathfrak{u}^-, X^t)$ as an $\mathfrak{r}^e$-module. In particular, we conclude $H_i(\mathfrak{u}^-, X^t)$ is finite-dimensional. Finally, in Section 4, by applying the Euler–Poincaré principle (Lemma 4.1) to Theorem 3.2, we obtain the combinatorial identities of Kac [4] and Macdonald [6] in Theorems 4.2 and 4.3. In particular, Theorem 4.3 is Kac's infinite-dimensional analogue of Weyl's character formula.

## 1. The Lie algebras of Kac and Moody

We let $\mathbb{Z}$ denote the set of integers. For $l \in \mathbb{Z}$, $l > 0$, we let $A = (A_{ij})_{i,j=1,\ldots,l}$, be an $l \times l$ matrix which satisfies the following conditions:

(i) $A_{ij} \in \mathbb{Z}$, for all $i, j$.
(ii) $A_{ii} = 2$, for all $i$. (1.1)
(iii) $A_{ij} \leq 0$, whenever $i \neq j$.
(iv) There exist positive, rational numbers $q_1, \ldots, q_l$ such that $\operatorname{diag}(q_1, \ldots, q_l)A$ is a symmetric matrix.

An integral matrix $A$ satisfying (i)–(iv) will be called a (*symmetrizable*) *Cartan matrix*.

We let $k$ be a field of char.0 and $\mathfrak{g} = \mathfrak{g}(A)$ be the Lie algebra over $k$ with $3l$ generators $h_i, e_i, f_i$ $(i = 1, \ldots, l)$ and relations

$$[h_i, h_j] = 0, \qquad [e_i, f_j] = \delta_{ij} h_i, \qquad [h_i, e_j] = A_{ij} e_j, \qquad [h_i, f_j] = -A_{ij} f_j,$$

for all $i, j = 1, \ldots, l$, and

$$(\operatorname{ad} e_i)^{-A_{ij}+1} e_j = 0, \qquad (\operatorname{ad} f_i)^{-A_{ij}+1} f_j = 0,$$

whenever $i \neq j$. We let $\mathfrak{h}$ denote the subalgebra of $\mathfrak{g}$ spanned by $h_1, \ldots, h_l$. For each $l$-tuple $(n_1, \ldots, n_l)$ of nonnegative (resp. nonpositive) integers, not all zero, we let $\mathfrak{g}(n_1, \ldots, n_l)$ be the subspace of $\mathfrak{g}$ spanned by the elements

$$[e_{i_1}, [e_{i_2}, \ldots, [e_{i_{r-1}} e_{i_r}] \ldots]] \qquad (\text{resp. } [f_{i_1}, [f_{i_2}, \ldots, [f_{i_{r-1}} f_{i_r}] \ldots]]),$$

where $e_j$ (resp. $f_j$) occurs $|n_j|$ times. Also, let $\mathfrak{g}(0, \ldots, 0) = \mathfrak{h}$ and $\mathfrak{g}(n_1, \ldots, n_l) = 0$, for any other $l$-tuple of integers. Then

$$\mathfrak{g} = \coprod_{(n_1, \ldots, n_l) \in \mathbb{Z}^l} \mathfrak{g}(n_1, \ldots, n_l).$$

This direct sum decomposition gives $\mathfrak{g}$ the structure of a $\mathbb{Z}^l$-graded Lie algebra, and we let $\mathfrak{r}_1$ be the unique graded ideal which is maximal among those graded ideals not intersecting the span of $h_i$, $e_i$, and $f_i$ $(1 \leq i \leq l)$. Then it is a theorem of Kac that $\mathfrak{r}_1 = 0$ (see [4]).

We let $D_i$ $(1 \leq i \leq l)$ be the $i$th-degree derivation of $\mathfrak{g}$; i.e., the derivation which acts on $\mathfrak{g}(n_1, \ldots, n_l)$ as scalar multiplication by $n_i$. Then $D_1, \ldots, D_l$ span an $l$-dimensional space $\mathfrak{d}_0$ of commuting derivations of $\mathfrak{g}$. If $\mathfrak{d} \subseteq \mathfrak{d}_0$ is a subspace, then $\mathfrak{d}$ may be regarded as an abelian Lie algebra acting on the $\mathfrak{d}$-module $\mathfrak{g}$ by derivations. We may thus form the semidirect product Lie algebra $\mathfrak{g}^e = \mathfrak{d} \times \mathfrak{g}$ (where e stands for "extended") with respect to this action. We set $\mathfrak{h}^e = \mathfrak{d} \oplus \mathfrak{h}$, which is an abelian subalgebra of $\mathfrak{g}^e$, and which acts on each subspace $\mathfrak{g}(n_1, \ldots, n_l)$ as scalar operators. We define $\alpha_1, \ldots, \alpha_l \in (\mathfrak{h}^e)^*$, the dual space of $\mathfrak{h}^e$, by the conditions

$$[h, e_i] = \alpha_i(h)e_i, \qquad h \in \mathfrak{h}^e, \qquad i = 1, \ldots, l.$$

We note that

$$\alpha_j(h_i) = A_{ij}, \qquad i, j = 1, \ldots, l.$$

From now on, we assume that $\mathfrak{d}$ is chosen so that $\alpha_1, \ldots, \alpha_l$ are linearly independent (for example, we could take $\mathfrak{d} = \mathfrak{d}_0$). If $A$ is nonsingular, we could of course take $\mathfrak{d} = 0$.

For $\phi \in (\mathfrak{h}^e)^*$, define

$$\mathfrak{g}^\phi = \{x \in \mathfrak{g} \,|\, [h, x] = \phi(h)x, \; h \in \mathfrak{h}^e\}.$$

It is clear that $e_i \in \mathfrak{g}^{\alpha_i}$ and $f_i \in \mathfrak{g}^{-\alpha_i}$ for each $i = 1, \ldots, l$, and that for all $(n_1, \ldots, n_l) \in \mathbb{Z}^l$, we have

$$\mathfrak{g}(n_1, \ldots, n_l) \subseteq \mathfrak{g}^{n_1\alpha_1 + \cdots + n_l\alpha_l}.$$

Moreover, since $\alpha_1, \ldots, \alpha_l$ are linearly independent, this inclusion is an equality, and the decomposition

$$\mathfrak{g} = \coprod_{(n_1, \ldots, n_l) \in \mathbb{Z}^l} \mathfrak{g}(n_1, \ldots, n_l)$$

coincides with the decomposition

$$\mathfrak{g} = \coprod_{\phi \in (\mathfrak{h}^e)^*} \mathfrak{g}^\phi.$$

We define the *roots* of $\mathfrak{g}$ (with respect to $\mathfrak{h}^e$) to be the nonzero elements $\phi$ of $(\mathfrak{h}^e)^*$ such that $\mathfrak{g}^\phi \neq 0$. A root is called *positive* if it is a nonnegative integral linear combination of $\alpha_1, \ldots, \alpha_l$. We let $\Delta$ be the set of roots, and $\Delta_+$ the set of positive roots. We set $\Delta_- = -\Delta_+$. Then $\Delta_-$ consists of roots (which we call *negative* roots) and $\Delta$ is the disjoint union

$$\Delta = \Delta_+ \cup \Delta_- .$$

The roots $\alpha_1, \ldots, \alpha_l$ are called *simple* roots. We have $\mathfrak{g}^0 = \mathfrak{h}$, and we have the direct sum decomposition

$$\mathfrak{g} = \mathfrak{h} \oplus \coprod_{\phi \in \Delta_+} \mathfrak{g}^\phi \oplus \coprod_{\phi \in \Delta_-} \mathfrak{g}^\phi,$$

with $\dim \mathfrak{g}^{-\phi} = \dim \mathfrak{g}^\phi$, for all $\phi \in \Delta$.

For each $i = 1, \ldots, l$, we define a linear automorphism $r_i$ of $(\mathfrak{h}^e)^*$ by the condition $r_i(\phi) = \phi - \phi(h_i)\alpha_i$, for all $\phi \in (\mathfrak{h}^e)^*$. We let $W$ denote the group of automorphisms generated by $\{r_i\}_{i=1,\ldots,l}$. We let $R \subseteq (\mathfrak{h}^e)^*$ be the linear span of the roots. Then each element of $W$ maps $R$ into itself, and moreover, one knows that the restriction homomorphism $w \mapsto w|_R$, $w \in W$, defines an isomorphism from $W$ onto its image in the group of linear automorphisms of $R$ (see e.g., [2, §2], and also [7, 10]). For a finite subset $\Phi$ of $\Delta$, we define

$$\langle \Phi \rangle = \sum_{\phi \in \Phi} \phi \qquad (\in R).$$

For $w \in W$, we set $\Phi_w = \Delta_+ \cap w\Delta_-$. One knows $\Phi_w$ is a finite subset of $\Delta_+$ (see [2]) and thus $\langle \Phi_w \rangle$ is defined.

Next, by (1.1), (iv), we have the symmetric matrix $\operatorname{diag}(q_1, \ldots, q_l)A$. We then define a symmetric bilinear form $\sigma$ on $R$ by the condition

$$\sigma(\alpha_i, \alpha_j) = q_i A_{ij}, \qquad i, j = 1, \ldots, l.$$

We then have $\sigma(\alpha_i, \alpha_i)/2 = q_i$, and we set $x_{\alpha_i} = q_i h_i = (\sigma(\alpha_i, \alpha_i)/2)h_i$, where, of course, $x_{\alpha_i}$ is an element of $\mathfrak{h}$. For $\phi \in R$, we have $\phi = \sum_{i=1}^l a_i \alpha_i$, $a_i \in k$, and we define $x_\phi$ in $\mathfrak{h}$ by $x_\phi = \sum_{i=1}^l a_i x_{\alpha_i}$. We may extend the symmetric form $\sigma$ on $R$ to a symmetric form $\sigma$ on $(\mathfrak{h}^e)^*$, satisfying the following condition: For all $\phi \in R$, $\lambda \in (\mathfrak{h}^e)^*$, we have

$$\sigma(\lambda, \phi) = \lambda(x_\phi).$$

We remark that the form $\sigma$ on $(\mathfrak{h}^e)^*$ is $W$-invariant (see [2, Prop. 2.10]).

We conclude this section with a few remarks and definitions concerning integral elements of $(\mathfrak{h}^e)^*$.

***Definition*** Let $\lambda \in (\mathfrak{h}^e)^*$. Then $\lambda$ is called *integral* if $\lambda(h_i) \in \mathbb{Z}$ for all $i = 1, \ldots, l$, and $\lambda$ is called *dominant integral* if $\lambda(h_i) \in \mathbb{Z}_+$ (the nonnegative integers) for all $i = 1, \ldots, l$. We let $P \subseteq (\mathfrak{h}^e)^*$ be the set of dominant integral elements.

We fix an element $\rho \in P$ such that $\rho(h_i) = 1$, for all $i = 1, \ldots, l$. One can then show (see [2, Prop. 2.5]) that

$$\langle \Phi_w \rangle = \rho - w \cdot \rho, \qquad w \in W.$$

## 2. Quasisimple modules

In the algebra $\mathfrak{g}^e$, we let $\mathfrak{n}^{\pm}$ be the subalgebra

$$\mathfrak{n}^{\pm} = \coprod_{\phi \in \Delta_{\pm}} \mathfrak{g}^{\phi},$$

and we simply write $\mathfrak{n}$ for $\mathfrak{n}^+$. If $X$ is a left $\mathfrak{h}^e$-module (e.g., if $X$ is a left $\mathfrak{g}^e$-module, then by restriction, $X$ inherits a left $\mathfrak{h}^e$-module structure) and if $\nu \in (\mathfrak{h}^e)^*$, then we define the *weight space* $X_\nu \subseteq X$ to be $\{x \in X \mid h \cdot x = \nu(h)x$, for all $x \in \mathfrak{h}^e\}$. We call $\nu$ a *weight* of $X$ if $X_\nu \neq 0$, and we call the nonzero elements of $X_\nu$ *weight vectors* with weight $\nu$. We call $X$ a *weight module* if it is a direct sum of its weight spaces.

A left $\mathfrak{g}^e$-module $X$ is called a *highest weight module* if it is generated by an $\mathfrak{n}$-invariant weight vector $x$, which we call a *highest weight vector*, and which is determined uniquely up to a nonzero scalar. The weight of $x$ is called the *highest weight* of $X$, and the weight space spanned by $x$ is called the *highest weight space* of $X$. $X$ is a weight module with finite-dimensional weight spaces.

***Definition*** Let $X$ be a $\mathfrak{g}^e$-module. Then we say $X$ is *quasisimple* if $X$ is a highest weight module with a highest weight vector $x$ such that there exists $n \in \mathbb{Z}_+$ with $f_i^n \cdot x = 0$, for all $i = 1, \ldots, l$ (see Kac [4]).

A quasisimple $\mathfrak{g}^e$-module turns out to be simple, and the set of equivalence classes of quasisimple $\mathfrak{g}^e$-modules is indexed (via highest weight) by the set $P$ of dominant integral linear forms (see [4, 2]).

## 3. Homology associated with $F$-parabolic subalgebras

We let $S \subseteq \{1, \ldots, l\}$ and we let $\mathfrak{g}_S \subseteq \mathfrak{g}$ denote the subalgebra generated by $\{h_i, e_i, f_i\}_{i \in S}$. We let

$$\Delta^S = \Delta \cap \coprod_{i \in S} \mathbb{Z}\alpha_i, \qquad \Delta_{\pm}{}^S = \Delta_{\pm} \cap \coprod_{i \in S} \mathbb{Z}\alpha_i,$$

and $\mathfrak{u}^{\pm} = \mathfrak{u}_S{}^{\pm} = \coprod_{\phi \in \Delta_{\pm} - \Delta_{\pm}{}^S} \mathfrak{g}^{\phi}$. We let $\mathfrak{r} = \mathfrak{r}_S = \mathfrak{g}_S + \mathfrak{h}$, and note that $\mathfrak{r}$, $\mathfrak{u}^{\pm}$, $\mathfrak{g}_S$, and $\mathfrak{p} = \mathfrak{r} \oplus \mathfrak{u}$ (where $\mathfrak{u} =_{df} \mathfrak{u}^+$), are subalgebras of $\mathfrak{g}$. For every $\mathfrak{d}$-invariant subalgebra $\mathfrak{t}$ of $\mathfrak{g}$, we let $\mathfrak{t}^e$ denote the subalgebra $\mathfrak{d} \oplus \mathfrak{t}$ of $\mathfrak{g}^e$. We say that $\mathfrak{p}$ (resp. $\mathfrak{p}^e = \mathfrak{r}^e \oplus \mathfrak{u}$) is an *F-parabolic subalgebra* of $\mathfrak{g}$ (resp. of $\mathfrak{g}^e$) if $\mathfrak{r}$ is finite-dimensional. In the case when $\mathfrak{g}$ is finite-dimensional, the algebra $\mathfrak{u}$ is the nilpotent radical of $\mathfrak{p}$. In general we may regard $\mathfrak{u}$ as a "pronilpotent" radical of $\mathfrak{p}$. Whenever convenient,we drop the subscript $S$.

We let $X$ denote a left quasisimple $\mathfrak{g}^e$-module. We let $T\colon \mathfrak{g}^e \to \mathfrak{g}^e$ denote the antiautomorphism

$$T(\xi) = -\xi, \qquad \xi \in \mathfrak{g}^e.$$

We denote by $X^t$ the right $\mathfrak{g}^e$-module whose underlying vector space is $X$ and whose (right) $\mathfrak{g}^e$ action is given by

$$v \cdot \xi = T(\xi)v, \qquad v \in X, \quad \xi \in \mathfrak{g}^e.$$

We now fix an $F$-parabolic subalgebra $\mathfrak{p}^e = \mathfrak{r}^e \oplus \mathfrak{u}$ of $\mathfrak{g}^e$, and by restriction, regard $X^t$ as a right $\mathfrak{u}^-$-module. We let

$$C_i = X^t \otimes_k \Lambda^i \mathfrak{u}^-, \qquad i \geq 0,$$

and we define $C_{-1}$ to be zero.

We then define

$$\partial_i\colon C_i \to C_{i-1}, \qquad i \geq 0,$$

by letting $\partial_0 = 0$, and

$$\begin{aligned}
&\partial_i(v \otimes \xi_1 \wedge \cdots \wedge \xi_i) \\
&\quad = \sum_{1 \leq j \leq i} (-1)^{j+1} v \cdot \xi_j \otimes \xi_1 \wedge \cdots \wedge \hat{\xi}_j \wedge \cdots \wedge \xi_i \\
&\qquad + \sum_{1 \leq j < k \leq i} (-1)^{j+k} v \otimes [\xi_j, \xi_k] \wedge \xi_1 \wedge \cdots \wedge \hat{\xi}_j \wedge \cdots \wedge \hat{\xi}_k \wedge \cdots \wedge \xi_i, \qquad (3.1)
\end{aligned}$$

where $i \geq 1$, and $v \in X^t$, $\xi_1, \ldots, \xi_i \in \mathfrak{u}^-$. (We extend $\partial_i$ to all of $C_i$, by linearity.) We let $H_i(\mathfrak{u}^-, X^t)$ denote the $i$th homology of the chain complex $\{C_i, \partial_i\}_i$.

Now $X$ is a left $\mathfrak{g}^e$-module, and hence, by restriction, a left $\mathfrak{r}^e$-module. With this same action of $\mathfrak{r}^e$, we regard $X^t$ as a *left* $\mathfrak{r}^e$-module. On the other hand, the algebra $\mathfrak{u}^-$ is normalized by $\mathfrak{r}^e$, and using the adjoint action of $\mathfrak{r}^e$ on $\mathfrak{u}^-$, we may regard $\mathfrak{u}^-$, and thus $\Lambda^i\mathfrak{u}^-$, as left $\mathfrak{r}^e$-modules. But then using the tensor product of the action of $\mathfrak{r}^e$ on $X^t$ and on $\Lambda\mathfrak{u}^-$, we may (and do) regard $X^t \otimes \Lambda^i\mathfrak{u}^- = C_i$ as a left $\mathfrak{r}^e$-module. Then a direct computation using definition (3.1) of $\partial_i$, shows that each $\partial_i$ is an $\mathfrak{r}^e$-module homomorphism. Hence each homology group $H_i(\mathfrak{u}^-, X^t)$ has an induced $\mathfrak{r}^e$-module structure.

Before describing the structure of $H_i(\mathfrak{u}^-, X^t)$ as an $\mathfrak{r}^e$-module we require a few further observations and notations. First, the group $W$ defined in Section 1 is a Coxeter group. More precisely, the pair $(W, \{r_i\}_{i=1,\ldots,l})$ is a *Coxeter system* in the sense of Bourbaki [1] (see [7, 10]). We say $S \subseteq \{1, \ldots, l\}$

is of *finite type* if the corresponding subalgebra $\mathfrak{r}_S \subseteq \mathfrak{g}$ is finite-dimensional. For $S \subseteq \{1, \ldots, l\}$, we let $W_S{}^1 \subseteq W$ be the set of all $w \in W$ such that $\Phi_w \subseteq \Delta_+ - \Delta_+{}^S$.

We let $P_S = \{\lambda \in (\mathfrak{h}^e)^* \,|\, \lambda(h_i) \in \mathbb{Z}_+$, for all $i \in S\}$. We now assume $S$ is of finite type, and we let $\mathfrak{h}_S$ denote the linear span of the $h_i$, $i \in S$. Then $\mathfrak{g}_S$ is a finite-dimensional, semisimple algebra, with Cartan subalgebra $\mathfrak{h}_S$. We let $\mathfrak{n}_S \subseteq \mathfrak{g}_S$ denote the subalgebra

$$\mathfrak{n}_S = \coprod_{\phi \in \Delta_+{}^S} \mathfrak{g}^\phi.$$

For each $\lambda \in P_S$, there corresponds a unique (up to isomorphism) finite-dimensional, irreducible $\mathfrak{r}^e$-module $M(\lambda)$. Namely, $M(\lambda)$ is determined by the conditions that as a $\mathfrak{g}_S$-module it is irreducible, and has highest weight space, relative to $\mathfrak{h}_S$ and $\mathfrak{n}_S$, stable with respect to $\mathfrak{h}^e$. Moreover, $\lambda$ is the resulting weight for the action of $\mathfrak{h}^e$ on this highest weight space.

If $w \in W$, then $l(w)$, the *length* of $w$, is defined to be the minimal integer $k$ such that $w$ has an expression $w = r_{i_1} \cdots r_{i_k}$, $1 \leq i_1, \ldots, i_k \leq l$. Finally, we note that for $\lambda \in P$ and $w \in W_S{}^1$, we have that $w(\lambda + \rho) - \rho \in P_S$ (see [2, Props. 8.3 and 8.4]). We are now prepared to describe $H_j(\mathfrak{u}^-, X^t)$ as an $\mathfrak{r}^e$-module (where $S \subseteq \{1, \ldots, l\}$ is a fixed subset of finite type, $\mathfrak{u}^- = \mathfrak{u}_S{}^-$, $\mathfrak{r}^e = \mathfrak{r}_S{}^e$, etc.):

***Theorem 3.2*** *Let $X$ be a (left) quasisimple $\mathfrak{g}^e$-module with highest weight $\lambda \in P$. The space $H_j(\mathfrak{u}^-, X^t)$ is a finite-dimensional vector space over $k$, and as an $\mathfrak{r}^e$-module, is isomorphic to the direct sum*

$$\coprod_{w \in W_S{}^1,\, l(w) = j} M(w(\lambda + \rho) - \rho).$$

*If $w, w' \in W_S{}^1$, then the corresponding constituents $M(w(\lambda + \rho) - \rho)$ and $M(w'(\lambda + \rho) - \rho)$ are isomorphic $\mathfrak{r}^e$-modules if and only if $w = w'$.*

*Remark* There are analogous results for cohomology; see [2, Th. 8.6]. Also see [2] for the proof of Theorem 3.2.

## 4. Combinatorial identities

We let $(\mathfrak{h}^e)_{\mathbb{Z}}{}^*$ denote the set of integral linear forms; i.e., the set of all $\lambda \in (\mathfrak{h}^e)^*$ such that $\lambda(h_i) \in \mathbb{Z}$, for all $i = 1, \ldots, l$. We let $\mathfrak{A}$ denote the abelian group of all (possibly infinite) formal integral combinations of elements of $(\mathfrak{h}^e)_{\mathbb{Z}}{}^*$. We let $e(\lambda) \in \mathfrak{A}$ denote the element corresponding to $\lambda \in (\mathfrak{h}^e)_{\mathbb{Z}}{}^*$, and $\mathfrak{A}^f$ the integral group algebra of $(\mathfrak{h}^e)_{\mathbb{Z}}{}^*$, so that $\mathfrak{A}^f \subseteq \mathfrak{A}$, and $\mathfrak{A}^f$ consists of the finite, formal, integral combinations of the $e(\lambda)$, $\lambda \in (\mathfrak{h}^e)_{\mathbb{Z}}{}^*$. In $\mathfrak{A}^f$ we have $e(\lambda)e(\mu) = e(\lambda + \mu)$, for all $\lambda$, $\mu \in (\mathfrak{h}^e)_{\mathbb{Z}}^*$. We will allow ourselves to multiply elements of $\mathfrak{A}$ whenever the product is well defined.

Let $\mathfrak{C}$ denote the category of $\mathfrak{h}^e$-modules $X$ such that $X$ has a direct sum decomposition

$$X = \coprod_{\lambda \in (\mathfrak{h}^e)_{\mathbb{Z}}^*} X_\lambda,$$

and each $X_\lambda$ is finite-dimensional (see Section 2 for the definition of $X_\lambda$). The morphisms of this category are the $\mathfrak{h}^e$-module maps.

If $X$ is an object of $\mathfrak{C}$, then we can define $\chi(X) \in \mathfrak{A}$, the *formal character* of $X$. Namely,

$$\chi(X) = \sum_{\lambda \in (\mathfrak{h}^e)_{\mathbb{Z}}^*} (\dim X_\lambda) e(\lambda).$$

We let

$$\cdots \longrightarrow C_{j+1} \xrightarrow{\partial_{j+1}} C_j \xrightarrow{\partial_j} \cdots \xrightarrow{\partial_1} C_0 \longrightarrow 0$$

be a chain complex in $\mathfrak{C}$ (so that $\partial_j \circ \partial_{j+1} = 0$, for $j = 1, 2, \ldots$), and we let $H_j$ denote the $j$th homology group of this complex. Then $H_j \in \mathfrak{C}$. We call this chain complex *admissible* in case $\coprod_j C_j \in \mathfrak{C}$. The following lemma is trivial to prove:

***Lemma 4.1*** (Euler–Poincaré) *Let $C_*$ be an admissible chain complex in $\mathfrak{C}$. Then $\coprod_j H_j$ is an object of $\mathfrak{C}$ and*

$$\sum_{j \in \mathbb{Z}_+} (-1)^j \chi(C_j) = \sum_{j \in \mathbb{Z}_+} (-1)^j \chi(H_j).$$

If we apply the Euler–Poincaré principle in this form, together with Theorem 3.2, to the complex $\{C_i, \partial_i\}_i$ defined in Section 3 ($C_i = X^t \otimes_k \Lambda^i \mathfrak{u}^-$), then we obtain, when $S$ is null and $X = k$, the trivial one-dimensional $\mathfrak{g}^e$-module:

***Theorem 4.2*** (Kac [4]) *We have the formal identity*

$$\prod_{\phi \in \Delta_+} (1 - e(-\phi))^{\dim \mathfrak{g}^\phi} = \sum_{w \in W} (\det w) e(w\rho - \rho) = \sum_{w \in W} (\det w) e(-\langle \Phi_w \rangle).$$

For an arbitrary quasisimple $\mathfrak{g}^e$-module with highest weight $\lambda \in P$, we have:

***Theorem 4.3*** (Kac [4]) *We have the formal identity*

$$\chi(X)\left(\sum_{w \in W} \det w \; e(w \cdot \rho)\right) = \sum_{w \in W} \det w \; e(w \cdot (\lambda + \rho)).$$

The following remarks are in order: Theorem 4.2 follows from the Euler–Poincaré principle and Theorem 3.2, in the case when $X = k$ is the trivial one-dimensional $\mathfrak{g}^e$-module. Theorem 4.3 then follows from Theorem 4.2, the Euler–Poincaré principle, and Theorem 3.2 for general $X$. Also, Macdonald's identities [6] comprise the special case of Theorem 4.2 when $\mathfrak{g}$ is a Euclidean Lie algebra (see [8, 9, 11]).

## References

1. N. Bourbaki, "Groupes et algèbres de Lie," Chapters 4–6. Hermann, Paris, 1968.
2. H. Garland and J. Lepowsky, Lie algebra homology and the Macdonald–Kac formulas, *Invent. Math.* **34** (1976), 37–76.
3. V. G. Kac, Simple irreducible graded Lie algebras of finite growth (in Russian), *Izv. Akad. Nauk. SSSR* **32** (1968), 1323–1367 [English transl.: *Math. USSR–Izv.* **2** (1968), 1271–1311].
4. V. G. Kac, Infinite-dimensional Lie algebras and Dedekind's $\eta$-function (in Russian), *Funktional Anal. i. Priložen.* **8** (1974), 77–78 [English transl.: *Functional Anal. Appl.* **8** (1974), 68–70].
5. B. Kostant, Lie algebra cohomology and the generalized Borel–Weil theorem, *Ann. of Math.* **74** (1961), 329–387.
6. I. G. Macdonald, Affine root systems and Dedekind's $\eta$-function, *Invent. Math.* **15** (1972), 91–143.
7. R. V. Moody, A new class of Lie algebras, *J. Algebra* **10** (1968), 211–230.
8. R. V. Moody, Euclidean Lie algebras, *Canad. J. Math.* **21** (1969), 1432–1454.
9. R. V. Moody, Macdonald identities and Euclidean Lie algebras, *Proc. Amer. Math. Soc.* **48** (1975), 43–52.
10. R. V. Moody and K. L. Teo, Tits' systems with crystallographic Weyl groups, *J. Algebra* **21** (1972), 178–190.
11. D.-N. Verma, Review of I. G. Macdonald's paper, "Affine root systems and Dedekind's $\eta$-function." *Math Revs.* **50** (1975), 1371–1374 (MR 9996).

H.G. partially supported by National Science Foundation Grant No. MPS71-03469.
J.L. partially supported by National Science Foundation Grants MPS71-03469 and MPS72-05055 A03, and a Yale University Junior Faculty Fellowship.

AMS (MOS) 1970 subject classifications: 18H25, 17B65, 17B10, 05A19

# *The characters of reductive p-adic groups*

*Harish-Chandra*

*The Institute for Advanced Study*

The motivation for the results which I am going to discuss comes from the Lefschetz principle, which says that whatever is true for real groups should also be true for $p$-adic groups. So let me begin by describing some known results in the real case.

Let $G$ be a compact, connected, real semisimple Lie group and $\mathfrak{g}$ its Lie algebra. Fix a Cartan subalgebra $\mathfrak{h}$ and define Fourier transforms on $\mathfrak{g}$ and $\mathfrak{h}$ as follows:

$$\hat{f}(Y) = \int_{\mathfrak{g}} e^{iB(Y,X)} f(X)\, dX \qquad (Y \in \mathfrak{g}, \quad f \in \mathscr{C}(\mathfrak{g})),$$

$$\hat{g}(H') = \int_{\mathfrak{h}} e^{iB(H',H)} g(H)\, dH \qquad (H' \in \mathfrak{h}, \quad g \in \mathscr{C}(\mathfrak{h})).$$

Here $B$ is the Killing form of $\mathfrak{g}$ and $\mathscr{C}(\mathfrak{g})$ and $\mathscr{C}(\mathfrak{h})$ the corresponding Schwartz spaces. It is convenient to assume that the Euclidean measures $dX$ and $dH$ on $\mathfrak{g}$ and $\mathfrak{h}$ respectively are self-dual. This means that

$$f(0) = \int_{\mathfrak{g}} \hat{f}(X)\, dX, \qquad g(0) = \int_{\mathfrak{h}} \hat{g}(H)\, dH$$

for all $f \in \mathscr{C}(\mathfrak{g})$ and $g \in \mathscr{C}(\mathfrak{h})$. Put

$$\phi_f(H) = \pi(H)\int_G f(x.H)dx \qquad (H \in \mathfrak{h},\ f \in \mathscr{C}(\mathfrak{g})).$$

Here $x.H = Ad(x)H$, $dx$ is the normalized Haar measure on $G$ and

$$\pi = \prod_{\alpha > 0} \alpha$$

where $\alpha$ runs over all positive roots of $(\mathfrak{g}, \mathfrak{h})$ (under some order). Then $f \mapsto \phi_f$ is a continuous mapping of $\mathscr{C}(\mathfrak{g})$ into $\mathscr{C}(\mathfrak{h})$ and the following theorem establishes a relation between this mapping and the Fourier transform [3, §5].

***Theorem 1*** *Let* $r = \frac{1}{2}\dim \mathfrak{g}/\mathfrak{h}$. *Then*

$$\phi_{\hat{f}} = i^r(\phi_f)^\wedge \qquad \textit{for } f \in \mathscr{C}(\mathfrak{g}).$$

***Corollary***

$$\int_G e^{iB(x.H,\, H')}\,dx = \frac{c\sum_{s\in W}\varepsilon(s)e^{iB(sH,\, H')}}{\pi(H)\pi(H')} \qquad \textit{for } H, H' \in \mathfrak{h}.$$

Here $W$ is the Weyl group of $(\mathfrak{g}, \mathfrak{h})$ and $c$ a constant given by

$$c = i^r[W]^{-1}\langle \pi, \pi\rangle$$

in the notation of [3, p. 90]. ($[W]$ denotes the order of $W$.) The striking resemblance between the above corollary and Weyl's character formula is obvious.

Now let us drop the assumption that $G$ is compact. Let $A$ be the Cartan subgroup of $G$ corresponding to $\mathfrak{h}$. Fix an invariant measure $dx^*$ on $G/A$ and let $\mathfrak{h}'$ be the set of all points $H \in \mathfrak{h}$ where $\pi(H) \neq 0$. Put

$$\phi_f(H) = \pi(H)\int_{G/A} f(x.H)\,dx^* \qquad (H \in \mathfrak{h}')$$

for $f \in \mathscr{C}(\mathfrak{g})$. This integral is well defined and $\phi_f$ is a $C^\infty$ function on $\mathfrak{h}'$. Fix $H_0 \in \mathfrak{h}'$ and consider the mapping

$$T : f \to \phi_f(H_0) \qquad (f \in \mathscr{C}(\mathfrak{g})).$$

Then $T$ is a tempered and $G$-invariant distribution on $\mathfrak{g}$.

Let $\mathfrak{g}_c$ be the complexification of $\mathfrak{g}$ and $S(\mathfrak{g}_c)$ the symmetric algebra over $\mathfrak{g}_c$. For $X \in \mathfrak{g}$, define a translation-invariant differential operator $\partial(X)$ on $\mathfrak{g}$ by

$$(\partial(X)f)(Y) = \left\{\frac{d}{dt} f(Y + tX)\right\}_{t=0} \qquad (Y \in \mathfrak{g})$$

for $f \in C^\infty(\mathfrak{g})$. Then $\partial$ can be extended (uniquely) to an isomorphism of $S(\mathfrak{g}_c)$ onto the algebra of all translation-invariant differential operators on $\mathfrak{g}$. (These are also known as differential operators with constant coefficients.) We may identify $\mathfrak{g}$ with its dual by means of the Killing form. Then $S(\mathfrak{g}_c)$ becomes the algebra of all polynomial functions on $\mathfrak{g}$.

Consider the orbit $H_0{}^G = G.H_0$ and let $\mu$ denote the invariant measure on this orbit so that

$$\mu(f) = \int_{G/A} f(x.H_0)\, dx^* \qquad (f \in \mathscr{C}(\mathfrak{g})).$$

Then $\mu$ is a tempered measure on $\mathfrak{g}$ and

$$T(f) = \pi(H_0)\mu(\hat{f}) = \pi(H_0)\hat{\mu}(f)$$

where $\hat{\mu}$ is the Fourier transform of $\mu$. This shows that $T = \pi(H_0)\hat{\mu}$.

Let $I(\mathfrak{g}_c)$ be the algebra of all $G$-invariants in $S(\mathfrak{g}_c)$. It is obvious that

$$p\mu = p(H_0)\mu \qquad (p \in I(\mathfrak{g}_c)).$$

Taking Fourier transforms, we get

$$p(H_0)\hat{\mu} = (p\mu)^\wedge = \partial(p^*)\hat{\mu}$$

where $p^*$ is the polynomial function $X \mapsto p(-iX)$ $(X \in \mathfrak{g})$. Therefore $T$ is an invariant distribution on $\mathfrak{g}$ which satisfies the differential equations

$$\partial(p)T = p(iH_0)T \qquad (p \in I(\mathfrak{g}_c)).$$

But then by a general theorem [4, p. 11], $T$ is a locally summable function on $\mathfrak{g}$ which is analytic on the regular set $\mathfrak{g}'$.

Now assume for a moment that $A$ is compact. Let $W$ denote the Weyl group of the pair $(\mathfrak{g}_c, \mathfrak{h}_c)$ and $W_G$ the subgroup of those elements which have a representative in $G$. Then on $\mathfrak{h}'$, $T$ is given by a formula of the form [6, §16]

$$T(H) = \pi(H)^{-1} \sum_{s \in W} \varepsilon(s) c_s e^{iB(sH_0, H)} \qquad (H \in \mathfrak{h}').$$

Here $c_s$ are constants such that $c_{ts} = c_s$ for $t \in W_G$. The analogy with the compact case is clear.

But what does this have to do with the theory of characters? Let $\mathfrak{g}_0$ be the set of all $X \in \mathfrak{g}$ such that $|\mathrm{Im}\ \lambda| < \pi = 3.14\ldots$ for every eigenvalue $\lambda$ of ad $X$. Then $\mathfrak{g}_0$ is an open $G$-invariant neighborhood of zero in $\mathfrak{g}$. Put $G_0 = \exp \mathfrak{g}_0$. Then $G_0$ is an open subset of $G$ and the exponential mapping defines an analytic diffeomorphism of $\mathfrak{g}_0$ onto $G_0$. Put

$$\xi(X) = |\det\{(e^{\mathrm{ad} X/2} - e^{-\mathrm{ad} X/2})/\mathrm{ad}\ X\}|^{1/2} \qquad (X \in \mathfrak{g}_0).$$

Then under the exponential mapping the Haar measure $dx$ on $G_0$ is related to the Euclidean measure $dX$ on $\mathfrak{g}_0$ by the formula

$$dx = \xi(X)^2\, dX.$$

For $\phi \in C^\infty(\mathfrak{g}_0)$, define $f_\phi \in C^\infty(G_0)$ by

$$f_\phi(\exp X) = \xi(X)^{-1}\phi(X) \qquad (X \in \mathfrak{g}_0).$$

Then

$$\int \phi_1 \phi_2\, dX = \int f_{\phi_1} \cdot f_{\phi_2}\, dx$$

for $\phi_1 \in C^\infty(\mathfrak{g}_0)$ and $\phi_2 \in C_c^{\ \infty}(\mathfrak{g}_0)$. Given a distribution $T$ on $G_0$, define a distribution $\tau_T$ on $\mathfrak{g}_0$ by

$$\tau_T(\phi) = T(f_\phi) \qquad (\phi \in C_c^{\ \infty}(\mathfrak{g}_0)).$$

Now $G$ operates on $G_0$ by conjugation and it is obvious that if $T$ is $G$-invariant, so is $\tau_T$.

Let $\mathfrak{Z}$ be the algebra of all differential operators on $G$ which commute with both left and right translations of $G$. Suppose $\Theta$ is an invariant distribution on $G_0$ and $z \in \mathfrak{Z}$. Then what is the relation between $\tau_\Theta$ and $\tau_{z\Theta}$? The answer is quite simple [5, p. 477]. There exists an algebra isomorphism

$$z \mapsto p_z$$

from $\mathfrak{Z}$ to $I(\mathfrak{g}_c)$ such that

$$\tau_{z\Theta} = \partial(p_z)\tau_\Theta.$$

Now suppose $\Theta$ is an eigendistribution of $\mathfrak{Z}$. Then

$$z\Theta = \chi(z)\Theta \qquad (z \in \mathfrak{Z})$$

where $\chi$ is a homomorphism of $\mathfrak{Z}$ into $\mathbf{C}$. Then if we write $\theta = \tau_\Theta$ and $\chi(p_z) = \chi(z)$, we get

$$\partial(p)\theta = \chi(p)\theta \qquad (p \in I(\mathfrak{g}_c)).$$

This shows that $\theta$ is an invariant eigendistribution of $\partial(I(\mathfrak{g}_c))$ on $\mathfrak{g}_0$.

The characters of $G$ are invariant eigendistributions of $\mathfrak{Z}$. Hence, in order to find them, one is led by the above procedure to the construction of invariant eigendistributions of $\partial(I(\mathfrak{g}_c))$ on $\mathfrak{g}_0$. But, as we have seen, the Fourier transform of the invariant measure on an orbit gives a solution to this problem. This suggests that there is an intimate connection between the multiplicative Fourier analysis on the group $G$ and the additive Fourier analysis on its Lie algebra $\mathfrak{g}$. In fact, the characters of the discrete series are actually constructed precisely by the method outlined above. Moreover, quite similar results have been discovered by Kirillov for nilpotent groups [10].

We would now like to apply the same philosophy in the $\mathfrak{p}$-adic case. So it is natural to expect that the characters are closely related to the Fourier transforms of the invariant measures on orbits.

Let $\Omega$ be a $\mathfrak{p}$-adic field. I shall have to assume that $\Omega$ has characteristic zero. We use the terminology of [7]. Let $\mathbf{G}$ be a connected reductive $\Omega$-group and $G$ the subgroup of all $\Omega$-rational points in $\mathbf{G}$. Then $G$, with its usual topology, is a locally compact, totally disconnected, unimodular group. Let $\mathfrak{g}$ be the Lie algebra of $G$. Then $\mathfrak{g}$ is a vector space over $\Omega$ of finite dimension and $G$ operates on $\mathfrak{g}$ by means of the adjoint representation. If $\omega$ is any subset of $\mathfrak{g}$, we denote by $J(\omega)$ the space of all $G$-invariant distributions $T$ on $\mathfrak{g}$ such that $\operatorname{Supp} T \subset \operatorname{Cl} \omega^G$. (Here $\omega^G$ is the orbit of $\omega$ under $G$ and Cl denotes closure.) Let $L$ be a lattice (i.e. an open, compact, additive subgroup) in $\mathfrak{g}$. Then $C_c(\mathfrak{g}/L) \subset C_c^\infty(\mathfrak{g})$. For any distribution $T$ on $\mathfrak{g}$, let $j_L T$ denote the restriction of $T$ on $C_c(\mathfrak{g}/L)$. The following remarkable theorem of Howe [8, §4] plays a very important role in harmonic analysis.

***Theorem 2*** (Howe) *Suppose $\omega$ is a compact subset in $\mathfrak{g}$ and $L$ a lattice in $\mathfrak{g}$. Then* $\dim j_L J(\omega) < \infty$.

In many proofs this theorem makes up for the absence of differential operators in the $\mathfrak{p}$-adic case.

Fix a nontrivial character $\chi$ of the additive group of $\Omega$. Also fix a nondegenerate, symmetric, $G$-invariant, bilinear form $B$ on $\mathfrak{g}$ (with values in $\Omega$). For $f \in C_c^\infty(\mathfrak{g})$, define

$$\hat{f}(Y) = \int_{\mathfrak{g}} \chi(B(Y, X)) f(X)\, dX \qquad (Y \in \mathfrak{g}).$$

Here $dX$ is a Haar measure on the additive group of $\mathfrak{g}$. Then $f \mapsto \hat{f}$ is a bijective mapping of $C_c^\infty(\mathfrak{g})$ onto itself. If $T$ is a distribution on $\mathfrak{g}$, its Fourier transform $\hat{T}$ is given by $\hat{T}(f) = T(\hat{f})$ $(f \in C_c^\infty(\mathfrak{g}))$.

Let $l$ be the (absolute) rank of $G$. If $t$ is an indeterminate, let $\eta(X)$ denote the coefficient of $t^l$ in $\det(t\text{-ad } X)$ $(X \in \mathfrak{g})$. Then $\eta$ is a polynomial function on $\mathfrak{g}$ which is not identically zero. Let $\mathfrak{g}'$ be the set of all $X \in \mathfrak{g}$ such that $\eta(X) \neq 0$.

***Theorem 3*** *Let $\omega$ be a compact subset of $\mathfrak{g}$ and $T$ a distribution in $J(\omega)$. Then $\hat{T}$ is a locally summable function on $\mathfrak{g}$ which is locally constant on $\mathfrak{g}'$. Moreover $|\eta|_{\mathfrak{p}}^{1/2} T$ is locally bounded on $\mathfrak{g}$.*

The analogy with the real case [4] is clear.

Fix $X_0 \in \mathfrak{g}$ and consider the orbit $\mathfrak{o} = X_0{}^G$. Let $G_{X_0}$ denote the stabilizer of $X_0$ in $G$. Then $G_{X_0}$ is unimodular and therefore $G/G_{X_0}$ has an invariant measure $dx^*$. By a theorem of Deligne and Rao [11], there exists a measure $\mu_{\mathfrak{o}}$ on $\mathfrak{g}$ given by

$$\mu_{\mathfrak{o}}(f) = \int_{G/G_{X_0}} f(x.X_0)\, dx^* \qquad (f \in C_c(\mathfrak{g})).$$

Since $\mu_{\mathfrak{o}} \in J(X_0)$, we conclude from Theorem 3 that $\hat{\mu}_{\mathfrak{o}}$ is a function.

Let $\pi$ be an irreducible representation of $G$ on a Hilbert space $\mathfrak{H}$. $\pi$ is said to be admissible if, for every open subgroup $K$ of $G$, $\dim \mathfrak{H}_K < \infty$. Here $\mathfrak{H}_K$ is the space of all vectors in $\mathfrak{H}$ which are left fixed by $K$. Suppose $\pi$ is admissible. For $f \in C_c^\infty(G)$ define

$$\pi(f) = \int_G f(x)\pi(x)\, dx$$

where $dx$ is the Haar measure on $G$. Then $\pi(f)$ is an operator of finite rank. Put

$$\Theta_\pi(f) = \operatorname{tr} \pi(f) \qquad (f \in C_c^\infty(G)).$$

Then $\Theta_\pi$ is an invariant distribution on $G$ which is called the character of $\pi$.

Let $D(x)$ denote the coefficient of $t^l$ in $\det(t + 1 - \mathrm{Ad}(x))$ $(x \in G)$ and $G'$ the set of all $x \in G$ where $D(x) \neq 0$.

***Theorem 4*** *Suppose $\pi$ is irreducible and admissible. Then $\Theta_\pi$ is a locally summable function on $G$ which is locally constant on $G'$. Moreover $|D|_{\mathfrak{p}}^{1/2}\Theta_\pi$ is locally bounded on $G$.*

It is possible to describe the behavior of $\Theta_\pi$ around singular points. We give here the result for $x = 1$. Recall that $\mathfrak{g} \subset gl(n, \Omega)$ for some $n \geq 1$. Let $\mathcal{N}$ be the set of all elements of $\mathfrak{g}$ which are nilpotent (as matrices). Then $\mathcal{N}$ is the union of a finite number of $G$-orbits which we call the nilpotent orbits.

***Theorem 5*** *We can choose an open neighborhood $U$ of zero in $\mathfrak{g}$ and, for each nilpotent orbit $\mathfrak{o}$, a complex constant $c_{\mathfrak{o}}$ such that*

$$\Theta_\pi(\exp X) = \sum_{\mathfrak{o}} c_{\mathfrak{o}} \hat{\mu}_{\mathfrak{o}}(X) \qquad (X \in U).$$

The constants $c_{\mathfrak{o}}$ are unique and we denote them by $c_{\mathfrak{o}}(\pi)$. In particular we consider the constant $c_0(\pi)$ corresponding to the nilpotent orbit $\mathfrak{o} = \{0\}$.

Let $Z$ denote the maximal split torus contained in the center of $G$. Fix a Haar measure $dx^*$ on $G/Z$. Let $\pi$ be an irreducible unitary representation of $G$ on $\mathfrak{H}$. As usual we say that $\pi$ is square-integrable if

$$\int_{G/Z} |(\phi, \pi(x)\psi)|^2 \, dx^* < \infty$$

for $\phi$, $\psi \in \mathfrak{H}$. Then there exists a positive number $d(\pi)$, called the formal degree of $\pi$, such that

$$\int_{G/Z} |(\phi, \pi(x)\psi)|^2 \, dx^* = d(\pi)^{-1} |\phi|^2 |\psi|^2$$

for all $\phi$, $\psi \in \mathfrak{H}$. A square-integrable representation is always admissible. It is called supercuspidal if we can choose $\phi$, $\psi \neq 0$ in $\mathfrak{H}$ such that the function

$$x \mapsto (\phi, \pi(x)\psi)$$

on $G$ has compact support mod $Z$.

**Theorem 6** *There exists a real number $c \neq 0$ such that*

$$c_0(\pi) = cd(\pi)$$

*for every supercuspidal representation $\pi$.*

It is natural to believe that this relation actually holds for all square-integrable $\pi$. Let $l_0 = \dim A_0/Z$ where $A_0$ is a maximal split torus of $G$. Then if $\pi_0$ is the Steinberg representation [7, §15], we expect that

$$c = (-1)^{l_0} d(\pi_0)^{-1}$$

**Theorem 7** *We can choose $Q \neq 0$ in $\mathbf{Z}$ such that $Qc_0(\pi) \in \mathbf{Z}$ for every supercuspidal $\pi$.*

**Corollary** *It is possible to normalize the Haar measure $dx^*$ on $G/Z$ so that $d(\pi)$ is an integer for every supercuspidal $\pi$.*

This has been proved for $GL(n, \Omega)$ by Howe [9] for arbitrary characteristic.

Bernshtein has shown (without any assumption on the characteristic) that every irreducible unitary representation of $G$ is admissible [1, 2]. In our case this result is an easy consequence of the above corollary.

## References

1. I. N. Bernshtein, All reductive $p$-adic groups are tame, *J. Functional Anal.* **8** (1974), 91–93.
2. P. Cartier, Les representations des groupes reductifs $p$-adiques et leurs caractères, *Sem. Bourbaki* (1975), No. 471.
3. Harish-Chandra, Differential operators on a semisimple Lie algebra, *Amer. J. Math.* **79** (1957), 87–120.
4. Harish-Chandra, Invariant eigendistributions on a semisimple Lie algebra, *Inst. Hautes Études Sci. Publ. Math.* **27** (1965), 5–54.
5. Harish-Chandra, Invariant eigendistributions on a semisimple Lie group, *Trans. Amer. Math. Soc.* **119** (1965), 457–508.
6. Harish-Chandra, Discrete series for semisimple Lie groups I, *Acta Math.* **113** (1965), 241–318.
7. Harish-Chandra, Harmonic analysis on reductive $p$-adic groups, *in* "Harmonic Analysis on Homogeneous Spaces," pp. 167–192. Amer. Math. Soc., Providence, Rhode Island, 1973.
8. R. Howe, Two conjectures about reductive $p$-adic groups, *in* "Harmonic Analysis on Homogeneous Spaces," pp. 377–380. Amer. Math. Soc., Providence, Rhode Island, 1973.
9. R. Howe, The Fourier transform and germs of characters (case of $Gl_n$ over a $p$-adic field), *Math. Ann.* **208** (1974), 305–322.
10. C. C. Moore, Representations of solvable and nilpotent groups and harmonic analysis on nil and solvmanifolds, *in* "Harmonic Analysis on Homogeneous Spaces." Amer. Math. Soc., Providence, Rhode Island, 1973.
11. R. R. Rao, Orbital integrals in reductive groups, *Ann. of Math.* **96** (1972), 505–510.

AMS (MOS) 1970 subject classification: 22E50

# *Basic constructions in group extension theory*

G. Hochschild

University of California
Berkeley

## 1. Introduction

Let $G$ be a group, $V$ a $G$-module, and $K$ a normal subgroup of $G$. From the cohomology theory of abstract groups, one has the following exact sequence of homomorphisms of cohomology groups:

$$(0) \to H^1(G/K, V^K) \to H^1(G, V) \to H^1(K, V)^G \to H^2(G/K, V^K)$$
$$\to H_K{}^2(G, V) \to H^1(G/K, H^1(K, V)) \to H^3(G/K, V^K).$$

In the category of abstract groups, $H^2(G/K, V^K)$ may be identified with the group $\text{Ext}_{G/K}(V^K)$ of equivalence classes of group extensions with kernel $V^K$ and image $G/K$, while $H_K{}^2(G, V)$ becomes the group $\text{Ext}_{G,K}(V)$ of equivalence classes of $K$-split extensions with kernel $V$ and image $G$. In this way, the major part of the exact sequence becomes a theorem concerning group extensions. It seems to be the custom not to spell this out in pure terms, but to regard it as subordinate to the cohomology theory. However, in many categories of groups with additional structure, the appropriate adaptation of cohomology theory is subject to severe limitations and involves considerable technical difficulties, while the adaptation of much group extension theory is easy.

In particular, it is rewarding to reexamine the above sequence entirely within the context of group extensions. In order to secure the validity of the results in various important categories of groups, one must take special care to free all the definitions and proofs from eventually inadmissible operations involving group algebras, cross sections or factor sets. Because of this, a certain number of basic constructions become dominant, and the results express exactness properties of these constructions, most of which remain significant also for group extensions with non-abelian kernels.

In order to outline the range of validity of the results, we begin with the description of a general setting for *legitimacy*, to which we subsequently adhere. This consists of a category of groups built over an auxiliary category of *spaces*, subject to suitable assumptions. It is designed to comprise, at least, the most important categories of groups admitting a strong group extension theory, such as locally compact separable topological groups, affine or general algebraic groups over an algebraically closed field of characteristic 0, and Lie groups with countable component groups.

The literature contains much of the substance of what is presented here; both, in the form of specialized fragments, and in the form of categorical generalizations.† The absence of a bibliography here must be interpreted so as to imply that no claim of true originality is made for any particular isolated portion of what follows. Our principal aim is to provide a self-contained and elementary exposition that can serve as a basis for a variety of past and future applications.

## 2. Preliminaries

We suppose there is given a category $\mathscr{S}$ of *spaces*, i.e., sets with some extra structure, like topological spaces, algebraic varieties, etc. The morphisms of $\mathscr{S}$ are certain set maps, and we assume, of course, that the composition of morphisms is that of set maps. A *subspace* of an object $A$ of $\mathscr{S}$ is an object $B$ of $\mathscr{S}$ such that the underlying set of $B$ is contained in that of $A$, the set injection $B \to A$ is a morphism of $\mathscr{S}$, and, for every morphism $f: X \to A$ of $\mathscr{S}$ for which $f(X) \subset B$, the set map $X \to B$ given by $f$ is a morphism of $\mathscr{S}$. Every 1-point subset of an object $A$ of $\mathscr{S}$ is assumed to be the underlying set of a subspace of $A$, and every set map from an object of $\mathscr{S}$ to a 1-point object of $\mathscr{S}$ is assumed to be a morphism of $\mathscr{S}$. We assume that $\mathscr{S}$ admits finite direct products, and that

† The discovery of the *reduction* map of Section 7 is due to Louis Auslander, who defined it (with the use of cross sections) in an unpublished manuscript dating back many years.

these, when viewed in the category of sets, are the usual Cartesian products.

We say that a subset $P$ of an object $A$ of $\mathcal{S}$ is *closed in* $A$ if it has the following property: for every point $x$ of the complement of $P$ in $A$, there is an object $D$ in $\mathcal{S}$ and a pair $(f, g)$ of morphisms $A \to D$ such that $f(p) = g(p)$ for every point $p$ of $P$, but $f(x) \neq g(x)$. Owing to the existence of finite direct products, this defines a topology on each object of $\mathcal{S}$. Our assumptions concerning 1-point sets imply that this is a $T_1$-topology. We assume that if $A$ is an object of $\mathcal{S}$ and $B$ is a closed subspace of $A$, then every closed subset of $B$ is closed in $A$. It is easy to see that all morphisms of $\mathcal{S}$ are continuous.

As an auxiliary notion, we introduce the category $\mathcal{S}_g$ of *$\mathcal{S}$-groups*. The objects of $\mathcal{S}_g$ are objects $A$ of $\mathcal{S}$, equipped with a group structure such that the group composition $A \times A \to A$ and the group inversion $A \to A$ are morphisms of $\mathcal{S}$. The morphisms of $\mathcal{S}_g$ are the abstract group homomorphisms that are also morphisms of $\mathcal{S}$.

The category $\mathcal{G}$ of groups with which we shall deal is to be a full subcategory of $\mathcal{S}_g$, i.e., the objects of $\mathcal{G}$ are objects of $\mathcal{S}_g$, and a map $f: U \to V$ from one object of $\mathcal{G}$ to another is a morphism of $\mathcal{G}$ if and only if it is a morphism of $\mathcal{S}_g$. We make several assumptions about $\mathcal{G}$, as follows.

By a *regular subgroup* of an object $A$ of $\mathcal{G}$ we mean an object $B$ of $\mathcal{G}$ such that $B$ is a subspace of $A$ and the injection $B \to A$ is a morphism of $\mathcal{G}$. We assume that every closed abstract subgroup of an object $A$ of $G$ is a regular subgroup of $A$.

Next, we assume that if $K$ is a closed normal subgroup of an object $G$ of $\mathcal{G}$, then the factor group $G/K$ is defined as an object of $\mathcal{G}$ and is such that the canonical map $\pi: G \to G/K$ is a morphism of $\mathcal{G}$ and, for every morphism $\mu: G \to H$ of $\mathcal{G}$ whose kernel contains $K$, the unique abstract group homomorphism $\mu': G/K \to H$ satisfying $\mu' \circ \pi = \mu$ is a morphism of $\mathcal{G}$.

The only significantly restrictive assumption we must make is that, in $\mathcal{G}$, every bijective morphism is an isomorphism.

If $\mathcal{C}$ is any category and $U$ is an object of $\mathcal{C}$, we denote by $\mathrm{Aut}_{\mathcal{C}}(U)$ the group of all $\mathcal{C}$-automorphisms of $U$. Suppose that $G$ is an object of $\mathcal{S}_g$ and $A$ is an object of $\mathcal{S}$. We say that $A$ is a *G-space* if there is given a homomorphism of abstract groups $\gamma: G \to \mathrm{Aut}_{\mathcal{S}}(A)$ such that the map $G \times A \to A$ sending each $(g, a)$ onto $\gamma(g)(a)$ is a morphism of $\mathcal{S}$. If, moreover, $A$ has an $\mathcal{S}$-group structure, and $\gamma$ sends $G$ into $\mathrm{Aut}_{\mathcal{S}_g}(A)$, then we say that $A$ is a *G-group*.

Suppose that $A$ and $G$ are objects of $\mathcal{G}$ and that we are given a group homomorphism $\gamma: G \to \mathrm{Aut}_{\mathcal{G}}(A)$ with which $A$ becomes a $G$-group. Then we can construct the *semidirect product* $(A \cdot G)_\gamma$ over the object $A \times G$ of $\mathcal{S}$, defining the group composition by $(a, g)(a', g') = (a\gamma(g)(a'), gg')$. It is easy to see that this makes $(A \cdot G)_\gamma$ into an object of $\mathcal{S}_{\mathbf{g}}$. We have the four $\mathcal{S}$-morphisms in the diagram

$$A \underset{\alpha}{\overset{\beta}{\leftrightarrows}} (A \cdot G)_\gamma \underset{\rho}{\overset{\sigma}{\leftrightarrows}} G$$

where $\alpha(a) = (a, 1)$, $\beta(a, g) = a$, $\rho(a, g) = g$, and $\sigma(g) = (1, g)$. Of these, $\alpha$, $\rho$, and $\sigma$ are also group homomorphisms. Our last assumption of $\mathcal{G}$ is that $(A \cdot G)_\gamma$ is always an object of $\mathcal{G}$. Note that this is simply a strengthening of the assumption that $\mathcal{G}$ should admit direct products. We shall permit ourselves to make the identifications $a = (a, 1)$ and $g = (1, g)$, by which $A$ and $G$ become closed subgroups of $(A \cdot G)_\gamma$.

## 3. Group extensions

If $A$ and $G$ are groups, then a *group extension* of $G$ by $A$ is a surjective morphism $\sigma\colon E \to G$ whose kernel coincides with $A$, and which is such that the induced morphism $E/A \to G$ is an isomorphism. Because we assume that, in our category $\mathcal{G}$, every bijective morphism is an isomorphism, a group extension of $G$ by $A$ in $\mathcal{G}$ is simply a surjective $\mathcal{G}$-morphism $\sigma\colon E \to G$ with kernel $A$. We denote such a group extension by $[A, E, G]_\sigma$. If $K$ is a regular subgroup of $G$, we say that $[A, E, G]_\sigma$ is *K-split* if there is a morphism $\rho\colon K \to E$ such that $\sigma \circ \rho$ is the identity map on $K$.

As an example of the use of our general assumptions, let us consider the situation just described in the case where $K$ is closed in $G$. Then, clearly, the inverse image $\sigma^{-1}(K)$ is a closed subgroup of $E$, and therefore is also a regular subgroup of $E$. The composite with $\sigma$ of the injection map $\sigma^{-1}(K) \to E$ is therefore a morphism $\sigma^{-1}(K) \to G$. Since the image of this morphism is contained in the regular subgroup $K$ of $G$, it defines a morphism $\sigma'\colon \sigma^{-1}(K) \to K$. An element $x$ of $\sigma^{-1}(K)$ belongs to $\rho(K)$ if and only if $(\rho \circ \sigma')(x) = x$. This shows that $\rho(K)$ is closed in $\sigma^{-1}(K)$. Since $\sigma^{-1}(K)$ is a closed subspace of $E$, this implies that $\rho(K)$ is closed in $E$. Thus, $\rho(K)$ is a regular subgroup of $E$, and it is now clear that $\rho$ defines an isomorphism $K \to \rho(K)$, by means of which we may identify $K$ with the closed subgroup $\rho(K)$ of $E$.

In particular, if $K = G$, we may identify $G$ with $\rho(G)$. Now, in any case, $A$ is an $E$-group with respect to the homomorphism $\gamma_{A,E}\colon E \to \mathrm{Aut}_{\mathcal{G}}(A)$ defined by $\gamma_{A,E}(e)(a) = eae^{-1}$. The composite $\gamma_{A,E} \circ \rho$ makes $A$ into a $G$-group, and it is now clear that $E$ may be identified with the semidirect product $(A \cdot G)_\tau$, where $\tau = \gamma_{A,E} \circ \rho$.

We say that two group extensions $[A, E, G]_\sigma$ and $[A, F, G]_\tau$ are *equivalent* if there is an isomorphism $\eta\colon E \to F$ such that $\eta(a) = a$ for every $a$ in $A$, and $\tau \circ \eta = \sigma$.

From now on, all groups and morphisms will be understood to belong to the category $\mathcal{G}$, unless a statement to the contrary is made. The groups $\mathrm{Aut}_{\mathcal{G}}(A)$ are *not* assumed to belong to $\mathcal{G}$.

***Proposition 3.1*** *Suppose there are given a group extension* $[A, E, G]_\sigma$, *a morphism* $\alpha: A \to B$, *and an E-group structure* $\gamma: E \to \mathrm{Aut}_{\mathcal{G}}(B)$ *for* $B$, *such that*

(1) $\gamma(a)(b) = \alpha(a)b\alpha(a)^{-1}$ *for all* $a$ *in* $A$ *and* $b$ *in* $B$, *and*
(2) $\gamma(e)(\alpha(a)) = \alpha(eae^{-1})$ *for all* $a$ *in* $A$ *and* $e$ *in* $E$.

*Then there is one and, to within equivalence, only one group extension* $[B, F, G]_\tau$ *for which there is a morphism* $\eta: E \to F$ *satisfying* $\eta_A = \alpha$, $\tau \circ \eta = \sigma$, *and* $\gamma_{B,F} \circ \eta = \gamma$.

*Proof* Consider the subset $D$ of $(B \cdot E)_\gamma$ consisting of the elements $(\alpha(a), a^{-1})$, with $a$ in $A$. It follows immediately from (1) that $D$ is an abstract subgroup of $(B \cdot E)_\gamma$ and that the elements of $D$ commute with the elements of $B$. It follows from (2) that $D$ is stable under the conjugation action of $E$, so that $D$ is normal in $E$. Much as in the discussion preceding the statement of our proposition, one sees that $D$ is closed in $(B \cdot E)_\gamma$, so that we can form the factor group $F = (B \cdot E)_\gamma/D$ in $\mathcal{G}$.

The map $(B \cdot E)_\gamma \to G$ sending each $(b, e)$ onto $\sigma(e)$ is a surjective morphism whose kernel contains $D$, so that it induces a surjective morphism $\tau: F \to G$. The kernel of $\tau$ is the canonical image of $B$ in $F$. Noting that $B \cap D = (1)$, we see that the canonical map from $B$ to the kernel of $\tau$ is an isomorphism. Using this for identifying the kernel of $\tau$ with $B$, we obtain a group extension $[B, F, G]_\tau$ (the middle group $F$ here is not really identical with the above group $F$, but results from it by replacing the canonical images of the elements of $B$ with the corresponding elements of $B$; in order to keep the notation within reasonable bounds, we shall always make this kind of abuse of notation). The map $\eta: E \to F$ defined by $\eta(e) = (1, e)D$ is a morphism satisfying the requirements of the proposition.

Now suppose that $[B, F', G]_{\tau'}$ and $\eta': E \to F'$ also satisfy the requirements of the proposition. Consider the map $\mu: (B \cdot E)_\gamma \to F'$ defined by $\mu(b, e) = b\eta'(e)$. Clearly, $\mu$ is an $\mathcal{S}$-morphism. From $\gamma_{B,F'} \circ \eta' = \gamma$, it follows that $\mu$ is a group homomorphism. Thus, $\mu$ is a morphism of $\mathcal{G}$. Clearly, $\mu$ is surjective, and it is easy to verify that the kernel of $\mu$ is precisely $D$. Finally, the isomorphism $(B \cdot E)_\gamma/D \to F'$ induced by $\mu$ is an equivalence from $[B, F, G]_\tau$ to $[B, F', G]_{\tau'}$. This completes the proof of Proposition 3.1.

We shall call the construction of Proposition 3.1 the *kernel shift* corresponding to $\alpha, \gamma$. When the context is limited to group extensions with abelian kernels and fixed image $G$, then the datum $\gamma$ comes naturally from a $G$-module structure of $B$ such that $\alpha$ is a morphism of $G$-modules. The kernel shifts then make $\mathrm{Ext}_G$ into a functor.

It is worth noting that kernel shifts can arise also in a very different way, such as in the following example. Suppose that $E$ is a real Lie group, that $A$ is the connected component of the identity in $E$, and that $G$ is the group $E/A$ of connected components of $E$. Using Lie algebra theory, one obtains a natural construction of the *universal complexification* $\alpha: A \to B$ of the real analytic group $A$ to a complex analytic group $B$ having the appropriate universal mapping property. An $E$-group structure $\gamma$ of $B$ satisfying the conditions of Proposition 3.1 is then obtained in a natural way from the adjoint representation of $E$. The corresponding kernel shift gives the universal complexification of $E$.

Now suppose we are given an exact sequence of morphisms

$$(1) \to A \underset{\alpha}{\to} B \underset{\beta}{\to} C \to (1),$$

a group extension $[B, F, G]_\tau$, and an $F$-group structure $\delta: F \to \mathrm{Aut}_{\mathscr{G}}(C)$ such that $\beta$, $\delta$, $[B, F, G]_\tau$ satisfy the (transcribed) conditions (1) and (2) of Proposition 3.1. Let $[C, P, G]_\rho$ be the group extension obtained from the kernel shift corresponding to $\beta$, $\delta$.

First, let us consider the situation where $[B, F, G]_\tau$ is obtained from an extension $[A, E, G]_\sigma$ by the kernel shift of Proposition 3.1. Then, if $\eta: E \to F$ is as in Proposition 3.1, we have an $E$-group structure $\delta \circ \eta$ for $C$, and $[A, E, G]_\sigma$, $\beta \circ \alpha$, $\delta \circ \eta$ satisfy the conditions of Proposition 3.1, so that we can carry out the kernel shift corresponding to $\beta \circ \alpha$, $\delta \circ \eta$. Now it follows from the unicity part of Proposition 3.1 that $[C, P, G]_\rho$ is equivalent to the extension obtained from $[A, E, G]_\sigma$ by making this kernel shift. Since $\beta \circ \alpha$ is the trivial map, this shows that $[C, P, G]_\rho$ is $G$-split.

Conversely, suppose that $[C, P, G]_\rho$ is $G$-split, so that $P$ may be identified with a semidirect product $(C \cdot G)_\varepsilon$. We shall show that there is an extension $[A, E, G]_\sigma$ and an $E$-group structure $\gamma: E \to \mathrm{Aut}_{\mathscr{G}}(B)$ such that the given extension $[B, F, G]_\tau$ is equivalent to the result of applying the corresponding kernel shift to $[A, E, G]_\sigma$. By assumption, there is a morphism $\mu: F \to (C \cdot G)_\varepsilon$ such that $\mu_B = \beta$, $\rho \circ \mu = \tau$, and $\gamma_{C,P} \circ \mu = \delta$. Put $E = \mu^{-1}(G)$, so that $E$ is a closed subgroup of $F$, and the restriction $\tau_E: E \to G$ of $\tau$ is a surjective morphism. The kernel of $\tau_E$ consists of the elements $b$ of $B$ for which $\mu(b) = 1$, i.e., for which $\beta(b) = 1$. Thus, the kernel of $\tau_E$ coincides with $\alpha(A)$. Since $\alpha(A)$ is closed in $B$, we may identify it with $A$ by means of $\alpha$. In this way, we obtain a group extension $[A, E, G]_{\tau_E}$, and the injection map $\eta: E \to F$ is a morphism satisfying the requirements of Proposition 3.1. Therefore, it follows from the unicity part of Proposition 3.1 that our given extension $[B, F, G]_\tau$ is equivalent to the extension of Proposition 3.1.

## 4. Crossed homomorphisms

Let $V$ and $G$ be groups, and suppose we are given a $G$-group structure $\gamma: G \to \mathrm{Aut}_{\mathscr{G}}(V)$ for $V$. Since $\gamma$ will remain fixed in our discussion, we suppress it in our notation, writing $g \cdot v$ for $\gamma(g)(v)$. Also, we shall denote the semidirect product $(V \cdot G)_\gamma$ simply by $V \cdot G$. The surjective morphism $V \cdot G \to G$ sending each $(v, g)$ onto $g$ will be denoted by $\mu$. A *crossed homomorphism* from $G$ to $V$ is an $\mathscr{S}$-morphism $f: G \to V$ such that $f(xy) = f(x)\, x{\cdot}f(y)$ for all $x$ and $y$ in $G$. This defines an $\mathscr{S}$-morphism $f^0: G \to V \cdot G$, where $f^0(x) = (f(x), x)$, and the formal requirement for $f$ means simply that $f^0$ is a group homomorphism. Thus, the correspondence associating $f^0$ with $f$ is a bijection from the set of crossed homomorphisms to the set of those morphisms $G \to V \cdot G$ whose composites with $\mu$ coincide with the identity map on $G$.

We denote the set of crossed homomorphisms from $G$ to $V$ by $Z(G, V)$. For an element $f$ in $Z(G, V)$, we denote the corresponding subgroup $f^0(G)$ of $V \cdot G$ by $G_f$. Note that $G_f$ is a closed subgroup of $V \cdot G$, and that $f^0$ defines an isomorphism $G \to G_f$.

Let us consider an exact sequence of $G$-groups and $G$-group morphisms

$$(1) \to A \underset{\alpha}{\to} B \underset{\beta}{\to} C \to (1).$$

Let $f$ be an element of $Z(G, C)$. We construct a group extension of $G$ by $A$ as follows. Let $\beta_1$ be the morphism $B \cdot G \to C \cdot G$ sending each $(b, g)$ onto $(\beta(b), g)$. Put $E = \beta_1^{-1}(G_f)$. We have the isomorphism $f^{0-1}$: $G_f \to G$, and we define the morphism $\sigma: E \to G$ as the restriction to $E$ of $f^{0-1} \circ \beta_1$. Then the kernel of $\sigma$ may be identified with the kernel of $\beta$, i.e., with $\alpha(A)$. Identifying $\alpha(A)$ with $A$ by means of $\alpha$, we obtain a group extension $[A, E, G]_\sigma$. We call this extension the *sequential extension determined by the given exact $G$-group sequence and the crossed homomorphism $f$.*

***Proposition 4.1*** *The above sequential extension is $G$-split if and only if $f = \beta \circ g$, with some $g$ in $Z(G, B)$. A group extension $[A, F, G]_\tau$ is equivalent to a sequential extension if and only if there is a morphism $\eta: F \to B \cdot G$ such that $\eta_A = \alpha$ and $\rho \circ \eta = \tau$, where $\rho$ is the canonical morphism $B \cdot G \to G$.*

*Proof* First, consider the above sequential extension $[A, E, G]_\sigma$. Suppose that $f = \beta \circ g$, with $g$ in $Z(G, B)$. Then the morphism $g^0$: $G \to B \cdot G$ actually sends $G$ into $E$, and thus defines a $G$-split of $[A, E, G]_\sigma$. Conversely, if $\gamma: G \to E$ is a $G$-split of our extension, then $\gamma(x)$ is of the form $(g(x), x)$, and the map $g: G \to B$ so defined is an element of $Z(G, B)$ such that $\beta \circ g = f$.

Clearly, the injection $E \to B \cdot G$ is a morphism $\eta$ as described in the proposition. Now consider an extension $[A, F, G]_\tau$ for which there is such a morphism $\eta: F \to B \cdot G$. Then the kernel in $F$ of the morphism $\beta_1 \circ \eta$: $F \to C \cdot G$ coincides with $A$, so that we obtain a morphism $\delta: G \to C \cdot G$ such that $\delta \circ \tau = \beta_1 \circ \eta$. Clearly, $\delta$ is of the form $f^0$, with a certain $f$ in $Z(G, C)$. Moreover, it is seen directly that $\eta(F) = \beta_1^{-1}(G_f) = E$, and that $\eta$ is injective. The required equivalence is the isomorphism $F \to E$ defined by $\eta$. This completes the proof of Proposition 4.1.

Let us return to a single $G$-group $V$ and the corresponding semidirect product $V \cdot G$. For an element $v$ of $V$, we denote the inner automorphism effected by $v$ in $V \cdot G$ by $v^0$, so that $v^0(x) = vxv^{-1}$. Let $V^0$ denote the subgroup of $\mathrm{Aut}_{\mathscr{G}}(V \cdot G)$ consisting of these inner automorphisms $v^0$. We say that two elements $f$ and $g$ of $Z(G, V)$ are *equivalent*, $f \sim g$, if $V^0 \circ f^0 = V^0 \circ g^0$. This relation may also be expressed by saying that there is an element $v$ in $V$ such that $f(x) = v^{-1}g(x)\ x{\cdot}v$ for every $x$ in $G$. The set of equivalence classes of elements of $Z(G, V)$ is denoted $H(G, V)$.

## 5. Transgression

Let $G$ be a group, $K$ a normal closed subgroup of $G$. Let $V$ be a $G$-group. Via the injection $K \to G$, we have then also the structure of a $K$-group on $V$, and we may identify $V \cdot K$ with a closed normal subgroup of $V \cdot G$. We regard $K$ as a $G$-group by the homomorphism $\alpha: G \to \mathrm{Aut}_{\mathscr{G}}(K)$, where $\alpha(x)(y) = xyx^{-1}$. Similarly, we view $V \cdot K$ as a $G$-group by $\beta: G \to \mathrm{Aut}_{\mathscr{G}}(V \cdot K)$, where $\beta(x)(u) = (1, x)u(1, x)^{-1}$. For an element $f$ in $Z(K, V)$ and $x$ in $G$, we define the element $x \cdot f$ of $Z(K, V)$ such that $(x \cdot f)^0 = \beta(x) \circ f^0 \circ \alpha(x^{-1})$. This amounts to defining $(x \cdot f)(y) = x \cdot f(x^{-1}yx)$. In this way, we have a permutation representation of $G$ on the set $Z(K, V)$. This representation preserves equivalence and therefore induces a permutation representation of $G$ on the set $H(K, V)$. The kernel of this last representation contains $K$. In fact, if $y$ belongs to $K$, we have $(y \cdot f)^0 = (f(y)^{-1})^0 \circ f^0$, i.e., $(y \cdot f)(z) = f(y)^{-1} f(z)\ z{\cdot}f(y)$.

Now let $f$ be an element of $Z(K, V)$, and let $K_f$ be the corresponding closed subgroup of $V \cdot K$. We are interested in the normalizer $N_f$ of $K_f$ in $V \cdot G$. Let $\rho$ be the canonical morphism $V \cdot G \to G$. The important fact for us here is that $\rho(N_f) = G$ *if and only if the equivalence class of $f$ in $H(K, V)$ is fixed under the action of $G$*. This is seen as follows. First, suppose that the equivalence class of $f$ is $G$-fixed. Let $x$ be an element of $G$. Then there is an element $v_x$ in $V$ such that $x \cdot f(x^{-1}yx) = v_x^{-1} f(y)\ y{\cdot}v_x$ for all $y$ in $K$. Now one verifies directly that, in $V \cdot G$, one has

$$(v_x, x)(f(y), y)(v_x, x)^{-1} = (f(xyx^{-1}), xyx^{-1})$$

so that $(v_x, x)$ belongs to $N_f$. Conversely, if $\rho(N_f) = G$, then, for every $x$ in $G$, there is an element $v_x$ in $V$ such that $(v_x, x)$ belongs to $N_f$, and it is seen from this that the equivalence class of $f$ is $G$-fixed by reading the (omitted) verification backwards.

Now suppose the equivalence class of $f$ is $G$-fixed. Then the canonical morphism $\rho$: $V \cdot G \to G$ yields a surjective morphism $\delta_f$: $N_f/K_f \to G/K$ in the natural way. The kernel of $\delta_f$ consists of the elements $(v, 1)K_f$, with $v$ in $V$ and $(v, 1)$ in $N_f$. This condition is equivalent to the condition $y \cdot v = f(y)^{-1}vf(y)$ for every $y$ in $K$. The elements of $V$ satisfying this condition clearly constitute a closed subgroup $V_f^K$ of $V$, which we may identify with its canonical image in $N_f/K_f$, i.e., with the kernel of $\delta_f$. This defines a group extension $[V_f^K, N_f/K_f, G/K]_{\delta_f}$, which we call the *transgression extension defined by* $f$.

**Proposition 5.1** *The above transgression extension is split if and only if* $f$ *is the restriction to* $K$ *of an element of* $Z(G, V)$.

*Proof* First, suppose that $f$ is the restriction to $K$ of a crossed homomorphism $g$: $G \to V$. Then the morphism $g^0$: $G \to V \cdot G$ actually sends $G$ into $N_f$ and $K$ into $K_f$. The morphism $G/K \to N_f/K_f$ naturally induced by $g^0$ is clearly a split of our transgression extension.

The converse will be proved with the aid of the following construction. Let $\sigma$: $N_f \to N_f/K_f$ and $\pi$: $G \to G/K$ be the canonical morphisms. Consider the fibered product $P_{\delta \circ \sigma, \pi}(N_f, G)$, i.e., the closed subgroup of $N_f \times G$ consisting of the elements $(a, x)$ with $(\delta \circ \sigma)(a) = \pi(x)$, where we have written $\delta$ for $\delta_f$: $N_f/K_f \to G/K$. We define a map $f'$: $P_{\delta \circ \sigma, \pi}(N_f, G) \to V \cdot G$ by setting

$$f'((v, y), x) = (v, y)(f(y^{-1}x), y^{-1}x) = (v\ yf(y^{-1}x), x)$$

(note that $y^{-1}x$ belongs to $K$, by virtue of the definition of the fibered product). Refering to Section 2, we see that $f'$ is an $\mathcal{S}$-morphism. Since $(v, y)$ belongs to $N_f$, we have, for every $t$ in $K$,

$$(f(t), t)(v, y) = (v, y)(f(y^{-1}ty), y^{-1}ty).$$

From this, we see that $f'$ is a group homomorphism. The kernel of $f'$ consists of the elements $((f(y), y), 1)$ with $y$ in $K$, i.e., it is the image in $P_{\delta \circ \sigma, \pi}(N_f, G)$ of the map from $K_f$ sending each $a$ onto $(a, 1)$. Since this image is closed, and since the map is an injective morphism $K_f \to P_{\delta \circ \sigma, \pi}(N_f, G)$, we may use this map for identifying $K_f$ with the kernel of $f'$. Now $f'$ induces an injective morphism $f''$: $P_{\delta, \pi}(N_f/K_f, G) \to V \cdot G$ in the natural way.

Now suppose that our transgression extension has a split $\tau$: $G/K \to N_f/K_f$. This yields a morphism

$$\tau^*: G \to P_{\delta, \pi}(N_f/K_f, G), \qquad \text{where} \quad \tau^*(x) = (\tau(xK), x).$$

It is clear from the definition of $f''$ that $f''(\tau^*(x)) = (g(x), x)$, with $g(x)$ in $V$. But this implies that $g$ is an element of $Z(G, V)$ and $g^0 = f'' \circ \tau^*$. If $x$ belongs to $K$, then $\tau^*(x) = (K_f, x)$ and

$$f''(\tau^*(x)) = f'((1, 1), x) = (f(x), x)$$

so that $g(x) = f(x)$. This completes the proof of Proposition 5.1.

## 6. Inflation

Let $\pi\colon G \to G/K$ be as in Section 5, let $V$ be a group, and let $W$ be a regular subgroup of $V$. Suppose there is given a group extension $[W, F, G/K]_\delta$ and a group homomorphism $\gamma\colon P_{\delta,\pi}(F, G) \to \mathrm{Aut}_{\mathscr{G}}(V)$ such that the following conditions are satisfied:

(1) $\gamma(w, 1)(v) = wvw^{-1}$ for all $w$ in $W$ and $v$ in $V$,
(2) $\gamma(u, x)(w) = uwu^{-1}$ for all $w$ in $W$ and $(u, x)$ in $P_{\delta,\pi}(F, G)$,
(3) $W$ is precisely the $\gamma(1, K)$-fixed part $V^{\gamma(1, K)}$ of $V$.

We construct a certain group extension in two stages, as follows. First, we note that the projection morphism $F \times G \to G$ restricts to a surjective morphism $\sigma\colon P_{\delta,\pi}(F, G) \to G$ whose kernel $(W, 1)$ may be identified with $W$. In this way, we obtain a group extension $[W, E, G]_\sigma$, where $E = P_{\delta,\pi}(F, G)$. Because of (1) and (2), we can carry out the kernel shift of Proposition 3.1, starting from $[W, E, G]_\sigma$, the injection $W \to V$ and the $E$-group structure $\gamma$ of $V$. The result is the group extension $[V, T, G]_\tau$, where $T = (V \cdot E)_\gamma/D$, with $D$ the closed normal subgroup consisting of the elements $(w, (w^{-1}, 1))$ corresponding to the elements $w$ of $W$, and $\tau$ is obtained in the natural way from $\sigma$. This group extension $[V, T, G]_\tau$ is called the *inflation of* $[W, F, G/K]_\sigma$ *with respect to* $\gamma$. Observe that this is a $K$-split extension, by the morphism $K \to T$ sending each $y$ onto $(1, (1, y))D$. Actually, we have not used condition (3) in the construction. This comes into play with the following two propositions.

***Proposition 6.1*** *Let $f$ be in $Z(K, V)$. and let $[V_f{}^K, N_f/K_f, G/K]_\delta$ be the corresponding transgression extension. Let $f''\colon P_{\delta,\pi}(N_f/K_f, G) \to V \cdot G$ be the morphism of the proof of Proposition* 5.1, *and let $\gamma = \gamma_{V, V\cdot G} \circ f''$. Then, with $W = V_f{}^K$, the inflation conditions* (1)–(3) *are satisfied, and the inflation of the transgression extension with respect to $\gamma$ is $G$-split.*

*Proof* For $w$ in $W$, we have $f''(w, 1) = w$, whence (1) is satisfied. Now let $(u, x)$ be an element of $P_{\delta,\pi}(N_f/K_f, G)$. Write $u = (v, y)K_f$. Then $f''(u, x)$ belongs to the coset $(v, y)K_f = u$, whence (2) is satisfied.

If $y$ is an element of $K$, then $f''(1, y) = (f(y), y)$. Since $V_f^K$ coincides with the centralizer of $K_f$ in $V$, it follows that (3) is satisfied.

Now let us write $E$ for $P_{\delta,\pi}(N_f/K_f, G)$, and define the map $\eta: (V \cdot E)_\gamma \to V \cdot G$ by $\eta(v, e) = vf''(e)$. Then $\eta$ is an $\mathscr{S}$-morphism, and we see from $\gamma = \gamma_{V, V \cdot G} \circ f''$ that $\eta$ is a group homomorphism. Clearly, $\eta$ is surjective. The kernel of $\eta$ consists of the elements $(v, e)$ for which $f''(e) = v^{-1}$. Let $((w, y), x)$ be a representative for $e$ in $P_{\delta \circ \sigma, \pi}(N_f, G)$, where $\sigma$ is the canonical morphism $N_f \to N_f/K_f$. Then we have $f'((w, y), x) = v^{-1}$, whence $x = 1$, $y$ belongs to $K$ and $w\, y{\cdot}f(y^{-1}) = v^{-1}$. Hence $(w, y)K_f = (v^{-1}, 1)K_f$. Moreover, since $(v^{-1}, 1)$ belongs to $N_f$, we have that $v$ belongs to $V_f^K$. Taking account of the identification of $V_f^K$ with its canonical image in $N_f/K_f$, we have $e = v^{-1}$, so that our element $(v, e)$ of the kernel of $\eta$ is $(v, v^{-1})$, with $v$ in $V_f^K$. Conversely, it is clear that all such elements belong to the kernel of $\eta$. Thus, the kernel of $\eta$ is precisely the group $D$ of the inflation construction, i.e., if the inflated extension is $[V, T, G]_\tau$, then $T = (V \cdot E)_\gamma/D$. Therefore, $\eta$ induces an isomorphism $T \to V \cdot G$ in the natural way, and it is easy to see that this is an equivalence from $[V, T, G]_\tau$ to the semidirect product extension $V \cdot G$ of $G$ by $V$. This completes the proof of Proposition 6.1.

***Proposition 6.2*** *If the inflation $[V, T, G]_\tau$ of a group extension $[W, F, G/K]_\delta$ with respect to a group homomorphism $\gamma: P_{\delta,\pi}(F, G) \to \mathrm{Aut}_{\mathscr{G}}(V)$ is G-split, then $[W, F, G/K]_\delta$ is equivalent to a transgression extension $[V_f^K, N_f/K_f, G/K]_{\delta_f}$, where $f$ is a crossed homomorphism $K \to V$ belonging to the G-group structure of $V$ obtained from a G-split of $[V, T, G]_\tau$.*

*Proof* Let $\mu: G \to T$ be a split of the inflated extension. Recall that there is also a $K$-split $\rho: K \to T$, where $\rho(y) = (1, (1, y))D$. Define the $\mathscr{S}$-morphism $f: K \to V$ by $f(y) = \rho(y)\mu(y^{-1})$. Let us make $V$ into a $G$-group by $\gamma_{V,T} \circ \mu: G \to \mathrm{Aut}_{\mathscr{G}}(V)$. Then we see directly from these definitions that $f$ belongs to $Z(K, V)$. We show that the equivalence class of $f$ is $G$-fixed, as follows.

For $x$ in $G$ and $y$ in $K$, we have

$$\begin{aligned} x \cdot f(y) = \mu(x)f(y)\mu(x^{-1}) &= \mu(x)\rho(y)\mu(y^{-1})\mu(x^{-1}) \\ &= \mu(x)\rho(y)\mu(x^{-1})\mu(xyx^{-1})^{-1}. \end{aligned}$$

Let us choose a representative $(v, (a, x))$ for $\mu(x)$ in $(V \cdot P_{\delta,\pi}(F, G))_\gamma$, so that $\mu(x) = (v, (a, x))D$, $v$ is in $V$ and $a$ is an element of $F$ such that $\delta(a) = \pi(x)$. Then a short computation gives

$$\mu(x)\rho(y)\mu(x^{-1}) = v\rho(xyx^{-1})v^{-1},$$

whence

$$\begin{aligned} x \cdot f(y) &= v\rho(xyx^{-1})v^{-1}\mu(xyx^{-1})^{-1} \\ &= vf(xyx^{-1})\mu(xyx^{-1})v^{-1}\mu(xyx^{-1})^{-1} \\ &= vf(xyx^{-1})\, xyx^{-1} \cdot v^{-1}, \end{aligned}$$

showing that $x \cdot f \sim f$.

For $v$ in $V$ and $y$ in $K$, we have that $\gamma(1, y)(v)$ is equal to the conjugate $\rho(y)v\rho(y)^{-1}$, taken in $T$. When this is written out in terms of $f(y)$ and $\mu(y)$, it is seen that $\gamma(1, y)(v) = f(y)(y \cdot v)f(y)^{-1}$. Therefore, the inflation assumption (3) that $W = V^{\gamma(1, K)}$ means that $W = V_f{}^K$.

Now consider the $\mathcal{S}$-morphism $\mu'$: $V \cdot G \to T$ defined by $\mu'(v, x) = v\mu(x)$. By virtue of our definition of the $G$-group structure of $V$, this map $\mu'$ is actually a group homomorphism. Since $\mu$ is a $G$-split of the extension $[V, T, G]_\tau$, it is now clear that $\mu'$ is an isomorphism. With $y$ in $K$, we have $\mu'(f(y), y) = f(y)\mu(y) = \rho(y)$, so that $\mu'(K_f) = \rho(K_f)$.

We may identify $P_{\delta, \pi}(F, G)$ with its canonical image in $T$, and it is clear that it then lies in the normalizer of $\rho(K)$. Conversely, suppose that an element $(v, (a, x))D$ of $T$ normalizes $\rho(K)$. Explicitly, this means that, for every $y$ in $K$, we have that

$$(v, (a, x))(1, (1, y))(\gamma(a, x)^{-1}(v^{-1}), (a^{-1}, x^{-1}))D$$

belongs to $\rho(K)$, which gives

$$(v\gamma(1, xyx^{-1})(v^{-1}), (1, xyx^{-1}))D = (1, (1, xyx^{-1}))D,$$

whence $v\gamma(1, xyx^{-1})(v^{-1}) = 1$, so that $v$ belongs to $V^{\gamma(1, K)} = W$, and $(v, (a, x))D = (1, (va, x))D \in P_{\delta, \pi}(F, G)$. Thus, the normalizer of $\rho(K)$ in $T$ coincides with $P_{\delta, \pi}(F, G)$.

Together with what we have seen above concerning $\mu'$, this shows that $\mu'^{-1}(P_{\delta, \pi}(F, G))$ is precisely the normalizer $N_f$ of $K_f$ in $V \cdot G$, and that the restriction of $\mu'$ to $N_f$ induces an isomorphism $N_f/K_f \to P_{\delta, \pi}(F, G)/\rho(K)$. Clearly, this last group may be identified with $F$, via the projection morphism $F \times G \to F$. In this way, we obtain an isomorphism $N_f/K_f \to F$, which is evidently an equivalence from the transgression extension $[V_f{}^K, N_f/K_f, G/K]_{\delta_f}$ to $[W, F, G/K]_\delta$. This completes the proof of Proposition 6.2.

## 7. Abelian kernels

Let $G$ be a group, and let $A$, $B$, $C$ be abelian groups with a $G$-group structure. Thus, $A$, $B$, $C$ are $G$-modules in the category $\mathcal{G}$, in the usual sense. The cohomology sets $H(G, A)$, etc., now become the usual 1-dimensional cohomology *groups*.

Now let

$$(0) \rightarrow A \underset{\alpha}{\rightarrow} B \underset{\beta}{\rightarrow} C \rightarrow (0)$$

be an exact sequence of morphisms of $G$-modules. Then $\alpha$ and $\beta$ induce group homomorphisms between the cohomology groups in the natural fashion, and the resulting sequence

$$H(G, A) \rightarrow H(G, B) \rightarrow H(G, C)$$

is, almost evidently, exact.

The sequential extension construction of Proposition 4.1 is easily seen to induce a map from $H(G, C)$ to the set $\mathrm{Ext}_G(A)$ of equivalence classes of group extensions of $G$ by $A$ compatible with the given $G$-group structure of $A$, in the evident sense. By means of the well-known Baer product of group extensions with abelian kernels, $\mathrm{Ext}_G(A)$ is endowed with the structure of an abelian group, and our map $H(G, C) \rightarrow \mathrm{Ext}_G(A)$ is a group homomorphism. By the first part of Proposition 4.1, the above exact sequence, when augmented with this homomorphism, remains exact.

In Proposition 3.1, an $E$-group structure $\gamma$ for $B$ satisfying the requirements (1) and (2) is now automaticaly available as the composite of $\sigma$ with the given $G$-module structure of $B$. It is easy to verify that, with this, the kernel shift yields a group homomorphism $\mathrm{Ext}_G(A) \rightarrow \mathrm{Ext}_G(B)$. It follows from the second part of Proposition 4.1 that the sequence $H(G, C) \rightarrow \mathrm{Ext}_G(A) \rightarrow \mathrm{Ext}_G(B)$ is exact.

Finally, the discussion of the kernel shift at the end of Section 3 proves that the sequence

$$\mathrm{Ext}_G(A) \rightarrow \mathrm{Ext}_G(B) \rightarrow \mathrm{Ext}_G(C)$$

is exact. Putting all this together, we have the usual six-term exact sequence

$$H(G, A) \rightarrow H(G, B) \rightarrow H(G, C) \rightarrow \mathrm{Ext}_G(A) \rightarrow \mathrm{Ext}_G(B) \rightarrow \mathrm{Ext}_G(C).$$

Now let $K$ be a closed normal subgroup of $G$, and let $V$ be a $G$-module. There is no difficulty in verifying that the following familiar inflation-restriction sequence for the 1-dimensional cohomology groups is exact:

$$(0) \rightarrow H(G/K, V^K) \rightarrow H(G, V) \rightarrow H(K, V)^G.$$

It is easy to see that the construction of the transgression extension of Section 5 respects equivalence so as to induce a map $H(K, V)^G \rightarrow \mathrm{Ext}_{G/K}(V^K)$. Taking this for granted, we show that this map is a group homomorphism.

Let $f$ and $g$ be elements of $Z(K, V)$ representing elements of $H(K, V)^G$. We consider the Baer product of the extensions $[V^K, N_f/K_f, G/K]_{\delta_f}$ and

$[V^K, N_g/K_g, G/K]_{\delta_g}$. This is defined as $P_{\delta_f, \delta_g}(N_f/K_f, N_g/K_g)/D$, where $D$ is the closed normal subgroup consisting of the elements $(w, -w)$ with $w$ in $V^K$, together with the surjective morphism $\delta$ from this group to $G/K$ that is defined by $\delta_f$ *or* $\delta_g$, and the identification of the kernel of $\delta$ with $V^K$, via the morphism sending each element $w$ of $V^K$ onto $(w, 1)D$. In order to compare this extended group with $N_{f+g}/K_{f+g}$, we make a construction using the semidirect product $V \cdot G$, as follows.

Let $\rho$: $V \cdot G \to G$ denote the canonical morphism. Let $\mu$ denote the surjective morphism $P_{\rho,\rho}(V \cdot G, V \cdot G) \to V \cdot G$ sending each $((u, x), (v, x))$ onto $(u + v, x)$. Clearly, $\mu(P_{\rho,\rho}(K_f, K_g)) = K_{f+g}$. We wish to show that $\mu(P_{\rho,\rho}(N_f, N_g)) = N_{f+g}$. In doing this, it will be convenient to use the following cohomological notation: if $w$ is an element of $V$, let $d(w)$ denote the element of $Z(K, V)$ defined by $d(w)(y) = y \cdot w - w$. Then we have $(u, x)$ in $N_f$ if and only if $d(u) = x \cdot f - f$, and similarly for $g$ or $f + g$ in the place of $f$. Hence, if $(u, x)$ belongs to $N_f$ and $(v, x)$ belongs to $N_g$, then $d(f + g) = x \cdot (f + g) - (f + g)$, so that $(u + v, x)$ belongs to $N_{f+g}$. This proves that $\mu(P_{\rho,\rho}(N_f, N_g)) \subset N_{f+g}$. Now suppose that $(w, x)$ is any element of $N_{f+g}$. There is an element $v_x$ in $V$ such that $d(v_x) = x \cdot g - g$, because the cohomology class of $g$ is $G$-fixed. We have $d(w) = x \cdot (f + g) - (f + g)$, whence $d(w - v_x) = x \cdot f - f$. Now $(w - v_x, x)$ belongs to $N_f$ and $(v_x, x)$ belongs to $N_g$, and $\mu((w - v_x, x), (v_x, x)) = (w, x)$. This completes the proof that $\mu(P_{\rho,\rho}(N_f, N_g)) = N_{f+g}$.

Now $P_{\rho,\rho}(K_f, K_g)$ is a closed normal subgroup of $P_{\rho,\rho}(N_f, N_g)$, and it is clear from the definitions of $\delta_f$ and $\delta_g$ (in Section 5) that the factor group may be identified with $P_{\delta_f, \delta_g}(N_f/K_f, N_g/K_g)$. From what we have just shown concerning $\mu$, it is now clear that $\mu$ induces a surjective morphism

$$\mu': P_{\delta_f, \delta_g}(N_f/K_f, N_g/K_g) \to N_{f+g}/K_{f+g}.$$

Evidently, the kernel of $\mu'$ contains the group $D$ figuring in the above definition of the Baer product. On the other hand, every element of the kernel of $\mu'$ can be written $((u, x)K_f, (v, x)K_g)$, with $(u + v, x)$ in $K_{f+g}$. This implies that $x$ belongs to $K$ and $u + v = f(x) + g(x)$. Therefore, our kernel element is equal to $((u', 1)K_f, (-u', 1)K_g)$, where $u' = u - f(x)$. Moreover, since $(u', 1)$ belongs to $N_f$, the element $u'$ must belong to $V^K$. Taking account of the identifications of $V^K$ with its canonical images in $N_f/K_f$ and $N_g/K_g$, we see that this means that our kernel element belongs to $D$. Thus, the kernel of $\mu'$ is precisely $D$.

Therefore, $\mu'$ induces an isomorphism $\mu''$ from the Baer product of the transgression extensions corresponding to $f$ and $g$ to $N_{f+g}/K_{f+g}$. It is easy to see that $\mu''$ has the formal properties to make it an equivalence of extensions of $G/K$ by $V^K$. The conclusion is that the map $H(K, V)^G \to \mathrm{Ext}_{G/K}(V^K)$ is indeed a group homomorphism. By Proposition 5.1, the resulting sequence

$$H(G, V) \to H(K, V)^G \to \mathrm{Ext}_{G/K}(V^K)$$

is exact.

Next, let us consider the group $\mathrm{Ext}_{G,K}(V)$ of equivalence classes of *K-split* extensions of $G$ by $V$. Noting the simplifications coming from the commutativity of the kernels, it is easy to see that the inflation construction of Section 6 yields a group homomorphism $\mathrm{Ext}_{G/K}(V^K) \to \mathrm{Ext}_{G,K}(V)$, it being understood that the required module structures of $V$ are always taken to be those coming naturally from the given $G$-module structure of $V$. Then it follows immediately from Propositions 6.1 and 6.2 that the sequence (transgression-inflation)

$$H(K, V)^G \to \mathrm{Ext}_{G/K}(V^K) \to \mathrm{Ext}_{G,K}(V)$$

is exact. Altogether, we have the exact sequence

$$(0) \to H(G/K, V^K) \to H(G, V) \to H(K, V)^G \to \mathrm{Ext}_{G/K}(V^K) \to \mathrm{Ext}_{G,K}(V).$$

The action of $G$ by abstract group automorphisms on $H(K, V)$ factors through $G/K$, so that we can speak of crossed homomorphisms $G/K \to H(K, V)$, in the sense of the category of abstract groups. As a warning of our thus leaving the category $\mathscr{G}$, we use the notation $H'(G/K, H(K, V))$ for the group of equivalence classes of crossed homomorphisms $G/K \to H(K, V)$. We wish to define a certain homomorphism $\mathrm{Ext}_{G,K}(V) \to H'(G/K, H(K, V))$, which will extend the above exact sequence.

Let $[V, E, G]_\sigma$ be a group extension representing an element of $\mathrm{Ext}_{G,K}(V)$. Since this is $K$-split, we may identify the inverse image of $K$ in $E$ with $V \cdot K$ in such a way that $\sigma(v, y) = y$ for every $y$ in $K$ and $v$ in $V$. Now $V \cdot K$ is normal in $E$, so that it has an $E$-group structure $\gamma_{V \cdot K, E}$. On the other hand, $V \cdot K$ has a $G$-group structure $\rho: G \to \mathrm{Aut}_{\mathscr{G}}(V \cdot K)$, where $\rho(x)(v, y) = (x \cdot v, xyx^{-1})$. For every $e$ in $E$, define the $\mathscr{G}$-automorphism $f(e)$ of $V \cdot K$ by $f(e) = \gamma_{V \cdot K, E}(e) \circ \rho(\sigma(e^{-1}))$.

Now let $A$ denote the abelian subgroup of $\mathrm{Aut}_{\mathscr{G}}(V \cdot K)$ consisting of the automorphisms that leave the elements of $V$ fixed and induce the indentity map on $(V \cdot K)/V$. Let $V^0$ denote the subgroup of $A$ consisting of the inner automorphisms $v^0$ effected by the elements $v$ of $V$. Then the factor group $A/V^0$ may be identified with $H(K, V)$ in the evident way.

It is seen immediately that $f(e)$ belongs to $A$ for every $e$ in $E$. Moreover, we have

$$f(e_1e_2) = f(e_1) \circ \rho(\sigma(e_1)) \circ f(e_2) \circ \rho(\sigma(e_1))^{-1},$$

which means that $f$ is a crossed homomorphism $E \to A$ with respect to the $E$-module structure of $A$ obtained from $\rho \circ \sigma$ and the conjugation action of $\rho(G)$ on $A$. Now put $g(e) = f(e)V^0$. Then $g$ is a crossed homomorphism $E \to A/V^0$ with respect to the $E$-module structure of $A/V^0$ induced by that of $A$.

Next, we observe that $f(V \cdot K) \subset V^0$, which implies that $g(e)$ depends only on the coset $e(V \cdot K)$. On the other hand, the $E$-module structure of $A/V^0$ factors via $\sigma$ to a $G$-module structure of $A/V^0$ which, through the identification $A/V^0 = H(K, V)$, coincides with the usual $G$-module structure of $H(K, V)$. This, in turn, we know to factor through $G/K$, via the canonical morphism $\pi\colon G \to G/K$. Therefore, $g$ defines a crossed homomorphism $h\colon G/K \to H(K, V)$ such that $h \circ \pi \circ \sigma = g$.

Now one verifies in the straightforward fashion that a different choice of the identification of $V \cdot K$ with the inverse image of $K$ in $E$ leads to an equivalent crossed homomorphism, and then that the cohomology class of $h$ depends only on the equivalence class of $[V, E, G]_\sigma$. Thus, the above construction yields a map

$$r\colon \mathrm{Ext}_{G,K}(V) \to H'(G/K, H(K, V)).$$

It is not difficult to verify that this map $r$ is a group homomorphism. We call it the *reduction*.

Taking the truth of these last assertions for granted, we shall now prove that the sequence

$$\mathrm{Ext}_{G/K}(V^K) \to \mathrm{Ext}_{G,K}(V) \underset{r}{\to} H'(G/K, H(K, V))$$

is exact.

Start with a group extension $[V^K, F, G/K]_\delta$. The inflated extension is $[V, E, G]_\sigma$, where (cf. Section 6) $E = (V \cdot P_{\delta,\pi}(F, G))/D$. An identification of $V \cdot K$ with $\sigma^{-1}(K)$ is obtained from the map sending each element $(v, y)$ of $V \cdot K$ onto $(v, (1, y))D$. Write an arbitrary element $e$ of $E$ in the form $e = (v, (u, x))D$. Then $\delta(u) = \pi(x) = xK$, and $\sigma(e) = x$. Now, in the notation used in defining $r$, we have

$$f(e) = \gamma_{V \cdot K, E}(e) \circ \rho(\sigma(e^{-1})) = \gamma_{V \cdot K, E}(e) \circ \rho(x^{-1}),$$

and $\gamma_{V \cdot K, E}(e) = v^0 \circ \rho(x)$, so that $f(e) = v^0$. This shows that the image of the inflation map is contained in the kernel of the reduction map $r$.

Now let $[V, E, G]_\sigma$ be an extension representing an element of the kernel of $r$. In the notation used in defining $r$, this means that there is an element $\mu$ in $A/V^0$ such that, for every $e$ in $E$, we have $g(e) = \mu^{-1}\sigma(e)\cdot\mu$. Choose a representative $u$ of $\mu$ in $A$. Then $f(e)$ belongs to the coset

$$u^{-1}\rho(\sigma(e))u\rho(\sigma(e^{-1}))V^0,$$

whence $\gamma_{V \cdot K, E}(e)$ belongs to $u^{-1}\rho(\sigma(e))uV^0$. Now let us replace the original imbedding $i: V \cdot K \to E$ with $i \circ u$. This changes the above $\gamma_{V \cdot K, E}$ to $u \circ \gamma_{V \cdot K, E} \circ u^{-1}$. Thus, we can choose the identification of $V \cdot K$ with $\sigma^{-1}(K)$ in such a way that $\gamma_{V \cdot K, E}(e)$ belongs to $\rho(\sigma(e))V^0$ for every $e$ in $E$.

Now let $N$ denote the normalizer of $K = (1, K)$ in $E$. We claim that $\sigma(N) = G$. In order to see this, let $x$ be an element of $G$, and choose an element $e$ in $E$ such that $\sigma(e) = x$. Next, choose an element $v$ in $V$ such that $\gamma_{V \cdot K, E}(e) = v^0 \circ \rho(x)$. Then $v^{-1}e$ lies in $N$ and $\sigma(v^{-1}e) = x$.

We have $V \cap N = V^K$, so that $\sigma$ induces a group extension $[V^K, N, G]_{\sigma'}$, from which we obtain a group extension $[V^K, N/K, G/K]_{\sigma''}$ in the evident way. Finally, we show that the inflation of this last extension is equivalent to $[V, E\ G]_\sigma$.

Let $[V, F, G]_\tau$ be the inflated extension, so that $F = (V \cdot P_{\sigma'', \pi}(N/K, G))/D$. Consider the morphism $\eta$: $V \cdot N \to F$, where $\eta(v, n) = (v, (nK, \sigma(n)))D$. This is surjective, and its kernel is the group $D_1$ consisting of the elements $(w, -w)$, with $w$ in $V^K$. Therefore, $\eta$ induces an isomorphism $\eta_1$: $(V \cdot N)/D_1 \to F$. On the other hand, the group composition of $E$ yields a surjective morphism $\alpha$: $V \cdot N \to E$ whose kernel is also $D_1$, so that $\alpha$ induces an isomorphism $\alpha_1$: $(V \cdot N)/D_1 \to E$. Now it is clear that $\eta_1 \circ \alpha_1^{-1}$ is an equivalence from the group extension $[V, E, G]_\sigma$ to the inflation $[V, F, G]_\tau$ of $[V^K, N/K, G/K]_{\sigma''}$.

We have established the six-term exact sequence

$$(0) \to H(G/K, V^K) \to H(G, V) \to H(K, V)^G \to \mathrm{Ext}_{G/K}(V^K) \\ \to \mathrm{Ext}_{G, K}(V) \to H'(G/K, H(K, V)).$$

## 8. Cohomology

In the category of abstract groups, the above exact sequence can be extended with one more term. The last group of the sequence is now the ordinary compound cohomology group $H^1(G/K, H^1(K, V))$. There is a group homomorphism $H^1(G/K, H^1(K, V)) \to H^3(G/K, V^K)$, with which one obtains the seven-term exact sequence exhibited in our introduction. Since its origin from the spectral sequence of $(G, K)$ is somewhat obscure, we shall sketch a direct cochain construction yielding this homomorphism.

Given an element $u$ of $H^1(G/K, H^1(K, V))$, we choose a 1-cocycle $\mu$ for $G/K$ in $H^1(K, V)$ representing $u$. For each element $xK$ of $G/K$, choose a 1-cocycle $f(xK)$ for $K$ in $V$ whose cohomology class is $\mu(xK)$, making $f(K) = 0$. Since $\mu$ is a cocycle, we have, for all elements $x$ and $y$ in $G$ and all elements $t$ in $K$,

$$(x \cdot f(yK) - f(xyK) + f(xK))(t) = t \cdot g(x, y) - g(x, y),$$

where $g(x, y)$ lies in $V$. Evidently, we may choose these such that $g(x, yt) = g(x, y)$ whenever $t$ lies in $K$, and $g(1, y) = 0 = g(x, 1)$. A direct computation, using that $f(yK)$ is a cocycle, shows that, for all $x$, $y$ in $G$ and $s$, $t$ in $K$,

$$t \cdot g(xs, y) - g(xs, y) = t \cdot h(s, x, y) - h(s, x, y),$$

where

$$h(s, x, y) = g(x, y) + x \cdot (f(yK)(s)).$$

Now make a coset decomposition $G = \bigcup x_\alpha K$, with $x_1 = 1$. Define $g'(x_\alpha s, y) = h(s, x_\alpha, y)$. Clearly, $g'$ satisfies the same original condition as $g$. Moreover, with $s$, $t$ *in* $K$,

$$\begin{aligned} g'(x_\alpha st, y) &= g(x_\alpha, y) + x_\alpha \cdot (f(yK)(st)) \\ &= g(x_\alpha, y) + x_\alpha \cdot (f(yK)(s)) + x_\alpha s \cdot (f(yK)(t)) \\ &= g'(x_\alpha s, y) + x_\alpha s \cdot (f(yK)(t)). \end{aligned}$$

Now we write $g$ for $g'$. Then $g$ satisfies the original condition and, for all $x$, $y$ in $G$ and $t$ in $K$,

$$g(xt, y) = g(x, y) + x \cdot (f(yK)(t)).$$

We consider the coboundary $dg$. Let $x$, $y$, $z$ be in $G$ and $t$ in $K$. Clearly, $dg(x, y, zt) = dg(x, y, z)$. Next, one verifies directly that $dg(y, t, z) = 0$. Using this and that $d^2 = 0$, one now shows that $dg(x, yt, z) = dg(x, y, z)$. It follows that

$$dg(xt, y, z) = x \cdot dg(t, y, z) + dg(x, y, z).$$

Next,

$$dg(t, y, z) = t \cdot g(y, z) - g(ty, z) + g(t, yz) - g(t, y)$$

and

$$\begin{aligned} g(ty, z) = g(y(y^{-1}ty), z) &= g(y, z) + y \cdot (f(zK)(y^{-1}ty)) \\ &= t \cdot g(y, z) + f(yzK)(t) - f(yK)(t) \\ &= t \cdot g(y, z) + g(t, yz) - g(t, y), \end{aligned}$$

whence $dg(t, y, z) = 0$ and $dg(xt, y, z) = dg(x, y, z)$.

Thus, $dg(x, y, z)$ depends only on the cosets $xK$, $yK$, $zK$. It follows that $dg$ takes its values in $V^K$. Now it is clear that $dg$ defines an element of $H^3(G/K, V^K)$. This is the image of the given element $u$ of $H^1(G/K, H^1(K, V))$. We omit the verification that this construction indeed yields the required group homomorphism

$$t\colon H^1(G/K, H^1(K, V)) \to H^3(G/K, V^K)$$

such that the sequence

$$\mathrm{Ext}_{G,K}(V) \underset{r}{\to} H^1(G/K, H^1(K, V)) \underset{t}{\to} H^3(G/K, V^K)$$

is exact. These facts are known from the spectral sequence associated with the pair $(G, K)$ and the $G$-module $V$. In order to see that the map obtained from the spectral sequence coincides with the above $t$, one must examine the underlying filtered complex of (nonhomogeneous) cochains for $G$ in $V$.

# *On the hyperalgebra of a semisimple algebraic group*

*J. E. Humphreys*

*University of Massachusetts*
*Amherst*

Let $G$ be a simply connected semisimple algebraic group over an algebraically closed field $K$ of characteristic $p > 0$. The purpose of this paper is to develop some information about the "hyperalgebra" of $G$ and certain of its finite-dimensional subalgebras, along lines suggested by D.-N. Verma (cf. [13, §6; 7]). In this framework we obtain a quick representation-theoretic proof of the main step in W. J. Haboush's proof of the Mumford conjecture [4].

## 1. The hyperalgebra

**1.1** The Lie algebra $\mathfrak{g}$ of $G$ has a basis $\{X_\alpha, \alpha \in \Phi; H_i, 1 \leq i \leq l\}$ derived from a Chevalley basis of the corresponding complex semisimple Lie algebra $\mathfrak{g}_{\mathbf{C}}$, where $\Phi$ is the root system and $l$ the rank (cf. [1; 6, §25; 12]). Denote by $m$ the number of positive roots, so $\dim \mathfrak{g} = l + 2m$. Kostant introduced a $\mathbf{Z}$-form $U_{\mathbf{Z}}$ of the universal algebra of $\mathfrak{g}_{\mathbf{C}}$ (cf. [1, 6, 9, 12] for details), the subring with 1 generated by all $X_\alpha{}^t/t!$ ($t \in \mathbf{Z}^+$, $\alpha \in \Phi$). This turns out to have a $\mathbf{Z}$-basis adapted from the usual Poincaré–Birkhoff–Witt basis, consisting of elements

$$\prod_{\alpha>0} X_{-\alpha}^{a(\alpha)}/a(\alpha)! \prod_i \binom{H_i}{b(i)} \prod_{\alpha>0} X_\alpha^{c(\alpha)}/c(\alpha)!,$$

where

$$\binom{H}{b} = \frac{H(H-1)(H-2)\cdots(H-b+1)}{b!}.$$

Call $U_K = U_{\mathbf{Z}} \otimes K$ the *hyperalgebra* of $G$. It inherits from $U_{\mathbf{Z}}$ a Hopf algebra structure and a basis manufactured from the elements

$$X_{\alpha,a} = (X_\alpha{}^a/a!) \otimes 1, \qquad H_{i,b} = \binom{H_i}{b} \otimes 1.$$

Note that $\mathfrak{g}$ can be viewed as a subspace of $U_K$; but it does not generate $U_K$. In Kostant's original treatment [9], $U_{\mathbf{Z}}$ is used to define a $\mathbf{Z}$-form of the (say, simply connected) algebraic group $G_{\mathbf{C}}$ associated with $\mathfrak{g}_{\mathbf{C}}$. The idea is to define the algebra of polynomial functions $\mathbf{Z}[G_{\mathbf{C}}]$ as a Hopf algebra of "representative functions" on $U_{\mathbf{Z}}$. In the setting of [3, II, §4, no. 5, 6], $U_K$ then becomes identified with an algebra of distributions on $K[G]$, in a way compatible with the usual action of $\mathfrak{g}$ on functions. Moreover, $G$ acts on $U_K$ in a way compatible with its adjoint action on $\mathfrak{g}$.

It is important to observe that $U_K$ acts naturally on any $G$-module (by which we mean a finite-dimensional rational $G$-module), in a way compatible with the derived action of $\mathfrak{g}$ [1, 5.13; 3]. (Verma [13] suggests, conversely, that all finite-dimensional $U_K$-modules should be liftable to $G$-modules. We shall not need to go in this direction.)

## 2. The algebra $\mathbf{u}_r$ and its representations

In this section we define, for each positive integer $r$, a subalgebra $\mathbf{u}_r$ of $U_K$, and investigate its structure and representations.

**2.1** Set $q = p^r$. Define $\mathbf{u}_r$ to be the subalgebra (with 1) of $U_K$ generated by all $X_{\alpha,t}$, where $\alpha \in \Phi$ and $0 \le t < q$.

***Proposition*** *A basis for* $\mathbf{u}_r$ *consists of the elements*

$$\prod_{\alpha>0} X_{-\alpha,a(\alpha)} \prod_i H_{i,b(i)} \prod_{\alpha>0} X_{\alpha,c(\alpha)},$$

*where the positive roots are ordered in some fixed way and* $0 \le a(\alpha)$, $b(i)$, $c(\alpha) < q$. *In particular,* $\dim \mathbf{u}_r = q^{\dim \mathfrak{g}}$.

*Proof* Inspection of the proof of Kostant's theorem [6, §26; 9; 12, §2] shows that all $H_{i,b}$ $(b < q)$ are obtainable from the given generators of $\mathbf{u}_r$, so the exhibited elements of $U_K$ do lie in $\mathbf{u}_r$, and are of course linearly independent over $K$. It remains to show that their $K$-span $\mathbf{v}$ is closed under multiplication. The commutation rules developed in the proof of Kostant's theorem reduce matters to showing that products of the form $X_{\alpha,a}X_{\alpha,b}$ or $H_{i,a}H_{i,b}$ lie in $\mathbf{v}$ if $a, b < q$ but $a + b \geq q$. The first of these is 0 in $U_K$, thanks to the identity over $\mathbf{Z}$:

$$\frac{T^a}{a!}\frac{T^b}{b!} = \binom{a+b}{b}\frac{T^{a+b}}{(a+b)!} \tag{$*$}$$

($T$ an indeterminate). This implies further that $X_{\alpha,a}$ is nilpotent, $a > 0 : (X_{\alpha,a})^p = 0$.

To study the subalgebra of $\mathbf{u}_r$ generated by the $H_{i,a}$ for fixed $i$, where $a < q$, we can suppress the index $i$ and just write $H_a$. From the proof of Kostant's theorem (cf. [6, Lemma 26.1]) we have a formal identity over $\mathbf{Z}$ for each $a, b \in \mathbf{Z}$, $T$ being an indeterminate:

$$\binom{T}{a}\binom{T}{b} = \sum_{c=0}^{a+b} n_c\binom{T}{c}, \tag{$**$}$$

for some $n_c \in \mathbf{Z}$. We assert that the $H_a$ $(a < q)$ all lie in the $\mathbf{F}_p$-subalgebra of $\mathbf{u}_r$ generated by the $H_{p^i}$ $(0 \leq i \leq r-1)$. The proof is a recursion, based on ($**$). First,

$$\binom{T}{1}^2 = 2\binom{T}{2} + \binom{T}{1},$$

which shows (if $p > 2$) that $H_2$ is an $\mathbf{F}_p$-linear combination of $H_1$ and $(H_1)^2$. In turn,

$$\binom{T}{2}\binom{T}{1} = 3\binom{T}{3} + 2\binom{T}{2},$$

which shows (if $p > 3$) that $H_3$ lies in the $\mathbf{F}_p$-subalgebra generated by $H_1$. This continues through $H_{p-1}$. Next,

$$\binom{T}{1}\binom{T}{p} = (p+1)\binom{T}{p+1} + n\binom{T}{p} + \cdots.$$

So $H_{p+1}$ lies in the subalgebra generated by $H_1$ and $H_p$; similarly for $H_{p+2}, \ldots, H_{2p-1}$. In general write

$$\binom{T}{a-p^c}\binom{T}{p^c} = \binom{a}{p^c}\binom{T}{a} + \text{lower terms.}$$

If $p^c$ is the largest power of $p$ dividing $a$, then $p \nmid \binom{a}{p^c}$. So induction may be used.

Next we assert that $(H_a)^p = H_a$ for arbitrary $a > 0$. From (**) we have

$$\binom{T}{a}^p = n_{ap}\binom{T}{ap} + \cdots + n_1\binom{T}{1} \qquad \text{(there is no constant term).}$$

Successive substitution of 1, 2, ... for $T$ shows that $n_1 = n_2 = \cdots = n_{a-1} = 0$, while $n_a = 1$. Setting $T = a + 1$, we get

$$\binom{a+1}{a}^p \equiv \binom{a+1}{a} \equiv n_{a+1} + \binom{a+1}{a} \qquad (\text{mod } p),$$

so $n_{a+1} \equiv 0 \pmod p$. Similarly, all higher coefficients are divisible by $p$.

The preceding paragraph, along with (**), shows that arbitrary monomials in the $H_{p^i}$ $(0 \le i \le r-1)$ belong to the $\mathbf{F}_p$-linear span of the $H_a$ $(a < q)$. Combined with the earlier argument, this shows that the subalgebra of $\mathbf{u}_r$ generated by these $H_a$ has the asserted dimension $q$. QED

Note that when $r = 1$, $\mathbf{u}_r$ may be identified with the restricted universal enveloping algebra $\mathbf{u}$ of $\mathfrak{g}$ (as remarked by Verma). The algebra $\mathbf{u}_r$ is in many ways analogous to the group algebra over $K$ of the finite Chevalley group $G(\mathbf{F}_q)$; this is emphasized in [7] when $r = 1$.

Note too that $G$ acts naturally on $\mathbf{u}_r$ as a group of algebra automorphisms, since the action of $G$ on $U_K$ respects the filtration induced by the standard Poincaré–Birkhoff–Witt basis.

**2.2** Let $X$ be the weight lattice associated with $\Phi$ (essentially the character group of a maximal torus of $G$), $X^+$ the set of dominant weights, and $X_q$ the subset of $X^+$ having coordinates, relative to the fundamental dominant weights $\lambda_1, \ldots, \lambda_l$, between 0 and $q - 1$.

Denote by $\mathbf{n}_r^-$, $\mathbf{h}_r$, $\mathbf{n}_r^+$ the respective subalgebras with 1 of $\mathbf{u}_r$ generated by the elements $X_{-\alpha, a}$, $H_{i, b}$, $X_{\alpha, c}$ $(\alpha > 0)$. Proposition 2.1 identifies $\mathbf{u}_r$ with $\mathbf{n}_r^- \otimes \mathbf{h}_r \otimes \mathbf{n}_r^+$. The weights $\lambda \in X_q$ induce $K$-algebra homomorphisms $\mathbf{h}_r \to K$ (also denoted $\lambda$); if $\lambda = \sum c_i\lambda_i$, $\lambda(H_{i,a})$ is the residue class in $\mathbf{F}_p$ of $\binom{c_i}{a}$. It is clear that these $q^l$ homomorphisms are all distinct, since $\lambda(H_{i,a}) = 0$ whenever $c_i < a$ while $\lambda(H_{i,a}) = 1$ if $c_i = a$. On the other hand, the proof of Proposition 2.1 shows that any $K$-algebra homomorphism $f\colon \mathbf{h}_r \to K$ is completely determined by its values at the elements $H_{i,1}, H_{i,p}, \ldots, H_{i,p^{r-1}}$. Moreover, these values lie in $\mathbf{F}_p$, since each of these elements equals its $p$th power, so there are at most $q^l$ distinct homomorphisms. It follows that all of them arise from weights $\lambda \in X_q$.

The behavior of finite-dimensional $\mathbf{u}_r$-modules is entirely analogous to that of $\mathbf{u}$-modules (cf. [1, 6.6]). The nilpotent elements $X_{\alpha,a}$ yield nilpotent operators, while the $H_{i,b}$ (equal to their $p$th powers) yield commuting semisimple operators. In particular, a $\mathbf{u}_r$-module is a direct sum of weight spaces relative to $\mathbf{h}_r$, the weights corresponding to the elements of $X_q$, and some weight vector is a maximal vector (killed by all $X_{\alpha, a}$, $\alpha > 0$). Straightforward imitation of the arguments in [1] shows that (up to isomorphism) the irreducible $\mathbf{u}_r$-modules correspond 1–1 with their highest weights $\lambda \in X_q$. To construct an irreducible $\mathbf{u}_r$-module of highest weight $\lambda$, one can imitate the construction of the modules $Z_\lambda$ in [5] (analogues for $\mathbf{u}$ of Verma modules), obtaining "universal" $\mathbf{u}_r$-modules of dimension $q^m$ generated by a maximal vector of weight $\lambda \in X_q$; denote such a module by $Z_{\lambda, r}$. $Z_{\lambda, r}$ is induced from a suitable 1-dimensional module for $\mathbf{b}_r = \mathbf{h}_r \otimes \mathbf{n}_r^+$ and, viewed as an $\mathbf{n}_r^-$-module, is isomorphic to $\mathbf{n}_r^-$. Passing to a quotient then yields the desired irreducible $\mathbf{u}_r$-module.

**2.3** The irreducible $\mathbf{u}_r$-modules discussed in Section 2.2 are in fact derived from $G$-modules. Recall that the dominant weights $\lambda \in X^+$ index the irreducible $G$-modules $M_\lambda$ [1, 2, 7, 11, 12].

***Proposition*** *If $\lambda \in X_q$, then $M_\lambda$ remains irreducible as a $\mathbf{u}_r$-module. If $\lambda \in X^+ - X_q$, write $\lambda = \mu_0 + p\mu_1 + \cdots + p^t\mu_t$ ($\mu_i \in X_p$, $\mu_t \neq 0$, $t \geq r$). Then $M_\lambda$ as $\mathbf{u}_r$-module is isomorphic to a direct sum of copies of $M_\mu$, where $\mu = \mu_0 + p\mu_1 + \cdots + p^{r-1}\mu_{r-1}$.*

*Proof* Let $\lambda \in X_q$. Straightforward imitation of the proof of [1, 6.4] shows that $M_\lambda$ is irreducible as a $\mathbf{u}_r$-module. Next let $\lambda \notin X_q$, with $\lambda$ written as indicated. Steinberg's twisted tensor product theorem [1, 7.2; 11; 12] says that

$$M_\lambda \cong M_{\mu_0} \otimes M_{\mu_1}^{(p)} \otimes \cdots \otimes M_{\mu_t}^{(p^t)}.$$

The action of $\mathbf{h}_r$ on the highest weight space of $M_\lambda$ is given by the indicated weight $\mu \in X_q$, so $M_\mu$ occurs as a $\mathbf{u}_r$-composition factor of $M_\lambda$. But an argument like that in [5, p. 73] shows that $\mathbf{u}_r$ acts completely reducibly on an irreducible $G$-module, with all its irreducible summands being isomorphic; so in fact $M_\lambda$ becomes a direct sum of copies of $M_\mu$, as asserted. QED

**2.4** The *Steinberg module* $\mathrm{St}_r = M_{(q-1)\delta}$ ($\delta$ is the half-sum of positive roots) is known [11] to have dimension $q^m$, so it must coincide with the cyclic module $Z_{(q-1)\delta, r}$ of Section 2.2. Moreover, all other $M_\lambda$ ($\lambda \in X_q$) have strictly smaller dimension.

***Proposition*** $\mathrm{St}_r$ *is a projective $\mathbf{u}_r$-module, but no other $M_\lambda$ ($\lambda \in X_q$) is projective.*

*Proof* The first assertion is proved just as in the case $r = 1$ (cf. [7, §5]), based on the fact that $\mathrm{St}_r = Z_{(q-1)\delta, r}$. Reasoning like that used for $\mathbf{u}$ in [5] shows that a projective $\mathbf{u}_r$-module must have dimension divisible by $q^m$, so the second assertion follows from the above remark on $\dim M_\lambda$. QED

## 3. Tensor products

Here we assemble a few observations about tensor products.

**3.1** If $M$, $N$ are $G$-modules, we obtain a derived action of $\mathbf{u}_r$ on $M \otimes N$, describable explicitly via the Hopf algebra structure. The following general principle, pointed out to the author by Sweedler [7, App. T] has been known for some time in the context of group algebras of finite groups or restricted universal enveloping algebras. We shall need it in the case of $\mathbf{u}_r$.

***Lemma*** *Let $M$, $N$ be modules for a Hopf algebra (with antipode) over $K$, with $N$ projective. Then $M \otimes N$ is also projective.*

**3.2** Since $\mathrm{St}_r$ is a projective $\mathbf{u}_r$-module (Section 2.4), it follows from Section 3.1 that $M \otimes \mathrm{St}_r$ is also projective ($M$ arbitrary), hence a direct sum of indecomposable projective modules (PIMs). The PIMs of $\mathbf{u}_r$ are indexed by the weights in $X_q$.

Suppose $M$ is a $G$-module. Since $G$ acts naturally on $\mathbf{u}_r$ as a group of automorphisms (Section 2.1), the action of $G$ on $M \otimes \mathrm{St}_r$ will permute the various PIMs for $\mathbf{u}_r$ occurring in $M \otimes \mathrm{St}_r$. Observe that $M \otimes \mathrm{St}_r$ is a direct sum of two $G$-submodules: the sum of all $\mathbf{u}_r$-summands isomorphic to $\mathrm{St}_r$ and the sum of all other $\mathbf{u}_r$-summands. This follows from the fact that $\mathrm{St}_r$ is the only projective $\mathbf{u}_r$-module which is simultaneously irreducible (Section 2.4).

**3.3** The possible $G$-composition factors of a tensor product $M \otimes M_\lambda$ ($M$ arbitrary) can be determined to some extent. It is known (cf. [8, p. 131]) that these composition factors are those of various $G$-modules $\overline{V}_{\mu+\lambda}$ gotten from irreducible $\mathfrak{g}_{\mathbf{C}}$-modules by reduction modulo $p$, where $\mu$ ranges over the weights of $M$ and where $\mu + \lambda$ is dominant. The highest weights of composition factors of $M \otimes \mathrm{St}_r$ are therefore among the dominant weights of the form $\nu + (q-1)\delta$, where $\nu \le \mu$ for some weight $\mu$ of $M$ in the usual partial ordering ($\nu$ not necessarily dominant). Let $P$ be the set of all such weights $\nu$. Since only finitely many dominant weights lie below a given weight in the partial ordering, it is clear that we can choose $q = p^r$ sufficiently large so that $P$ contains no nonzero dominant weight of the form $q\pi$. Having chosen $r$ in this way (relative to $M$), consider a $G$-composition factor of $M \otimes \mathrm{St}_r$ having highest weight $\nu + (q-1)\,\delta$, $\nu = \sum c_i \lambda_i$ in $P$. Write

$$\nu + (q-1)\,\delta = \sum d_i \lambda_i + q \sum e_i \lambda_i, \qquad \text{where} \qquad 0 \le d_i < q, \quad 0 \le e_i,$$

so that $c_i + q - 1 = d_i + q e_i$. In order that $\sum d_i \lambda_i = (q-1)\,\delta$, it is necessary and sufficient that $c_i = q e_i$ for all $i$, but our choice of $r$ allows this to happen only when $\nu = 0$. In other words, taking Section 2.3 into account, a $G$-composition factor of $M \otimes \mathrm{St}_r$ which for $\mathbf{u}_r$ becomes a direct sum of copies of $\mathrm{St}_r$ must already be isomorphic (as a $G$-module) to $\mathrm{St}_r$.

## 4. Completely reducible $G$-modules

**4.1** For use in the next section, we note a simple (and more-or-less known) fact.

***Lemma*** *Let $S$ be a $G$-module all of whose composition factors have the same highest weight $\lambda$. Then $S$ is completely reducible.*

*Proof* The weights $\mu$ of $M_\lambda$ all satisfy $\mu \leq \lambda$, so it is clear that each vector of weight $\lambda$ in $S$ is a maximal vector (fixed by the maximal unipotent subgroup of $G$ corresponding to the positive roots) and that these vectors generate $S$ as $G$-module. In turn the submodule generated by a vector of weight $\lambda$ involves $\lambda$ with multiplicity 1, therefore involves $M_\lambda$ just once as a composition factor. It follows that $S$ is a direct sum of copies of $M_\lambda$. QED

## 5. Mumford's conjecture

**5.1** Here we shall indicate a representation-theoretic proof of Mumford's conjecture, using the algebra $\mathbf{u}_r$ for sufficiently large $r$. In the original proof by Haboush (cf. [4, 10]), the main step is as follows: Given a $G$-module $M$ containing a 1-dimensional $G$-invariant subspace $L$, find a $G$-module homomorphism $\varphi\colon M \to \mathrm{End}(\mathrm{St}_r)$ for some $r$, so that $\varphi(L) \neq 0$. (The pullback of the "determinant" function will then define a homogeneous $G$-invariant hypersurface in $M$ not containing $L$, whose existence was conjectured by Mumford.)

An equivalent problem is to find a $G$-homomorphism $M \otimes \mathrm{St}_r \to \mathrm{St}_r$ which is nonzero (hence 1–1) on $L \otimes \mathrm{St}_r$ ($\cong \mathrm{St}_r$). This amounts to showing that $L \otimes \mathrm{St}_r$ is a *direct summand* of $M \otimes \mathrm{St}_r$. In general, the $G$-summands of a tensor product such as $M \otimes \mathrm{St}_r$ are extremely difficult to describe. But here we can exploit the algebra $\mathbf{u}_r$: $M \otimes \mathrm{St}_r$ is a $U_K$-module (Section 1.1), hence a $\mathbf{u}_r$-module. Since $\mathrm{St}_r$ is a projective $\mathbf{u}_r$-module (Section 2.4), so is the tensor product (Section 3.1). Moreover, $G$ permutes the $\mathbf{u}_r$-summands isomorphic to $\mathrm{St}_r$, whose sum is a $G$-summand (Section 3.2), call it $S$. Some of the $G$-composition factors of $S$ might be larger than $\mathrm{St}_r$, having highest weights congruent to $(q-1)\,\delta$ modulo $qX$ (cf. Section 2.3). But if $r$ is chosen sufficiently large, only the highest weight $(q-1)\,\delta$ can actually occur in $S$ (Section 3.3). In this case, $S$ is completely reducible as a $G$-module (Section 4.1). Therefore $L \otimes \mathrm{St}_r$ splits off as a direct summand of $S$, hence also of $M \otimes \mathrm{St}_r$. QED

## References

1. A. Borel, "Properties and Linear Representations of Chevalley Groups" (Lect. Notes in Math. **131**), pp. 1–55. Springer-Verlag, New York–Heidelberg–Berlin, 1970.
2. A. Borel, Linear representations of semi-simple algebraic groups, *Proc. Symp. Pure Math.* **29**, (1975), 421–440.
3. M. Demazure and P. Gabriel, "Groupes Algébriques," Vol. I. Masson, Paris; North-Holland Publ., Amsterdam, 1970.
4. W. J. Haboush, Reductive groups are geometrically reductive, *Ann of Math.* **102** (1975), 67–83.
5. J. E. Humphreys, Modular representations of classical Lie algebras and semisimple groups, *J. Algebra* **19** (1971), 51–79.
6. J. E. Humphreys, "Introduction to Lie Algebras and Representation Theory." Springer-Verlag, New York–Heidelberg–Berlin, 1972.
7. J. E. Humphreys, "Ordinary and Modular Representations of Chevalley Groups" (Lect. Notes in Math. **528**). Springer-Verlag, New York–Heildelberg–Berlin, 1976.
8. J. C. Jantzen, Zur Charakterformel gewisser Darstellungen halbeinfacher Gruppen und Lie-Algebren, *Math. Z.* **140** (1974), 127-149.
9. B. Kostant, Groups over **Z**, *Proc. Symp. Pure Math.* **9** (1966), 90–98.
10. C. S. Seshadri, Theory of moduli, *Proc. Symp. Pure Math.* **29** (1975), 263–304.
11. R. Steinberg, Representations of algebraic groups, *Nagoya Math. J.* **22** (1963), 33–56.
12. R. Steinberg, Lectures on Chevalley Groups, Yale Univ. Math. Dept., 1968.
13. D.-N. Verma, Role of affine Weyl groups in the representation theory of algebraic Chevalley groups and their Lie algebras, *in* "Lie Groups and Their Representations" (I. M. Gel'fand, ed.), pp. 653–705. Halsted, New York, 1975.

Research was partially supported by a National Science Foundation grant. I am grateful to D.-N. Verma for pointing out the usefulness of the algebras $\mathbf{u}_r$, and to W. J. Haboush for some helpful discussions.

AMS (MOS) 1970 subject classification: 20G05

# *A notion of regularity for differential local algebras*

*Joseph Johnson*

*Rutgers University*

## Introduction

Let $k$ be a differential field and $A$ a differential local algebra of finitely generated type over $k$ without zero divisors. Roughly speaking, $A$ should be "regular" when there exists a family of local generators $x_1, \ldots, x_n$ of $A$ and relations $F_1(x) = \cdots = F_n(x) = 0$, where the $F_j$ are differential polynomials over $k$ in differential indeterminates $y_1, \ldots, y_n$ that satisfy the following:

(1) With respect to some ranking of $y_1, \ldots, y_n$, $F_1, \ldots, F_p$ is a characteristic set for the prime differential ideal $\{F \in k\{y\} : F(x) = 0\}$.

(2) If $u_j$ is the leader of $F_j$ and $I_j$ its initial, then $I_j(x)$ and $(\partial F_j / \partial u_j)(x)$ are units of $A$.

Use of this approach will lead us into computational problems of a substantial nature even if we only want to prove desirable basic properties of $A$ such as $\bigcap_{d>0} \mathfrak{m}^d = 0$, where $\mathfrak{m}$ is the maximal ideal of $A$. The problem is of course that characteristic sets are fairly complicated and should not be considered when easier alternatives are available. In this paper an alternative approach is developed, which in many respects simplifies and makes more conceptual the study of differential local algebras.

For the most part Ritt developed differential algebra for ordinary differential rings. For partial differential algebra the notion of ranking was developed so we could have a tool for partial differential rings which is as fine for them as the notion of order is for ordinary differential rings. It is my belief that for most purposes the notion of order alone is quite enough if we consider in the right way how elements of order $< r$ are related to elements of order $\leq r$. If for instance we want to consider a differential algebra $A = k\{x_1, \ldots, x_n\}$ and if $A_r \subset A$ is the $k$-subalgebra of $A$ generated by all derivatives of order $\leq r$ of $x_1, \ldots, x_n$, we should consider $A_r$ as a finitely generated $A_{r-1}$-algebra and take fullest advantage of all that commutative algebra can tell us about such things. That is roughly the program followed here.

In this paper the role that separants play in differential algebra is entirely subsumed by the study of differential modules of Kähler differentials. Hence Section 1 is devoted to sketching the part of that theory which we need. In Section 2 is a definition of a regular differential local algebra over $k$ that involves a choice of local generators but no choice of rankings. In Section 3 it is shown that the rough definition for regularity given above implies regularity if we insist that the ranking in the rough definition be an orderly one. The theorem there is more general, however, since no assumption needs to be made about the initials. It is also shown by an example that in nonzero characteristic, regularity is not a phenomenon that can always be expressed by the nonvanishing of separants. In Section 5 we show roughly that if $A$ is a finitely generated differential integral domain over $k$, then the set of all $p$ in Spec $A$ such that $R_p$ is regular is open. To prove the theorem of Section 5 we use Section 4 which contains a theory of regularity for graded modules over a polynomial ring. In Section 6 it is shown that for differential fields regularity amounts to separability.

## 1. Some preliminary facts about Kähler differentials

It will be assumed that the reader knows the contents of [1] which is a very brief and readable reference on the theory of Kähler differentials. That material together with Section 1 of [2] is the only background that will be assumed on that theory, without which this paper could not have been written. This section assembles some other known material on Kähler differentials for which suitable references are lacking. In addition, a few notational conventions are introduced.

Let $A \to B \to C$ be a given sequence of ring homomorphisms. We will usually use the very simple (and incomplete) notation $d(b)$ (or just $db$) for $d_{B/A}(b) \in \Omega_{B/A}$ and $1_C \otimes d_{B/A}(b) \in C \otimes_B \Omega_{B/A}$. If the context does not make clear which is meant we will be careful to state that $db \in \Omega_{B/A}$ or $db \in C \otimes_B \Omega_{B/A}$ so that confusion will not arise.

When $C = B/I$, with $I$ an ideal of $B$, the kernel of $C \otimes_B \Omega_{B/A} \to \Omega_{C/A}$ is generated as a $C$-module by $d(I) = \{dx : x \in I\}$. It follows that the kernel of $\Omega_{B/A} \to \Omega_{C/A}$ is generated as a $B$-module by $d(I) \subset \Omega_{B/A}$. Suppose now that $C$ is any $A$-algebra generated by a family $(x_\lambda)_{\lambda \in \Lambda}$. Let $B = A[X_\lambda]_{\lambda \in \Lambda}$ be a polynomial algebra over $A$ and write $C = B/I$ in the obvious way. It is clear that $(dx_\lambda)_{\lambda \in \Lambda}$ generates $\Omega_{C/A}$. Also, if $(F_h)_{h \in H}$ generates $I$, the relations $F_h \, dX_\lambda = 0$ and $dF_h = 0$ ($h \in H$, $\lambda \in \Lambda$) comprise a defining set of relations for the $dx_\lambda$. If the reader has an easy familiarity with this way of describing a module of Kähler differentials by generators and relations, that which is written here will be much clearer.

Whenever a local ring $A$ is given we will always associate with it the following notation:

$\mathfrak{m}$ = the maximal ideal of $A$

$K = A/\mathfrak{m}$

$P(A) = \mathbf{S}_K(\mathfrak{m}/\mathfrak{m}^2)$ the symmetric algebra over $K$ on the $K$-vector space $\mathfrak{m}/\mathfrak{m}^2$

$G(A) = \sum_{d \geq 0} \mathfrak{m}^d/\mathfrak{m}^{d+1}$ the associated graded algebra of $A$

$\tau_A : P(A) \to G(A)$ the unique $K$-algebra homomorphism that extends the identity mapping of $\mathfrak{m}/\mathfrak{m}^2$

Let $B$ be any ring, $S$ a multiplicatively closed subset of $B$, and $R$ a subring of $B$. Then $S^{-1}R$ will be written instead of $(S \cap R)^{-1}R$. Similarly, if $\mathfrak{p}$ is a prime ideal of $B$, we will write $R_\mathfrak{p}$ for $R_{(\mathfrak{p} \cap R)}$.

***Lemma 1*** *Let $A$ be a local ring and $k$ a subfield of $A$ such that $K$ is a separable extension of $k$. Then the sequence*

$$0 \to \mathfrak{m}/\mathfrak{m}^2 \xrightarrow{i} K \otimes_A \Omega_{A/k} \to \Omega_{K/k} \to 0$$

*is exact.*

The only part of this lemma not already established in [1] is the injectivity of $i$. We will show this first with the following additional hypotheses:

1. There exist $x_1, \ldots, x_n \in A$ such that $A = k[x_1, \ldots, x_n]_\mathfrak{m}$, i.e., $A$ is finitely generated as a local algebra over $k$.
2. $\mathfrak{m}^2 = 0$.

We will need

***Lemma 2*** *Let $A$ be finitely generated as a local algebra over $k$ and suppose $\mathfrak{m}^2 = 0$. If $K$ is a separable extension of $k$, then $A$ contains a subfield which is mapped isomorphically onto $K$.*

Choose $x_1, \ldots, x_n \in A$ so that $x_1 + \mathfrak{m}, \ldots, x_n + \mathfrak{m}$ is a separating transcendence basis for $K$ over $k$. We may replace $k$ by $k(x_1, \ldots, x_n)$ and so assume that $K$ is separably algebraic over $k$. Then $K = k(x)$, where $x \in K$ satisfies $F(x) = 0$ with $F$ an irreducible polynomial over $k$ that has no multiple roots. If $y \in A$ and $w \in \mathfrak{m}$, $F(y + w) = F(y) + F'(y)w$. If $y$ is a preimage of $x$ and $w = -F(y)/F'(y)$, then $w \in \mathfrak{m}$ and $F(y + w) = 0$. It follows that $k[y + w]$ is mapped isomorphically onto $K$.

From Lemma 2 and our assumptions it follows that we have an isomorphism of vector spaces over $k$, $A \approx K \oplus \mathfrak{m}$. It is easy to see that $\mathfrak{m}$ is an $A$-module and that multiplication in $A$ obeys the rule $(\alpha_1, \xi_1)(\alpha_2, \xi_2) = (\alpha_1\alpha_2, \alpha_1\xi_2 + \alpha_2\xi_1)$ whenever $\alpha_1, \alpha_2 \in K$ and $\xi_1, \xi_2 \in \mathfrak{m}$. Let $p_2$ be the projection of $A$ onto $\mathfrak{m}$. Then $p_2 \in \operatorname{Der}_k(A, \mathfrak{m})$ so there is a unique $A$-linear map $\lambda\colon \Omega_{A/k} \to \mathfrak{m}$ with $\lambda \circ d_{A/k} = p_2$. Now $i = d_{A/k} \mid \mathfrak{m}$ so $\lambda \circ i = (p_2 \mid \mathfrak{m}) = \mathrm{id}_{\mathfrak{m}}$ so $i$ is injective.

We now need to justify assumptions 1 and 2. For assumption 2 we observe that the sequence of Lemma 1 does not change if we replace $A$ by $A/\mathfrak{m}^2$, so it is clear we can asssme that $A$ satisfies assumption 2. For assumption 1 let $A'$ be any local $k$-subalgebra of $A$ which is (locally) finitely generated. Let $\mathfrak{m}' = \mathfrak{m} \cap A'$ be the maximal ideal of $A'$ and let $K' = A'/\mathfrak{m}'$. Then the sequence of Lemma 1 is $\varinjlim_{A}$ of the sequences

$$0 \to \mathfrak{m}'/\mathfrak{m}'^2 \to K' \otimes_{A'} \Omega_{A'/k} \to \Omega_{K'/k} \to 0$$

so it must be exact if these are.

The following simple lemma is contained in Lemma 5.

***Lemma 3*** *If $k$ is a field and $K$ a finitely generated and separable extension of $k$, then* $\dim_K \Omega_{K/k} = \operatorname{tr\,deg}_k K$.

***Lemma 4*** *Let $A$ be a finitely generated local integral domain over a field $k$. Assume that $K/k$ is separable. Then*

(1) $\dim_K K \otimes_A \Omega_{A/k} \geq \operatorname{tr\,deg}_k A$ *with equality if and only if $A$ is regular;*
(2) *$A$ is regular if and only if $\Omega_{A/k}$ is a free $A$-module of rank equal to* $\operatorname{tr\,deg}_k A$.

We know that $\dim_K \mathfrak{m}/\mathfrak{m}^2 \geq \operatorname{tr\,deg}_k A - \operatorname{tr\,deg}_k K$ with equality exactly when $K$ is regular. Thus from Lemmas 1 and 3

$$\dim_K K \otimes_A \Omega_{A/k} = \dim_K \mathfrak{m}/\mathfrak{m}^2 + \operatorname{tr\,deg}_k K \geq \operatorname{tr\,deg}_k A$$

with equality if and only if $A$ is regular. Statement (2) follows from (1), Nakayama's lemma, and Lemma 5:

***Lemma 5*** *Let $L$ be a finitely generated field extension of $k$. Then* $\operatorname{tr\,deg}_k L \leq \dim_L \Omega_{L/k}$ *with equality if and only if $L$ is separable over $k$.*

Let $x_1, \ldots, x_n$ generate $L$ and set $p = \dim_L \Omega_{L/k}$. There exist polynomial relations $f_j(x_1, \ldots, x_n) = 0$ over $k$ with $j = p + 1, \ldots, n$ such that the matrix $(\partial f_j/\partial x_i)_{1 \le i \le n,\, p+1 \le j \le n}$ (which has coefficients in $L$) is of rank $n - p$. It follows that $n - p$ of $x_1, \ldots, x_n$ are separably algebraic over the others. Hence $\operatorname{tr\,deg}_k L \le p$ and equality implies that $L$ is separable over $k$. Conversely if $L$ is separable over $k$ and $x_1, \ldots, x_p$ is a separating transcendance basis for $L$ over $k$, then $dx_1, \ldots, dx_p$ is a basis for $\Omega_{L/k}$.

***Corollary*** *If $A$ is a regular local algebra of finitely generated type over the field $k$ and if $K$ is a separable extension of $k$, then the quotient field of $A$ is a separable extension of $k$.*

In Lemma 6 for any polynomial $F$, $D_i F$ will represent its partial derivative with respect to the $i$th variable.

***Lemma 6*** *Let $A$ be a local integral domain of finitely generated type over a field $k$. Let $B$ be a local integral domain containing $A$ with maximal ideal $\mathfrak{n}$ such that $\mathfrak{m} \subset \mathfrak{n}$ and with $B$ locally generated over $A$ by elements $x_1, \ldots, x_n$. Let $L$ be a separable extension of $k$ and $\phi$: $B \to L$ a $k$-algebra homomorphism with $\phi(\mathfrak{n}) = 0$. The following are then equivalent if $A$ is regular:*

(1) *$L \otimes_A \Omega_{A/k} \to L \otimes_B \Omega_{B/k}$ is injective and $B$ is regular;*

(2) *for some $0 \le p \le n$ there are polynomial relations $F_{p+1}(x) = \cdots = F_n(x) = 0$ on $x_1, \ldots, x_n$ with coefficients in $A$ and $I \subset \{1, \ldots, n\}$ of cardinality $n - p$ such that the $x_j$ with $j \notin I$ are algebraically independent over $A$ and such that* $\det(D_i F_j)_{i \in I,\, j = p+1, \ldots, n} \notin \mathfrak{n}$.

Assume that (1) holds and set $p = \dim_L L \otimes_B \Omega_{B/A}$. We can rearrange $x_1, \ldots, x_n$ so that $dx_1, \ldots, dx_p$ is a basis (over $L$) of $L \otimes_B \Omega_{B/A}$. Since $B$ is regular, Lemma 4 shows that $\operatorname{tr\,deg}_k B = \dim_L L \otimes_B \Omega_{B/k}$. Also we have $\dim_L L \otimes_A \Omega_{A/k} = \operatorname{tr\,deg}_k A$. By (1) and Th. 1.6 of [1] we have an exact sequence

$$0 \to L \otimes_A \Omega_{A/k} \to L \otimes_B \Omega_{B/k} \to L \otimes_B \Omega_{B/A} \to 0 \qquad (*)$$

so $\dim_L L \otimes_B \Omega_{B/k} = \operatorname{tr\,deg}_k A + p$. By the definition of $p$ there exist polynomial relations $F_{p+1}(x) = \cdots = F_n(x) = 0$ on $x_1, \ldots, x_n$ with coefficients in $A$ such that $\det(D_i F_j(x))_{p<i,\, j \le n} \notin \mathfrak{n}$. If $x_1, \ldots, x_p$ are algebraically dependent over $A$, as $B$ is separably algebraic over $A[x_1, \ldots, x_p]$, $\operatorname{tr\,deg}_k B < \operatorname{tr\,deg}_k A + p$ so $B$ is not regular. This contradiction shows that $x_1, \ldots, x_p$ are algebraically independent over $A$.

Now assume (2). Then the quotient field of $B$ is separable over the quotient field of $A$, and $\operatorname{tr\,deg}_A B = p$. Also (2) implies that any element of $\operatorname{Der}_k(A, L)$ extends to an element of $\operatorname{Der}_k(B, L)$. Hence we have an exact sequence $(*)$. Now by (2) $\dim_L L \otimes_B \Omega_{B/A} \le p$, so $\dim_L L \otimes_B \Omega_{B/k} \le \operatorname{tr\,deg}_k A + \operatorname{tr\,deg}_A B = \operatorname{tr\,deg}_k B$. That shows $B$ is regular so (1) holds.

## 2. Regular differential algebras

Let $\Delta$ be a set with $m$ elements $\delta_1, \ldots, \delta_m$. A differential ring with $\Delta$ as its set of (commuting) derivation operators will be called a $\Delta$-ring. Similarly we shall speak of $\Delta$-modules and $\Delta$-ideals preferring to write $\Delta$ in those places where formerly we would have written "differential." For ordinary differential algebra ($m = 1$), however, we will often stick to our old ways.

Let $k$ be a $\Delta$-ring, $A$ any $\Delta$-algebra over $k$, and $S$ the set of units of $A$. We will say $A$ is *of finitely generated type* (*over* $k$) if there is a finite subset $x_1, \ldots, x_n$ of $A$ (for some $n$) so that $A = S^{-1}k\{x_1, \ldots, x_n\}$. In that case such a set $x_1, \ldots, x_n$ will be called a *set of quasi-generators* (*of* $A$). Given such a set we shall always define $A_r = S^{-1}k[\Theta_r x_1 \cdots \Theta_r x_n]$ where

$$\Theta_r = \{\delta_1^{r_1} \cdots \delta_m^{r_m} : r_1 + \cdots + r_m \leq r, 0 \leq r_1, \ldots, r_m\}.$$

For $r \in \mathbb{Z}$ with $r < 0$ we shall set $A_r = S^{-1}(k \cdot 1_A)$. When $A$ is also a local ring we shall call $A$ a $\Delta$-*local algebra* (*over* $k$) even if the maximal ideal $\mathfrak{m}$ of $A$ is not a $\Delta$-ideal of $A$. Any set $x_1, \ldots, x_n$ of quasi-generators of $A$ will be called a set of *local generators*. When a set of local generators is given we shall define $\mathfrak{m}_r = \mathfrak{m} \cap A_r$ and $K_r = A_r/\mathfrak{m}_r$ for all $r$ in $\mathbb{Z}$. Of course it should be noted that $K$ need not be a $\Delta$-field.

For the rest of this section $k$ will be any $\Delta$-field and $A$ a $\Delta$-local algebra of finitely generated type over $k$. We will say what it means for $A$ to be regular by introducing three axioms.

***Axiom 1*** *$\tau_A$: $P(A) \to G(A)$ is an isomorphism.*

When $m = 0$ (and so $\Delta = \phi$) this axiom is quite enough by itself. The following example suggests that when $m > 0$ we will need more.

***Example*** Let $k$ be an ordinary differential field and $k\langle y \rangle$ a differential field extension of $k$ with $y, y'$ algebraically independent over $k$ and $y'' = y'/y$. If we let $B = k\{y\}$, then $B_0 = k[y]$ and $B_r = k[y, y'/y^{r-1}]$ if $r > 0$. Clearly $By$ is a prime differential ideal of $B$, and in fact $B/By = k$. Let $A = B_{By}$. Since $y' = y^{r-1}(y'/y^{r-1}) \in By^{r-1}$ for all $r > 1$, $y' \in \bigcap_{d \geq 0} \mathfrak{m}^d$. Hence $\bigcap_{d \geq 0} \mathfrak{m}^d \neq (0)$ although, since $\mathfrak{m}^d = Ay^d \neq \mathfrak{m}^{d+1}$, condition (1) is satisfied.

The second axiom requires the choice of a set of local generators for $A$. I do not know at present any way to get around this.

***Axiom 2*** *There exists a finite set of local generators such that*

$$\phi_r: K \otimes_{A_r} \Omega_{A_r/k} \to K \otimes_A \Omega_{A/k}$$

*is injective for all $r \geq 0$.*

Of course in Axiom 2, $A_r$ is defined using the set of local generators in the manner explained at the beginning of this section.

Finally, we add an axiom which could perhaps be done away with but without which our theory would certainly be a good deal more complicated.

***Axiom 3*** *K is a separable extension of k.*

When $A$ satisfies Axioms 1 and 3, any finite set of local generators for $A$ such that Axiom 2 holds will be called a *regular set of local generators* (*for* $A$). When $A$ satisfies Axioms 1–3, we will say that $A$ is *regular*. Some properties of regular $\Delta$-algebras will now be derived.

***Proposition 1*** *Let $A$ be a $\Delta$-local algebra with a given regular set of local generators. Then*

(1) *$A_r$ is a regular local ring for every $r$;*
(2) *$A$ is an integral domain, and its quotient field is separable over $k$;*
(3) $\bigcap_{d \geq 0} \mathfrak{m}^d = 0$.

To see this note that in the following diagram one can conclude (in the order indicated) that 1 is injective, 2 and 3 are isomorphisms, and 4 is injective, the injectivity of 1 being a consequence of Axiom 2 and Lemma 1 of Section 1:

$$\begin{array}{ccc} P(A_r) & \xrightarrow{3} & G(A_r) \\ {\scriptstyle 1}\downarrow & & \downarrow{\scriptstyle 4} \\ P(A) & \xrightarrow[\tau_A]{2} & G(A) \end{array}$$

From this and some well-known facts about regular local rings, (1) and (2) are immediate, separability being a consequence of Lemmas 4 and 5 of Section 1. If $x \in \mathfrak{m}$ and $x \neq 0$, then for some $r$, $x \in \mathfrak{m}_r{}^d$. By increasing $d$ we can assume that $x \notin \mathfrak{m}_r^{d+1}$. Then by the injectivity of 4, $x \notin \mathfrak{m}^{d+1}$, so $x \notin \bigcap_{d \geq 0} \mathfrak{m}^d$ proving (3).

***Corollary*** *If $A$ is a differential local algebra with a given set of local generators, then*

$$\dim_K(K \otimes_{A_r} \Omega_{A_r/k}) \geq \operatorname{tr\,deg}_k A_r$$

*for all $r \in \mathbb{Z}$, with equality if the set of local generators is regular.*

This is immediate from Lemma 4 of Section 1 because

$$K \otimes_{A_r} \Omega_{A_r/k} \approx K \otimes_{K_r} (K_r \otimes_{A_r} \Omega_{A_r/k}).$$

The inequality in this corollary can sometimes be helpful. Let us illustrate its use in considering the system of differential equations

$$\delta_1{}^2\eta = \eta\, \delta_1\, \delta_2\, \eta = \delta_2{}^2\eta \qquad (*)$$

since later on we will need this example anyway. Let $m = 2$, let $k$ be a $\Delta$-field, and let $B$ be any $\Delta$-algebra over $k$ without zero divisors and generated by a single element $\eta$ that satisfies $(*)$. If $1 \notin [\eta]$, then $[\eta]$ is prime and we can let $A = B_{[\eta]}$. Now $k = K$ and we find that for every $r$, $\dim_k k \otimes_{A_r} \Omega_{A_r/k} \leq 4$. Now by the corollary, either $\operatorname{tr}\deg_k A_2 < 4$ or every element of $\Theta_3\, \eta$ is algebraically dependent over $A_2$. This encourages us to seek additional relations of order $\leq 3$ on $\eta$. We find that any $\eta$ that satisfies $(*)$ also satisfies

$$(1 - \eta^2)\, \delta_2{}^3\eta = (1 - \eta^2)\, \delta_1{}^2\, \delta_2\, \eta = (\delta_2\, \eta + \eta\, \delta_1\eta)\, \delta_1\, \delta_2\, \eta$$

and

$$(1 - \eta^2)\, \delta_1{}^3\eta = (1 - \eta^2)\, \delta_1\, \delta_2{}^2\eta = (\delta_1\eta + \eta\, \delta_2\, \eta)\, \delta_1\, \delta_2\, \eta,$$

a very explicit form of the observation about $\Theta_3\, \eta$ that was made before. We can also compute $\delta_1{}^2\, \delta_2{}^2\eta$ twice using these equations. We find that as $k\{\eta\}$ is an integral domain,

$$2(1 - \eta^2)\eta((\delta_1\, \eta)^2 - (\delta_2\eta)^2)\, \delta_1\, \delta_2\eta = 0.$$

If char $k \neq 2$, then necessarily $\delta_1{}^2\eta$, $\delta_1\, \delta_2\, \eta$, $\delta_2{}^2\eta$ are all zero. This shows for instance that the only meromorphic functions $f(t_1, t_2)$ that satisfy

$$\frac{\partial^2 f}{\partial t_1{}^2} = f\, \frac{\partial^2 f}{\partial t_1\, \partial t_2} = \frac{\partial^2 f}{\partial t_2{}^2}$$

are $f(t_1, t_2) = a + bt_1 + ct_2$ with $a, b, c \in \mathbb{C}$. When char $k = 2$, there does exist a $\Delta$-algebra $k\{\eta\}$ such that $\delta_1{}^2\eta = \eta\, \delta_1\, \delta_2\eta = \delta_2{}^2$ with $\eta_1$, $\delta_1\eta$, $\delta_2\eta$, $\delta_1\, \delta_2\, \eta$ algebraically independent over $k$. Let $C = k[\eta, \delta_1\, \eta, \delta_2\, \eta, \delta_1\, \delta_2\, \eta] \times [1/(1 - \eta^2)]$. Clearly $C$ is a $\Delta$-ring and $[\eta]$ is a prime $\Delta$-ideal of $C$.

Before going on to the next section let us note the following criterion for regularity. It is more useful than the original definition for the proofs that will follow.

***Proposition 2*** *Let $A$ be a differential local algebra over a differential field $k$ with a given set of local generators $x_1, \ldots, x_n$. If*

(1) *$K$ is a separable extension of $k$;*
(2) *$A_r$ is a regular local ring for all $r \geq 0$;*
(3) *for all $r \geq 0$, $\phi_r$: $K \otimes_{A_r} \Omega_{A_r/k} \to K \otimes_A \Omega_{A/k}$ is injective,*

*then $x_1, \ldots, x_n$ is a regular set of local generators of $A$.*

Clearly it is only necessary to show that when (1)–(3) are satisfied, it follows that $\tau_A\colon P(A) \to G(A)$ is injective. For each $r \geq 0$ consider the commutative diagram

$$\begin{array}{ccc} P(A_r) & \xrightarrow{\tau_{A_r}} & G(A_r) \\ \downarrow & & \downarrow \\ P(A) & \xrightarrow{\tau_A} & G(A) \end{array}$$

Since

$$P(A) = \varinjlim_r P(A_r) \qquad \text{and} \qquad G(A) = \varinjlim_r G(A_r)$$

and since by (1) $\tau_{A_r}$ is injective for every $r$, it follows that $\tau_A$ is injective.

It should be noted that when $A$ is regular there may well be a finite set of local generators for $A$ which is not regular. For instance, if $y$ is a $\Delta$-indeterminate and $\Delta = \{\delta\}$ (so $m = 1$), $A = k\{y\}_{[y]}$ is regular but $y, (\delta y)^2$ is not a regular set of local generators. This is because for each $r > 0$,

$$d(2\,\delta y\,\delta^{r+1}y) \in \ker(K \otimes_{A_r} \Omega_{A_r/k} \to K \otimes_A \Omega_{A/k})$$

but is nonzero. It remains an interesting question whether Axiom 2 can be replaced by one that does not involve a choice of local generators.

## 3. A sufficient condition for regularity

Let $k$ be a $\Delta$-field, $y_1, \ldots, y_n$ $\Delta$-indeterminates over $k$, and $\mathfrak{Q}$ a prime $\Delta$-ideal of $R = k\{y_1, \ldots, y_n\}$. Let $\mathfrak{P}$ be any prime ideal of $R$ with $\mathfrak{P} \supset \mathfrak{Q}$, let $A = R_{\mathfrak{p}}/\mathfrak{Q}R_{\mathfrak{p}}$, and let $x_1, \ldots, x_n$ be the images of $y_1, \ldots, y_n$ respectively in $A$. In this section a sufficient condition will be given for the set $x_1, \ldots, x_n$ of local generators of $A$ to be regular. For this purpose let us suppose that an orderly ranking of $y_1, \ldots, y_n$ is given. Then if $F_1 < \cdots < F_h$ is a characteristic set of $\mathfrak{Q}$ and $u_j$ is the leader of $F_j$ for $j = 1, \ldots, h$, the sequence $u_1 < \cdots < u_h$ depends only on the ranking. We will assume in the theorem that follows that $K$ (which in this case is the quotient field of $R/\mathfrak{P}$) is a separable extension of $k$ and that $u_1, \ldots, u_h$ are as just defined.

***Theorem*** *Suppose there exist elements $F_1, \ldots, F_h$ of $\mathfrak{Q}$ with $u_j$ the leader of $F_j$ and such that $\partial F_j/\partial u_j \notin \mathfrak{P}$ for $j = 1, \ldots, h$. Then $x_1, \ldots, x_n$ is a regular set of local generators of $A$.*

By Proposition 2 of Section 2 we need to show that $A_r$ is regular and $K \otimes_{A_r} \Omega_{A_r/k} \to K \otimes_A \Omega_{A/k}$ is injective for all $r \geq 0$. Now $K \otimes_A \Omega_{A/k}$ is the direct limit of the $K \otimes_{A_r} \Omega_{A_r/k}$. Hence it will suffice to show that if $A_{r-1}$ is regular, then $A_r$ is regular and $K \otimes_{A_{r-1}} \Omega_{A_{r-1}/k} \to K \otimes_{A_r} \Omega_{A_r/k}$ is injective.

For brevity let us call each $u_j$ a *leader* $(1 \leq j \leq h)$. Also if $v \in R_{\mathfrak{P}}$, let $\bar{v}$ denote its image in $A$. Let $r \in \mathbb{N}$ and let $v_1, \ldots, v_g$ be those elements of $\Theta y$ $(= \Theta y_1 \cup \cdots \cup \Theta y_n)$ whose order is $r$ but which are not derivatives of any leader of order $< r$.

**Lemma** *Let $u \in \Theta y$ be of order $r$ and also a derivative of some leader whose order is $< r$. Then $\bar{u} = \bar{P}/\bar{Q}$, where $P, Q \in R$, order $Q < r$, each element of $\Theta y$ that occurs in $P$ is $< u$, and $Q \notin \mathfrak{P}$. Also if $v \in \Theta y$ occurs in $P$ is of order $r$, $v$ is not a derivative of some leader of order $< r$.*

If the lemma is false, choose the least $u$ for which the conclusion does not hold. Write $u = \theta u_j$ where $\theta \in \Theta$ and order $u_j < r$. Then $\theta F_j = (\partial F_j/\partial u_j)u - P'$, where each derivative that occurs in $P'$ is $< u$. Clearly $\bar{u} = \bar{P}'/\bar{Q}'$, where $Q' = \partial F_j/\partial u_j$. It may happen that for some $l$, order $u_l < r$ and that there is a $v$ in $\Theta u_l$ with order $v = r$ such that $v$ occurs in $P'$. We will call such a $v$ *unfortunate*. Let $V$ be the set of unfortunates. If $v \in V$, then $v < u$ and so (by the way $u$ was chosen) we may write $\bar{v} = \bar{P}_v/\bar{Q}_v$ with $P_v$ and $Q_v$ as in the conclusion of the lemma. Replace every unfortunate $v$ in $P'$ by $P_v/Q_v$ and let $P''$ be the result of those substitutions. Let $H = \prod_{v \in V} Q_v$. If $l$ is large enough, $H^l P'' \in R_r$. If $P = H^l P''$ and $Q = H^l Q'$, $\bar{u} = \bar{P}/\bar{Q}$. Since $P$ and $Q$ satisfy the conclusion of the lemma this contradicts the way $u$ was chosen.

From the lemma it is clear that $\bar{v}_1, \ldots, \bar{v}_g$ is a set of local generators for $A_r$ as a local algebra over $A_{r-1}$. Let us assume that $u_\alpha, \ldots, u_\beta$ are the leaders among $v_1, \ldots, v_g$ and that $v_1, \ldots, v_e$ are what remain. If $j = \alpha, \ldots, \beta$ it can be that $F_j$ involves elements of $\Theta y$ of order $r$ other than $v_1, \ldots, v_g$. However, any such element $u$ of $\Theta y$ must be a derivative of a leader of order $< r$. We can use the lemma to replace $u$ by $P_u/Q_u$ as in the lemma. Let $H_j$ be the product of all the $Q_u$ that are needed. If $d$ is large enough and $F_j{}'$ is what results from $F_j$ after all the replacements are made, then $G_j = H_j{}^l F_j{}' \in R$ if $l$ is large enough. Now $G_j \in \mathfrak{Q}$ also and has leader $u_j$. It is easy to see that $\partial G_j/\partial u_j \notin \mathfrak{P}$. Let $P_j \in A_{r-1}[v_1, \ldots, v_g]$ be the result of substituting $\bar{v}$ for $v$ in $G_j$ for every $v \in \Theta_{r-1} y$. Now $v_1, \ldots, v_e$ are algebraically independent over $A_{r-1}$. Let $\lambda_{ij}$ be what results from $\partial P_i/\partial u_j$ when $\bar{v}_1, \ldots, \bar{v}_g$ are substituted for $v_1, \ldots, v_g$ $(\alpha \leq i, j \leq \beta)$. Now $\det(\lambda_{ij})_{\alpha \leq i,\, j \leq \beta} \notin \mathfrak{m}$. By Lemma 6 of Section 1, $A_r$ is regular and

$$K \otimes_{A_{r-1}} \Omega_{A_{r-1}/k} \to K \otimes_{A_r} \Omega_{A_r/k}$$

is injective.

It is probable that when the characteristic of $k$ is zero some kind of a converse to this theorem exists. That is, if $x_1, \ldots, x_n$ is a regular set of local generators of $A$ we can probably replace $x_1, \ldots, x_n$ by $z_1, \ldots, z_n$ where $z_i = \sum_{j=1}^{n} a_{ij} z_j$ with the $a_{ij}$ in $k$ and $\delta_1, \ldots, \delta_m$ by $e_i = \sum_{j=1}^{m} c_{ij} \delta_j$ with the $c_{ij}$ constants of $k$ and det $c \neq 0$ and then be able to assert the converse.†

However, when the characteristic of $k$ is not zero, there is no such converse. As an example consider the case char $k = 2$ and the $\Delta$-algebra $C = k[\eta, \delta, \eta, \delta_2 \eta, \delta_1 \delta_2 \eta][1/(1 - \eta^2)]$ mentioned in Section 2. Let $A = C_{[\eta]}$. Now clearly $\eta$ is a regular set of local generators for $A$. No matter what orderly ranking we use on a single $\Delta$-indeterminate $y$, we will have $u_1 = \delta_1 \delta_2 y$. Also any $F_1$ we can choose will be a multiple of $y\, \delta_1 \delta_2 y - \delta_1{}^2 y$ or $y\, \delta_1 \delta_2 y - \delta_2{}^2 y$ (depending on which ranking we use), and no matter how we change $\Delta$, $\partial F/\partial u_1$ will obstinately remain in $[y]$. This example suggests we should try approaches to the study of regular $\Delta$-local algebras that do not require a converse to this theorem. We will therefore study $\Omega_{A/k}$ more closely in Section 5 by taking advantage of the filtration given to it by the system of local generators. It should after all be apparent by now that looking for "separants" that do not vanish is just a way of looking for relations on the generators $dx_1, \ldots, dx_n$ of $\Omega_{A/k}$.

It seems a good idea to point out that to prove $A$ satisfies Axiom 2 in the definition of regularity by some sort of calculation with differential polynomials (starting with the hypothesis of our theorem) would be a very tedious exercise, perhaps even an impossible one.

## 4. Regular graded modules

This section contains a theory which will be used in proving the theorem of Section 5. To start we recall a basic concept of algebra.

Let $T$ be any ring with identity. A left $T$-module $M$ is called *finitely presented* if there is an exact sequence of left $T$-modules $0 \to L \to F \to M \to 0$ with $L$ finitely generated and free. We have the following well-known lemmas.

***Lemma 1*** *Let $M$ be a finitely presented left $T$-module. Then if $0 \to L \to F \to M \to 0$ is an exact sequence of left $T$-modules, $L$ is finitely generated if $F$ is.*

This is proven in [3, p. 37]. The following is left to the reader.

***Lemma 2*** *Let $0 \to M' \to M \to M'' \to 0$ be an exact sequence of left $T$-modules. Then if $M'$ and $M''$ are finitely presented, $M$ is also.*

† *Note added in proof:* An example shows that some modification of this "converse" statement is necessary.

For what follows $m$ will be a fixed natural number and $k$ a fixed field with infinitely many elements. We will let $X_1, \ldots, X_m$ be a given set of indeterminates. If $A$ is any commutative ring, set $S_A = A[X_1, \ldots, X_m]$. Whenever $\mathfrak{p} \in \operatorname{Spec} A$ we will write $S_\mathfrak{p}$ for $S_{A_\mathfrak{p}} = A_\mathfrak{p}[X_1, \ldots, X_m]$.

Let $A$ be a commutative ring and $M$ a graded $S_A$-module. Call $M$ *regular* if $M_d$ is a free $A$-module for every $d$ in $\mathbb{Z}$. If $\mathfrak{p} \in \operatorname{Spec} A$, call $\mathfrak{p}$ a *regular point for* $M$ if the graded $S_\mathfrak{p}$-module $M_\mathfrak{p}$ is regular; that is, if for every $d$ in $\mathbb{Z}$, $(M_d)_\mathfrak{p}$ is a free $A_\mathfrak{p}$-module. We will let $\operatorname{Reg} M$ denote the $\mathfrak{p}$ in $\operatorname{Spec} A$ such that $\mathfrak{p}$ is a regular point for $M$. The following is our fundamental theorem concerning this notion.

***Theorem*** *Let $A$ be an integral domain with $k \subset A$ and $M$ a finitely generated graded $S_A$-module. If $\mathfrak{p} \in \operatorname{Reg} M$, then for some $f$ in $A \backslash \mathfrak{p}$, $M[1/f]$ is a finitely presented and regular graded $S_{A[1/f]}$-module.*

It should be noted that this theorem when properly phrased resembles (3.4.6) of [4] by Raynaud and Gruson. However, a proof based on [4] does not seem easier than the direct proof given here.

The proof of the theorem will involve induction on $m$. Part (ii) of the following lemma establishes the special case $m = 0$ (in which case $S_A = A$). We will use the following notation:

$K_\mathfrak{p} = A_\mathfrak{p}/\mathfrak{p}A_\mathfrak{p}$ the residue class field of $A_\mathfrak{p}$

$m_\mathfrak{p} = \mathfrak{p}A_\mathfrak{p}$

$\bar{M}_\mathfrak{p} = K_\mathfrak{p} \otimes_A M = M_\mathfrak{p}/\mathfrak{p}M_\mathfrak{p}$ for any $A$-module $M$

***Lemma 3*** *Let $A$ be an integral domain, $M$ a finitely generated $A$-module, $\mathfrak{p} \in \operatorname{Spec} A$. Then:*

(i) *$M_\mathfrak{p}$ is free if and only if $\dim_{K_\mathfrak{p}} \bar{M}_\mathfrak{p} = \operatorname{rk}_A M$;*

(ii) *if $M_\mathfrak{p}$ is free, then for some $f$ in $A \backslash \mathfrak{p}$, $M[1/f]$ is a free $A[1/f]$-module;*

(iii) *if $0 \to L \to F \to M \to 0$ is an exact sequence of $A$-modules with $F$ finitely generated free and $M_\mathfrak{p}$ a free $A_\mathfrak{p}$-module, then for some $f$ in $A \backslash \mathfrak{p}$, $L[1/f]$ is a finitely generated free $A[1/f]$-module.*

Let $Q$ be the quotient field of $A$ and suppose that $M_\mathfrak{p}$ is a free $A_\mathfrak{p}$-module. Then $\dim_{K_\mathfrak{p}}(\bar{M}_\mathfrak{p}) = \dim_Q(Q \otimes_{A_\mathfrak{p}} M_\mathfrak{p})$, so $\dim_{K_\mathfrak{p}} \bar{M}_\mathfrak{p} = \operatorname{rk}_A M$. Conversely, assume $n = \dim_{K_\mathfrak{p}} M_\mathfrak{p} = \operatorname{rk}_A M$. Choose $x_1, \ldots, x_n$ in $M$ so that their images $\bar{x}_1, \ldots, \bar{x}_n$ in $\bar{M}_\mathfrak{p}$ form a basis for $\bar{M}_\mathfrak{p}$ over $K_\mathfrak{p}$. By Nakayama's lemma, $x_1, \ldots, x_n$ (or rather their images in $M_\mathfrak{p}$) generate $M_\mathfrak{p}$. They are linearly independent over $Q$ and so must also be linearly independent over $A_\mathfrak{p} \subset Q$.

For (ii) choose a surjective map $\phi\colon F \to M$, where $F$ is an $A$-module free on generators $e_1, \ldots, e_n$, and let $e_i' = \phi(e_i)$. If $d = \operatorname{rk}_A M$, we may assume that $e_1', \ldots, e_d'$ generate $M_{\mathfrak{p}}$ over $A_{\mathfrak{p}}$. Then for $j = d+1, \ldots, n$ there exists $a_j \in A \backslash \mathfrak{p}$ such that $a_j e_j' \in Ae_1' + \cdots + Ae_d'$. If $f = a_{d+1} \cdots a_n$, then $e_j' \in A[1/f]e_1' + \cdots + A[1/f]e_d'$, so $M[1/f]$ is a free $A[1/f]$-module. For (iii) we can assume by (ii) that $M$ is free. Then the sequence splits, so $L$ is a direct summand of $F$ and so must be finitely generated. Now $\operatorname{rk}_A L = \operatorname{rk}_A F - \operatorname{rk}_A M$ and also $\dim_{K_{\mathfrak{p}}} \bar{L}_{\mathfrak{p}} = \dim_{K_{\mathfrak{p}}} \bar{F}_{\mathfrak{p}} - \dim_{K_{\mathfrak{p}}} \bar{M}_{\mathfrak{p}}$ which implies that $\operatorname{rk}_A L = \dim_{K_{\mathfrak{p}}} \bar{L}$. By (i) and (ii) $f$ must exist.

In proving the theorem let us fix $\mathfrak{p}$ in Reg $M$ and set $K = K_{\mathfrak{p}}$, $\bar{M} = (\bar{M}_{\mathfrak{p}})$, etc., that is, generally omitting $\mathfrak{p}$ as a subscript. Also if $x \in M$, its image in $\bar{M}$ will be denoted as $\bar{x}$. Now $\bar{M}$ is a graded $S_K$-module, and if we fix a set $x_1, \ldots, x_n$ of homogeneous generators for $M$, $\bar{x}_1, \ldots, \bar{x}_n$ will be a set of homogeneous generators for $\bar{M}$. Let $d$ be the rank of $\bar{M}$. Then by renumbering we can assume that $\bar{x}_1, \ldots, \bar{x}_d$ are linearly independent over $S_K$. Let $L = S_A x_1 + \cdots + S_A x_d$ and $N = M/L$. It follows immediately from the next lemma that $\mathfrak{p}$ is a regular point of $L$ and $N$.

**Lemma 4** *Let $P$ be a finitely generated $A$-module, $\mathfrak{p} \in \operatorname{Spec} A$, $u_1, \ldots, u_k \in P$. If $P_{\mathfrak{p}}$ is a free $A_{\mathfrak{p}}$-module and $\bar{u}_1, \ldots, \bar{u}_k \in \bar{P}$ are linearly independent over $K$, then $u_1, \ldots, u_k$ are linearly independent over $A_{\mathfrak{p}}$ and $(P/Au_1 + \cdots + Au_k)$ is a free $A_{\mathfrak{p}}$-module.*

Indeed there exist $u_{h+1}, \ldots, u_n$ in $P$ such that $\bar{u}_1, \ldots, \bar{u}_n$ is a basis for $\bar{P}$ over $K$. Then $u_1, \ldots, u_n$ is a free basis for $P_{\mathfrak{p}}$ over $A_{\mathfrak{p}}$ and Lemma 4 follows.

To prove the theorem it will be enough to show it holds for $L$ and $N$. That is clear because of Lemma 2 and because if in a short exact sequence of modules the end terms are free, then the middle term is free also.

Proving the theorem for $L$ is the simpler. Indeed, by Lemma 4 the family $\{X_1^{d_1} \cdots X_m^{d_m} x_j : 1 \leq j \leq d, d_1, \ldots, d_m \in \mathbb{N}\}$ is free over $A_{\mathfrak{p}}$ and hence free over $A$ also. Hence $L$ is free on the homogeneous generators $x_1, \ldots, x_d$ so it is regular and finitely presented. The proof for $N$ is more difficult and requires induction on $m$.

It should be noted that the theorem involves the indeterminates $X_1, \ldots, X_m$ only to the extent that they are used to grade $S_A$. Hence any $m$ elements of $k[X_1, \ldots, X_m]$ which are homogeneous of degree 1 and linearly independent over $k$ may be used instead. These remarks should make clear the statement of the next lemma.

**Lemma 5** *Let $P$ be a finitely generated $S_K$-module and suppose that no element of $P$ is free over $S_K$. Then $X_1, \ldots, X_m$ may be chosen so that $P$ is finitely generated as a $K[X_1, \ldots, X_{m-1}]$-module.*

We shall defer the proof of this and also the much needed

***Lemma 6*** *Suppose that $N$ is an $A[X_1, \ldots, X_m]$-module and is finitely presented as an $A[X_1, \ldots, X_{m-1}]$-module. Then $N$ is finitely presented as an $A[X_1, \ldots, X_m]$-module.*

Let us see how these two lemmas imply the theorem. Set $S_K' = K[X_1, \ldots, X_{m-1}]$, $S_A' = A[X_1, \ldots, X_{m-1}]$, and $S_\mathfrak{p}' = A_\mathfrak{p}[X_1, \ldots, X_{m-1}]$. Then Lemma 5 implies that for some choice of $X_1, \ldots, X_m$, $N_\mathfrak{p}$ is finitely generated as an $S_\mathfrak{p}'$-module. Indeed, in Lemma 5 set $P = \bar{N}$, and let $X_1, \ldots, X_m$ be chosen so that the conclusion of the lemma holds. If $x_1, \ldots, x_h$ is any set of homogeneous elements of $N$ such that $\bar{x}_1, \ldots, \bar{x}_h$ generate $P = \bar{N}$ as an $S_K'$-module, then $x_1, \ldots, x_h$ generate $N_\mathfrak{p}$ as an $S_\mathfrak{p}'$-module. Now let $y_1, \ldots, y_n$ generate $N$ as an $S_A$-module and also $N_\mathfrak{p}$ as an $S_\mathfrak{p}'$-module. Let $H$ be the $S_A'$-submodule of $N$ generated by $y_1, \ldots, y_n$. Then $f \in A \backslash \mathfrak{p}$ may be chosen so that $fX_m y_j \in H$ for $j = 1, \ldots, n$. Then $X_m$ maps the $S_A'[1/f]$-submodule $H[1/f]$ of $N[1/f]$ into itself. Hence $H[1/f] = N[1/f]$ and so $N[1/f]$ is finitely generated as an $S_A'[1/f]$-module. By induction on $m$ we know that a better choice of $f$ in $A \backslash \mathfrak{p}$ will make $N[1/f]$ regular as an $S_A'[1/f]$-module and also finitely presented as an $S_A'[1/f]$-module. Clearly for such a choice of $f$, $N[1/f]$ is regular as an $S_A[1/f]$-module. Also by Lemma 6, $N[1/f]$ will be finitely presented as an $S_A[1/f]$-module.

We now come to the proof of Lemma 5. Let $\sigma$ be a nonsingular $m \times m$-matrix over $k$ and define $Y_1, \ldots, Y_m$ by $X_i = \sum_{j=1}^m \sigma_{ij} Y_j$ for $i = 1, \ldots, m$. It will be shown that there is a nonzero polynomial $H(\sigma) = H(\sigma_{1m}, \ldots, \sigma_{mm})$ over $k$ in the entries of the last column of $\sigma$ such that if $H(\sigma) \neq 0$, then $P$ is finitely generated as a $K[Y_1, \ldots, Y_{m-1}]$-module. Now this last statement follows easily by induction on the number $n$ of generators of $P$ (as an $S_K$-module) once it is known for $n = 1$. Hence we will assume that $P$ is generated by a single homogeneous element $x$.

By hypothesis there is a nonzero $F$ in $S_K$ with $Fx = 0$. We can assume $F$ is homogeneous of some degree $d > 0$, and we will write $F(\sigma)$ for $F(\sigma_{1m}, \ldots, \sigma_{mm})$. Then

$$F(X) = F(\sigma) Y_m{}^d + G(Y), \qquad (*)$$

where $G$ is homogeneous of degree $d$ in $Y_1, \ldots, Y_m$ and is free of $Y_m{}^d$. Now write $F = u_1 H_1 + \cdots + u_N H_N$ where $u_1, \ldots, u_N \in K$ are linearly independent over $k$ and $H_1, \ldots, H_N$ are nonzero elements of $k[X_1, \ldots, X_m]$. Pick any $i = 1, \ldots, N$ and write $H(\sigma) = H_i(\sigma_{1m}, \ldots, \sigma_{mm})$. Then if $H(\sigma) \neq 0$, $F(\sigma) \neq 0$ since the entries of $\sigma$ lie in $k$. Let $\tilde{P}$ be the $K[Y_1, \ldots, Y_{m-1}]$-submodule of $P$ generated by $x, Y_m x, \ldots, Y_m^{d-1} x$ and assume $H(\sigma) \neq 0$. Then from $(*)$ since $Fx = 0$ and $F(\sigma) \neq 0$ we get $Y_m{}^d x \in \tilde{P}$ so $Y_m \tilde{P} \subset \tilde{P}$. As $x \in \tilde{P}$ and $S_K \tilde{P} \subset \tilde{P}$, $\tilde{P} = P$ and $P$ is generated as a $K[Y_1, \ldots, Y_{m-1}]$-module by $x, Y_m x, \ldots, Y_m^{d-1} x$. This proves Lemma 5.

To prove Lemma 6 construct a commutative diagram of solid arrows:

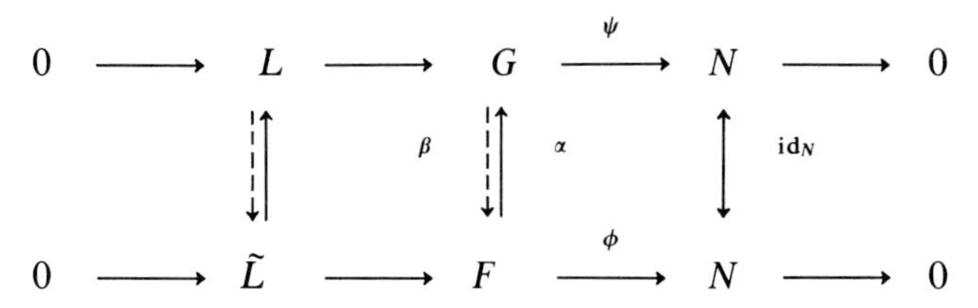

The top row is any exact sequence of $S_A'$-modules with $G$ finitely generated free, say on generators $e_1, \ldots, e_n$. Now in the category of $S_A$-modules (instead of $S_A'$-modules) let $F$ be free on generators $e_1, \ldots, e_n$ also and map $F$ to $N$ by defining $\phi(e_i) = \psi(e_i)$ for $i = 1, \ldots, n$. Let $L = \ker \psi$ and $\tilde{L} = \ker \phi$.

We can write $X_m \psi(e_i) = \sum_{j=1}^{n} f_{ij} \psi(e_j)$ for $i = 1, \ldots, n$, where the $f_{ij} \in S_A'$. Extend the $S_A'$-module structure on $G$ to an $S_A$-module structure by defining $X_m e_i = \sum_{j=1}^{n} f_{ij} e_j$ for $i = 1, \ldots, n$. Then $\psi$ is $S_A$-linear and $L$ is an $S_A$-submodule of $G$. Let $\alpha$ be the unique $S_A$-linear map of $F$ to $G$ such that $\alpha(e_i) = e_i$ for $i = 1, \ldots, n$. The diagram is completed by letting the left-hand vertical arrow be induced by $\alpha$. It is a commutative diagram of $S_A$-modules.

With respect to the induced $S_A'$-module structures we have an $S_A'$-linear map $\beta\colon G \to F$ defined by $\beta(e_i) = e_i$ for $i = 1, \ldots, n$. Clearly $\alpha\beta = \mathrm{id}_G$ so $\beta$ identifies $G$ with an $S_A'$-submodule of $F$ (which is very easy to describe) and $L$ with an $S_A'$-submodule of $\tilde{L}$.

For $i = 1, \ldots, n$ let $u_i$ be the element $X_m e_i - \sum_{j=1}^{n} f_{ij} e_j$ of $\tilde{L}$ and let $H$ be the $S_A$-submodule of $F$ generated by $u_1, \ldots, u_n$. We need to show that $\tilde{L}$ is finitely generated. For that it will suffice to show that $F = H + \beta(G)$, for then it will follow that $\tilde{L} = H + \beta(L)$ which certainly is finitely generated since $L$ is. The relations $X_m e_i = u_i + \sum_{j=1}^{n} f_{ij} e_j$ show $X_m(\beta(G)) \subset H + \beta(G)$, so $H + \beta(G)$ is an $S_A$-submodule of $F$ and contains $e_1, \ldots, e_n$. Hence $F = H + \beta(G)$.

***Theorem 2*** *Let $x_1, \ldots, x_n$ be a set of homogeneous generators for the graded $S_A$-module $M$ and suppose that $\mathfrak{p} \in \operatorname{Spec} A$ is regular for $M$. Then for some choice of $X_1, \ldots, X_m$ it is possible to choose $f \in A \backslash \mathfrak{p}$ and subsets $x_{01}, \ldots, x_{0n_0}; x_{11}, \ldots, x_{1n_1}; \ldots; x_{m1}, \ldots, x_{mn_m}$ of $x_1, \ldots, x_n$ so that the family*

$$\{X_1^{r_1} \cdots X_l^{r_l} x_{lh} : 1 \leq l \leq m, r_1, \ldots, r_l \in \mathbb{N}, h = 1, \ldots, n_l\}$$

*is a free basis for $M[1/f]$ as an $A[1/f]$-module.*

This is certainly true when $m = 0$, so assume $m > 0$. We will freely use the notation of the preceding proof. Rearranging $x_1, \ldots, x_n$ causes no problems, so we will assume that $d = \operatorname{rank} \bar{M}$ and that $\bar{x}_1, \ldots, \bar{x}_d$ are linearly independent over $S_K$. Set $n_m = d$ and let $x_{ml}, \ldots, x_{mn_m}$ be $x_1, \ldots, x_d$. Let $N$ be defined as before. We can choose $X_1, \ldots, X_m$ and $f \in A \backslash \mathfrak{p}$ so that $N[1/f]$ is generated as an $S'_{A[1 \backslash f]}$-module by $j(x_{d+1}), \ldots, j(x_n)$ where $j: M \to N$. We can replace $A$ by $A[1/f]$, i.e., we can assume to begin with that $N$ is generated as an $S_A'$-module by $j(x_{d+1}), \ldots, j(x_n)$. Now $\mathfrak{p}$ is regular for $N$, so by induction on $m$ we can choose $f$ in $A$, rechoose $X_1, \ldots, X_{m-1}$, and choose subsets $x_{01}, \ldots, x_{0n_0}; \ldots; x_{m-1,1}, \ldots, x_{m-1,n_{m-1}}$ of $x_{d+1}, \ldots, x_n$ so that the family

$$\{X_1^{r_1} \cdots X_l^{r_l} j(x_{lh}) : 1 \leq l \leq m-1, r_1, \ldots, r_l \geq 0, h = 1, \ldots, n_l\}$$

is a free basis for $N[1/f]$ as an $A[1/f]$-module. It is then clear that $f$ and the $x_{lh}$, $0 \leq l \leq m$, $1 \leq h \leq n_l$, that we have chosen do satisfy the conclusion of Theorem 2.

It should be remarked that if $k_0$ is any infinite subfield of $k$, the choice of $X_1, \ldots, X_m$ referred to in Theorem 2 can as well be made by taking linear combinations over $k_0$ of $X_1, \ldots, X_m$.

## 5. The openness of the set of regular points

Let $A$ be a $\Delta$-algebra over the $\Delta$-field $k$ and $x_1, \ldots, x_n$ a given set of quasi-generators for $A$. For convenience we will set $(\Omega'_{A/k})_r = A \otimes_{A_r} \Omega_{A_r/k}$, and let $(\Omega_{A/k})_r$ be the image of $(\Omega'_{A/k})_r$ in $\Omega_{A/k}$. Then the $(\Omega_{A/k})_r$ give an excellent filtration of $\Omega_{A/k}$ which is a left module over $\mathscr{D}_A$ (cf. [2]). As is well known (cf. [7]) gr $\mathscr{D}_A$ is a polynomial ring $S_A$ over $A$ in the (indeterminates) $x_1, \ldots, x_n$, where $x_i = \delta_i + A \in \mathrm{gr}_1 \mathscr{D}_A$. Let $M_{A/k}$ be the graded $S_A$-module $\mathrm{gr}(\Omega_{A/k})$. Whenever the context makes clear what $k$ and $A$ we are considering, these thereby redundant subscripts will be deleted from the notation; e.g., we will write $\Omega_r'$ or $(\Omega_A')_r$ in place of $(\Omega'_{A/k})_r$ and $M_A$ or $M$ in place of $M_{A/k}$. Consonant with this convention, $M_r$ will then denote $(M_{A/k})_r$ (i.e., the part homogeneous of degree $r$).

***Proposition*** *Suppose that $A$ is a $\Delta$-local algebra over $k$ and that its residue class field $K$ is a separable extension of $k$. If $x_1, \ldots, x_n$ is a regular set of local generators of $A$, then conditions* (i) *and* (ii) *are satisfied:*

(i) *for every $r$ the map $\Omega_r' \to \Omega_r$ is an isomorphism;*
(ii) *$M$ is a regular graded $S_A$-module.*

*Conversely if* (i) *and* (ii) *are satisfied and if the quotient field of $A$ is a separable extension of $k$, then $x_1, \ldots, x_n$ is a regular set of local generators of $A$.*

To prove the first part assume that $x_1, \ldots, x_n$ is a regular set of local generators of $A$. By Lemma 4 of Section 1 and Proposition 1 of Section 2 we have for every $r$ that $\Omega_{A_r/k}$ is a finitely generated free $A_r$-module of rank equal to $\dim_{K_r} K_r \otimes_{A_r} \Omega_{A_r/k}$. Also if $r \leq s$, we know that $K \otimes_{A_r} \Omega_{A_r/k} \to K \otimes_{A_s} \Omega_{A_s/k}$

is injective. Hence $K \otimes_A \Omega_r' \to K \otimes_A \Omega_s'$ is injective and so by Lemma 4 of Section 4, $\Omega_r'$ is a direct summand of $\Omega_s'$ and $\Omega_s'/\Omega_r'$ is free. As $\Omega = \varinjlim_r \Omega_r'$, the map $\Omega_r' \to \Omega$ is injective for every $r$. This establishes (i) and also (ii) since clearly $\Omega_r/\Omega_{r-1}$ is a free $A$-module for every $r$.

To prove the converse statement in the proposition, a simple lemma is needed.

***Lemma 1*** *Let $U \subset V$ be commutative rings, $L$ a $U$-module generated by $d$ elements. If $V \otimes_U L$ is a free $V$-module on $d$ generators, $L$ is a free $U$-module on $d$ generators.*

The proof of this lemma is routine. Now suppose the quotient field of $A$ is separable over $k$ and that (i) and (ii) of the proposition are satisfied. As each $M_r = \Omega_r/\Omega_{r-1}$ is a finitely generated free $A$-module and since $\Omega_r = 0$ if $r < 0$, it follows that $\Omega_r = \Omega_r'$ is a finitely generated free $A$-module for every $r$ and also that if $r \leq s$, $\Omega_r$ is a direct summand of $\Omega_s$. For the moment fix an $r$ in $\mathbb{Z}$ and let $d = \dim_K K \otimes_A \Omega_r$. Then also $d = \dim_{K_r} K_r \otimes_{A_r} \Omega_{A_r/k}$, so $\Omega_{A_r/k}$ is generated by $d$ elements as an $A_r$-module and $\Omega_r = A \otimes_{A_r} \Omega_{A_r/k}$ is an $A$-module free on $d$ generators. By Lemma 2, $\Omega_{A_r/k}$ is free on $d$ elements. Using the fact that the quotient field of $A_r$ is separable over $k$ we can combine Lemmas 4 and 5 of Section 1 to conclude that $A_r$ is a regular local ring. Also since $\Omega_r$ is a direct summand of $\Omega_s$ if $s \geq r$, $K \otimes_A \Omega_r \to K \otimes_A \Omega_s$ is always injective. Since $K \otimes_A \Omega$ is the direct limit of the $K \otimes_A \Omega_s$, $K \otimes_A \Omega_r \to K \otimes_A \Omega$ is injective too. It follows from Proposition 2 of Section 2 that $x_1, \ldots, x_n$ is a regular set of local generators for $A$.

***Corollary*** *Let $A$ be a $\Delta$-algebra over an infinite $\Delta$-field $k$. Suppose $x_1, \ldots, x_n$ is a set of generators of $A$ and $\mathfrak{p}$ a prime ideal of $A$ such that $x_1, \ldots, x_n$ is a regular set of local generators for $A_{\mathfrak{p}}$. Then $\Delta$ may be replaced by $m$ linear combinations of $\delta_1, \ldots, \delta_m$ over the constants of $k$ so that the following is true. There is an $f$ in $A \backslash \mathfrak{p}$ and there are subsets $x_{01}, \ldots, x_{0n_0}$; $x_{11}, \ldots, x_{1n_1}; \ldots; x_{m1}, \ldots, x_{mn_m}$ of $x_1, \ldots, x_n$ such that if*

$$B = \{\delta_1^{r_1} \cdots \delta_l^{r_l}\, dx_{lh} : 0 \leq l \leq m,\ 1 \leq h \leq n_l,\ r_1, \ldots, r_l \in \mathbb{N}\},$$

*then $B$ is a free basis over $A[1/f]$ of $\Omega_{A[1/f]/k}$.*

This is almost immediate from the proposition and Theorem 2 of Section 4. We do, however, have to let $k_0$ be the constants of $k$ and use the remark at the end of Section 4. More precisely, we apply that theorem to $M = \mathrm{gr}(\Omega_{A/k})$. By the definitions we have a canonical inclusion $M_0 \subset \Omega_{A/k}$ and the elements $dx_1, \ldots, dx_n$ of $\Omega_{A/k}$ are in $M_0$. We can use $dx_1, \ldots, dx_n$ as the set of homogeneous generators mentioned in Theorem 2 of Section 4 and make a choice of $X_1, \ldots, X_m$ over $k_0$ (instead of over $k$) so that the conclusion of that theorem holds. We make the modification of $\Delta$ which is analogous to the modification we made of $X_1, \ldots, X_m$. We may as well replace $A$ by $A[1/f]$, that is, assume $f = 1$. We use the $x_{lh}$ and the indexing given to us by the conclusion of Theorem 2 of Section 4. To make it clear that the set $B$ we obtain in this way has the required properties, define $B_r$ for $r \in \mathbb{N}$ to be the elements of $B$ with $r_1 + \cdots + r_l \leq r$. Now the elements of $B_r \backslash B_{r-1}$ are linearly independent over $A$ mod$(\Omega_{A/k})_{r-1}$. If we use induction on $r$ to assume that the elements of $B_{r-1}$ are linearly independent over $A$, since $B_{r-1} \subset (\Omega_{A/k})_{r-1}$, it follows that the elements of $B_r$ are linearly independent over $A$. Hence it is clear that the elements of $B$ are linearly independent over $A$. A similar induction shows that $\Omega_{A/k}$ is generated as an $A$-module by $B$.

If $R$ is a $\Delta$-ring containing the $\Delta$-field $k$ and if $\mathfrak{p} \in \operatorname{Spec} R$, call $\mathfrak{p}$ *separable* if the quotient field of $R/\mathfrak{p}$ is a separable extension of $k$. Denote by $\operatorname{Sep}\operatorname{Spec}_k R$ the set of all such $\mathfrak{p}$, and consider it as a topological space by using the relative topology from $\operatorname{Spec} R$.

***Theorem*** *Let $k$ be an infinite differential field and $A$ a differential integral domain containing $k$. Suppose that a set $x_1, \ldots, x_n$ of generators of $A$ as a differential algebra over $k$ is given. Then the set of all $\mathfrak{p}$ in* Sep Spec$_k$ *$A$ such that $x_1, \ldots, x_n$ is a regular set of generators for $A_\mathfrak{p}$ is open.*

Notice that the wording of this theorem obscures a difficulty that is very real. Namely, it is possible that there exists no $\mathfrak{p}$ at all such that $x_1, \ldots, x_n$ is a regular set of generators for $A_\mathfrak{p}$. It is shown in Section 6 that for all sufficiently large $r$ in $\mathbb{N}$, $\Theta_r x_1 \cup \cdots \cup \Theta_r x_n$ is a regular set of generators for $A_{(0)}$ if $A_{(0)}$ is a separable extension of $k$.

Let us show first that our theorem is a consequence of the following lemma.

***Lemma 2*** *Let $\mathfrak{p} \in$ Sep Spec$_k$ $A$ and suppose that $x_1, \ldots, x_n$ is a regular set of generators for $A_\mathfrak{p}$. Then for some $f$ in $A \backslash \mathfrak{p}$, $\Omega_r'[1/f] \to \Omega_r[1/f]$ is an isomorphism for every $r$.*

Let $\mathfrak{p} \in \operatorname{Sep}\operatorname{Spec}_k A$ and assume that $x_1, \ldots, x_n$ is a regular set of generators for $A_\mathfrak{p}$. We need to show that for some $f$ in $A \backslash \mathfrak{p}$, $x_1, \ldots, x_n$ is a regular set of local generators for every $\mathfrak{q}$ in $\operatorname{Sep}\operatorname{Spec}_k A$ such that $f \notin \mathfrak{q}$. Assuming Lemma 2 allows us to assume in effect that $\Omega_r' \to \Omega_r$ is an isomorphism for every $r$.

By the proposition, $M_\mathfrak{p}$ is a regular graded $S_{A_\mathfrak{p}}$-module. By Theorem 1 of Section 4 there is an $f$ in $A \backslash \mathfrak{p}$ such that if $\mathfrak{q} \in \operatorname{Spec} A$ and $f \notin \mathfrak{q}$, $M_\mathfrak{q}$ is a regular graded $S_{A_\mathfrak{q}}$-module. Now the existence of a regular $\mathfrak{p} \in \operatorname{Spec} A$ implies that the quotient field of $A$ is separable over $k$ (corollary to Lemma 5 of Section 1). Hence we can again apply the proposition to conclude that $x_1, \ldots, x_n$ is a regular set of local generators for $A_\mathfrak{q}$. Hence it remains only to prove Lemma 2.

Let $i^r\colon \Omega'_{r-1} \to \Omega_r'$ be the natural map and let $M' = \bigoplus_{r \in \mathbb{Z}} \Omega_r'/i^r\Omega'_{r-1}$. Now $M'$ has a natural structure of graded $S_A$-module such that the canonical map $M' \to M$ is $S_A$-linear. Let us assume that for the moment and show how Lemma 2 follows. First $M_\mathfrak{p}' \to M_\mathfrak{p}$ is an isomorphism by the proposition, and we can choose $\mathfrak{p} \in A\backslash\mathfrak{p}$ such that $M'[1/f]$ and $M[1/f]$ are regular. Now $M'[1/f] \to M[1/f]$ is surjective so its kernel $L$ is regular. Since $L_\mathfrak{p} = 0$, $L = 0$ and $M'[1/f] \approx M[1/f]$. If we replace $A$ by $A[1/f]$ we can, to begin with, assume $M' \approx M$. Let us show by induction on $r$ that $\Omega_r'$ is free and $i^r$ is injective for every $r$. We have a commutative diagram

$$\begin{array}{ccc}
\Omega'_{r-1} & \xrightarrow{i^r} & \Omega_r' \\
\downarrow & & \downarrow \\
(\Omega'_{r-1})_\mathfrak{p} & \longrightarrow & (\Omega_r')_\mathfrak{p} \\
\wr\wr & & \wr\wr \\
(\Omega_{r-1})_\mathfrak{p} & \longrightarrow & (\Omega_r)_\mathfrak{p}
\end{array}$$

and we can assume that $\Omega'_{r-1}$ is free. Then the left and bottom arrows are injections, so $i^r$ must be an injection too. Because $\Omega_r'/i^r\Omega'_{r-1}$ is free, $\Omega_r'$ is free also. Since we have now shown that each $i^r$ is an injection and since $\Omega = \lim_r \Omega_r'$, $\Omega_r' \to \Omega$ is injective. Thus the proof of Lemma 2 is complete once we establish the assertion concerning $M'$.

To show that $M'$ has the asserted $S_A$-module structure, consider for any given $r$ and $i = 1, \ldots, m$ the commutative diagram

$$\begin{array}{ccccccccc}
0 & \longrightarrow & I_{r-1} & \longrightarrow & A_{r-1} \otimes_k A_{r-1} & \xrightarrow{\mu_{r-1}} & A_{r-1} & \longrightarrow & 0 \\
 & & \downarrow{\scriptstyle \tau_i} & & \downarrow{\scriptstyle \sigma_i} & & \downarrow{\scriptstyle \delta_i} & & \\
0 & \longrightarrow & I_r & \longrightarrow & A_r \otimes_k A_r & \xrightarrow{\mu_r} & A_r & \longrightarrow & 0
\end{array}$$

where $\mu_r$ is the multiplication map and $I_r = \ker \mu_r$. We let $\sigma_i = \delta_i \otimes \iota + \iota \otimes \delta_i$, where $\iota\colon A_{r-1} \to A_r$ is the inclusion. Let $\xi = \sum_{j=1}^p a_j \otimes b_j$ be an arbitrary element of $A_{r-1} \otimes_k A_{r-1}$. Then

$$\xi = \left(\sum_{j=1}^{p} a_j b_j\right) \otimes 1 + \sum_{j=1}^{p} (a_j \otimes 1)(1 \otimes b_j - b_j \otimes 1).$$

If $\mu_{r-1}(\xi) = 0$, $\xi = \sum_{j=1}^p (a_j \otimes 1)(1 \otimes b_j - b_j \otimes 1)$ and an easy calculation shows $\sigma_i(\xi) \in \ker \mu_r = I_r$. Hence $\sigma_i$ induces a map $\tau_i\colon I_{r-1} \to I_r$. Now $\sigma_i(uv) = (\sigma_i u)(\phi v) + (\phi u)(\sigma_i v)$, where $\phi\colon A_{r-1} \otimes_k A_{r-1} \to A_r \otimes_k A_r$ is $\iota \otimes \iota$

and $\tau_i(uv)$ has a similar expression. Hence $\tau_i(I_{r-1}^2) \subset I_r{}^2$, so we have a map $\delta_i: \Omega_{A_{r-1}/k} \to \Omega_{A_r/k}$ because $\Omega_{A_r/k} = I_r/I_r{}^2$. Define $\delta_i: \Omega'_{r-1} \to \Omega_r{}'$ by $\delta_i(u \otimes w) = \delta_i(a) \otimes i^r w + a \otimes \delta_i w$ for $a \in A$, $w \in \Omega_{A_{r-1}/k}$. This is well defined and induces a $A$-linear map $X_i: \Omega'_{r-1}/i^{r-1}\Omega'_{r-2} \to \Omega_r{}'/i^r\Omega'_{r-1}$. This gives an action of $X_i$ on $M'$. If $1 \leq i, j \leq m$, the actions of $\delta_i$ and $\delta_j$ on $A$ commute and it follows that the actions of $X_i$ and $X_j$ on $M'$ commute also. This gives the desired $S_A$-module structure on $M'$.

## 6. Regularity for Δ-fields

In [6, p. 115] the proof of Th. 6 contains a proof of the following result.

***Theorem 5*** *Let $k$ be a $\Delta$-field, $K$ a $\Delta$-field extension of $k$ generated by a finite set $E \subset K$. Then there is a natural number $s'$ such for every natural number $s \geq s'$, the canonical map $\phi_s$: $K \otimes_{K_s} \Omega_{K_s/k} \to \Omega_{K/k}$ is injective.*

Before we explain how this is so we should note its import for the theory of regularity studied in the previous sections. Consider the special case of Axiom 2 (Section 2) in which $K = A$. This theorem says Axiom 2 is verified automatically by the set of local generators $\Theta_N E$ ($N \in \mathbb{N}$) if $N$ is sufficiently large. When $K = A$, Axiom 1 is automatic. Hence (cf. Axiom 3) for such $N$ in $\mathbb{N}$, $\Theta_N E$ is a regular set of local generators for $K$ over $k$ if and only if $K$ is a separable extension of $k$.

Let $\eta_1, \ldots, \eta_n$ be the elements of $E$ and let $y_1, \ldots, y_n$ be a set of $\Delta$-indeterminates. Let $R = k\{y_1, \ldots, y_n\}$ and for each $f$ in $R$ let $f(\eta)$ denote the result of substituting $\eta_1, \ldots, \eta_n$ differentially in $f$ for $y_1, \ldots, y_n$ respectively. If $F$ is any subset of $R$, let $F(\eta) = \{f(\eta) : f \in F\}$. Let $\mathfrak{p}$ be $\{f \in R : f(\eta) = 0\}$. Then $\mathfrak{p}$ is a prime $\Delta$-ideal of $R$. Let $\mathbf{A}$ be a characteristic set of $\mathfrak{p}$ with respect to some fixed orderly ranking of $y_1, \ldots, y_n$. Let $U = \Theta y_1 \cup \cdots \cup \Theta y_n$ and for any $r$ in $\mathbb{N}$ and $H \subset U$ let $H_r = H \cap (\Theta_r y_1 \cup \cdots \cup \Theta_r y_n)$. For any $F$ in $R$, $u_F$ will denote the leader of $F$. Let $V = \{v \in U : v \notin \Theta u_A \text{ for any } A \text{ in } \mathbf{A}\}$. In [6] it is shown that there is $r$ in $\mathbb{N}$ and $V' \subset V_r$ such that if $B = (V \backslash V_r) \cup V'$, then the elements of $B(\eta)$ are separately algebraically independent over $k$ and such that every element of $K$ is separably algebraic over $k(B(\eta))$. It follows that $d_{K/k}(B(\eta))$ is a basis of the $K$-vector space $\Omega_{K/k}$.

Because $V_r$ is finite we can choose a natural number $s' > r$ such that every element of $V_r(\eta)$ is separably algebraic over $k(B_{s'}(\eta))$. Then for every natural number $s \geq s'$, every element of $V_s(\eta)$ is separably algebraic over $k(B_s(\eta))$. It is easy to show that whenever $s \in \mathbb{N}$, $K_s$ is separably algebraic over $k(V_s(\eta))$. Hence if also $s \geq s'$, $K_s$ is separably algebraic over $k(B_s(\eta))$. Thus $d_{K_s/k}(B_s(\eta))$ generates $\Omega_{K_s/k}$ as a vector space over $K_s$. On the other hand, $d_{K/k}(B_s(\eta))$ is linearly independent over $K$. Hence $\phi_s$ is an injection if $s \geq s'$.

Most though not all of the theorem on p. 115 of [6] can be derived from Theorem 5. As an example we have:

***Corollary 1*** *Let the hypotheses be as in Theorem 5 and let* $\mathfrak{r}$ *be an indeterminate. Then there is an element* $\omega$ *of* $\mathbb{Q}(\mathfrak{r})$ *such that for all sufficiently large natural numbers* $r$, $\omega(r)$ *is the inseparability degree of* $K_r$ *over* $k$.

Indeed let $(\Omega_{K/k})_r$ be $K \cdot d_{K/k}(\Theta_r E)$. By the theorem on p. 241 of [7] there is $\omega$ in $\mathbb{Q}(\mathfrak{r})$ such that for all sufficiently large natural numbers $r$, $\omega(r) = \dim_K(\Omega_{K/k})_r$. By Theorem 5, $(\Omega_{K/k})_r \approx K \otimes_{K_r} \Omega_{K_r/k}$ if $r$ is sufficiently large. Since $\dim_K(K \otimes_{K_r} \Omega_{K_r/k}) = \dim_{K_r} \Omega_{K_r/k}$ is the inseparability degree of $K_r$ over $k$, Corollary 1 follows.

The following is an application of Theorem 5 which does not seem to be contained in the theorem of [6] already mentioned.

***Corollary 2*** *In addition to the hypotheses of Theorem 5, assume that* $K$ *is a separable extension of* $k$. *There is a natural number* $N$ *such that if* $r \geq N$, $K_r$ *is a separable extension of* $K_{r-1}$.

Let $N = s' + 1$, where $s'$ is the number given by Theorem 5. For any $r$ in $\mathbb{N}$ we have a commutative diagram

$$\begin{array}{ccc} K_r \otimes_{K_{r-1}} \Omega_{K_{r-1}/k} & \xrightarrow{\;i\;} & \Omega_{K_r/k} \\ \downarrow{\scriptstyle\alpha} & & \downarrow \\ K \otimes_{K_{r-1}} \Omega_{K_{r-1}/k} & \xrightarrow[\;\beta\;]{} & \Omega_{K/k} \end{array}$$

If $r \geq N$, $\beta$ is an injection, and $\alpha$ is an injection for any $r$. It follows that if $r \geq N$, the sequence

$$0 \to K_r \otimes_{K_{r-1}} \Omega_{K_{r-1}/k} \to \Omega_{K_r/k} \to \Omega_{K_r/K_{r-1}} \to 0$$

is exact. Abbreviate this sequence as $0 \to A \to B \to C \to 0$. Then $\dim_{K_r} B = \text{tr deg}_k K_r$ and $\dim_K A = \text{tr deg}_k K_{r-1}$. Hence $\dim_{K_r} C = \text{tr deg}_{K_{r-1}} K_r$. It follows from Lemma 5 of Section 1 that $K_r$ is a separable extension of $K_{r-1}$ if $r \geq N$.

Actually the hypotheses of Corollary 2 imply $K_r$ is a pure extension of $K_{r-1}$ if $r$ is sufficiently large.

Corollary 2 allows us to state the following improvement of the theorem on p. 244 of [7].

***Theorem*** *Let* $G$ *and* $F$ *be* $\Delta$*-fields and assume that* $G$ *is a separable extension of* $F$. *Suppose we have an excellent filtration* $(G_r)_{r \in \mathbb{Z}}$ *of* $G$ *over* $F$. *Then there exists a polynomial* $\chi(\mathfrak{r})$ *with rational coefficients such that for all sufficiently large* $r$ *in* $\mathbb{Z}$, $\chi(r)$ *is the transcendance degree of* $G_r$ *over* $F$.

## References

1. Altman, A., and Kleiman, S., "Introduction to Grothendieck Duality Theory" (Lecture Notes in Math., **146**), pp. 102–108. Springer-Verlag, Berlin, 1970.
2. Johnson, J., Kähler differentials and differential algebra, *Ann. of Math.* **89** (1969), 92–98.
3. Bourbaki, N., "Algèbre Commutative," Chapter I, §2, p. 37 (Lemma 9).
4. Raynaud, M., and Gruson, L., Critères de platitude et projectivité, *Invent. Math.* **13** (1971), 1–89.
5. Johnson, J., Kähler differentials and differential algebra in arbitrary characteristic, *Trans. Amer. Math. Soc.* **192** (1974), 201–208.
6. Kolchin, E. R., "Differential Algebra and Algebraic Groups." Academic Press, New York, 1973.
7. Johnson, J., Differential dimension polynomials and a fundamental theorem on differential modules, *Amer. J. Math.* **91** (1969), 239–248.

AMS (MOS) 1970 subject classification: 12H05

# *The Engel–Kolchin theorem revisited*

*Irving Kaplansky*

*The University of Chicago*

## 1. Introduction

A matrix $A$ is *unipotent* if $A - I$ is nilpotent, where $I$ is the identity matrix. In 1948 Kolchin [4, pp. 775–776] proved his splendid globalization of Engel's theorem: any multiplicative semigroup of unipotent matrices can be put in simultaneous triangular form.

In this note I shall discuss several generalizations and analogues of Kolchin's theorem.

## 2. A second unification of Kolchin's theorem and Levitzki's theorem

A theorem proved by Levitzki [5] is a kind of companion: any multiplicative semigroup of nilpotent matrices can be put in simultaneous triangular form. It is to be noted that Levitzki's theorem works not just over a field but over a division ring. In [3, Th. H, p. 137] I presented a theorem which unified Kolchin's theorem and the field case of Levitzki's theorem; the assumption I made was that every matrix has the form scalar plus nilpotent (the scalar varying with the matrix). Here is a different unification.

***Theorem 1*** *Let $S$ be a multiplicative semigroup of linear transformations on an $n$-dimensional vector space over a field. Assume that, for a certain* *fixed integer $r$ ($0 \leqq r \leqq n$), every member of $S$ has $r$ zeros and $n - r$ ones as its characteristic roots. Then $S$ can be put in simultaneous triangular form.*

*Remark* We cannot allow $r$ to vary. Consider, for example, the semigroup consisting of 0 and all matrix units.

*Proof* Algebraic closure of the ground field may be harmlessly assumed as usual. Suppose first that $S$ acts irreducibly. All the members of $S$ have the same trace (to wit, $r$). It follows that $S$ consists of just one linear transformation, and the given vector space (say $V$) is one-dimensional. Thus we may assume a decomposition $V = V_1 \oplus V_2$ into subspaces invariant under $S$.

The only difficulty in the proof is the possibility that the hypothesis that $r$ is fixed may be lost when $S$ is restricted to $V_i$. We proceed to grapple with this.

Let $T_1$ and $T_2$ be members of $S$. Assume that the number of zero characteristic roots in $T_j$, when restricted to $V_1$, is $s_j$. We suppose that $s_1 < s_2$ and shall reach a contradiction. Now $V_1$ can be split into a direct sum of two subspaces invariant under $T_1$ on the first of which $T_1$ is unipotent and on the second nilpotent. By replacing $T_1$ by a power, we can make $T_1$ zero on the second of these subspaces. In this way we can arrange that the nilpotent parts of both $T_1$ and $T_2$ on both $V_1$ and $V_2$ are zero. Now let us examine $T_1 T_2$. It inherits from $T_2$ at least $s_2$ zero characteristic roots on $V_1$ and from $T_1$ at least $r - s_1$ on $V_2$. This gives a total of at least $s_2 + r - s_1$, a contradiction since $s_2 + r - s_1 > r$.

## 3. Two more theorems

The linear transformations in Theorem 1 all have the same trace. This leads us to formulate another theorem which, however, requires characteristic 0.

***Theorem 2*** *Let $S$ be a multiplicative semigroup of linear transformations on a finite-dimensional vector space over a field of characteristic* 0. *Assume that all the members of $S$ have the same trace. Then $S$ can be put in simultaneous triangular form.*

*Proof* The plan is to reduce the problem to the case treated in Theorem 1. Let $a_1, \ldots, a_n$ be the characteristic roots of a member of $S$. We have $a_1{}^m + \cdots + a_n{}^m = c$ for all $m$, where $c$ is the fixed value of the trace. We simply forget any $a_i$'s which are zero, and the rest of the proof is covered in a lemma.

***Lemma*** *Let $a_1, \ldots, a_r$ be nonzero elements in a field of characteristic zero. Assume that $a_1^m + \cdots + a_r^m$ equals a fixed element $c$ for $m = 1, \ldots, r+1$. Then each $a_i = 1$.*

*Proof* Let

$$x^r + t_1 x^{r-1} + \cdots + t_{r-1}x + t_r$$

be the polynomial with roots $a_1, \ldots, a_r$. Put $x = a_i$ and add; the result is

$$c + t_1 c + \cdots + t_{r-1}c + rt_r = 0.$$

Next multiply the polynomial by $x$ and again set $x = a_i$ and add:

$$c + t_1 c + \cdots + t_{r-1}c + ct_r = 0.$$

Hence $rt_r = ct_r$ and $r = c$.

It is standard that the elements $a_1, \ldots, a_r$ are determined if one knows the sums $\sum a_i^m$ with $m = 1, \ldots, r$ (this uses characteristic 0). Since we have $\sum a_i^m = r$, and since the $r$-ple $1, \ldots, 1$ is a solution, we deduce that each $a_i = 1$, as required.

*Remarks* 1. Characteristic 0 is vital in Theorem 2. For instance, in characteristic 2 take the 4 by 4 matrices

$$\begin{pmatrix} A & 0 \\ 0 & A \end{pmatrix},$$

where $A$ ranges over all 2 by 2 matrices.

2. By a brutal discussion I have checked that Theorem 2 holds for any characteristic for matrices of size $\leqq 3$.

I conclude this section with a theorem that does not unify anything; it is a modest generalization of the characteristic 0 case of Kolchin's theorem.

***Theorem 3*** *Let $S$ be a multiplicative semigroup of linear transformations on a finite-dimensional vector space over a field of characteristic* 0. *Assume that each member of $S$ has characteristic roots $\pm 1$. Then $S$ can be put in simultaneous triangular form.*

*Proof* Algebraic closure of the field and irreducibility of $S$ are available as usual. Only a finite number of traces occur, and then Burnside's argument shows $S$ to be finite. Then the square of every member of $S$ is the identity (this uses characteristic 0), and $S$ is commutative.

Theorem 3 fails for 3 by 3 matrices in characteristic $\neq 0$: take the permutation matrices in characteristic 3.

## 4. Lie and Jordan analogues

The original Engel theorem of course referred to Lie algebras of nilpotent matrices. One can weaken the closure assumption to just the hypothesis of closure under the Lie bracket. There is a generalization to scalar-plus-nilpotent matrices that does not seem to have been noted in the literature. As one easily sees, characteristic 0 is indispensable here. The proof follows a standard pattern, but is included for completeness.

***Theorem 4*** *Let $S$ be a set of linear transformations on a finite-dimensional vector space $V$ over a field of characteristic* 0. *Assume that each member of $S$ has the form scalar plus nilpotent* (*the scalar varying with the element*) *and that $S$ is closed under the operation $AB - BA$. Then $S$ can be put in simultaneous triangular form.*

*Proof* We can assume that $V$ is irreducible under $S$. Let $S_0$ be the set of nilpotent elements of $S$ (alternatively, those of trace 0). $S_0$ is closed under commutation and hence admits joint triangular form. Let $W$ be the subspace annihilated by all of $S_0$; $W$ is not 0 and we shall prove that $W$ is invariant under $S$. Given $w \in W, T \in S$, and $U \in S_0$, we must prove $UTw = 0$. Now $w$ is annihilated by $U$ and also by $UT - TU$, since $UT - TU \in S_0$. Hence $UTw = 0$, and $W$ is invariant. We deduce that $W = V$, so that $S$ is commutative.

There is a Jordan algebra analogue [6]. It does not apply to sets closed just under $AB + BA$, as appears from the 2 by 2 matrices $e_{12}$, $e_{21}$, and $I$. For sets of nilpotent matrices Jacobson has given a neat unification of the semigroup, Lie, and Jordan theorems [2, pp. 31–35]. See also [1]. In the scalar-plus-nilpotent case we have seen that one needs characteristic assumptions and/or closure under addition. I have experimented with several formulations of unified theorems, but they seem too artificial to be worth pursuing.

## 5. The infinite case

Beyond the theorems discussed here, there lies an uncharted region where finiteness assumptions are dropped. Here is an example of the kind of statement that can be proposed (I do not say "conjectured," feeling that to be too daring): in a ring with unit, let $S$ be a multiplicative semigroup consisting of unipotent elements; then the nilpotent parts of the members of $S$ generate a nil subring. A special case is the assertion that the Engel–Kolchin theorem works over a division ring.

One fact to be noted is that McCrimmon's theorem is valid in the infinite-dimensional case. This should not occasion too much surprise, since the Jordan case ranks just below the associative case in good behavior, and there is a corresponding (very easy) associative result. In this vein, here is a final exercise for the reader. Let $R$ be a ring with unit in which every element $x$ satisfies at least one of the following four conditions: $x$ has a right inverse, $x$ has a left inverse, $1 + x$ has a right inverse, $1 + x$ has a left inverse. Then $R$ is local, i.e. it is a division ring modulo its Jacobson radical.

*Note added in proof* (January 25, 1977) Mochizuki (*Notices AMS* (*Jan. 1977*), A-2) has announced that the Engel–Kolchin theorem is valid over a division ring of characteristic 0.

## References

1. K. Hunter, Nilpotence of nil subrings implies more general nilpotence, *Arch. Math.* **18** (1967), 136–139.
2. N. Jacobson, "Lie Algebras." Wiley (Interscience), New York, 1962.
3. I. Kaplansky, "Fields and Rings," 2nd ed. Univ. of Chicago Press, Chicago, Illinois, 1972.
4. E. Kolchin, On certain concepts in the theory of algebraic matric groups, *Ann. of Math.* **49** (1948), 774–789.
5. J. Levitzki, Über nilpotente Unterringe, *Math. Ann.* **105** (1931), 620–627.
6. K. McCrimmon, Jordan algebras of degree 1, *Bull. Amer. Math. Soc.* **70** (1964), 702.

AMS (MOS) 1970 subject classification: 15A21

# ***Prime differential ideals in differential rings***

*William F. Keigher*

*University of Tennessee*

Throughout this paper, all rings are commutative with identity and all ring homomorphisms preserve the identity. Also, all differential rings are ordinary, i.e., possess a single derivation operator which is suppressed from the notation. If $A$ is a differential ring and $x \in A$, then $x^{(n)}$ denotes the $n$th derivative of $x$; we note that $x^{(0)} = x$. A subset of a differential ring is called differential if it contains the derivative of each of its elements.

## 1. Special differential rings

Let $A$ be a differential ring and let $X$ be a subset of $A$. We define a subset $X_{\#}$ of $A$ by $X_{\#} = \{x \mid x^{(n)} \in X \text{ for all } n \geq 0\}$. Some of the properties of the operator $(\ )_{\#}$ are given in the following.

***Proposition 1.1*** *Let $A$ and $B$ be differential rings and let $f$: $A \to B$ be a differential ring homomorphism.*

(i) *For any $X \subset A$, $X_{\#} \subset X$ and $(X_{\#})_{\#} = X_{\#}$.*
(ii) *For any $X \subset A$, $X_{\#} = X$ if and only if $X$ is a differential subset of $A$.*
(iii) *For any $X$, $Y \subset A$ with $X \subset Y$, $X_{\#} \subset Y_{\#}$.*

(iv) *For any* $\{X_i\}_{i \in I}$ *with* $X_i \subset A$, $(\bigcap_I X_i)_\# = \bigcap_I (X_i)_\#$ *and* $\bigcup_I (X_i)_\# \subset (\bigcup_I X_i)_\#$.

(v) *For any* $X, Y \subset A$, $(X + Y)_\# \supset X_\# + Y_\#$ *and* $X_\# \cdot Y_\# \subset (X \cdot Y)_\#$.

(vi) *For any* $X \subset A$ *and* $Y \subset B$, $f^{-1}(Y_\#) = f^{-1}(Y)_\#$ *and* $f(X_\#) \subset f(X)_\#$ *with equality if* $f$ *is injective.*

*Proof* The proof is elementary and follows immediately from the definitions.

It follows from Proposition 1.1 that for any subset $X$ of a differential ring $A$, $X_\#$ is a differential subset. Moreover, the union and the intersection of any family of differential subsets is again a differential subset, and finite sums and products of differential subsets are differential. Also, direct and inverse images of differential subsets under a differential homomorphism are differential.

There is another related operator defined on subsets of a differential ring which will be of some value to us at a later time. For any subset $X$ of a differential ring $A$, define $X_\infty$ by $X_\infty = \{x^{(n)} | x \in X, n \geq 0\}$. Then we have the following.

***Proposition 1.2*** *Let* $A$ *and* $B$ *be differential rings and let* $f\colon A \to B$ *be a differential ring homomorphism.*

(i) *For any* $X \subset A$, $X \subset X_\infty$ *and* $(X_\infty)_\infty = X_\infty$.

(ii) *For any* $X \subset A$, $X_\infty = X$ *if and only if* $X$ *is a differential subset of* $A$.

(iii) *For any* $X, Y \subset A$ *with* $X \subset Y$, $X_\infty \subset Y_\infty$.

(iv) *For any* $\{X_i\}_{i \in I}$ *with* $X_i \subset A$, $(\bigcup_I X_i)_\infty = \bigcup_I (X_i)_\infty$ *and* $(\bigcap_I X_i)_\infty \subset \bigcap_I (X_i)_\infty$.

(v) *For any* $X, Y \subset A$, $(X + Y)_\infty \subset X_\infty + Y_\infty$ *and* $(X \cdot Y)_\infty \subset X_\infty \cdot Y_\infty$.

(vi) *For any* $X \subset A$ *and* $Y \subset B$, $f(X)_\infty = f(X_\infty)$ *and* $f^{-1}(Y)_\infty \subset f^{-1}(Y_\infty)$.

(vii) *For any* $X \subset A$, $(X_\infty)_\# = X_\infty$ *and* $(X_\#)_\infty = X_\#$.

We observe that since the collection of differential subsets of a differential ring $A$ forms a complete lattice, we may consider the differential subsets as being either the open or the closed sets for a topology on $A$. When one considers them as the open sets for a topology, it is clear that $(\ )_\#$ is the interior operator for that topology and any differential ring homomorphism is then continuous and open with respect to these topologies. Similarly, if one considers the differential subsets as closed, then $(\ )_\infty$ is the closure operator and any differential ring homomorphism is continuous and closed.

If $A$ is a differential ring and $X$ is a subset of $A$, let $X^* = \{x \in X |$ there exists $y \in X$ with $xy = 1\}$. Thus if $X$ is a subring of $A$, $X^*$ is the set of units in $X$. The following lemma will prove useful.

***Lemma 1.3*** *Let $A$ be a differential ring and $B$ a subring of $A$. Then $(B_{\#})^* = B_{\#} \cap B^*$.*

*Proof* Clearly $(B_{\#})^* \subset B_{\#} \cap B^*$, so let $x \in B$ be such that $x^{(n)} \in B$ for each $n \geq 0$, and suppose that $xy = 1$ for some $y \in B$. We wish to show that $y^{(n)} \in B$ for each $n \geq 0$. We may assume $n \geq 1$ and that for each $k < n$, $y^{(k)} \in B$. Then by Leibnitz's rule [5, p. 9] we have

$$0 = (xy)^{(n)} = xy^{(n)} + \sum_{k=1}^{n} {}^nC_k\, x^{(k)} y^{(n-k)},$$

so that

$$y^{(n)} = -y\left(\sum_{k=1}^{n} {}^nC_k\, x^{(k)} y^{(n-k)}\right) \in B.$$

Using this lemma we now show that $(\ )_{\#}$ preserves a certain amount of the algebraic structure the subset $X$ may possess.

***Proposition 1.4*** *Let $A$ be a differential ring and let $X$ be a subset of $A$. If $X$ is any of the following, the same is true of $X_{\#}$.*

(i) *A subring of $A$.*
(ii) *An ideal in $A$.*
(iii) *A semilocal subring of $A$.*
(iv) *A local subring of $A$.*
(v) *A subfield of $A$.*

*Proof* Assertion (i) follows immediately from the definitions. For (ii), suppose $a \in A$ and $x \in X_{\#}$. Then by Leibnitz's rule we have

$$(ax)^{(n)} = \sum_{k=0}^{n} {}^nC_k\, a^{(k)} x^{(n-k)}.$$

Since each $x^{(n-k)} \in X$ and $X$ is an ideal in $A$, $(ax)^{(n)} \in X$ and hence $ax \in X_{\#}$. For (iii), let $X$ be a semilocal subring of $A$ [1, p. 117] and let $\mathfrak{m}_1, \ldots, \mathfrak{m}_k$ denote the distinct maximal ideals of $X$, so that $\bigcup_{i=1}^{k} \mathfrak{m}_i = X - X^*$. Then by Lemma 1.3 we have

$$\begin{aligned}(X_{\#})^* = X_{\#} \cap X^* &= X_{\#} \cap \left(X - \bigcup_{i=1}^{k} \mathfrak{m}_i\right)\\ &= X_{\#} - \bigcup_{i=1}^{k} (X_{\#} \cap \mathfrak{m}_i) = X_{\#} - \bigcup_{i=1}^{k} \mathfrak{m}_i',\end{aligned}$$

where $\mathfrak{m}_i' = X_{\#} \cap \mathfrak{m}_i$ is a prime ideal in $X_{\#}$ for $i = 1, \ldots, k$. Hence if $S = (X_{\#})^*$, then $S^{-1}X_{\#}$ is a semilocal ring [1, Prop. 17, p. 117] and if $\mathfrak{M}_1$,

..., $\mathfrak{M}_r$ are the distinct maximal elements (with respect to inclusion) among the $\mathfrak{m}_i'$, then $S^{-1}\mathfrak{M}_1, \ldots, S^{-1}\mathfrak{M}_r$ are the distinct maximal ideals of $S^{-1}X_\#$. But since $S = (X_\#)^*$, $S^{-1}X_\# \cong X_\#$ [1, p. 78] and $S^{-1}\mathfrak{M}_i \cong \mathfrak{M}_i$, so that $X_\#$ is semilocal and $\mathfrak{M}_1, \ldots, \mathfrak{M}_r$ are the distinct maximal ideals of $X_\#$.

The assertion in (iv) now follows by taking $k = 1$ in the proof of (iii). Finally, (v) follows from (iv) by taking the maximal ideal of $X$ to be 0.

Recall from [5] that a Ritt algebra is a differential ring which contains the rational numbers. The operator $(\ )_\#$ defined on subsets of a Ritt algebra preserves even more structure as shown in the following.

***Proposition 1.5*** *Let $A$ be a Ritt algebra and let $X$ be a subset of $A$. If $X$ is any of the following, the same is true of $X_\#$.*

(i) *A prime ideal in $A$.*
(ii) *A radical ideal in $A$.*

*Proof* For (i), it follows from Proposition 1.4 that $X_\#$ is an ideal in $A$, so suppose that $a \notin X_\#$ and $b \notin X_\#$. Then there exist integers $m, n \geq 0$ such that $a^{(m)} \notin X$, $b^{(n)} \notin X$ and for all $k < m$ and $l < n$, $a^{(k)} \in X$ and $b^{(l)} \in X$. Consider now

$$(ab)^{(m+n)} = \sum_{k=0}^{m+n} {}^{m+n}C_k\, a^{(k)} b^{(m+n-k)}.$$

For $k < m$, ${}^{m+n}C_k\, a^{(k)} b^{(m+n-k)} \in X$, while for $k > m$, i.e., for $m + n - k < n$, ${}^{m+n}C_k\, a^{(k)} b^{(m+n-k)} \in X$. Now for $k = m$, $a^{(m)} b^{(n)} \notin X$ since $X$ is prime, and since $A$ is a Ritt algebra ${}^{m+n}C_m\, a^{(m)} b^{(n)} \notin X$. Hence $(ab)^{(m+n)} \notin X$, so that $X_\#$ is prime.

To prove (ii), note that any radical ideal in $A$ is an intersection of prime ideals in $A$ and conversely. Since $(\ )_\#$ preserves intersections by Proposition 1.1 and prime ideals by (i), $(\ )_\#$ preserves radical ideals as well.

We will call a differential ring $A$ special if for each prime ideal $\mathfrak{p}$ in $A$, $\mathfrak{p}_\#$ is also prime in $A$. Any Ritt algebra is special by the above proposition, and any differentially simple differential domain (e.g., any differential field) is special. If $A$ is any ring equipped with the trivial derivation (i.e., $x^{(n)} = 0$ for all $x \in A$ and $n \geq 1$), then $A$ is special. Special differential rings arise in other ways as in the following.

***Proposition 1.6*** *Let $A$ be a special differential ring and $f\colon A \to B$ a surjective differential ring homomorphism. Then $B$ is also special.*

*Proof* The surjective $f$ induces a one-to-one correspondence between prime ideals $\mathfrak{q}$ in $B$ and prime ideals $\mathfrak{p}$ in $A$ containing the kernel of $f$ via

$\mathfrak{p} = f^{-1}(\mathfrak{q})$ and $\mathfrak{q} = f(\mathfrak{p})$. Hence if $\mathfrak{q}$ is a prime ideal in $B$, then we have $\mathfrak{q}_\# = f(f^{-1}(\mathfrak{q}_\#)) = f(f^{-1}(\mathfrak{q})_\#)$. But since $A$ is special, $f^{-1}(\mathfrak{q})_\#$ is prime in $A$, and hence $\mathfrak{q}_\#$ is prime in $B$.

Recall that if $S$ is a multiplicative subset of a differential ring $A$, then the ring of fractions $S^{-1}A$ is a differential ring via $(a/s)^{(1)} = (sa^{(1)} - as^{(1)})/s^2$ [2, p. 198; 6, p. 64].

***Lemma 1.7*** *Let $A$ be a differential ring, $S$ a multiplicative subset of $A$, and $\mathfrak{p}$ a prime ideal in $A$ disjoint from $S$. Then in the differential ring of fractions $S^{-1}A$ we have $(S^{-1}\mathfrak{p})_\# = S^{-1}\mathfrak{p}_\#$.*

*Proof* Let $a/s \in S^{-1}\mathfrak{p}_\#$. We prove by induction that $(a/s)^{(n)} \in S^{-1}\mathfrak{p}$. For $n = 0$ this is immediate, so assume $(a/s)^{(k)} \in S^{-1}\mathfrak{p}$ for $k < n$, where $n \geq 1$. Since $(a/s)(s/1) = a/1$ in $S^{-1}A$, we have

$$(a/s)^{(n)} = \left[(a/1)^{(n)} - \sum_{k=0}^{n-1} {}^nC_k(a/s)^{(k)}(s/1)^{(n-k)}\right](1/s), \qquad (*)$$

and since $(x/1)^{(n)} = x^{(n)}/1$ for all $n \geq 0$ and $x \in A$, the result follows. On the other hand, if we assume that $(a/s)^{(n)} \in S^{-1}\mathfrak{p}$ for each $n \geq 0$, then again by $(*)$ we see that $a^{(n)}/1 \in S^{-1}\mathfrak{p}$. Since $\mathfrak{p}$ is prime and disjoint from $S$, it follows that $a^{(n)} \in \mathfrak{p}$, so that $a/s \in S^{-1}\mathfrak{p}_\#$.

***Proposition 1.8*** *Let $A$ be a special differential ring and $S$ a multiplicative subset of $A$. Then $S^{-1}A$ is also special.*

*Proof* This follows immediately from Lemma 1.7, since there is a one-to-one correspondence between prime ideals in $S^{-1}A$ and prime ideals in $A$ disjoint from $S$ [1, Prop. 11, p. 90].

***Corollary 1.9*** *A differential ring $A$ is special if and only if $A_\mathfrak{p}$ is special for each prime ideal $\mathfrak{p}$ in $A$.*

*Proof* If $A$ is special, then so is each $A_\mathfrak{p}$ by Proposition 1.8. Conversely, let $\mathfrak{p}$ be a prime ideal in $A$, let $\phi\colon A \to A_\mathfrak{p}$ denote the canonical differential ring homomorphism and let $S = A - \mathfrak{p}$. Then since $\mathfrak{p} = \phi^{-1}(S^{-1}\mathfrak{p})$, we see that $\mathfrak{p}_\# = \phi^{-1}((S^{-1}\mathfrak{p})_\#)$ by Proposition 1.1, and since $A_\mathfrak{p}$ is special, $(S^{-1}\mathfrak{p})_\#$ is prime in $A_\mathfrak{p}$. Hence $\mathfrak{p}_\#$ is prime in $A$ and $A$ is special.

***Proposition 1.10*** *Let $A_1, \ldots, A_n$ be any finite family of differential rings and let $A = A_1 \times \cdots \times A_n$ denote their product. Then $A$ is special if and only if each $A_i$ is special.*

*Proof* If $A$ is special, then so is each $A_i$ by Proposition 1.6. Conversely, suppose that $\mathfrak{p}$ is a prime ideal in $A$, and let $\pi_i\colon A \to A_i$ denote the canonical projections, $i = 1, \ldots, n$. Then $\pi_k(\mathfrak{p}) = \mathfrak{p}_k$ is a prime ideal in $A_k$ for some $k$, $1 \leq k \leq n$, and $\pi_j(\mathfrak{p}) = A_j$ for $j \neq k$. Clearly $\mathfrak{p}_\# = \pi_k^{-1}(\mathfrak{p}_{k\#})$, and since $A_k$ is special, $\mathfrak{p}_\#$ is prime in $A$ and $A$ is special.

Recall from [2] that a *d-MP* ring $A$ is a differential ring in which the radical of a differential ideal is again a differential ideal, and that this is equivalent to each of the following:

(i) Prime ideals minimal over differential ideals are differential ideals.
(ii) If $\mathfrak{I}$ is a differential ideal of $A$ and $S$ is a multiplicative subset of $A$ disjoint from $\mathfrak{I}$, then ideals maximal among differential ideals which contain $\mathfrak{I}$ and are disjoint from $S$ are prime.

***Proposition 1.11*** *A differential ring A is special if and only if it is a d-MP ring.*

*Proof* Assume first that $A$ is special, let $\mathfrak{I}$ be a differential ideal in $A$, and let $\mathfrak{p}$ be a prime ideal in $A$ which is minimal among prime ideals containing $\mathfrak{I}$. Then by Proposition 1.1 we see that $\mathfrak{I} = \mathfrak{I}_\# \subset \mathfrak{p}_\# \subset \mathfrak{p}$, and since $\mathfrak{p}_\#$ is a prime ideal, $\mathfrak{p}_\# = \mathfrak{p}$ and $\mathfrak{p}$ is a differential ideal. Hence $A$ is a *d-MP* ring by (i). Conversely, suppose that $A$ is a *d-MP* ring and let $\mathfrak{p}$ be a prime ideal in $A$. Then $S = A - \mathfrak{p}$ is a multiplicative subset of $A$ and 0 is a differential ideal disjoint from $S$. Also, $\mathfrak{p}_\#$ is a differential ideal in $A$ disjoint from $S$, and if $\mathfrak{I}$ is a differential ideal in $A$ with $\mathfrak{p}_\# \subset \mathfrak{I}$ and $\mathfrak{I} \cap S = \phi$, then $\mathfrak{I} \subset \mathfrak{p}$ and hence $\mathfrak{I} = \mathfrak{I}_\# \subset \mathfrak{p}_\#$. It follows that $\mathfrak{p}_\#$ is maximal among differential ideals disjoint from $S$. Hence $\mathfrak{p}_\#$ is prime by (ii), and $A$ is special.

If $A$ is a differential ring which is not necessarily special, one might ask whether anything can be said of the differential ideal $\mathfrak{p}_\#$ where $\mathfrak{p}$ is a prime ideal in $A$. Recall first that an ideal $\mathfrak{I}$ in a ring $A$ is prime if and only if there is a multiplicative subset $S$ of $A$ such that $\mathfrak{I}$ is maximal among ideals disjoint from $S$ [4, Th. 1, p. 1]. With this in mind we make the following definition.

Let $A$ be a differential ring. A differential ideal $\mathfrak{I}$ in $A$ will be called quasi-prime if there is a multiplicative subset $S$ of $A$ such that $\mathfrak{I}$ is maximal among differential ideals disjoint from $S$. Clearly every prime differential ideal is quasi-prime, and every quasi-prime ideal is prime if and only if $A$ is special. Also, every quasi-prime ideal is primary [6, p. 63]. Quasi-prime ideals arise from prime ideals as in the following.

***Proposition 1.12*** *Let A be a differential ring. Then for any prime ideal* $\mathfrak{p}$ *in A,* $\mathfrak{p}_\#$ *is quasi-prime. Moreover, for any quasi-prime ideal* $\mathfrak{q}$ *in A, there is a prime ideal* $\mathfrak{p}$ *in A such that* $\mathfrak{p}_\# = \mathfrak{q}$.

*Proof* Let $\mathfrak{p}$ be a prime ideal in $A$ and let $S = A - \mathfrak{p}$. Clearly $\mathfrak{p}_\#$ is a differential ideal disjoint from $S$, and if $\mathfrak{I}$ is any differential ideal disjoint from $S$, then $\mathfrak{I} \subset \mathfrak{p}$, so that $\mathfrak{I} = \mathfrak{I}_{\#} \subset \mathfrak{p}_{\#}$. Hence $\mathfrak{p}_{\#}$ is maximal among differential ideals disjoint from $S$. Now let $\mathfrak{q}$ be a quasi-prime ideal in $A$ and let $S$ be a multiplicative subset of $A$ such that $\mathfrak{q}$ is maximal among differential ideals disjoint from $S$. Then there exists a prime ideal $\mathfrak{p}$ in $A$ such that $\mathfrak{q} \subset \mathfrak{p}$ and $\mathfrak{p} \cap S = \phi$ [1, Cor. 2, p. 92]. Hence $\mathfrak{q} = \mathfrak{q}^{\#} \subset \mathfrak{p}^{\#}$ and $\mathfrak{p}_{\#} \cap S = \phi$, so that $\mathfrak{q} = \mathfrak{p}_{\#}$.

It follows from Proposition 1.12 that for any differential ring $A$ there is a surjection from the set of prime ideals in $A$ to the set of quasi-prime ideals in $A$. In particular, when $A$ is special, even more can be said about this surjection.

## 2. The prime differential spectrum of a differential ring

For any differential ring $A$, Spec($A$) will denote the set of prime ideals in (the underlying ring of) $A$ with the Zariski topology [1, p. 125; 7, p. 18]. The set of prime differential ideals in $A$ will be denoted by $\mathrm{Spec}_{\mathrm{D}}(A)$ and will be called the prime differential spectrum of $A$. As a topological space, it has the subspace topology from Spec($A$), so that the closed sets in $\mathrm{Spec}_{\mathrm{D}}(A)$ are of the form $V_0(E) = V(E) \cap \mathrm{Spec}_{\mathrm{D}}(A)$, where $E$ is a subset of $A$. It follows that the sets $D_0(f) = D(f) \cap \mathrm{Spec}_{\mathrm{D}}(A)$, where $f \in A$, form an open basis for the topology.

For any subset $E$ of $A$, we define the differential radical of $E$, denoted by $\mathrm{r}_{\mathrm{D}}(E)$, to be the intersection of all differential prime ideals in $A$ containing $E$. If $\mathrm{r}(E)$ denotes the radical of the ideal in $A$ generated by $E$ [1, p. 94; 7, p. 18], then clearly $E \subset \mathrm{r}(E) \subset \mathrm{r}_{\mathrm{D}}(E)$ and $\mathrm{r}_{\mathrm{D}}(\mathrm{r}_{\mathrm{D}}(E)) = \mathrm{r}_{\mathrm{D}}(E)$. An ideal $\mathfrak{I}$ in $A$ will be called a differential radical ideal if $\mathfrak{I} = \mathrm{r}_{\mathrm{D}}(\mathfrak{I})$. Clearly every differential radical ideal is a differential ideal equal to its own radical, i.e., a radical differential ideal [5, p. 13], and it follows from [5, Th. 2.1, p. 13] that every radical differential ideal is a differential radical ideal.

If $Y$ is any subset of $\mathrm{Spec}_{\mathrm{D}}(A)$, let $J_{\mathrm{D}}(Y)$ denote the intersection of all prime differential ideals in $A$ which belong to $Y$. Clearly $J_{\mathrm{D}}(Y)$ is a differential ideal in $A$, and the mapping $Y \rightsquigarrow J_{\mathrm{D}}(Y)$ is order-reversing with respect to the partial ordering by inclusion in $\mathrm{Spec}_{\mathrm{D}}(A)$ and $A$. Moreover, $J_{\mathrm{D}}(\phi) = A$ and $J_{\mathrm{D}}(\bigcup_I Y_i) = \bigcap_I J_{\mathrm{D}}(Y_i)$ for every family $\{Y_i\}_{i \in I}$ of subsets of $\mathrm{Spec}_{\mathrm{D}}(A)$. We have the following

***Proposition 2.1*** *Let $A$ be a differential ring, $E$ a subset of $A$, and $Y$ a subset of* $\mathit{Spec}_{\mathrm{D}}(A)$.

(i) *$V_0(E)$ is closed in* $\mathrm{Spec}_{\mathrm{D}}(A)$ *and $J_{\mathrm{D}}(Y)$ is a differential radical ideal in $A$.*

(ii) *$J_{\mathrm{D}} V_0(E)$ is the differential radical of $E$ and $V_0 J_{\mathrm{D}}(Y)$ is the closure of $Y$ in* $\mathrm{Spec}_{\mathrm{D}}(A)$.

(iii) *The mappings $J_D$ and $V_0$ are inverse order-reversing bijections between the set of closed subsets of* $\mathrm{Spec}_D(A)$ *and the set of differential radical ideals in A.*

(iv) *The restriction of $V_0$ to the set of prime differential ideals in A is an order-reversing bijection between* $\mathrm{Spec}_D(A)$ *and the closed irreducible subsets of* $\mathrm{Spec}_D(A)$ (*i.e., every closed irreducible subset of* $\mathrm{Spec}_D(A)$ *admits a unique generic point*).

*Proof* Both assertions in (i) and the first assertion in (ii) follow from the definitions, while the second assertion of (ii) follows from the corresponding assertion in $\mathrm{Spec}(A)$. The assertion in (iii) follows immediately from (ii), while the proof of the assertion in (iv) corresponds exactly to the proof of Proposition 14 [1, p. 129].

Let $A$, $B$ be differential rings and $f: A \to B$ a differential ring homomorphism. Then $f$ induces a continuous mapping ${}^a f$: $\mathrm{Spec}(B) \to \mathrm{Spec}(A)$ via ${}^a f(y) = f^{-1}(j_y)$, where $j_y$ denotes the prime ideal corresponding to $y \in \mathrm{Spec}(B)$. It follows from Proposition 1.1 that ${}^a f$ restricts to give a continuous mapping also denoted by ${}^a f$, namely ${}^a f$: $\mathrm{Spec}_D(B) \to \mathrm{Spec}_D(A)$. Clearly if $g: B \to C$ is another differential ring homomorphism, then ${}^a(gf) = {}^a f \cdot {}^a g$, so that $\mathrm{Spec}_D$ is a contravariant functor from the category of differential rings and differential ring homomorphisms to the category of topological spaces and continuous mappings.

In particular, if $f$: $A \to B$ is a surjective differential ring homomorphism with kernel $\mathfrak{I}$, then ${}^a f$: $\mathrm{Spec}_D(B) \to \mathrm{Spec}_D(A)$ is a homeomorphism of $\mathrm{Spec}_D(B)$ onto the closed subspace $V_0(\mathfrak{I})$ of $\mathrm{Spec}_D(A)$. Also, if $S$ is a multiplicative subset of $A$ and $\phi: A \to S^{-1}A$ is the canonical differential ring homomorphism, then ${}^a\phi$ is a homeomorphism of $\mathrm{Spec}_D(S^{-1}A)$ onto the subspace of $\mathrm{Spec}_D(A)$ consisting of those prime differential ideals in $A$ disjoint from $S$.

We can say much more about the structure of $\mathrm{Spec}_D(A)$ for special differential rings, as in the following.

***Theorem* 2.2** *For any special differential ring A there is a natural continuous open retraction* $\rho_A$: $\mathrm{Spec}(A) \to \mathrm{Spec}_D(A)$.

*Proof* Since $A$ is special, we know that for each prime ideal $\mathfrak{p}$ in $A$, $\mathfrak{p}_\#$ is a prime differential ideal in $A$; this defines $\rho_A$. If $D_0(f)$ is any basic open set in $\mathrm{Spec}_D(A)$, then $\rho_A^{-1}(D_0(f)) = \bigcup_{n \geq 0} D(f^{(n)})$. To see this, note that for any prime ideal $\mathfrak{p}$ in $A$, $f \notin \mathfrak{p}_\#$ if and only if $f^{(n)} \notin \mathfrak{p}$ for some $n \geq 0$. It follows that $\rho_A$ is continuous. Now $(\mathfrak{p}_\#)_\# = \mathfrak{p}_\#$ by Proposition 1.1, so that $\rho_A$ is the identity on $\mathrm{Spec}_D(A)$ and hence is a retraction. From the definitions it is clear that $\rho_A(D(f)) = D_0(f)$, and hence $\rho_A$ is open. Finally, $\rho_A$ is natural in the sense that if $A$ and $B$ are special differential rings and

$f\colon A \to B$ is any differential ring homomorphism, there is a commutative diagram of continuous maps

$$\begin{array}{ccc} \mathrm{Spec}(B) & \xrightarrow{\ {}^a f\ } & \mathrm{Spec}(A) \\ \downarrow{\scriptstyle \rho_B} & & \downarrow{\scriptstyle \rho_A} \\ \mathrm{Spec}_{\mathrm{D}}(B) & \xrightarrow{\ {}^a f\ } & \mathrm{Spec}_{\mathrm{D}}(A) \end{array}$$

This is immediate, for if $x \in \mathrm{Spec}(B)$, then $\rho_A \cdot {}^a f(x) = f^{-1}(j_x)_\# = f^{-1}((j_x)_\#) = {}^a f \cdot \rho_B(x)$, where the second equation follows from Proposition 1.1.

***Proposition 2.3*** *Let $A$ be a special differential ring and let $\rho$ denote the retraction $\rho_A$: $\mathrm{Spec}(A) \to \mathrm{Spec}_{\mathrm{D}}(A)$.*

(i) *For any basic open set $D(f)$ in $\mathrm{Spec}(A)$, $\rho(D(f)) = D_0(f)$.*
(ii) *For any closed set $V_0(E)$ in $\mathrm{Spec}_{\mathrm{D}}(A)$, $\rho^{-1}(V_0(E)) = V(E_\infty)$.*
(iii) *The basic open sets $D_0(f)$ in $\mathrm{Spec}_{\mathrm{D}}(A)$ are quasi-compact. In particular, $\mathrm{Spec}_{\mathrm{D}}(A)$ is quasi-compact.*
(iv) *For any subset $E$ of $A$, $\mathfrak{r}_{\mathrm{D}}(E_\#) \subset \mathfrak{r}(E)_\#$.*
(v) *$\mathrm{Spec}_{\mathrm{D}}(A)$ is dense in $\mathrm{Spec}(A)$.*
(vi) *An open set in $\mathrm{Spec}_{\mathrm{D}}(A)$ is quasi-compact if and only if its complement is of the form $V_0(\mathfrak{I})$ where $\mathfrak{I}$ is a differentially finitely generated differential ideal in $A$.*

*Proof* The assertion in (i) follows immediately from the definitions. For (ii), if $\mathfrak{p}$ is a prime ideal in $A$ with $E_\infty \subset \mathfrak{p}$, then from Proposition 1.1 and 1.2 we see that $E \subset E_\infty = (E_\infty)_\# \subset \mathfrak{p}_\#$, so that $V(E_\infty) \subset \rho^{-1}(V_0(E))$. The inclusion $\rho^{-1}(V_0(E)) \subset V(E_\infty)$ is proved similarly. For the assertion in (iii), the basic open sets $D(f)$ in $\mathrm{Spec}(A)$ are quasi-compact [1, Prop. 12, p. 128]. Since the continuous image of a quasi-compact set is quasi-compact, $D_0(f)$ is quasi-compact by (i). For (iv) we have $\mathfrak{r}(E)_\# = (\bigcap\{\mathfrak{p} \mid E \subset \mathfrak{p},\ \mathfrak{p} \text{ prime in } A\})_\# = \bigcap\{\mathfrak{p}_\# \mid E \subset \mathfrak{p},\ \mathfrak{p} \text{ prime in } A\} \supset \bigcap\{\mathfrak{p}_\# \mid E_\# \subset \mathfrak{p}_\#,\ \mathfrak{p}_\# \text{ prime in } A\} = \mathfrak{r}_{\mathrm{D}}(E_\#)$. In order to verify (v), it is enough to show that the intersection of all prime differential ideals in $A$ equals the nilradical of $A$, i.e., the intersection of all prime ideals in $A$. This, however, follows from (iv) by taking $E = \{0\}$. Finally, to verify (vi), let $U \subset \mathrm{Spec}_{\mathrm{D}}(A)$ be open and quasi-compact, and cover $U$ by basic open sets $D_0(f_i)$, $f_i \in A$, $i \in I$. Then there exists a finite subset $J \subset I$ with $U = \bigcup_{j \in J} D_0(f_j)$. Hence $V = \mathrm{Spec}_{\mathrm{D}}(A) - U = \bigcap_{j \in J} V_0(f_j) = V_0(\mathfrak{F})$ where $\mathfrak{F}$ denotes the differential ideal in $A$ generated by $\{f_j\}_{j \in J}$. Conversely, if $\mathfrak{I} = [a_1, \ldots, a_n]$ is a differentially finitely generated differential ideal in $A$, then $V_0(\mathfrak{I}) = \bigcap_{i=1}^n V_0(a_i)$, so that $U = \mathrm{Spec}_{\mathrm{D}}(A) - V_0(\mathfrak{I}) = \bigcup_{i=1}^n D_0(a_i)$ is a finite union of quasi-compacts and hence is itself quasi-compact.

Recall from [3] that a topological space $X$ is called spectral if it is $T_0$ and quasi-compact, the quasi-compact open subsets are closed under finite intersection and form an open basis, and every nonempty irreducible closed subset has a generic point. Moreover, a continuous map of spectral spaces is called spectral if the inverse image of quasi-compact open sets are quasi-compact. It is well known, e.g., [3, p. 43], that for every ring $A$, Spec($A$) is a spectral space, and for every ring homomorphism $f$, ${}^af$ is a spectral map. It is also true that every spectral space (resp., spectral map) is the image of a ring (resp., ring homomorphism) via the functor Spec [3, Th. 6, p. 51], so that spectral spaces and maps are exactly the image of Spec. Spectral spaces and maps also arise from special differential rings and differential ring homomorphisms as in the following.

***Theorem 2.4*** *If $A$ is a special differential ring,* $\mathrm{Spec}_D(A)$ *is a spectral space. If $f: A \to B$ is a differential ring homomorphism between special differential rings, ${}^af$ is a spectral map.*

*Proof* Suppose first that $A$ is special. Then $\mathrm{Spec}_D(A)$ is a subspace of Spec($A$) and hence is $T_0$, and by Proposition 2.3 $\mathrm{Spec}_D(A)$ is quasi-compact. Also by Proposition 2.3 the quasi-compact open subsets of $\mathrm{Spec}_D(A)$ are those whose complements are closed sets of the form $V_0(\mathfrak{I})$, where $\mathfrak{I}$ is a differentially finitely generated differential ideal. Hence if $U_1, \ldots, U_n$ are open and quasi-compact in $\mathrm{Spec}_D(A)$, then

$$\bigcap_{i=1}^{n} U_i = \mathrm{Spec}_D(A) - \bigcup_{i=1}^{n} V_0(\mathfrak{I}_i) = \mathrm{Spec}_D(A) - V_0(\mathfrak{I}_1 \cdot \cdots \cdot \mathfrak{I}_n)$$

is again quasi-compact. Since the basic open sets $D_0(f)$ are quasi-compact by Proposition 2.3, the quasi-compact opens form a basis. Finally, every closed irreducible subset of $\mathrm{Spec}_D(A)$ has a unique generic point by Proposition 2.1. Hence $\mathrm{Spec}_D(A)$ is spectral.

Now let $A$ and $B$ be special, $f$: $A \to B$ a differential ring homomorphism and $U$ quasi-compact open in $\mathrm{Spec}_D(A)$. Then we have $V = \mathrm{Spec}_D(A) - U = V_0(\mathfrak{I})$, where $\mathfrak{I} = [a_1, \ldots, a_n]$, $a_i \in A$, by Proposition 2.3. Hence ${}^af^{-1}(V) = {}^af^{-1}(V_0(\mathfrak{I})) = V_0(f(\mathfrak{I}))$, where $f(\mathfrak{I}) = [f(a_1), \ldots, f(a_n)]$, so that ${}^af^{-1}(U) = \mathrm{Spec}_D(A) - V_0(f(\mathfrak{I}))$ is quasi-compact by Proposition 2.3.

***Corollary 2.5*** *Let $A$ and $B$ be special differential rings and $f$: $A \to B$ a differential ring homomorphism. Then there exist rings $A'$ and $B'$ and a ring homomorphism $f'$: $A' \to B'$ such that* $\mathrm{Spec}_D(A)$ *is homeomorphic to* Spec($A'$), $\mathrm{Spec}_D(B)$ *is homeomorphic to* Spec($B'$), *and ${}^af$ is equivalent to ${}^af'$.*

## References

1. N. Bourbaki, "Algèbre Commutative" (Élém. Math. **27**). Hermann, Paris, 1961.
2. H. Gorman, Radical regularity in differential rings, *Canad. J. Math.* **23** (1971), 197–201.
3. M. Hochster, Prime ideal structure in commutative rings, *Trans. Amer. Math. Soc.* **142** (1969), 43–60.
4. I. Kaplansky, "Commutative Rings." Allyn and Bacon, Boston, 1970.
5. I. Kaplansky, An introduction to differential algebra, *Actualitiés Sci. Indust.* **1251** (1957), 9–63.
6. E. Kolchin, "Differential Algebra and Algebraic Groups." Academic Press, New York, 1973.
7. I. Macdonald, "Algebraic Geometry." Benjamin, New York, 1968.
8. O. Zariski and P. Samuel, "Commutative Algebra," Vols. I and II. Van Nostrand-Reinhold, Princeton, New Jersey, 1958 and 1960.

AMS (MOS) 1970 subject classifications: 12H05, 13A15

# *Constrained cohomology*

*J. Kovacic*

*Brooklyn College*
*CUNY*

## Introduction

In [2] Kolchin developed a cohomology theory for strongly normal extensions, but he did so only for dimensions 0 and 1. For many applications (e.g., the classification of central differentially simple differential algebras) higher dimensional cohomology groups are required. This cohomology theory suffers from serious limitations. If it were applied to the problem of classifying algebras, the additional hypothesis would have to be made that the field of constants of the base field be algebraically closed. A second crucial limitation is its inapplicability to the problem of classifying principal homogeneous spaces for differential algebraic groups; thus Kolchin developed the far more general "constrained" cohomology theory in [4], but he did so only for dimensions 0 and 1. In the present paper we begin the extension of this latter theory to higher dimensions.

Our definition of cochain is somewhat complicated by the fact that a normal extension is not necessarily the union of finitely generated *normal* extensions (as is the case in the theory of fields). Definitions and basic properties are found in Section 1. It is important to note that the generality of the present theory allows the target groups to be arbitrary commutative differential algebraic groups.

In Section 2 we show that if the normal extension happens to be algebraic and the target group happens to be an algebraic group, then the constrained cohomology groups are equal to the usual Galois cohomology groups.

If a short exact sequence of differential algebraic groups is given, and a differential rational cross section exists, then there is an associated long exact cohomology sequence. This is described in Section 3.

In Section 4 we show that the constrained cohomology groups are trivial whenever the base field is algebraically closed and the target group is an algebraic group. This fact, together with the inflation–restriction sequence, is a powerful tool for the computation of cohomology groups. We are unable, unfortunately, to obtain the inflation–restriction sequence in complete generality. The special case in which we can do so (see Sections 5 and 6) does allow us to show that the cohomology groups are isomorphic to the usual Galois cohomology groups in case the target group is an algebraic group.

## Notation

Throughout this paper $\mathscr{F}$ denotes a differential field of characteristic zero with set of derivation operators $\mathbf{\Delta}$, and $\mathscr{F}^{\dagger}$ denotes a fixed constrained closure of $\mathscr{F}$ in some universal differential field $\mathscr{U}$. $\mathscr{N}$ always denotes a normal extension of $\mathscr{F}$ in $\mathscr{F}^{\dagger}$. This means that every ($\mathbf{\Delta}$-, i.e. differential) isomorphism of $\mathscr{N}$ into $\mathscr{F}$ has image in $\mathscr{N}$ (however, not necessarily equal to $\mathscr{N}$). (For a discussion of constrained extensions see Kolchin [3].)

We recall (Kolchin [4]) that for any finitely ($\mathbf{\Delta}$-) generated extension $\mathscr{G}$ of $\mathscr{F}$ in $\mathscr{F}^{\dagger}$, $\mathrm{Iso}(\mathscr{G}/\mathscr{F})$ (the set of $\mathbf{\Delta}$-isomorphisms of $\mathscr{G}$ over $\mathscr{F}$ into $\mathscr{U}$) has the structure of an affine $\mathbf{\Delta}$-$\mathscr{F}$-set in which every element is generic over $\mathscr{F}$. Indeed, if $\mathscr{G} = \mathscr{F}\langle\eta_1, \ldots, \eta_n\rangle$ and $C \in \mathscr{F}\{y_1, \ldots, y_n\}$ is a constraint for $(\eta_1, \ldots, \eta_n)$ over $\mathscr{F}$, we let $\eta_0 = C(\eta_1, \ldots, \eta_n)^{-1}$ and observe that $\mathrm{Iso}(\mathscr{G}/\mathscr{F})$ may be identified with the $\mathbf{\Delta}$-$\mathscr{F}$-locus of $(\eta_0, \ldots, \eta_n)$.

Throughout this paper $G$ denotes a commutative $\mathbf{\Delta}$-$\mathscr{F}$-group which we shall write additively.

## 1. Constrained cohomology groups

***Definition*** For each positive integer $n$ we denote by $\mathfrak{M}^n(\mathscr{N}/\mathscr{F}, G)$ the set of mappings $f$ of $\mathrm{Aut}(\mathscr{N}/\mathscr{F})^n$ into $G$ with the property that there is a finitely generated extension $\mathscr{G}$ of $\mathscr{F}$ and an everywhere defined $\mathbf{\Delta}$-$\mathscr{G}$-mapping $f_{\mathscr{G}}$ of $\mathrm{Iso}(\mathscr{G}/\mathscr{F})^n$ into $G$ with $f(\sigma) = f_{\mathscr{G}}(\sigma\,|\,\mathscr{G})$ for $\sigma = (\sigma_1, \ldots, \sigma_n) \in \mathrm{Aut}(\mathscr{N}/\mathscr{F})^n$. $\mathfrak{M}^0(\mathscr{N}/\mathscr{F}, G)$ is defined to be $G_{\mathscr{N}}$.

If $f \in \mathfrak{M}^n(\mathscr{N}/\mathscr{F}, G)$ and $\mathscr{G}$ is as above, then $f_{\mathscr{G}}$ is unique because the image of $\mathrm{Aut}(\mathscr{N}/\mathscr{F})$ in $\mathrm{Iso}(\mathscr{G}/\mathscr{F})$ (via restriction) is $\mathbf{\Delta}$-dense. We say that $\mathscr{G}$ exhibits the rationality of $f$. Any larger finitely generated extension $\mathscr{H}$ of $\mathscr{F}$ in $\mathscr{N}$ also exhibits the rationality of $f$; indeed, $f_{\mathscr{H}}$ may be defined by $f_{\mathscr{H}}(\sigma) = f_{\mathscr{G}}(\sigma \mid \mathscr{G})$.

If $f, g \in \mathfrak{M}^n(\mathscr{N}/\mathscr{F}, G)$, then $f + g$ (defined by adding images in $G$) also is in $\mathfrak{M}^n(\mathscr{N}/\mathscr{F}, G)$; if $\mathscr{G}$ and $\mathscr{H}$ exhibit the rationality of $f$ and $g$, then the composite $\mathscr{G}\mathscr{H}$ exhibits the rationality of $f + g$. With this law of composition, $\mathfrak{M}^n(\mathscr{N}/\mathscr{F}, G)$ is a commutative group (with identity being the constant mapping 0).

For any mapping $f$ of $\mathrm{Aut}(\mathscr{N}/\mathscr{F})^n$ into $G$, we denote by $f^{\#}$ the mapping of $\mathrm{Aut}(\mathscr{N}/\mathscr{F})^n$ into $G$ defined by the formula $f^{\#}(\sigma_1, \ldots, \sigma_n) = f(\sigma_1, \sigma_1^{-1}\sigma_2, \ldots, \sigma_{n-1}^{-1}\sigma_n)$.

***Definition*** A mapping $f$ of $\mathrm{Aut}(\mathscr{N}/\mathscr{F})^n$ into $G$ is called a (nonhomogeneous) *n-cochain* if $f^{\#}$ is in $\mathfrak{M}^n(\mathscr{N}/\mathscr{F}, G)$. The set of $n$-cochains, together with the law of composition $(f, g) \mapsto f + g$, is a group that is denoted by $\mathscr{C}^n(\mathscr{N}/\mathscr{F}, G)$.

Evidently $\mathscr{C}^0(\mathscr{N}/\mathscr{F}, G) = \mathfrak{M}^0(\mathscr{N}/\mathscr{F}, G) = G_{\mathscr{N}}$ and $\mathscr{C}^1(\mathscr{N}/\mathscr{F}, G) = \mathfrak{M}^1(\mathscr{N}/\mathscr{F}, G)$.

If $f \in \mathfrak{M}^n(\mathscr{N}/\mathscr{F}, G)$ and if $\mathscr{G}$ exhibits the rationality of $f^{\#}$, then

$$\begin{aligned} f(\sigma_1, \ldots, \sigma_n) &= f^{\#}(\sigma_1, \sigma_1\sigma_2, \ldots, \sigma_1 \cdots \sigma_n) \\ &= f^{\#}_{\mathscr{G}}(\sigma_1 \mid \mathscr{G}, \ldots, \sigma_1 \cdots \sigma_n \mid \mathscr{G}) \in G_{\mathscr{G}\sigma_1\mathscr{G} \cdots (\sigma_1 \cdots \sigma_n)\mathscr{G}} \subset G_{\mathscr{N}}. \end{aligned}$$

This allows us to define the $n$th coboundary operator $\partial^n$ by the usual formula:

$$\begin{aligned} \partial^n f(\sigma_1, \ldots, \sigma_{n+1}) = \sigma_1 f(\sigma_2, \ldots, \sigma_{n+1}) &\\ + \sum_{i=1}^{n} (-1)^i f(\sigma_1, \ldots, \sigma_i\sigma_{i+1}, \ldots, \sigma_{n+1}) &\\ + (-1)^{n+1} f(\sigma_1, \ldots, \sigma_n). & \end{aligned}$$

***Proposition 1*** *The image of $\partial^n$ is contained in $\mathscr{C}^{n+1}(\mathscr{N}/\mathscr{F}, G)$.*

This will be proven as a corollary of Propositions 3 and 4.

The usual computations show that

$$0 \longrightarrow \mathscr{C}^0(\mathscr{N}/\mathscr{F}, G) \xrightarrow{\partial^0} \mathscr{C}^1(\mathscr{N}/\mathscr{F}, G) \xrightarrow{\partial^1} \cdots$$

is a complex of commutative groups. Therefore we define the cohomology groups $\mathscr{H}^n(\mathscr{N}/\mathscr{F}, G)$ to be $\mathscr{Z}^n(\mathscr{N}/\mathscr{F}, G)/\mathscr{B}^n(\mathscr{N}/\mathscr{F}, G)$, where $\mathscr{Z}^n(\mathscr{N}/\mathscr{F}, G) = \mathrm{Ker}\,\partial^n$ is the group of $n$-cocycles and $\mathscr{B}^n(\mathscr{N}/\mathscr{F}, G) = \mathrm{Im}\,\partial^{n-1}$ is the group of $n$-coboundaries ($\mathscr{B}^0(\mathscr{N}/\mathscr{F}, G) = 0$).

If $\mathscr{N}$ and $\mathscr{F}$ and $G$ are fixed in a discussion, we allow ourselves to write $\mathscr{H}^n$, $\mathscr{C}^n$, etc., in place of $\mathscr{H}^n(\mathscr{N}/\mathscr{F}, G)$, $\mathscr{C}^n(\mathscr{N}/\mathscr{F}, G)$, etc.

It follows that $\mathscr{H}^0 = G_{\mathscr{F}}$, that an element of $\mathscr{Z}^1$ is a crossed homomorphism, and that an element of $\mathscr{B}^1$ is a principal crossed homomorphism. Observe also that, for $n = 0, 1$, $\mathscr{H}^n(\mathscr{N}/\mathscr{F}, G) = H_{\Delta}{}^n(\mathscr{N}/\mathscr{F}, G)$ as defined by Kolchin [4].

We shall deal almost exclusively with nonhomogeneous cochains, primarily because of their utility in applications. However, in order to justify the somewhat peculiar definition of nonhomogeneous cochains given above, we shall introduce homogeneous cochains.

***Definition*** An element $c$ of $\mathfrak{M}^{n+1}$ is called a *homogeneous n-cochain* if $\tau c(\sigma) = c(\tau\sigma)$ whenever

$$\tau \in \operatorname{Aut}(\mathscr{N}/\mathscr{F}) \qquad \text{and} \qquad \sigma = (\sigma_0, \ldots, \sigma_n) \in \operatorname{Aut}(\mathscr{N}/\mathscr{F})^{n+1}.$$

The group of homogeneous $n$-cochains is denoted by $\mathscr{C}^{*n}(\mathscr{N}/\mathscr{F}, G)$.

***Proposition 2*** *Let $c \in \mathfrak{M}^{n+1}$ and let $\mathscr{G}$ exhibit the rationality of $c$. Then $c \in \mathscr{C}^{*n}$ if and only if $c_{\mathscr{G}}$ (which is given to be a $\mathbf{\Delta}$-$\mathscr{G}$-mapping) is a $\mathbf{\Delta}$-$\mathscr{F}$-mapping.*

*Proof* Assume that $c \in \mathscr{C}^{*n}$ and let $\tau \in \operatorname{Aut}(\mathscr{N}/\mathscr{F})$. Then $c_{\mathscr{G}}$ is a $\mathbf{\Delta}$-$\mathscr{F}$-mapping if and only if $\tau(c_{\mathscr{G}}) = c_{\mathscr{G}}$ (see Kolchin [4]). But

$$\begin{aligned} \tau(c_{\mathscr{G}})(\sigma \mid \mathscr{G}) &= \tau c_{\mathscr{G}}(\tau^{-1}(\sigma \mid \mathscr{G})) \qquad \text{(definition of } \tau(c_{\mathscr{G}})) \\ &= \tau c_{\mathscr{G}}(\tau^{-1}\sigma \mid \mathscr{G}) = \tau c(\tau^{-1}\sigma) = c(\sigma) = c_{\mathscr{G}}(\sigma \mid \mathscr{G}). \end{aligned}$$

The converse is similar.

The homogeneous coboundary operator is defined in the usual way:

$$\partial^{*n} c(\sigma_0, \ldots, \sigma_{n+1}) = \sum_{i=0}^{n+1} (-1)^i c(\sigma_0, \ldots, \hat{\sigma}_i, \ldots, \sigma_{n+1}),$$

where the caret ($\hat{\ }$) indicates deletion.

***Definition*** For each mapping $f$ of $\operatorname{Aut}(\mathscr{N}/\mathscr{F})^n$ into $G_{\mathscr{N}}$ we define a mapping $f^*$ of $\operatorname{Aut}(\mathscr{N}/\mathscr{F})^{n+1}$ into $G$ by the formula

$$f^*(\sigma_0, \ldots, \sigma_n) = \sigma_0\, f(\sigma_0^{-1}\sigma_1, \sigma_1^{-1}\sigma_2, \ldots, \sigma_{n-1}^{-1}\sigma_n).$$

For each mapping $c$ of $\operatorname{Aut}(\mathscr{N}/\mathscr{F})^{n+1}$ into $G$ we define a mapping $c^0$ of $\operatorname{Aut}(\mathscr{N}/\mathscr{F})^n$ into $G$ by the formula

$$c^0(\sigma_1, \ldots, \sigma_n) = c(\mathrm{id}_{\mathscr{N}}, \sigma_1, \sigma_1\sigma_2, \ldots, \sigma_1 \cdots \sigma_n).$$

***Proposition 3*** (a) *If $f \in \mathscr{C}^n$, then $f^* \in \mathscr{C}^{*n}$.*

(b) *If $c \in \mathscr{C}^{*n}$, then $c^0 \in \mathscr{C}^n$.*

(c) *The mappings* $*$ and $^0$ *between $\mathscr{C}^n$ and $\mathscr{C}^{*n}$ are bijective and inverse to each other.*

(d) *The mappings* * *and* $^0$ *commute with* $\partial$ *and* $\partial^*$, *that is,* $\partial^* f^* = (\partial f)^*$ *and* $\partial c^0 = (\partial^* c)^0$.

*Proof* (a) Let $f \in \mathscr{C}^n$ and let $\mathscr{G}$ exhibit the rationality of $f^\#$. Define a mapping $f_{\mathscr{G}}{}^*\colon \operatorname{Iso}(\mathscr{G}/\mathscr{F})^{n+1} \to G$ by the formula

$$f_{\mathscr{G}}{}^*(\sigma_0, \ldots, \sigma_n) = \sigma_0(f_{\mathscr{G}}{}^\#)(\sigma_1, \ldots, \sigma_n).$$

$f_{\mathscr{G}}{}^*$ is an everywhere defined $\mathbf{\Delta}$-$\mathscr{G}$-mapping and, for $\sigma_0, \ldots, \sigma_n \in \operatorname{Aut}(\mathscr{N}/\mathscr{F})$,

$$\begin{aligned} f^*(\sigma_0, \ldots, \sigma_n) &= \sigma_0\, f(\sigma_0^{-1}\sigma_1, \sigma_1^{-1}\sigma_2, \ldots, \sigma_{n-1}^{-1}\sigma_n) \\ &= \sigma_0\, f^\#(\sigma_0^{-1}\sigma_1, \ldots, \sigma_0^{-1}\sigma_n) \\ &= \sigma_0\, f_{\mathscr{G}}{}^\#(\sigma_0^{-1}\sigma_1 | \mathscr{G}, \ldots, \sigma_0^{-1}\sigma_n | \mathscr{G}) \\ &= \sigma_0(f_{\mathscr{G}}{}^\#)(\sigma_1 | \mathscr{G}, \ldots, \sigma_n | \mathscr{G}) = f_{\mathscr{G}}{}^*(\sigma_0 | \mathscr{G}, \ldots, \sigma_n | \mathscr{G}). \end{aligned}$$

Moreover, if $\tau \in \operatorname{Aut}(\mathscr{N}/\mathscr{F})$, then

$$f^*(\tau\sigma_0, \ldots, \tau\sigma_n) = \tau\sigma_0\, f(\sigma_0^{-1}\sigma_1, \ldots, \sigma_{n-1}^{-1}\sigma_n) = \tau f^*(\sigma_0, \ldots, \sigma_n),$$

whence $f^* \in \mathscr{C}^{*n}$.

(b) Let $c \in \mathscr{C}^{*n}$ and let $\mathscr{G}$ exhibit the rationality of $c$. Define a mapping $c_{\mathscr{G}}^{0\#}\colon \operatorname{Iso}(\mathscr{G}/\mathscr{F})^n \to G$ by the formula

$$c_{\mathscr{G}}^{0\#}(\sigma_1, \ldots, \sigma_n) = c_{\mathscr{G}}(\mathrm{id}_{\mathscr{G}}, \sigma_1, \ldots, \sigma_n).$$

$c_{\mathscr{G}}^{0\#}$ is an everywhere defined $\mathbf{\Delta}$-$\mathscr{G}$-mapping and one easily sees that $c^{0\#}(\sigma) = c_{\mathscr{G}}^{0\#}(\sigma | \mathscr{G})$ for $\sigma \in \operatorname{Aut}(\mathscr{N}/\mathscr{F})^n$. Therefore $c^0 \in \mathscr{C}^n$.

Straightforward computations will establish parts (c) and (d).

***Proposition 4*** *Let* $c \in \mathscr{C}^{*n}$. *Then* $\partial^{*n} c \in \mathscr{C}^{*n+1}$.

*Proof* Let $\mathscr{G}$ exhibit the rationality of $c$. We may define $(\partial^* c)_{\mathscr{G}}$: $\operatorname{Iso}(\mathscr{G}/\mathscr{F})^{n+2} \to G$ by the formula

$$(\partial^* c)_{\mathscr{G}}(\sigma_0, \ldots, \sigma_{n+1}) = \sum_{i=0}^{n+1} (-1)^i c_{\mathscr{G}}(\sigma_0, \ldots, \hat{\sigma}_i, \ldots, \sigma_{n+1}).$$

Observe that Proposition 2 is a corollary to Propositions 3 and 4.

***Definition*** An $n$-cochain $f \in \mathscr{C}^n$ is said to be *normalized* if either $n = 0$ or else $f(\sigma_1, \ldots, \sigma_n) = 0$ whenever $\sigma_i = \mathrm{id}_{\mathscr{N}}$ for some $i$.

A homogeneous $n$-cochain $c$ is said to be normalized if either $n = 0$ or else $c(\sigma_0, \ldots, \sigma_n) = 0$ whenever $\sigma_i = \sigma_{i+1}$ for some $i$.

Evidently $*$, $^0$, $\partial$, $\partial^*$ preserve normalization.

***Proposition 5*** *Let* $f$ *be a cocycle. Then there is a coboundary* $g$ *such that* $f + g$ *is normalized.*

*Let f be a coboundary that is normalized. Then there is a cochain g that is normalized having the property* $\partial g = f$.

The proof is standard. See e.g. Mac Lane [5, VII, Sec. 6] and observe that all the constructed mappings do have the rationality properties required by the present theory.

## 2. Galois cohomology

In this section we assume that $\mathcal{N}$ is a normal *algebraic* extension of $\mathcal{F}$. In this case, every automorphism of $\mathcal{N}$ over $\mathcal{F}$ is a $\mathbf{\Delta}$-automorphism, so $\operatorname{Aut}(\mathcal{N}/\mathcal{F})$ is the Galois group of $\mathcal{N}$ over $\mathcal{F}$. This group, together with the Krull topology, is denoted by $G(\mathcal{N}/\mathcal{F})$. If $M$ is a commutative group, given the discrete topology, on which $G(\mathcal{N}/\mathcal{F})$ acts continuously, then $M$ is said to be a $G(\mathcal{N}/\mathcal{F})$-module. Evidently $G_{\mathcal{N}}$ is a $G(\mathcal{N}/\mathcal{F})$-module for any commutative $\mathbf{\Delta}$-$\mathcal{F}$-group $G$. We denote by $H^n(\mathcal{N}/\mathcal{F}, M)$ the cohomology computed with continuous cochains.

***Proposition 6*** *Let* $\mathcal{N}$ *be a normal algebraic extension of* $\mathcal{F}$ *in* $\mathcal{F}^\dagger$. *Then* $\mathscr{H}^n(\mathcal{N}/\mathcal{F}, G) = H^n(\mathcal{N}/\mathcal{F}, G_{\mathcal{N}})$.

*Proof* Consider any continuous mapping $f$: $G(\mathcal{N}/\mathcal{F})^n \to G_{\mathcal{N}}$; evidently $f^\#$ is also continuous. Therefore there is a finite algebraic extension $\mathcal{G}$ of $\mathcal{F}$ in $\mathcal{N}$ such that $f^\#(\sigma\tau) = f^\#(\sigma)$ for $\sigma = (\sigma_1, \ldots, \sigma_n) \in G(\mathcal{N}/\mathcal{F})^n$ and $\tau \in G(\mathcal{N}/\mathcal{G})$, i.e., $f^\#$ depends only on the cosets of $G(\mathcal{N}/\mathcal{G})$ in $G(\mathcal{N}/\mathcal{F})$. We let $f_{\mathcal{G}}{}^\#$: $\operatorname{Iso}(\mathcal{G}/\mathcal{F})^n \to G$ be defined by $f_{\mathcal{G}}{}^\#(\sigma) = f^\#(\sigma^*)$, where $\sigma^* | \mathcal{G} = \sigma$. Since $\operatorname{Iso}(\mathcal{G}/\mathcal{F})^n$ is finite and since each of its elements is rational over $\mathcal{G}$, $f_{\mathcal{G}}{}^\#$ is an everywhere defined $\mathbf{\Delta}$-$\mathcal{G}$-mapping. This shows that $f \in \mathscr{C}^n$.

Let $f \in \mathscr{C}^n$ and let $\mathcal{G}$ exhibit the rationality of $f^\#$. We may assume that $\mathcal{G}$ is a finite normal extension of $\mathcal{F}$. If $B$ is any subset of $G$ and if $f^{\#-1}(B)$ contains $\sigma = (\sigma_1, \ldots, \sigma_n)$, then $f^{\#-1}(B)$ contains the open neighborhood $(\sigma_1 \cdot G(\mathcal{N}/\mathcal{G}), \ldots, \sigma_n \cdot G(\mathcal{N}/\mathcal{G}))$ of $\sigma$. Therefore $f^\#$ is continuous and so is $f$.

## 3. Change of group

Let $\phi$: $H \to G$ be a $\mathbf{\Delta}$-$\mathcal{F}$-homomorphism of commutative $\mathbf{\Delta}$-$\mathcal{F}$-groups. Then there is a homomorphism $\mathscr{C}^n(\mathcal{N}/\mathcal{F}, H) \to \mathscr{C}^n(\mathcal{N}/\mathcal{F}, G)$ defined by $f \to \phi \circ f$; this homomorphism evidently commutes with $\partial$, and the induced homomorphism of $\mathscr{H}^n(\mathcal{N}/\mathcal{F}, H)$ into $\mathscr{H}^n(\mathcal{N}/\mathcal{F}, G)$ is denoted by $\phi^n$.

Now suppose that

$$0 \to H \to G \overset{\rho}{\to} G' \to 0$$

is an exact sequence of commutative $\mathbf{\Delta}$-$\mathscr{F}$-groups and that $\rho(G_{\mathscr{N}}) \supset G_{\mathscr{N}}'$. Let $\alpha' \in G_{\mathscr{F}}'$. Choose $\alpha \in G_{\mathscr{N}}$ such that $\rho\alpha = \alpha'$ and define $f_\alpha$: $\operatorname{Aut}(\mathscr{N}/\mathscr{F}) \to G$ by the formula $f_\alpha(\sigma) = \sigma\alpha - \alpha$. Evidently $f_\alpha \in \mathscr{Z}^1(\mathscr{N}/\mathscr{F}, G)$ (indeed, $\mathscr{F}\langle\alpha\rangle$ exhibits the rationality of $f_\alpha = f_\alpha{}^{\#}$). Moreover, $\rho(f_\alpha(\sigma)) = \sigma\alpha' - \alpha' = 0$, so $f_\alpha \in \mathscr{Z}^1(\mathscr{N}/\mathscr{F}, H)$. It is easy to check that if $\beta \in G_{\mathscr{N}}$ is such that $\rho\beta = \alpha'$, then $f_\beta$ is cohomologous to $f_\alpha$. Hence there is a homomorphism $\Delta^0$: $\mathscr{H}^0(\mathscr{N}/\mathscr{F}, G') \to \mathscr{H}^1(\mathscr{N}/\mathscr{F}, H)$ having the property that if $\alpha' \in G_{\mathscr{F}}'$ and $\alpha \in G_{\mathscr{N}}$ is such that $\rho\alpha = \alpha'$, then $f_\alpha \in \Delta^0(\alpha')$.

***Proposition 7*** *Let*

$$0 \to H \xrightarrow{\iota} G \xrightarrow{\rho} G' \to 0$$

*be an exact sequence of commutative* $\mathbf{\Delta}$-$\mathscr{F}$*-groups and suppose that* $\rho G_{\mathscr{N}} = G_{\mathscr{N}}'$. *Then the following sequence is exact:*

$$0 \longrightarrow \mathscr{H}^0(\mathscr{N}/\mathscr{F}, H) \xrightarrow{\iota^0} \mathscr{H}^0(\mathscr{N}/\mathscr{F}, G) \xrightarrow{\rho^0} \mathscr{H}^0(\mathscr{N}/\mathscr{F}, G')$$
$$\xrightarrow{\Delta^0} \mathscr{H}^1(\mathscr{N}/\mathscr{F}, H) \xrightarrow{\iota^1} \mathscr{H}^1(\mathscr{N}/\mathscr{F}, G) \xrightarrow{\rho^1} \mathscr{H}^1(\mathscr{N}/\mathscr{F}, G').$$

This is proven by Kolchin [4].

For $n > 0$ we are unable to obtain a connecting homomorphism $\Delta^n$ without assuming the existence of a cross section.

Let

$$0 \to H \to G \xrightarrow{\rho} G' \to 0$$

be an exact sequence of $\mathbf{\Delta}$-$\mathscr{F}$-groups. By a cross section for $\rho$ we shall mean an everywhere defined $\Delta$-$\mathscr{F}$-mapping $s$: $G' \to G$ such that $\rho \circ s = \mathrm{id}_{G'}$.

Assume that a cross section exists for $\rho$ and let $f \in \mathscr{Z}^n(\mathscr{N}/\mathscr{F}, G')$. Then $s \circ f$ is a mapping of $\operatorname{Aut}(\mathscr{N}/\mathscr{F})^n$ into $G$. If $\mathscr{G}$ exhibits the rationality of $f^{\#}$, we may define a mapping $(s \circ f)_{\mathscr{G}}{}^{\#}$ of $\operatorname{Iso}(\mathscr{G}/\mathscr{F})^n$ into $G$ by the formula $(s \circ f)_{\mathscr{G}}{}^{\#} = s \circ (f_{\mathscr{G}}{}^{\#})$. $(s \circ f)_{\mathscr{G}}{}^{\#}$ is an everywhere defined $\mathbf{\Delta}$-$\mathscr{G}$-mapping and therefore $s \circ f$ is in $\mathscr{C}^n(\mathscr{N}/\mathscr{F}, G)$. Because $\partial^n(s \circ f) = \partial^n f = 0$, $\partial^n(s \circ f) \in \mathscr{Z}^{n+1}(\mathscr{N}/\mathscr{F}, H)$. Moreover, if $f = \partial^{n-1}g$, then $\partial^n(s \circ f) = 0$, hence there is a homomorphism of $\mathscr{H}^n(\mathscr{N}/\mathscr{F}, G')$ into $\mathscr{H}^{n+1}(\mathscr{N}/\mathscr{F}, H)$ which is denoted by $\Delta^n$. We remark that $\Delta^n$ is independent of the choice of cross section.

***Proposition 8*** *Let*

$$0 \to H \xrightarrow{\iota} G \xrightarrow{\rho} G' \to 0$$

*be an exact sequence of commutative* $\mathbf{\Delta}$-$\mathscr{F}$*-groups and assume that there is a cross section for* $\rho$. *Then the following sequence is exact:*

$$0 \longrightarrow \cdots$$

$$\xrightarrow{\rho^{n-1}} \mathscr{H}^{n-1}(\mathscr{N}/\mathscr{F}, G') \xrightarrow{\Delta^{n-1}} \mathscr{H}^n(\mathscr{N}/\mathscr{F}, H) \xrightarrow{\iota^n} \mathscr{H}^n(\mathscr{N}/\mathscr{F}, G)$$

$$\xrightarrow{\rho^n} \mathscr{H}^n(\mathscr{N}/\mathscr{F}, G') \xrightarrow{\Delta^n} \cdots.$$

The proof is straightforward.

## 4. $\mathscr{F}$-Cohomology

In this section $A$ is an $\mathscr{F}$-set and $G$ is a commutative $\mathscr{F}$-group (as opposed to $\mathbf{\Delta}$-$\mathscr{F}$-set and group).

***Definition*** By an *n-cochain of A into G* is meant an $\mathscr{F}$-mapping of $A^{n+1}$ into $G$ (not necessarily everywhere defined). The group of $n$-cochains is denoted by $C^n(A, G)$. The coboundary operator is defined by the usual formula

$$\partial^n\phi(a_0, \ldots, a_{n+1}) = \sum_{i=0}^{n+1} (-1)^i \phi(a_0, \ldots, \hat{a}_i, \ldots, a_{n+1}),$$

where $\phi \in C^n(A, G)$ and $(a_0, \ldots, a_{n+1})$ is generic for $A^{n+1}$ over $\mathscr{F}$. $Z^n(A, G)$ is defined as usual.

***Proposition 9*** *Let* $\phi \in Z^n(A, G)$ *and suppose that* $\phi$ *is defined at* $(a_0, \ldots, a_n)$ *whenever* $a_i = a_{i+1}$ *for some i. Then* $\phi$ *is everywhere defined.*

*Proof* First note that if $\phi$ is defined at $(a_0, \ldots, \hat{a}_i, \ldots, a_{n+1})$ for all $i \neq i_0$, then $\phi$ is also defined at $(a_0, \ldots, \hat{a}_{i_0}, \ldots, a_{n+1})$ (because $\partial^n\phi = 0$).

Now let $(a_0, \ldots, a_n) \in A^{n+1}$. For $k = n, n-1, \ldots, 0$, choose $b_k$ generic for $A$ over $\mathscr{F}(a_0, \ldots, a_n, b_{k+1}, \ldots, b_n)$ with the property that $b_k$ specializes over $\mathscr{F}$ to $a_k$. We shall prove that $\phi$ is defined at $(a_0, \ldots, a_k, b_{k+1}, \ldots, b_n)$ by induction on $k$.

Let $k = 0$. If $i > 0$, then $(a_0, b_0, \ldots, \hat{b}_i, \ldots, b_n)$ specializes to $(a_0, a_0, \ldots, \hat{b}_i, \ldots, b_n)$, and hence $\phi$ is defined at $(a_0, b_0, \ldots, \hat{b}_i, \ldots, b_n)$. Moreover $\phi$ is defined at $(b_0, \ldots, b_n)$ (since $(b_0, \ldots, b_n)$ is generic for $A^{n+1}$). By the remark above, $\phi$ is defined at $(a_0, \hat{b}_0, b_1, \ldots, b_n)$.

Now let $k > 0$. If $i < k$, then $(a_0, \ldots, \hat{a}_i, \ldots, a_k, b_k, \ldots, b_n)$ specializes to $(a_0, \ldots, \hat{a}_i, \ldots, a_k, a_k, b_{k+1}, \ldots, b_n)$, and hence $\phi$ is defined at the former point. $\phi$ is defined at $(a_0, \ldots, \hat{a}_k, b_k, \ldots, b_n)$ by the induction assumption. Similarly, $\phi$ is defined at $(a_0, \ldots, a_k, b_k, \ldots, \hat{b}_i, \ldots, b_n)$ for $i > k$. By the remark above, $\phi$ is defined at $(a_0, \ldots, a_k, \hat{b}_k, \ldots, b_n)$. This completes the induction. Setting $k = n$, we obtain the proposition.

We denote by $C^n_{\text{ed}}(A, G)$, $Z^n_{\text{ed}}(A, G)$, etc., the group of everywhere defined $n$-cochains, -cocycles, etc.

***Proposition 10*** *If $\mathscr{F}$ is algebraically closed, then $H^n_{\text{ed}}(A, G) = 0$ for all $n > 0$.*

*Proof* Choose $\alpha \in A_{\mathscr{F}}$. Given $\phi \in Z^n_{\text{ed}}(A, G)$, define $\psi \in C^{n-1}_{\text{ed}}(A, G)$ by the formula $\psi(a_0, \ldots, a_{n-1}) = (-1)^n \phi(a_0, \ldots, a_{n-1}, \alpha)$. Then $\partial^{n-1}\psi = \phi$.

Recall [4] that if $A$ is a $\mathbf{\Delta}$-$\mathscr{F}$-subset of $\mathbf{G}_a{}^n$ ($\mathbf{G}_a$ being the one-dimensional additive group) and $s \in \mathbf{N}$, then $A^{(s)}$ is the $\mathbf{\Delta}$-$\mathscr{F}$-subset of $\mathbf{G}_a{}^k$, where $k = \binom{s+m}{m} n$ and $m = \text{card } \mathbf{\Delta}$, consisting of all elements $u^{(s)} = (\theta u_j)_{\theta \in \Theta(s),\, 1 \le j \le n}$ with $u = (u_1, \ldots, u_n) \in A$, and $A_s$ is the Zariski closure of $A^{(s)}$. If $\phi$ is a $\mathbf{\Delta}$-$\mathscr{F}$-mapping of $A^{n+1}$ into an $\mathscr{F}$-group $G$, then, for $s$ sufficiently big, there is an $\mathscr{F}$-mapping $\phi_s$ of $A_s^{n+1}$ into $G$ such that $\phi_s(u_0^{(s)}, \ldots, u_n^{(s)}) = \phi(u_0, \ldots, u_n)$ for all $(u_0, \ldots, u_n)$ generic for $A^{n+1}$ over $\mathscr{F}$.

***Lemma*** *Let $G$ be a commutative $\mathscr{F}$-group (as opposed to $\mathbf{\Delta}$-$\mathscr{F}$-group), and let $c \in \mathscr{Z}^{*n}(\mathscr{N}/\mathscr{F}, G)$ be normalized. Suppose that $\mathscr{G}$ exhibits the rationality of $c$ and set $A = \text{Iso}(\mathscr{G}/\mathscr{F})$. Fix $s$ sufficiently big so that $(c_{\mathscr{G}})_s$ exists. Then $(c_{\mathscr{G}})_s \in Z^n_{\text{ed}}(A_s, G)$. If $(c_{\mathscr{G}})_s \in B^n_{\text{ed}}(A_s, G)$, then $c \in \mathscr{B}^{*n}(\mathscr{N}/\mathscr{F}, G)$.*

*Proof* Fixing $i$, with $0 \le i \le n$, we let $B$ be the set of $(u_0, \ldots, u_n)$ in $A^{n+1}$ with $u_i = u_{i+1}$ and let $\iota$ be the inclusion of $B$ in $A^{n+1}$. Since $c$ is normalized, $c_{\mathscr{G}} \circ \iota = 0$. $c_{\mathscr{G}s}$ is defined on $B^{(s)} = B_s \cap A^{(s)n+1}$ so we may consider the generic composite of $c_{\mathscr{G}s}$ and $\iota_s$. This composite is equal to $(c_{\mathscr{G}} \circ \iota)_s$ which is 0. Thus $c_{\mathscr{G}s}$ is everywhere defined (and equal to 0) on $B_s$. This being so for any index $i$, it follows that, by Proposition 9, $c_{\mathscr{G}s}$ is everywhere defined and therefore $c_{\mathscr{G}s} \in Z^n_{\text{ed}}(A_s, G)$.

Now suppose that $c_{\mathscr{G}s} \in B^n_{\text{ed}}(A_s, G)$, say $c_{\mathscr{G}s} = \partial e$ with $e \in C^{n-1}_{\text{ed}}(A_s, G)$. Define a $\mathbf{\Delta}$-$\mathscr{F}$-mapping $d_{\mathscr{G}}$ of $A^{n-1}$ into $G$ by the formula $d_{\mathscr{G}}(u_0, \ldots, u_{n-1}) = e(u_0^{(s)}, \ldots, u_{n-1}^{(s)})$. Because $e$ is everywhere defined, so is $d_{\mathscr{G}}$. Finally, define $d \in \mathscr{C}^{*n-1}(\mathscr{N}/\mathscr{F}, G)$ by setting $d(\sigma_0, \ldots, \sigma_{n-1}) = d_{\mathscr{G}}(\sigma_0 | \mathscr{G}, \ldots, \sigma_{n-1} | \mathscr{G})$; clearly $c = \partial d$, which proves the lemma.

***Proposition 11*** *Let $G$ be a commutative $\mathscr{F}$-group (as opposed to a $\mathbf{\Delta}$-$\mathscr{F}$-group). If $\mathscr{F}$ is algebraically closed, then $\mathscr{H}^n(\mathscr{N}/\mathscr{F}, G) = 0$ for every $n > 0$.*

This is immediate from the lemma and Proposition 10.

## 5. Change of field

In this section $G$ is a commutative $\mathbf{\Delta}$-$\mathscr{F}$-group.

Let $\mathscr{N}$ and $\mathscr{N}'$ be normal extensions of $\mathscr{F}$ in $\mathscr{F}^\dagger$ with $\mathscr{N} \subset \mathscr{N}'$. Then every element of $\mathrm{Aut}(\mathcal{N}'/\mathcal{F})$ restricts to an element of $\mathrm{Aut}(\mathcal{N}/\mathcal{F})$. Given $f \in \mathscr{C}^n(\mathcal{N}/\mathcal{F}, G)$, we may define a mapping of $\mathrm{Aut}(\mathcal{N}'/\mathcal{F})^n$ into $G$ by the formula $\sigma \mapsto f(\sigma \mid \mathcal{N})$. This mapping is called the *inflation of* $f$ and is denoted by $\mathrm{infl}(f)$. Clearly $\mathrm{infl}(f) \in \mathscr{C}^n(\mathcal{N}'/\mathcal{F}, G)$ (in fact, if $\mathscr{G}$ exhibits the rationality of $f^{\#}$, then $\mathscr{G}$ also exhibits the rationality of $\mathrm{infl}(f)^{\#}$). Moreover, inflation commutes with $\partial$. The induced homomorphism of $\mathscr{H}^n(\mathcal{N}/\mathcal{F}, G)$ into $\mathscr{H}^n(\mathcal{N}'/\mathcal{F}, G)$ is denoted by $\mathrm{infl}^n$.

Now suppose that $\mathscr{U}$ is universal over $\mathcal{N}$. Let $f \in \mathscr{C}^n(\mathcal{N}'/\mathcal{F}, G)$. Because $\mathrm{Aut}(\mathcal{N}'/\mathcal{N})$ is a subgroup of $\mathrm{Aut}(\mathcal{N}'/\mathcal{F})$ we may define a mapping of $\mathrm{Aut}(\mathcal{N}'/\mathcal{N})^n$ into $G$ by restricting the domain of $f$. This mapping is called the *restriction of* $f$ and is denoted by $\mathrm{res}(f)$. Evidently $\mathrm{res}(f) \in \mathscr{C}^n(\mathcal{N}'/\mathcal{N}, G)$ (in fact if $\mathscr{G}$ exhibits the rationality of $f^{\#}$, then $\mathscr{G}\mathcal{N}$ exhibits the rationality of $\mathrm{res}(f)^{\#}$). Moreover, restriction commutes with $\partial$. The induced homomorphism of $\mathscr{H}^n(\mathcal{N}'/\mathcal{F}, G)$ into $\mathscr{H}^n(\mathcal{N}'/\mathcal{N}, G)$ is denoted by $\mathrm{res}^n$.

It is apparent that $\mathrm{res}^n \circ \mathrm{infl}^n = 0$.

***Proposition 12*** *Let $\mathcal{N}$ and $\mathcal{N}'$ be normal extension of $\mathcal{F}$ with $\mathcal{N} \subset \mathcal{N}'$, and suppose that $\mathscr{U}$ is universal over $\mathcal{N}$. Then the sequence*

$$0 \longrightarrow \mathscr{H}^1(\mathcal{N}/\mathcal{F}, G) \xrightarrow{\mathrm{infl}^1} \mathscr{H}^1(\mathcal{N}'/\mathcal{F}, G) \xrightarrow{\mathrm{res}^1} \mathscr{H}^1(\mathcal{N}'/\mathcal{N}, G)$$

*is exact.*

This is a special case of a result of Kolchin [4].

We shall next define an action of $\mathrm{Aut}(\mathcal{N}/\mathcal{F})$ on $\mathscr{H}^n(\mathcal{N}'/\mathcal{N}, G)$ and show that the image of $\mathrm{res}^n$ lies in the group of fixed points under this action.

Let $f$ be a mapping of $\mathrm{Aut}(\mathcal{N}'/\mathcal{N})^n$ into $G_{\mathcal{N}'}$ and let $\tau \in \mathrm{Aut}(\mathcal{N}'/\mathcal{F})$. We define a mapping $\tau \cdot f$ of $\mathrm{Aut}(\mathcal{N}'/\mathcal{N})^n$ into $G$ by the formula $(\tau \cdot f)(\sigma_1, \ldots, \sigma_n) = f(\tau^{-1}\sigma_1\tau, \ldots, \tau^{-1}\sigma_n\tau)$.

***Proposition 13*** *Let $f \in \mathscr{C}^n(\mathcal{N}'/\mathcal{N}, G)$ and $\tau \in \mathrm{Aut}(\mathcal{N}'/\mathcal{F})$. Then $\tau \cdot f \in \mathscr{C}^n(\mathcal{N}'/\mathcal{N}, G)$. Moreover $\tau \cdot (\partial f) = \partial(\tau \cdot f)$.*

*Proof* Let $\mathscr{G}$ exhibit the rationality of $f^{\#}$. Because $\mathrm{Iso}(\mathscr{G}/\mathcal{N})$ is a $\mathbf{\Delta}$-$\mathcal{N}$-subset of affine space, $\tau(\mathrm{Iso}(\mathscr{G}/\mathcal{N}))$ is defined and is a $\mathbf{\Delta}$-$\mathcal{N}$-set (since $\tau(\mathcal{N}) \subset \mathcal{N}$). Because $\mathcal{N}\langle \mathrm{id}_{\tau\mathscr{G}}\rangle = \tau\mathscr{G} = \tau(\mathcal{N}\langle \mathrm{id}_{\mathscr{G}}\rangle) = \mathcal{N}\langle \tau(\mathrm{id}_{\mathscr{G}})\rangle$, there is a unique $\mathbf{\Delta}$-$\mathcal{N}$-mapping $\phi$ of $\mathrm{Iso}(\tau\mathscr{G}/\mathcal{N})$ into $\tau(\mathrm{Iso}(\mathscr{G}/\mathcal{N}))$ such that $\phi(\mathrm{id}_{\tau\mathscr{G}}) = \tau(\mathrm{id}_{\mathscr{G}})$. The fact that every element of $\mathrm{Iso}(\tau\mathscr{G}/\mathcal{N})$ is generic over $\mathcal{N}$ implies that $\phi$ is everywhere defined.

Denote by $\mathrm{Iso}^{\dagger}(\tau\mathscr{G}/\mathcal{N})$ the set of isomorphisms of $\tau\mathscr{G}$ over $\mathcal{N}$ into $\mathcal{F}^{\dagger}$ (and therefore into $\mathcal{N}'$). $\mathrm{Iso}^{\dagger}(\tau\mathscr{G}/\mathcal{N})$ is the image of $\mathrm{Aut}(\mathcal{N}'/\mathcal{N})$ (via restriction). For any $\sigma \in \mathrm{Iso}^{\dagger}(\tau\mathscr{G}/\mathcal{N})$, $(\tau^{-1} \mid \sigma\tau\mathscr{G}) \circ \sigma \circ (\tau \mid \mathscr{G}) \in \mathrm{Iso}(\mathscr{G}/\mathcal{N})$. We claim that $\phi(\sigma) = \tau((\tau^{-1}|\sigma\tau\mathscr{G}) \circ \sigma \circ (\tau|\mathscr{G}))$. Indeed, the $\mathscr{N}$-specialization $\mathrm{id}_{\tau\mathscr{G}} \leftrightarrow \sigma$ induces an isomorphism $\tau\mathscr{G} = \mathscr{N}\langle \mathrm{id}_{\tau\mathscr{G}}\rangle \simeq \mathscr{N}\langle\sigma\rangle = \sigma\tau\mathscr{G}$ which is merely $\sigma$. Therefore

$$\begin{aligned}\phi(\sigma) &= \sigma(\phi(\mathrm{id}_{\tau\mathscr{G}})) = \sigma(\tau(\mathrm{id}_{\mathscr{G}})) = \tau((\tau^{-1}\sigma\tau)(\mathrm{id}_{\mathscr{G}}))\\ &= \tau(\tau^{-1}\sigma\tau \circ \mathrm{id}_{\mathscr{G}}) = \tau((\tau^{-1}|\sigma\tau\mathscr{G}) \circ \sigma \circ (\tau|\mathscr{G})).\end{aligned}$$

Now $f_{\mathscr{G}}{}^{\#}$ is an everywhere defined $\mathbf{\Delta}$-$\mathscr{G}$-mapping of $\mathrm{Iso}(\mathscr{G}/\mathscr{N})^n$ into $G$ and hence $\tau(f_{\mathscr{G}}{}^{\#}) \circ (\phi \times \cdots \times \phi)$ is an everywhere defined $\mathbf{\Delta}$-$\tau\mathscr{G}$-mapping of $\mathrm{Iso}(\tau\mathscr{G}/\mathscr{N})^n$ into $G$. For $\sigma \in \mathrm{Aut}(\mathscr{N}'/\mathscr{N})^n$, a direct computation shows that $(\tau \cdot f)^{\#}(\sigma) = \tau(f_{\mathscr{G}}{}^{\#})(\phi(\sigma|\tau\mathscr{G}))$. This shows that $\tau \cdot f \in \mathscr{C}^n(\mathscr{N}'/\mathscr{N}, G)$. The remainder of the proposition is straightforward.

***Proposition 14*** *Let* $\tau \in \mathrm{Aut}(\mathscr{N}'/\mathscr{N})$ *and let* $f \in \mathscr{Z}^n(\mathscr{N}'/\mathscr{N}, G)$. *Then* $\tau \cdot f - f \in \mathscr{B}^n(\mathscr{N}'/\mathscr{N}, G)$.

*Proof* For each $i = 1, \ldots, n$, we define a mapping $T_i^n f$ of $\mathrm{Aut}(\mathscr{N}'/\mathscr{N})^{n-1}$ into $G$ by the formula

$$T_i^n f(\sigma_1, \ldots, \sigma_{n-1}) = f(\sigma_1, \ldots, \sigma_{i-1}, \tau, \tau^{-1}\sigma_i\tau, \ldots, \tau^{-1}\sigma_{n-1}\tau),$$

and claim that $T_i^n f \in \mathscr{C}^{n-1}(\mathscr{N}'/\mathscr{N}, G)$.

First note that $(T_i^n f)^{\#}(\sigma_1, \ldots, \sigma_{n-1}) = f^{\#}(\sigma_1, \ldots, \sigma_{i-1}, \sigma_{i-1}\tau, \ldots, \sigma_{n-1}\tau)$. If $\mathscr{G}$ exhibits the rationality of $f^{\#}$, then $\mathscr{G}\tau\mathscr{G}$ exhibits the rationality of $(T_i^n f)^{\#}$. Indeed $(T_i^n f)^{\#}_{\mathscr{G}\tau\mathscr{G}}(\sigma)$ is defined to be $f_{\mathscr{G}}{}^{\#}(\sigma_1|\mathscr{G}, \ldots, \sigma_{i-1}|\mathscr{G}, \sigma_{i-1}\tau|\mathscr{G}, \ldots, \sigma_{n-1}\tau|\mathscr{G})$ for $\sigma = (\sigma_1, \ldots, \sigma_{n-1}) \in \mathrm{Iso}(\mathscr{G}\tau\mathscr{G}/\mathscr{N})^{n-1}$.

Let $g = -\sum_{i=1}^{n} (-1)^i T_i^n f \in \mathscr{C}^{n-1}(\mathscr{N}'/\mathscr{N}, G)$. Then

$$f - \tau \cdot f + \partial^{n-1} g = \sum_{i=1}^{n+1} (-1)^i T_i^{n+1}\, \partial^n f = 0,$$

which proves the proposition.

Propositions 13 and 14 allow us to define an action of $\mathrm{Aut}(\mathscr{N}/\mathscr{F})$ on the group $\mathscr{H}^n(\mathscr{N}'/\mathscr{N}, G)$. The fixed points of this action are denoted by $\mathscr{H}^n(\mathscr{N}'/\mathscr{N}, G)^{\mathscr{N}/\mathscr{F}}$.

***Proposition 15*** *The image of* $\mathrm{res}^n$ *is contained in* $\mathscr{H}^n(\mathscr{N}'/\mathscr{N}, G)^{\mathscr{N}/\mathscr{F}}$.

*Proof* Given $f \in \mathscr{Z}^n(\mathscr{N}'/\mathscr{F}, G)$ and $\tau \in \mathrm{Aut}(\mathscr{N}'/\mathscr{F})$, define

$$g = -\sum_{i=1}^{n} (-1)^i T_i^n f \in \mathscr{C}^{n-1}(\mathscr{N}'/\mathscr{F}, G).$$

Then $\tau \cdot f - f = \partial g$, so $\tau \cdot (\mathrm{res}(f)) - \mathrm{res}(f) = \partial(\mathrm{res}(g))$.

If $\mathscr{N}$ and $\mathscr{N}'$ are arbitrary normal extensions, we are unable to develop further the Hochschild–Serre inflation–restriction sequence. In particular, we are unable to define the "transgression." Under restrictive hypotheses, however, we can obtain the entire sequence, and we do so in the following section.

## 6. The Hochschild–Serre sequence

In this section we fix a normal extension $\mathcal{N}$ of $\mathcal{F}$ and denote by $\mathcal{N}_0$ the relative algebraic closure of $\mathcal{F}$ in $\mathcal{N}$; we note that $\mathcal{N}_0$ is a Galois extension of $\mathcal{F}$. The group $\text{Aut}(\mathcal{N}_0/\mathcal{F})$ together with the Krull topology is denoted by $G(\mathcal{N}_0/\mathcal{F})$. In this discussion all cochains are nonhomogeneous and are *assumed to be normalized*. Recall that $G(\mathcal{N}_0/\mathcal{F})$ acts on $\mathcal{H}^n(\mathcal{N}/\mathcal{N}_0, G)$, but that we have not defined an action of $G(\mathcal{N}_0/\mathcal{F})$ on $\mathcal{C}^n(\mathcal{N}/\mathcal{N}_0, G)$. Nevertheless, we shall write, by abuse of notation, $C^p(\mathcal{N}_0/\mathcal{F}, \mathcal{C}^q(\mathcal{N}/\mathcal{N}_0, G))$ for the group of continuous mappings (in the Krull and discrete topologies) of $G(\mathcal{N}_0/\mathcal{F})^p$ into $C^q(\mathcal{N}/\mathcal{N}_0, G)$. Our development will follow that of Hochschild and Serre [1, Chap. 2] closely. This reference will henceforth be abbreviated to H–S.

***Definition*** $K^{p,q}$ denotes the subgroup of $\mathcal{C}^{p+q}(\mathcal{N}/\mathcal{F}, G)$ consisting of those $f$ such that $f(\sigma, \tau) = f(\sigma, \tau')$ for all $\sigma \in \text{Aut}(\mathcal{N}/\mathcal{F})^q$ and $\tau$, $\tau' \in \text{Aut}(\mathcal{N}/\mathcal{F})^p$ which satisfy the condition $\tau' | \mathcal{N}_0 = \tau | \mathcal{N}_0$.

As usual, we define a spectral sequence by setting

$$Z_r^{p,q} = \{f \in K^{p,q} \mid \partial f \in K^{p+r,q-r+1}\}$$

and

$$E_r^{p,q} = Z_r^{p,q}/(\partial Z_{r-1}^{p-r+1,q+r-2} + K^{p+1,q-1}) \qquad (r \geq 1).$$

Our first goal is to compute $E_1^{p,q}$.

Fix, for each $\gamma \in G(\mathcal{N}_0/\mathcal{F})$, a representative $\gamma^* \in \text{Aut}(\mathcal{N}/\mathcal{F})$ (so that $\gamma^* | \mathcal{N}_0 = \gamma$) in such a way that $\text{id}_{\mathcal{N}_0}^* = \text{id}_{\mathcal{N}}$.

***Definition*** Let $f \in K^{p,q}$. Then $r_p f \in C^p(\mathcal{N}_0/\mathcal{F}, \mathcal{C}^q(\mathcal{N}/\mathcal{N}_0, G))$ is defined by $r_p f(\gamma_1, \ldots, \gamma_p)(\sigma_1, \ldots, \sigma_q) = f(\sigma_1, \ldots, \sigma_q, \gamma_1^*, \ldots, \gamma_p^*)$.

For fixed $\gamma$, we must show that $r_p f(\gamma) \in \mathcal{C}^q(\mathcal{N}/\mathcal{N}_0, G)$. We claim that if $\mathcal{G}$ exhibits the rationality of $f^\#$, then $\mathcal{G}' = \mathcal{N}_0 \mathcal{G} \gamma_1^* \mathcal{G} \cdots (\gamma_1^* \cdots \gamma_p^*)\mathcal{G}$ exhibits the rationality of $r_p f(\gamma)^\#$. Indeed, define a mapping of $\text{Iso}(\mathcal{G}'/\mathcal{N}_0)^q$ into $G$ by the formula $r_p f(\gamma)_{\mathcal{G}'}^\#(\sigma) = f_{\mathcal{G}}^{\#}(\sigma | \mathcal{G}, \gamma_1^* | \mathcal{G}, \ldots, \gamma_1^* \cdots \gamma_p^* | \mathcal{G})$, where $\sigma = (\sigma_1, \ldots, \sigma_q) \in \text{Iso}(\mathcal{G}'/\mathcal{N}_0)^q$. The straightforward computations are left to the reader. Since $r_p f(\gamma)$ is evidently normalized, it is an element of $\mathcal{C}^q(\mathcal{N}/\mathcal{N}_0, G)$. Moreover, $r_p f(\gamma) = r_p f(\gamma')$ whenever $\gamma | (\mathcal{G} \cap \mathcal{N}_0) = \gamma' | (\mathcal{G} \cap \mathcal{N}_0)$, hence $r_p f$ is continuous in the Krull and discrete topologies and $r_p f \in C^p(\mathcal{N}_0/\mathcal{F}, \mathcal{C}^q(\mathcal{N}/\mathcal{N}_0, G))$.

For $\sigma \in \mathrm{Aut}(\mathcal{N}/\mathcal{N}_0)^{q+1}$, $\partial^q(r_p f(\gamma))(\sigma) = (\partial^{p+q} f)(\sigma, \gamma^*) = r_p(\partial^{p+q} f)(\gamma)(\sigma)$ by a direct computation using the fact that $f \in K^{p,q}$ and that $f$ is normalized. We omit the details (see H–S). It follows that the homomorphism $r_p$ induces a homomorphism $r_p{}^1: E_1^{p,q} \to C^p(\mathcal{N}_0/\mathcal{F}, \mathcal{C}^q(\mathcal{N}/\mathcal{N}_0, G))$.

***Proposition 16*** *$r_p{}^1$ is an isomorphism.*

*Proof* We follow the construction given in H–S exactly and thus need address ourselves only to the question of the rationality.

Let $f \in Z_1^{p,q+1}$ and $h \in C^p(\mathcal{N}_0/\mathcal{F}, \mathcal{C}^q(\mathcal{N}/\mathcal{N}_0, G))$ be such that $r_p f(\gamma) = \partial^q(h(\gamma))$. We have replaced $q$ by $q + 1$ for convenience in the formulas which follow. The case $q = 0$ (which is thereby omitted) is trivial. We must find $g \in Z_0^{p,q} = K^{p,q}$ such that $f - \partial g \in K^{p+1,q}$.

The mapping $g$ is defined by Hochschild and Serre by the formula

$$g(\tau_1, \ldots, \tau_q, \mu_1, \ldots, \mu_p) = \lambda_1{}^*h(\mu \,|\, \mathcal{N}_0)(\phi_1, \ldots, \phi_q) + \sum_{i=1}^{q} (-1)^i f(\tau_1, \ldots, \tau_{i-1}, \lambda_i{}^*, \phi_i, \ldots, \phi_q, \mu),$$

where $\lambda_q = \tau_q \,|\, \mathcal{N}_0$, $\phi_q = \lambda_q^{*-1}\tau_q$, and, for $i < q$, $\lambda_i = (\tau_i \,|\, \mathcal{N}_0)\lambda_{i+1}$ and $\phi_i = \lambda_i^{*-1}\tau_i\lambda_{i+1}^*$.

Since

$$g^\#(\tau_1, \ldots, \tau_q, \mu_1, \ldots, \mu_p) = g(\tau_1, \tau_1^{-1}\tau_2, \ldots, \tau_{q-1}^{-1}\tau_q, \tau_q^{-1}\mu_1, \mu_1^{-1}\mu_2, \ldots, \mu_{p-1}^{-1}\mu_p),$$

we introduce the following notation.

Let $v = (\tau_q^{-1}\mu_1, \mu_1^{-1}\mu_2, \ldots, \mu_{p-1}^{-1}\mu_p)$, $\gamma_i = \tau_i \,|\, \mathcal{N}_0$, and $\sigma_i = \gamma_q^{*-1}\tau_i(\gamma_i^{-1}\gamma_q)^*$ $(i = 1, \ldots, q)$. Then $\lambda_q = \tau_{q-1}^{-1}\tau_q \,|\, \mathcal{N}_0 = \gamma_{q-1}^{-1}\gamma_q$, and

$$\phi_q = \lambda_q^{*-1}\tau_{q-1}^{-1}\tau_q = \sigma_{q-1}^{-1}\sigma_q.$$

In general, for $i = 2, \ldots, q - 1$, $\lambda_i = \gamma_{i-1}^{-1}\gamma_q$ (by induction) and $\phi_i = \sigma_{i-1}^{-1}\sigma_q$. Moreover, $\lambda_1 = (\tau_1 \,|\, \mathcal{N}_0)\lambda_2 = \gamma_q$, and $\phi_1 = \sigma_1$. We obtain the formula

$$\begin{aligned} g^\#(\tau_1, \ldots, \tau_q, \mu_1, \ldots, \mu_p) &= \gamma_q{}^*(h(v \,|\, \mathcal{N}_0)(\sigma_1, \sigma_1^{-1}\sigma_2, \ldots, \sigma_{q-1}^{-1}\sigma_q)) \\ &\quad + \sum_{i=1}^{q} (-1)^i f(\tau_1, \tau_1^{-1}\tau_2, \ldots, \tau_{i-2}^{-1}\tau_{i-1}, (\gamma_{i-1}^{-1}\gamma_q)^*, \sigma_{i-1}^{-1}\sigma_i, \ldots, \sigma_{q-1}^{-1}\sigma_q, v) \\ &= \gamma_q{}^*(h(v \,|\, \mathcal{N}_0)^\#(\sigma)) \\ &\quad + \sum_{i=1}^{q} (-1)^i f^\#(\tau_1, \ldots, \tau_{i-1}, \tau_{i-1}(\gamma_{i-1}^{-1}\gamma_q)^*, \ldots, \tau_q(\gamma_q^{-1}\gamma_q)^*, \mu). \quad (\star) \end{aligned}$$

We must find a field which exhibits the rationality of $g^*$. We do this in the following sequence of lemmas.

There is a finitely generated extension of $\mathcal{N}_0$ which exhibits the rationality of $h(v\,|\,\mathcal{N}_0)^*$ but also, by the next lemma, a finitely generated extension of $\mathcal{F}$ which serves the same role.

***Lemma 1*** *Let $h \in \mathcal{C}^q(\mathcal{N}/\mathcal{N}_0, G)$. Then there is a finitely generated extension $\mathcal{H}$ of $\mathcal{F}$ in $\mathcal{N}$ and an everywhere defined $\mathbf{\Delta}$-$\mathcal{H}$-mapping $h_{\mathcal{H}}{}^*$ of* $\mathrm{Iso}(\mathcal{H}/\mathcal{H} \cap \mathcal{N}_0)^q$ *into $G$ such that $h^*(\sigma) = h_{\mathcal{H}}{}^*(\sigma\,|\,\mathcal{H})$ $(\sigma \in \mathrm{Aut}(\mathcal{N}/\mathcal{N}_0)^q)$.*

*Proof* Let $\mathcal{G}\mathcal{N}_0$ exhibit the rationality of $h^*$, where $\mathcal{G}$ is a finitely generated extension of $\mathcal{F}$ in $\mathcal{N}$. Since $\mathcal{G}$ is a regular extension of $\mathcal{G} \cap \mathcal{N}_0$, $\mathrm{Iso}(\mathcal{G}/\mathcal{G} \cap \mathcal{N}_0)$ is $\mathbf{\Delta}$-$\mathcal{U}$-irreducible and a fortiori $\mathbf{\Delta}$-$\mathcal{N}_0$-irreducible. We define a $\mathbf{\Delta}$-$\mathcal{N}_0$-mapping $s$: $\mathrm{Iso}(\mathcal{G}/\mathcal{G} \cap \mathcal{N}_0) \to \mathrm{Iso}(\mathcal{G}\mathcal{N}_0/\mathcal{N}_0)$ such that $s(\mathrm{id}_{\mathcal{G}}) = \mathrm{id}_{\mathcal{G}\mathcal{N}_0}$. It is easy to see that $s(\sigma)$ is the unique (since $\mathcal{G}$ and $\mathcal{N}_0$ are linearly disjoint over $\mathcal{G} \cap \mathcal{N}_0$) element of $\mathrm{Iso}(\mathcal{G}\mathcal{N}_0/\mathcal{N}_0)$ whose restriction to $\mathcal{G}$ is $\sigma$. $h^{\#}_{\mathcal{G}\mathcal{N}_0} \circ (s \times \cdots \times s)$ is an everywhere defined $\mathbf{\Delta}$-$\mathcal{G}\mathcal{N}_0$-mapping of $\mathrm{Iso}(\mathcal{G}/\mathcal{G} \cap \mathcal{N}_0)^q$ into $G$. We now choose a finitely generated extension $\mathcal{H}$ of $\mathcal{G}$ so that $h^{\#}_{\mathcal{G}\mathcal{N}_0} \circ (s \times \cdots \times s)$ is a $\mathbf{\Delta}$-$\mathcal{H}$-mapping and then set $h_{\mathcal{H}}{}^* = h^{\#}_{\mathcal{G}\mathcal{N}_0} \circ (s \times \cdots \times s) \circ (\rho \times \cdots \times \rho)$, where $\rho$: $\mathrm{Iso}(\mathcal{H}/\mathcal{H} \cap \mathcal{N}_0) \to \mathrm{Iso}(\mathcal{G}/\mathcal{G} \cap \mathcal{N}_0)$ is the restriction.

We now continue with the proof of the proposition. Choose a finitely generated extension $\mathcal{G}_0$ of $\mathcal{F}$ in $\mathcal{N}$ which satisfies the following conditions.

(a) $\mathcal{G}_0$ exhibits the rationality of $f^*$.

(b) $\mathcal{F}_0 = \mathcal{G}_0 \cap \mathcal{N}_0$ is a Galois extension of $\mathcal{F}$.

(c) $\mathcal{F}_0$ exhibits the continuity of $h$ in the sense that $h(\mu) = h_0(\mu\,|\,\mathcal{F}_0)$ for some $h_0 \in C^p(\mathcal{F}_0/\mathcal{F}, \mathcal{C}^q(\mathcal{N}/\mathcal{N}_0, G))$.

(d) $\mathcal{G}_0$ exhibits the rationality of $h(\mu)^*$ (in the sense of Lemma 1) for every $\mu \in G(\mathcal{F}_0/\mathcal{F})^p$.

Set $\mathcal{G}$ equal to the compositum of the fields $\gamma^*\mathcal{G}_0$ $(\gamma \in G(\mathcal{F}_0/\mathcal{F}))$. We claim that $\mathcal{G}$ exhibits the rationality of $g^*$.

Since we shall define $g_{\mathcal{G}}{}^*$ separately on each of the $\mathbf{\Delta}$-$\mathcal{G}$-components of $\mathrm{Iso}(\mathcal{G}/\mathcal{F})^{p+q}$, we shall need the following lemma.

***Lemma 2*** *Let $\mathcal{G}$ be a finitely generated constrained extension of $\mathcal{F}$ and suppose that the relative algebraic closure $\mathcal{F}_0$ of $\mathcal{F}$ in $\mathcal{G}$ is a Galois extension of $\mathcal{F}$. Then the $\mathbf{\Delta}$-components of* $\mathrm{Iso}(\mathcal{G}/\mathcal{F})$ *are the sets* $\gamma(\mathrm{Iso}(\mathcal{G}/\mathcal{F}_0))$, *where $\gamma \in G(\mathcal{F}_0/\mathcal{F})$. Moreover, $\tau \in \gamma(\mathrm{Iso}(\mathcal{G}/\mathcal{F}_0))$ if and only if $\tau\,|\,\mathcal{F}_0 = \gamma$.*

*Proof* We prove the last sentence first. Fix $\gamma \in G(\mathcal{F}_0/\mathcal{F})$. Choose a generic point $\sigma$ of $\mathrm{Iso}(\mathcal{G}/\mathcal{F}_0)$ over $\mathcal{F}_0$ and let $\gamma'$ be an extension of $\gamma$ to $\mathcal{F}_0\sigma\mathcal{G}$. Then $\gamma' \circ \sigma = \gamma'(\sigma) \in \mathrm{Iso}(\mathcal{G}/\mathcal{F})$ and so specializes generically over $\mathcal{F}$ to any $\tau \in \mathrm{Iso}(\mathcal{G}/\mathcal{F})$. The associated isomorphism $\phi : \gamma'\sigma\mathcal{G} \simeq \tau\mathcal{G}$ over $\mathcal{F}$ is given by

$\phi = \tau \circ (\gamma'\sigma)^{-1}$. The condition that $\tau \in \gamma(\mathrm{Iso}(\mathscr{G}/\mathscr{F}_0))$ is equivalent to the condition that $\phi$ extends to an isomorphism of $\mathscr{F}_0 \cdot \gamma'\sigma\mathscr{G}$ over $\mathscr{F}_0$. Since $\mathscr{F}_0 \subset \gamma'\sigma\mathscr{G}$ ($\mathscr{F}_0$ is a Galois extension of $\mathscr{F}$), the above condition is equivalent to the condition that $\phi \mid \mathscr{F}_0 = \mathrm{id}_{\mathscr{F}_0}$. Clearly this condition is satisfied if and only if $\tau \mid \mathscr{F}_0 = \gamma$.

The sets $\gamma(\mathrm{Iso}(\mathscr{G}/\mathscr{F}_0))$ $(\gamma \in G(\mathscr{F}_0/\mathscr{F}))$ are evidently $\mathbf{\Delta}$-$\mathscr{F}_0$-irreducible and, by the above, disjoint. Their union, by the above, is $\mathrm{Iso}(\mathscr{G}/\mathscr{F})$. Hence they are the $\mathbf{\Delta}$-$\mathscr{F}_0$-components. But $\gamma(\mathrm{Iso}(\mathscr{G}/\mathscr{F}_0))$ has a generic point which is regular over $\mathscr{F}_0$ and so is $\mathbf{\Delta}$-$\mathscr{U}$-irreducible. This proves the lemma.

Because $\gamma(\mathrm{Iso}(\mathscr{G}/\mathscr{F}_0))$ is $\mathbf{\Delta}$-$\mathscr{U}$-irreducible, the $\mathbf{\Delta}$-$\mathscr{G}$-components of $\mathrm{Iso}(\mathscr{G}/\mathscr{F})^{p+q}$ are the $(p+q)$-fold products of the form $\gamma_1(\mathrm{Iso}(\mathscr{G}/\mathscr{F}_0)) \times \cdots$ with $\gamma_1, \ldots \in G(\mathscr{F}_0/\mathscr{F})$.

We return to the proof of the proposition. Fix a $\mathbf{\Delta}$-$\mathscr{G}$-component

$$X = \gamma_1(\mathrm{Iso}(\mathscr{G}/\mathscr{F}_0)) \times \cdots \times \gamma_p(\mathrm{Iso}(\mathscr{G}/\mathscr{F}_0)) \times \mu_1{}'(\quad) \times \cdots \times \mu_q{}'(\quad)$$

of $\mathrm{Iso}(\mathscr{G}/\mathscr{F})^{p+q}$. We shall define a mapping $f_X{}^{\#}$ of $X$ into $G$ such that (for any $i = 1, \ldots, q$)

$$\begin{aligned} &f_X{}^{\#}(\tau_1, \ldots, \tau_q, \mu_1, \ldots, \mu_p) \\ &\quad = f^{\#}_{\mathscr{G}_0}(\tau_1 \mid \mathscr{G}_0, \ldots, \tau_{i-1} \mid \mathscr{G}_0, \tau_{i-1}(\gamma_{i-1}^{-1}\gamma_q)^* \mid \mathscr{G}_0, \ldots) \end{aligned}$$

(cf. formula ($\star$)). The following lemma shows that $f_X{}^{\#}$ is an everywhere defined $\mathbf{\Delta}$-$\mathscr{G}$-mapping.

***Lemma 3*** *Let $\gamma \in G(\mathscr{F}_0/\mathscr{F})$. Then there is an everywhere defined $\mathbf{\Delta}$-$\mathscr{F}$-mapping $\phi_\gamma$: $\mathrm{Iso}(\mathscr{G}/\mathscr{F}) \to \mathrm{Iso}(\mathscr{G}_0/\mathscr{F})$ such that $\phi_\gamma(\tau) = \tau \circ \gamma^* \mid \mathscr{G}_0$.*

*Proof* Since $\mathscr{F}\langle \mathrm{id}_{\mathscr{G}} \rangle = \mathscr{G}$, $\gamma^*\mathscr{G}_0 = \mathscr{F}\langle \gamma^* \mid \mathscr{G}_0 \rangle$, we may define $\phi_\gamma$ by the formula $\phi_\gamma(\mathrm{id}_{\mathscr{G}}) = \gamma^* \mid \mathscr{G}_0$. For any $\tau \in \mathrm{Iso}(\mathscr{G}/\mathscr{F})$,

$$\phi_\gamma(\tau) = \phi_\gamma(\tau(\mathrm{id}_{\mathscr{G}})) = \tau(\tau^{-1}(\phi_\gamma)(\mathrm{id}_{\mathscr{G}})) = \tau(\phi_\gamma(\mathrm{id}_{\mathscr{G}})) = \tau(\gamma^* \mid \mathscr{G}_0) = \tau \circ \gamma^* \mid \mathscr{G}_0 .$$

This proves the lemma.

We shall now consider the first summand in formula ($\star$). Observe that the image of the mapping $\phi_\gamma$ of Lemma 3 lies in $\gamma(\mathrm{Iso}(\mathscr{G}_0/\mathscr{F}_0))$. Moreover,

$$\sigma_i = \gamma_q^{*-1}\tau_i(\gamma_i^{-1}\gamma_q)^* = \gamma_q^{*-1}(\tau_i(\gamma_i^{-1}\gamma_q)^*),$$

so

$$\gamma_q{}^*(h(v \mid \mathscr{N}_0)^{\#}_{\mathscr{G}_0}(\sigma_1, \ldots, \sigma_q)) = \gamma_q(h(v \mid \mathscr{N}_0)^{\#}_{\mathscr{G}_0})(\phi_{\gamma_1{}^{-1}\gamma_q}(\tau_1), \ldots).$$

This shows that $g \in \mathscr{C}^{p+q}(\mathscr{N}/\mathscr{F}, G)$. The proof that $g$ has the desired property (i.e., that $f - \partial g \in K^{p+1,q}$) may be found in H–S along with the remainder of the proof of the proposition.

***Proposition* 17** *$E_2^{p,q}$ is isomorphic to $H^p(\mathcal{N}_0/\mathcal{F}, \mathcal{H}^q(\mathcal{N}/\mathcal{F}, G))$.*

See H–S.

***Theorem*** *Let $\mathcal{N}$ be a normal extension of $\mathcal{F}$ in $\mathcal{F}^\dagger$ and let $\mathcal{N}_0$ be the relative algebraic closure of $\mathcal{F}$ in $\mathcal{N}$. Let $n \geq 1$ and suppose that $\mathcal{H}^k(\mathcal{N}/\mathcal{N}_0, G) = 0$ for $0 < k < n$. Then there is a homomorphism $\mathrm{tg}^{n+1}$ (called the transgression) such that the following sequence is exact:*

$$0 \longrightarrow \mathcal{H}^n(\mathcal{N}_0/\mathcal{F}, G) \xrightarrow{\mathrm{infl}^n} \mathcal{H}^n(\mathcal{N}/\mathcal{F}, G)$$

$$\xrightarrow{\mathrm{res}^n} \mathcal{H}^n(\mathcal{N}/\mathcal{N}_0, G)^{\mathcal{N}_0/\mathcal{F}}$$

$$\xrightarrow{\mathrm{tg}^{n+1}} \mathcal{H}^{n+1}(\mathcal{N}_0/\mathcal{F}, G) \xrightarrow{\mathrm{infl}^{n+1}} \mathcal{H}^n(\mathcal{N}/\mathcal{F}, G).$$

Given the above propositions, the proof is exactly that given in H–S.

***Corollary*** *Let $\mathcal{N}$ be a normal extension of $\mathcal{F}$ in $\mathcal{F}^\dagger$ which contains the algebraic closure $\mathcal{F}_{\mathbf{a}}$ of $\mathcal{F}$, and let G be a commutative $\mathcal{F}$-group (as opposed to $\Delta$-$\mathcal{F}$-group). Then $\mathcal{H}^n(\mathcal{N}/\mathcal{F}, G) \simeq H^n(\mathcal{F}_{\mathbf{a}}/\mathcal{F}, G_{\mathcal{F}_{\mathbf{a}}})$.*

Propositions 6 and 11.

## References


1. Hochschild, G., and Serre, J.-P., Cohomology of group extensions, *Trans. Amer. Math. Soc.* **74** (1953), 110–134.
2. Kolchin, E. R., Galois theory of differential fields, *Amer. J. Math.* **75** (1953), 753–824.
3. Kolchin, E. R., Constrained extensions of differential fields, *Adv. in Math.* **12** (1974), 141–170.
4. Kolchin, E. R., Differential algebraic groups (in preparation).
5. Mac Lane, S., "Homology." Academic Press, New York, 1963.


This work was partially supported by National Science Foundation Grant MPS 75-07569.

AMS (MOS) 1970 subject classification: 12H05

# *The integrability condition of deformations of CR structures*

*Masatake Kuranishi*
*Columbia University*

## Introduction

The parameterization of deformations of a complex manifold by type (1, 0)-valued differential forms of type (0, 1) and the representation of the integrability condition by differential equations on the forms were the keys to open the way to apply the theory of elliptic partial differential equations to deformation theory of complex manifolds (cf. [2]). This note shows that a similar parameterization and representation can be obtained for *CR* structures. The motivation for doing this is the program, suggested by H. Rossi, to regard deformation theory of isolated singularities as deformation theory of *CR* structures. This program was carried almost to its conclusion in [3] (an outline appears in [4]). As for the parameterization and integrability condition, a more intrinsic and simple formulation than appeared in [3] was given by Goldschmidt and Spencer [1] and independently by Akabori. The method developed here is different from theirs as well as from the original one (cf. [3]) and seems to show more clearly the nature of the integrability condition.

## 1. The integrability condition of $CR$ structures

Let $M$ be a $C^\infty$ manifold together with an embedding

$$i\colon M \to N \tag{1}$$

into a complex manifold $N$ of real codimension 1. The complex tangent vector space $\mathbb{C}T_xM$ to $M$ at a point $x$ may be identified by $di$ with a vector subspace of $\mathbb{C}T_{i(x)}N = T''_{i(x)}N + T'_{i(x)}N$, where $T''$ (resp. $T'$) indicates tangent vector bundles of type (0, 1) (resp. of type (1, 0)). Set

$${}^0T''_x(M, i) = \mathbb{C}T_xM \cap T''_{i(x)}N. \tag{2}$$

In other words, we have an exact sequence

$$0 \to {}^0T''_x(M, i) \to \mathbb{C}T_xM \to T'_{i(x)}N, \tag{3}$$

where the last arrow is the restriction to $\mathbb{C}T_xM$ of the projection of $\mathbb{C}T_{i(x)}N$ to $T'_{i(x)}N$ with kernel $T''_{i(x)}N$. The last arrow is surjective, i.e. for any nonzero $\xi \in T'_{i(x)}N$, there is $\eta \in T''_{i(x)}N$ such that $\eta + \xi \in \mathbb{C}T_xM$. Otherwise, $\bar{\xi} + \xi$ and $\sqrt{-1}(\bar{\xi} - \xi)$ generate over $\mathbb{R}$ a vector subspace of dimension 2 in the tangent vector space $T_{i(x)}N$ to $N$ at $i(x)$ lying outside of $T_xM$. This contradicts the assumption that the codimension of the embedding $i$ is one. Therefore it follows by (3) that ${}^0T''_x(M, i)$ is a vector subspace (over $\mathbb{C}$) of $\mathbb{C}T_xM$ of codimension $n$, where $n$ is the complex dimension of $N$. Thus we have a subbundle ${}^0T''(M, i)$ of fiber complex dimension $n - 1$ in the bundle $\mathbb{C}TM$ of complex tangent vectors to $M$, where $M$ is of real dimension $2n - 1$. This suggests that we introduce

***Definition 1*** A vector subbundle $E''$ over $\mathbb{C}$ of $\mathbb{C}TM$, where $M$ is a $C^\infty$ manifold of real dimension $2n - 1$, is called an almost $CR$ structure when the complex fiber dimension of $E''$ is $n - 1$ and $E'' \cap \overline{E''} = \{0\}$.

${}^0T''(M, i)$ is an almost $CR$ structure. However, not all almost $CR$ structures are obtained in this way as the following observation shows. Let $X$, $Y$ be sections of ${}^0T''(M, i)$ over an open subset $U$ of $M$. Still regarding $\mathbb{C}TM$ as a subset of $\mathbb{C}TN$ by $di$, $X$ and $Y$ may be extended to tangent vector fields of type (0, 1) on an open submanifold $U'$ containing $U$, say $X'$ and $Y'$, respectively. Then the bracket $[X', Y']$ is a tangent vector field of type (0, 1) and tangent to $M$ at each point in $U$. Therefore $[X, Y]$ is also a section of ${}^0T''(M, i)$.

***Definition 2*** An almost $CR$ structure $E'' \subseteq \mathbb{C}TM$ is called a $CR$ structure when it satisfies the following condition:

$$\text{If } X, Y \text{ are local sections of } E'', \text{ so is } [X, Y]. \tag{$*$}$$

The condition $(*)$ is often referred to as the integrability condition on the almost $CR$ structure $E''$. We sometimes refer to a $CR$ structure as an almost $CR$ structure satisfying the integrability condition or an integrable almost $CR$ structure. The majority of $CR$ structures are believed to be obtained locally by embedding into complex manifolds. However, there are examples of $CR$ structures which are not of this type (cf. Nirenberg [5]).

*Remark* We do not need the full assumption of $N$ being a complex manifold. What we need is the assumptions that we have the decomposition $\mathbb{C}T_{i(x)}N = T''_{i(x)}N + T'_{i(x)}N$ assigned for all $x$ in $M$ and that on a neighborhood of each $i(x)$ the decomposition can be extended to a decomposition induced by a complex structure. This will be the case if $i(M)$ disconnects $N$ into two components, one of which has a complex structure which extends locally to $M$ but cannot extend beyond $M$ globally.

We are going to rewrite the integrability condition in terms of differential forms instead of vector fields. We start by fixing notations. For a vector space $V$ over $\mathbb{C}$, $\Lambda^l V^*$ denotes the vector space of $l$-alternate multilinear (over $\mathbb{C}$) functions on $V$. Thus $V^* = \Lambda^1 V^*, \mathbb{C} = \Lambda^0 V^*$. $\Lambda V^* = \sum_l \Lambda^l V^*$ is an algebra, the multiplication being by the exterior product denoted by $\wedge$. Let $W$ be a vector subspace of $V$. We denote by $W^\perp$ the subspace of all $\phi \in V^*$ which vanish on $W$. When we wish to emphasize $V$, we write $(W^\perp, V)$ instead of $W^\perp$. The restriction map $r\colon V^* \to W^*$ induces the isomorphism

$$V^*/W^\perp \to W^*. \tag{4}$$

More generally, the restriction map $r\colon \Lambda^l V^* \to \Lambda^l W^*$ induces the isomorphism

$$\Lambda^l V^*/(W^\perp \wedge \Lambda^{l-1} V^*) \to \Lambda^l W^*. \tag{5}$$

For a vector bundle $F$ over $M$, $C^\infty(M, F)$ denotes the vector space of $C^\infty$ sections of $F$ over $M$. If $\theta$ is a differential form of degree $l$ on $M$, the evaluation of $\theta$ at $X_1, \ldots, X_l$ in $\mathbb{C}TM$ will be denoted either by $\theta(X_1, \ldots, X_l)$ or by $\langle \theta, X_1 \wedge \cdots \wedge X_l \rangle$.

Let $E''$ be an almost $CR$ structure on $M$. We define a differential operator

$$\bar\partial_{E''}\colon C^\infty(M, \mathbb{C}) \to C^\infty(M, (E'')^*) \tag{6}$$

as the composition of the exterior derivative $d\colon C^\infty(M, \mathbb{C}) \to C^\infty(M, \mathbb{C}T^*M)$ and the linear map $r\colon C^\infty(M, \mathbb{C}T^*M) \to C^\infty(M, (E'')^*)$ induced by the restriction map $\mathbb{C}T^*M \to (E'')^* = \mathbb{C}T^*M/(E'')^{\perp}$. We may try the same construction for differential forms of order 1. Thus we consider the diagram

$$\begin{array}{ccc} C^\infty(M, \mathbb{C}T^*M) & \longrightarrow & C^\infty(M, \Lambda^2\mathbb{C}T^*M) \\ \downarrow & & \downarrow \\ C^\infty(M, (E'')^*) & \dashrightarrow & C^\infty(M, \Lambda^2(E'')^*) \end{array} \tag{7}$$

where the horizontal undotted arrow is the exterior derivative and the two vertical arrows are induced by the restriction maps.

***Proposition 1*** *Let $E''$ be an almost CR structure on $M$. Then $E''$ is integrable if and only if there exists a dotted arrow in* (7) *which makes the diagram commutative.*

*Proof* It is easy to see that the existence of the arrow as above is equivalent to the following condition: If $\theta$ is a differential form of order 1 on an open $U$ of $M$ such that $\theta(L) = 0$ for any $L \in C^\infty(U, E'')$, then $d\theta(L, L') = 0$ for all $L, L' \in C^\infty(U, E'')$. Then clearly it is equivalent to the integrability condition because of the formula

$$2\, d\theta(L, L') = L \cdot \theta(L') - L' \cdot \theta(L) + \theta([L, L']), \qquad \text{QED} \tag{8}$$

This case of differential forms of degree 1 we considered is the critical one. In fact it follows easily by (5) and Proposition 1 that for a $CR$ structure $E''$ we have, for all integers $r \geqq 0$, a commutative diagram

$$\begin{array}{ccc} C^\infty(M, \Lambda^r\mathbb{C}T^*M) & \longrightarrow & C^\infty(M, \Lambda^{r+1}\mathbb{C}T^*M) \\ \downarrow & & \downarrow \\ C^\infty(M, \Lambda^r(E'')^*) & \longrightarrow & C^\infty(M, \Lambda^{r+1}(E'')^*) \end{array} \tag{9}$$

where the horizontal arrow on the top is the exterior derivative and the vertical arrows are induced by the restrictions. We denote by

$$\bar\partial_{E''} \tag{10}$$

the horizontal arrow at the bottom. It is clear by the construction that

$$\bar\partial_{E''} \circ \bar\partial_{E''} = 0. \tag{11}$$

***Definition 3*** The complex $\{C^\infty(M, \Lambda^r(E'')^*), \bar\partial_{E''}\}$ is called the complex induced by the $CR$ structure $E''$ on $M$.

Thus the existence of the induced complex is equivalent to the integrability condition on $E''$. For an almost $CR$ structure $E''$ there is no differential operator $\bar\partial_{E''}$ (except in the case $r = 0$). However, we may introduce it artificially by picking a cross section of the vertical arrow on the right in (9). We do this by picking a projection

$$j\colon \mathbb{C}TM \to E'' \tag{12}$$

($j(L) = L$ for all $L \in E''$). $j$ induces

$$j^*\colon C^\infty(M, \Lambda^r(E'')^*) \to C^\infty(M, \Lambda^r\mathbb{C}T^*M). \tag{13}$$

We denote by

$$\bar\partial_{E'',j}\colon C^\infty(M, \Lambda^r(E'')^*) \to C^\infty(M, \Lambda^{r+1}(E'')^*) \tag{14}$$

the composition of $j^*$, the exterior derivative, and the restriction map. It is clear by the construction that $\bar\partial_{E'',j} = \bar\partial_{E''}$ when $E''$ is a $CR$ structure. For a $C^\infty$ function $f$, $\bar\partial_{E''} f = \bar\partial_{E'',j} f$. Elements in $C^\infty(M, \Lambda^r(E'')^*)$ are referred to as differential forms of type $(0, r)_{E''}$. If $V$ is a vector bundle over $M$, we may also consider $V$-valued differential forms of type $(0, r)_{E''}$.

***Proposition 2*** *Let $E''$ be an almost CR structure and $j$ the projection of $\mathbb{C}TM$ onto $E''$ picked in* (12). *Then the following three statements are equivalent:*

(1) *$E''$ is a CR structure,*
(2) *$\{C^\infty(M, \Lambda^r(E'')^*), \bar\partial_{E'',j}\}$ forms a complex,*
(3) *for any $C^\infty$ function $f$ on $M$, $\bar\partial_{E'',j} \circ \bar\partial_{E''} f = 0$.*

*Proof* It is obvious that (1) implies (2) and (2) implies (3). It remains to show that (3) implies (1). Assume that (3) holds. Let $X$, $Y$ be sections of $E''$. To show that $[X, Y]$ is a section of $E''$, it is enough to show that $j([X, Y]) = [X, Y]$, i.e. $\langle df, [X, Y]\rangle = \langle j^*\, df, [X, Y]\rangle$ for all $C^\infty$ functions $f$ on $M$. Applying (8) to $\theta = j^*\, df$, we find that

$$\begin{aligned}\langle j^*\, df, [X, Y]\rangle &= 2\langle d(j^*\, df), X \wedge Y\rangle - X\, df(Y) + Y\, df(X)\\ &= -X\, df(Y) + Y\, df(X)\end{aligned}$$

because $\langle d(j^*\, df), X \wedge Y\rangle = \langle \bar\partial_{E'',j} \circ \bar\partial_{E''} f, X \wedge Y\rangle$. On the other hand, $\langle df, [X, Y]\rangle$ is also equal to $-X\, df(Y) + Y\, df(X)$, QED

***Proposition 3*** *Let $E''$ be an almost CR structure and $j$ a projection with kernel $K$ of $\mathbb{C}TM$ onto $E''$. Then there is a unique $K$-valued differential form of type $(0, 2)_{E''}$, say $P_{E'',j}$, such that for all $C^\infty$ functions $f$ on $M$*

$$\langle \bar\partial_{E'',j} \circ \bar\partial_{E''} f,\ X \wedge Y\rangle = \langle P_{E'',j}, X \wedge Y\rangle f \tag{15}$$

*for all $X$, $Y \in E''$ located at the same point.*

*Remark* $P_{E'',j}$ being a $K$-valued differential form of type $(0, 2)_{E''}$, $\langle P_{E'',j}, X \wedge Y \rangle$ is an element in $K \subseteq \mathbb{C}TM$ and hence it operates on $f$ as a partial differential operator.

*Proof* We show first that the map $\tau: f \mapsto \langle \bar{\partial}_{E'',j} \circ \bar{\partial}_{E''} f, X \wedge Y \rangle$ is induced by the operation of a complex tangent vector. Since $\tau$ is linear, it will be enough to show that $\tau(fg) = f(x_0)\tau(g) + g(x_0)\tau(f)$, where $x_0$ is the point where $X$ and $Y$ are located. The formula holds because

$$\begin{aligned}\bar{\partial}_{E'',j} \circ \bar{\partial}_{E''}(fg) &= \bar{\partial}_{E'',j} \circ (f\bar{\partial}_{E''} g + g\,\bar{\partial}_{E''} f) \\ &= f\,\bar{\partial}_{E'',j} \circ \bar{\partial}_{E''} g + g\,\bar{\partial}_{E'',j} \circ \bar{\partial}_{E''} f + \bar{\partial}_{E''} f \wedge \bar{\partial}_{E''} g + \bar{\partial}_{E''} g \wedge \bar{\partial}_{E''} f \\ &= f\,\bar{\partial}_{E'',j} \circ \bar{\partial}_{E''} g + g\,\bar{\partial}_{E'',j} \circ \bar{\partial}_{E''} f.\end{aligned}$$

Thus there is a unique $\mathbb{C}TM$-valued differential form $P_{E'',j}$ which satisfies (15). It remains to show that this $P_{E'',j}$ is actually $K$ valued. For this, it is enough to show the following: For any $C^\infty$ function $f$ on $M$ such that $Wf = 0$ for all $W$ in $K$ located at $x_0$, $\langle P_{E'',j}, X \wedge Y \rangle f = 0$. Since $\bar{\partial}_{E'',j} \circ \bar{\partial}_{E''} f = d \circ j^* \circ df$, by extending $X$ and $Y$ to a section in $E''$ we see that

$$\begin{aligned}-2\langle P_{E'',j}, X \wedge Y \rangle f &= 2\langle d \circ j^* \circ df, X \wedge Y \rangle \\ &= X\langle j^* \circ df, Y \rangle - Y\langle j^* \circ df, X \rangle + \langle j^* \circ df, [X, Y] \rangle \\ &= X\langle df, Y \rangle - Y\langle df, X \rangle + \langle df, j[X, Y] \rangle.\end{aligned}$$

The assumption on $f$ implies that $\langle df, j[X, Y] \rangle = \langle df, [X, Y] \rangle$ at $x_0$. Hence we see that, evaluating at $x_0$,

$$\langle P_{E'',j}, X \wedge Y \rangle = \langle d \circ j^* \circ df, [X, Y] \rangle = \langle d \circ df, [X, Y] \rangle = 0, \qquad \text{QED}$$

**Corollary** *Under the same assumption as in Proposition* 3, $E''$ *is integrable if and only if* $P_{E'',j}$ *vanishes.*

## 2. The integrability condition of deformations

We fix once and for all a $CR$ structure, denoted by ${}^0T''$, on $M$ induced by an embedding $i: M \to N$ into a complex manifold $N$ of codimension 1 together with a subbundle $F \subset TM$ of fiber real dimension 1 such that $({}^0T' = \overline{{}^0T''})$

$$\mathbb{C}TM = {}^0T'' + {}^0T' + \mathbb{C}F. \tag{16}$$

It is easy to see that we can always pick such $F$ when ${}^0T''$ is given. We identify $\mathbb{C}TM \subseteq \mathbb{C}TN|iM$ via $di$. Denote by

$$\pi'': \mathbb{C}TM \to {}^0T'' \tag{17}$$

the projection with respect to the decomposition (16). We regard

$$C^\infty(M, \Lambda^r((^0T'')^*)) \subset C^\infty(M, \Lambda^r(\mathbb{C}T^*M)) \tag{18}$$

via $\pi''$. Following the standard notation, we write $\bar{\partial}_b$ instead of $\bar{\partial}_{^0T''}$, and say type $(0, r)_b$ instead of type $(0, r)_{^0T''}$. In particular, a differential form of type $(0, 1)_b$ is a differential form which vanishes when evaluated at elements in $^0T' + \mathbb{C}F$. In this section we rewrite the material developed in Section 1 referring everything to $(^0T'', F)$.

***Definition 4*** An almost $CR$ structure $E''$ on $M$ is said to be of finite distance to $^0T''$ when $\pi''|E''$ is injective.

When this is the case, $E''$ can be regarded as the graph of a linear map

$$\phi_1: {}^0T'' \to {}^0T' + \mathbb{C}F, \tag{19}$$

i.e.

$$E'' = \{X - \phi_1(X) : X \in {}^0T''\}. \tag{20}$$

Thus we can parameterize almost $CR$ structures of finite distance to $^0T''$ by $(^0T' + \mathbb{C}F)$-valued differential forms of type $(0, 1)_b$. However, it will be nicer if we employ a bundle more natural than $^0T' + \mathbb{C}F$. To do this, we note that the last arrow, say $\rho'$, in the exact sequence (3) is surjective and $^\circ T' + \mathbb{C}F$ is complimentary to $^0T'' = {}^0T''(M, i)$. Therefore $\rho'$ induces an isomorphism of $^0T' + \mathbb{C}F$ onto $T'N|M$. Let

$$\tau: T'N|M \to {}^0T' + \mathbb{C}F \tag{21}$$

be the inverse mapping. Then there is a linear mapping

$$\phi: {}^0T'' \to T'N|M \tag{22}$$

such that

$$\phi_1 = \tau \circ \phi. \tag{23}$$

***Proposition 4*** *Let $E''$ be an almost $CR$ structure on $M$ of finite distance to $^0T''$. Then there is a unique $T'N|M$-valued differential form of type $(0, 1)_b$ such that*

$$E'' = \{X - \tau \circ \phi(X) : X \in {}^0T''\}. \tag{24}$$

*Conversely, for any sufficiently small $T'N|M$-valued differential form of type $(0, 1)_b$, formula (24) defines an almost $CR$ structure $E''$ of finite distance to $^0T''$.*

***Definition 5*** $E''$ defined by (24) in terms of $\phi$ will be called the almost $CR$ structure determined by $\phi$ relative to ${}^0T''$ and will be denoted by ${}^\phi T''$.

As for the projection $j$ in (12) we assume that $\phi$ is sufficiently small so that $\mathbb{C}TM = {}^\phi T'' + {}^0T' + \mathbb{C}F$ and we choose

$$j^\phi : \mathbb{C}TM \to {}^\phi T'' \tag{25}$$

with kernel ${}^0T' + \mathbb{C}F$. We transform the sequence of differential operators $\bar{\partial}_{{}^\phi T'', j^\phi}$ by means of $1 - \tau \circ \phi : {}^0T'' \to {}^\phi T''$. More explicitly, we introduce a sequence of differential operators $\bar{\partial}_b{}^\phi$ as the horizontal arrows at the bottom of the commutative diagrams:

$$\begin{array}{ccc} C^\infty(M, \Lambda^r(({}^\phi T'')^*)) & \longrightarrow & C^\infty(M, \Lambda^{r+1}(({}^\phi T'')^*)) \\ \big\downarrow & & \big\downarrow \\ C^\infty(M, \Lambda^r(({}^0 T'')^*)) & \longrightarrow & C^\infty(M, \Lambda^{r+1}(({}^0 T'')^*)) \end{array} \tag{26}$$

where the horizontal arrow on the top is $\bar{\partial}_{{}^\phi T'', j^\phi}$ and the vertical arrows are induced by $1 - \tau \circ \phi$. We introduce a $T'N|M$-valued differential form $P(\phi)$ of type $(0, 2)_b$ by

$$-\tau \circ \langle P(\phi), X \wedge Y \rangle = \langle P_{{}^\phi T'', j^\phi}, (1 - \tau\phi)X \wedge (1 - \tau\phi)Y \rangle. \tag{27}$$

By (15) we see easily that

$$(\tau \circ \langle P(\phi), X \wedge Y \rangle) f = -\langle \bar{\partial}_b{}^\phi \, \bar{\partial}_b{}^\phi f, X \wedge Y \rangle. \tag{28}$$

Proposition 2 and the Corollary to Proposition 3 are now rewritten as

***Proposition 5*** *Let $\phi$ be a sufficiently small $T'N|M$-valued differential form of type $(0, 1)_b$. Then the following statements are equivalent:*

(1) *${}^\phi T''$ is a $CR$ structure,*
(2) *$\{C^\infty(M, \Lambda^r({}^0T'')^*), \bar{\partial}_b{}^\phi\}$ form a complex,*
(3) *for any $C^\infty$ function $f$, $\bar{\partial}_b{}^\phi \, \bar{\partial}_b{}^\phi f = 0$,*
(4) *$P(\phi) = 0$.*

It will be interesting to see chart expressions of $\bar{\partial}_b{}^\phi$ and $P(\phi)$. To find them we use a chart $z = (z^1, \ldots, z^n)$ of the ambient complex manifold $N$. Let $\mathcal{U}$ be the domain of the chart and let $iM \cap \mathcal{U} = U$ be defined by

$$h = 0, \tag{29}$$

where $h$ is a $C^\infty$ real-valued function on $\mathcal{U}$ with nonvanishing gradient. We keep a choice of $h$ fixed throughout this paper. Since $F \subseteq TM$, we can pick a vector field $P'$ of type $(1, 0)$, nonvanishing everywhere on $U$, such that

$$P' - P'' \in iF, \qquad P'' = \overline{P'} \tag{30}$$

(provided we may shrink $\mathscr{U}$ if necessary). Since $P' - P''$ is tangent to $M$, $\langle dh, P'\rangle = \langle dh, P''\rangle$. Noting that $P'' = \overline{P'}$ and $h$ is real valued, it follows that $\langle dh, P'\rangle$ is real. If $\langle dh, P'\rangle = 0$ at a point, $dh$ vanishes on $\mathbb{C}TM$ as well as at one nonzero vector outside of $\mathbb{C}TM$. Since the codimension of $M$ in $N$ is 1, it then follows that the gradient of $h$ at the point is zero. Therefore we can normalize the choice of $P'$ by imposing the condition

$$\langle dh, P'\rangle = \langle dh, P''\rangle = 1 \tag{31}$$

on $\mathscr{U}$. We extend $P'$ to a vector field of type (0, 1) on $\mathscr{U}$ satisfying (31). We set

$$h_j = \partial h/\partial z^j, \qquad h_{\bar{j}} = \overline{h_j} = \partial h/\partial \bar{z}^j, \tag{32}$$

$$P' = p^j\, \partial/\partial z^j, \qquad P'' = p^{\bar{j}}\, \partial/\partial \bar{z}^j, \qquad p^{\bar{j}} = \overline{p^j}. \tag{33}$$

By (31),

$$p^j h_j = p^{\bar{j}} h_{\bar{j}} = 1. \tag{34}$$

We set

$$Z_j = \partial/\partial z^j - h_j P', \qquad Z_{\bar{j}} = \overline{Z_j} = \partial/\partial \bar{z}^j - h_{\bar{j}} P'' \tag{35}$$

$(j = 1, \ldots, n)$. Since $\langle dh, Z_j\rangle = 0$ by (32) and (34), it is clear that $Z_j$ (resp. $Z_{\bar{j}}$), when restricted to $U$, generate ${}^0T'|U$ (resp. ${}^0T''|U$). They satisfy the relation

$$p^j Z_j = p^{\bar{j}} Z_{\bar{j}} = 0. \tag{36}$$

Similarly, we introduce a generator of type $(0, 1)_b$ differential forms on $U$ as

$$Z^{\bar{k}} = i^*\, d\bar{z}^k - p^{\bar{k}} i^*\, d''h = i^*\, d\bar{z}^k + p^{\bar{k}} i^*\, d'h, \tag{37}$$

where $d''h$ (resp. $d'h$) is the type (0, 1)-part (resp. type (1, 0)-part) of $dh$. Clearly $\langle Z^{\bar{k}}, X\rangle = 0$ for all $X$ in ${}^0T'|U$ and $\langle Z^{\bar{k}}, P'' - P'\rangle = 0$ by (33) and (34). Therefore $Z^{\bar{k}}$ is consistent with the identification (18). By (34) and (37) we see easily that

$$h_{\bar{k}} Z^{\bar{k}} = 0. \tag{38}$$

Therefore any $\psi \in C^\infty(U, \Lambda^r(({}^0T'')^*))$ has the unique expression

$$\psi = \psi_{\bar{k}_1, \ldots, \bar{k}_r} Z^{\bar{k}_1} \wedge \cdots \wedge Z^{\bar{k}_r}, \qquad h_l\, \psi^{l\bar{k}_1 \cdots \bar{k}_{r-1}} = 0, \tag{39}$$

$$\psi_{\bar{k}_1, \ldots, \bar{k}_r} \text{ skew-symmetric in } (k_1, \ldots, k_r).$$

The projection $\rho'$ to type (0, 1) vectors maps $P' - P''$ to $P'$. Since $P' - P'' \in \mathbb{C}F$, it follows by the definition of $\tau$ in (21) that

$$\tau P' = P' - P''. \tag{40}$$

Since $Z_j \in {}^0T'$, $\tau Z_j = Z_j$. Hence by (34) we see that $\tau(\partial/\partial z^j) = \tau(Z_j + h_j P') = \tau Z_j + h_j \tau P' = Z_j + h_j(P' - P'') = \partial/\partial z^j - h_j P''$. Therefore

$$\tau(\partial/\partial z^j) = \partial/\partial z^j - h_j P''. \tag{41}$$

For simplicity of notation we set

$$\tau(\partial/\partial z^j) = \partial^\tau/\partial z^j. \tag{42}$$

Let $f$ be a differentiable function on $\mathscr{U}$. Then

$$\begin{aligned} df &= (\partial f/\partial \overline{z^k} - h_{\bar{k}} P'' f)\, d\overline{z^k} + (\partial f/\partial z^k - h_k P'' f)\, dz^k + (P'' f)\, dh \\ &= (\partial f/\partial z^k - h_{\bar{k}} P'' f)(d\overline{z^k} - p^{\bar{k}}\, d''h) + (\partial f/\partial z^k - h_k P'' f)\, dz^k + (P'' f)\, dh, \end{aligned}$$

i.e.

$$df = (Z_{\bar{k}} f) Z^{\bar{k}} + (\partial^\tau f/\partial z^k)\, dz^k + (P'' f)\, dh. \tag{43}$$

Therefore, for a differentiable function $u$ on $U$,

$$du = (Z_{\bar{k}} u) Z^{\bar{k}} + (\partial^\tau u/\partial z^l) i^*\, dz^l. \tag{44}$$

For $X \in {}^0T''$ we see that $X = (X - \tau\phi(X)) + \tau\phi(X)$, $X - \tau\phi(X) \in {}^\phi T''$, and that $\tau\phi(X) \in {}^0T' + \mathbb{C}F$. Hence by (25),

$$j^\phi = j - \tau \circ \phi \circ j. \tag{45}$$

Write

$$\phi = \phi_{\bar{k}}{}^l Z^{\bar{k}}\, \partial/\partial z^l, \qquad p^{\bar{k}} \phi_{\bar{k}}{}^l = 0. \tag{46}$$

Then for $L \in \mathbb{C}TM$ we see by (44), (45), (41), and the fact that $Z^{\bar{k}}$ annihilates vectors in ${}^0T' + \mathbb{C}F$, that

$$\langle du, j^\phi L\rangle = (Z_{\bar{k}} u)\langle Z^{\bar{k}}, L\rangle - (\partial^\tau u/\partial z^l)\phi_{\bar{k}}{}^l \langle Z^{\bar{k}}, L\rangle = (Z_{\bar{k}}{}^\phi u)\langle Z^{\bar{k}}, L\rangle$$

where

$$Z_{\bar{k}}{}^\phi = Z_{\bar{k}} - \phi_{\bar{k}}{}^l\, \partial^\tau/\partial z^l. \tag{47}$$

Since $Z^{\bar{k}} \circ (1 - \tau \circ \phi) = Z^{\bar{k}}$, it follows then that

$$\bar{\partial}_b{}^\phi u = (Z_{\bar{k}}{}^\phi u) Z^{\bar{k}}. \tag{48}$$

***Lemma 1*** $\bar{\partial}_b{}^\phi Z^l = \phi^j \wedge \bar{\partial}_b{}^\phi(h_j p^l)$, $\phi^j = \phi_{\bar{l}}{}^j Z^{\bar{l}}$.

*Proof* $Z^l \circ (1 - \tau\phi) = Z^l$. Hence

$$\langle \bar{\partial}_b{}^\phi Z^l, X \wedge Y\rangle = \langle dZ^l, (1 - \tau\phi)X \wedge (1 - \tau\phi)Y\rangle$$

for $X, Y \in {}^0T''$. Since $dZ^l = d(p^l h_k) \wedge di^* z^k$ by the second equality in (37), it follows that $\bar{\partial}_b{}^\phi Z^l = \theta_k \wedge \psi^k$, where $\theta_k = d(p^l h_k) \circ (1 - \tau\phi)$ and $\psi^k = di^* z^k \circ (1 - \tau\phi)$. Clearly, $\theta = \bar{\partial}_b{}^\phi(p^l h_k)$ and $\psi^k = -\phi^k$, QED

Since $\bar{\partial}_b{}^\phi(\theta \wedge \psi) = (\bar{\partial}_b{}^\phi\theta) \wedge \psi + (-1)^p\theta \wedge \bar{\partial}_b{}^\phi\psi$, where $\theta$ is of type $(0, p)_b$, we see immediately by Lemma 1 that

***Proposition 6*** *If*

$$\mu = \mu_{\bar{k}_1 \cdots \bar{k}_r} Z^{\bar{k}_1} \wedge \cdots \wedge Z^{\bar{k}_r} \in C^\infty(M, \Lambda^r({}^0T'')^*)$$

*with* $\mu_{\bar{k}_1 \cdots \bar{k}_{s-1} l \bar{k}_{s+1} \cdots \bar{k}_r} p^l = 0$ *for* $s = 1, \ldots, r$, *then*

$$\bar{\partial}_b{}^\phi \mu = (\bar{\partial}_b{}^\phi \mu_{\bar{k}_1 \cdots \bar{k}_r}) \wedge Z^{\bar{k}_1} \wedge \cdots \wedge Z^{\bar{k}_r}$$
$$+ \mu_{\bar{k}_1 \cdots \bar{k}_r}\left( \sum_{s=1}^{r} (-1)^{s+1} h_l \phi^l \wedge (\bar{\partial}_b{}^\phi p^{\bar{k}_s}) \wedge Z^{\bar{k}_1} \wedge \cdots \wedge Z^{\bar{k}_{s-1}} \right.$$
$$\left. \wedge Z^{\bar{k}_{s+1}} \wedge \cdots \wedge Z^{\bar{k}_r} \right).$$

In order to find a formula for $P(\phi)$ we need a number of commutator relations. Since they are obtained by direct calculations we list them without proof:

$$h_l \partial p^{\bar{l}}/\partial \bar{Z}^k = -P'' h_{\bar{k}}, \tag{49}$$

$$[Z_{\bar{k}}, Z_{\bar{l}}] = (h_{\bar{k}} \partial p^{\bar{i}}/\partial \bar{z}^l - h_{\bar{l}} \partial p^{\bar{i}}/\partial \bar{z}^k) Z_{\bar{i}}, \tag{50}$$

$$[Z_{\bar{k}}, Z_{\bar{l}}] \otimes (Z^{\bar{k}} \wedge Z^{\bar{l}}) = 0, \tag{51}$$

$$h_{\bar{k}} \partial p^{\bar{k}}/\partial z^j = -P'' h_j, \tag{52}$$

$$[\partial^\tau/\partial z^l, Z_{\bar{k}}] = (h_l \partial p^{\bar{i}}/\partial \bar{z}^k - h_{\bar{k}} \partial p^{\bar{i}}/\partial z^l) Z_{\bar{i}}, \tag{53}$$

$$[\partial^\tau/\partial z^l, \partial^\tau/\partial z^k] = (h_l \partial p^{\bar{i}}/\partial z^k - h_k \partial p^{\bar{i}}/\partial z^l) Z_{\bar{i}}. \tag{54}$$

***Lemma 2***

$$[Z_{\bar{k}}{}^\phi, Z_{\bar{l}}{}^\phi] \otimes (Z^{\bar{k}} \wedge Z^{\bar{l}}) = 2Z_{\bar{i}}{}^\phi \otimes (\bar{\partial}_b{}^\phi p^{\bar{i}} \wedge h_l \phi^l) - 2(\partial^\tau/\partial z^l) \otimes \bar{\partial}_b{}^\phi \phi^l.$$

*Proof* We see by (47) that

$$A = [Z_{\bar{k}}{}^\phi, Z_{\bar{l}}{}^\phi] \otimes (Z^{\bar{k}} \wedge Z^{\bar{l}})$$
$$= [Z_{\bar{k}}, Z_{\bar{l}}] \otimes (Z^{\bar{k}} \wedge Z^{\bar{l}}) - 2(Z_{\bar{k}}{}^\phi \phi_{\bar{l}}{}^i)(\partial^\tau/\partial z^i) \otimes (Z^{\bar{k}} \wedge Z^{\bar{l}})$$
$$+ 2[\partial^\tau/\partial z^i, Z_{\bar{k}}] \otimes (Z^{\bar{k}} \wedge \phi^i) + [\partial^\tau/\partial z^k, \partial^\tau/\partial z^l] \otimes (\phi^k \wedge \phi^l).$$

Hence by (51), (53), (54), and (38),

$$A = -2(\partial^\tau/\partial z^i) \otimes ((Z_{\bar{k}}{}^\phi \phi_{\bar{l}}{}^i) Z^{\bar{k}} \wedge Z^{\bar{l}}) + 2Z_{\bar{i}} \otimes (\partial p^{\bar{i}}/\partial z^k - \phi_{\bar{k}}{}^l\, \partial p^{\bar{i}}/\partial z^l) Z^{\bar{k}} \wedge h_j \phi^j.$$

Since $\phi_{\bar{k}}{}^{l}(\partial p^{\bar{j}}/\partial z^{l})Z^{\bar{k}} \wedge h_j \phi^j = \phi_{\bar{k}}{}^{l}(\partial^{\tau} p^{\bar{j}}/\partial z^{l})Z^{\bar{k}} \wedge h_j \phi^j$, it follows by (47) and Proposition 1 that

$$A = -2(\partial^{\tau}/\partial z^{i}) \otimes (\bar{\partial}_b{}^{\phi}\phi^{i} + \phi_{l}{}^{i}\,\bar{\partial}_b{}^{\phi} p^{l} \wedge h_j\phi^j) + 2Z_{\bar{l}} \otimes (Z_{\bar{k}}{}^{\phi} p^{\bar{l}}) \wedge Z^{k} \wedge h_j \phi^j$$

$$= -2(\partial^{\tau}/\partial z^{l}) \otimes \bar{\partial}_b{}^{\phi}\phi^{l} + 2Z_{\bar{l}}{}^{\phi} \otimes (\bar{\partial}_b{}^{\phi} p^{\bar{l}} \wedge h_j \phi^j), \qquad \text{QED}$$

***Proposition* 7** *For a* $C^{\infty}$ *function* $u$ *on* $M$, $\bar{\partial}_b{}^{\phi}\,\bar{\partial}_b{}^{\phi} u = -(\partial^{\tau} u/\partial z^{l})\,\bar{\partial}_b{}^{\phi}\phi^{l}$.

*Proof* Since $\bar{\partial}_b{}^{\phi} u = (Z_{\bar{l}}{}^{\phi} u)Z^{\bar{l}}$ and $p^{\bar{l}} Z_{\bar{l}}{}^{\phi} = 0$, it follows by Proposition 6 that

$$\bar{\partial}_b{}^{\phi}\,\bar{\partial}_b{}^{\phi} u = (Z_{\bar{k}}{}^{\phi} Z_{\bar{l}}{}^{\phi} u)Z^{\bar{k}} \wedge Z^{\bar{l}} + (Z_{\bar{l}}{}^{\phi} u)h_k \phi^k \wedge \bar{\partial}_b{}^{\phi} p^{\bar{l}}$$

$$= \tfrac{1}{2}([Z_{\bar{k}}{}^{\phi}, Z_{\bar{l}}{}^{\phi}]u)Z^{\bar{k}} \wedge Z^{\bar{l}} + (Z_{\bar{l}}{}^{\phi} u)h_k \phi^k \wedge \bar{\partial}_b{}^{\phi} p^{\bar{l}}.$$

Then our assertion follows by Lemma 2, QED

***Proposition* 8** *Let* $\phi$ *be a* $T'N|M$*-valued differential form of type* $(0, 1)_b$. *Write* $\phi = \phi^{l}\,\partial/\partial z^{l}$, *where* $\phi^{l}$ *is a scalar-valued differential form of type* $(0, 1)_b$. *Then*

$$P(\phi) = (\bar{\partial}_b{}^{\phi}\phi^{l})\,\partial/\partial z^{l}.$$

*Proof* This is an immediate corollary of Proposition 7, (28), and (42).

## References


1. H. Goldschmidt and D. C. Spencer, Submanifolds and overdetermined differential operators, *Acta Math.* **136** (1976), 103–239.
2. K. Kodaira, L. Nirenberg, and D. C. Spencer, On the existence of deformations of complex analytic structures, *Ann. of Math.* **68** (1958), 450–459.
3. M. Kuranishi, Deformations of isolated singularities and $\bar{\partial}_b$, preprint, Columbia Univ., New York.
4. M. Kuranishi, Application of $\bar{\partial}_b$ to deformation of isolated singularities, *Proc. Symp. Pure Math.* **30** (1976), 97–106.
5. L. Nirenberg, On a question of Hans Lewy, *Uspekhi Mat. Nauk* **29** (1974), 241–251.


This work was supported in part by National Science Foundation Grant GP 8988.

AMS (MOS) 1970 subject classifications: 14B05, 14D99, 32G05, 35N99, 58G99

# *Noetherian rings with many derivations*

*Hideyuki Matsumura*

*Nagoya University*

The purpose of this paper is to study derivations of noetherian rings containing a field. Generally speaking, if such a ring admits "sufficiently many" derivations into itself, then it has good properties, such as being "universally catenary" or "excellent." We obtained a theorem of this type already in [7, Th. 2.7]. Here we try to generalize the results of that paper. Our main result is Theorem 9, which shows that the class of regular rings containing the rational numbers in which the strong Jacobian condition (SJ) holds is closed under polynomial extensions as well as formal power series extensions. We also discuss the "universal finite module of differentials" of Bingener [1] and Scheja and Storch [14] in connection with our Jacobian conditions (Theorem 11). Lastly we discuss the case of characteristic $p$. In this case we do not have such a complete result as Theorem 9 yet, but we prove a few theorems, which can be applied to some interesting cases. We obtain, for instance, a new proof of Valabrega's theorem that $k[X_1, \ldots, X_n][[Y_1, \ldots, Y_m]]$ is excellent for any field $k$.

We have tried to make this paper self-contained, so that the knowledge of [7] is not necessary. The terminology and the results in our book [6] will be freely used. Let $A$ be a ring (always assumed to be commutative), and $M$ an $A$-module. We denote the set of derivations from $A$ to $M$ by $\mathrm{Der}(A, M)$, which is an $A$-module. We write $\mathrm{Der}(A)$ for $\mathrm{Der}(A, A)$. If $k$ is a subring of $A$, the set of derivations over $k$ is denoted by $\mathrm{Der}_k(A, M)$.

If $P$ is a prime ideal of $A$, then any derivation $D \in \mathrm{Der}(A)$ extends uniquely to a derivation of $A_P$. Similarly, if $I$ is an ideal of $A$ and $A^*$ is the $I$-adic completion of $A$, then $D$ extends uniquely to a derivation of $A^*$. These extensions will be denoted usually by the same letter $D$ if there is no confusion.

If $A$ is an integral domain with quotient field $K$, and if $M$ is a finitely generated $A$-module, then rank $M$ will mean $\mathrm{rank}_K(M \otimes K)$, which is equal to the maximal number of linearly independent elements in $M$.

Let $P$ be an ideal generated by $f_1, \ldots, f_n$, and let $D_1, \ldots, D_r \in \mathrm{Der}(A)$. Let $Q$ be a prime ideal containing $P$. We denote the matrix $(D_i f_j \bmod Q)$ by $J(f_1, \ldots, f_n; D_1, \ldots, D_r)(Q)$. The rank of this matrix does not depend on the choice of the system of generators of $P$, so we denote it by rank $J(P; D_1, \ldots, D_r)(Q)$. If $\Delta$ is a subset of $\mathrm{Der}(A)$, by rank $J(P; \Delta)(Q)$ we mean the supremum of rank $J(P; D_1, \ldots, D_r)(Q)$ when $\{D_1, \ldots, D_r\}$ runs over the set of all finite subsets of $\Delta$.

By a regular ring we mean a noetherian ring whose local rings are regular.

***Theorem 1*** *Let $(R, \mathfrak{m})$ be a regular local ring and $P$ be a prime ideal of height $r$. Then:*

(i) rank $J(P; \mathrm{Der}(R))(\mathfrak{m}) \leqq$ rank $J(P; \mathrm{Der}(R))(P) \leqq r$,

(ii) *if $D_1, \ldots, D_r \in \mathrm{Der}(R)$, $f_1, \ldots, f_r \in P$ and* $\det(D_i f_j) \notin \mathfrak{m}$ (i.e. rank $J(f_1, \ldots, f_r; D_1, \ldots, D_r)(\mathfrak{m}) = r$), *then $P = (f_1, \ldots, f_r)$ and $R/P$ is regular.*

*Proof* (i) The first inequality is trivial, while the second follows from the observation that, since $R_P$ is regular, $PR_P$ is generated by $r$ elements.

(ii) It is easy to see that $f_1, \ldots, f_r$ are linearly independent mod $\mathfrak{m}^2$ over $R/\mathfrak{m}$. The assertions follow from this.

***Theorem 2*** *Let $(A, \mathfrak{m})$ be a noetherian local domain of dimension $n$ containing the rational number field $\mathbb{Q}$. Let $k$ be a coefficient field of the completion $A^*$ of $A$, and let $\mathrm{Der}_k(A)$ denote the $A$-submodule of $\mathrm{Der}(A)$ consisting of derivations which, when extended to $A^*$, vanish on $k$. Then $\mathrm{Der}_k(A)$ is isomorphic to a submodule of $A^n$, and consequently*

$$\text{rank } \mathrm{Der}_k(A) \leqq \dim A.$$

*Proof* Take a system of parameters $x_1, \ldots, x_n$ of $A$. We claim that the map $\phi: \mathrm{Der}_k(A) \to A^n$ defined by $\phi(D) = (Dx_1, \ldots, Dx_n)$ is injective. Suppose that $D \in \mathrm{Der}_k(A)$ and that $Dx_1 = \cdots = Dx_n = 0$. The completion $A^*$ is finite over the subring $k[[x_1, \ldots, x_n]]$, on which $D$ vanishes. Let $a$ be an element of $A$. As an element of $A^*$ it satisfies polynomial relations $f(a) = 0$ with coefficients in $k[[x_1, \ldots, x_n]]$. Pick up such a polynomial $f(T)$ of the smallest degree. Then $0 = D(f(a)) = f'(a)Da$, and $f'(a) \neq 0$. Since $Da \in A$ and since the nonzero elements of $A$ are not zero divisors, we have $Da = 0$. QED

*Remark* In the unequal characteristic case, if $k$ is a coefficient ring of $A^*$, we get

$$\text{rank } \mathrm{Der}_k(A) \leqq \dim A - 1.$$

The above proof applies to this case with obvious modifications. In the case of prime characteristic the inequality does not hold in general, but cf. Theorem 13 as well as [8].

The next theorem will not be used in the sequel.

***Theorem 3*** *Let $A$ and $k$ be as in Theorem 2, and suppose that* rank $\mathrm{Der}_k(A) = \dim A$. *Then the completon $A^*$ is equidimensional and has no embedded primes. Consequently $A$ is universally catenary.*

*Proof* Let $P \in \mathrm{Ass}(A^*)$. Since $\mathbb{Q} \subset A^*$, any derivation $D \in \mathrm{Der}(A^*)$ maps $P$ into itself (cf. Seidenberg [15] or Matsumura [8]), thus inducing a derivation of $A^*/P$. Let $D_1, \ldots, D_n \in \mathrm{Der}(A)$ be linearly independent over $A$. Then there are $n$ elements $a_1, \ldots, a_n \in A$ such that $\det(D_i a_j) \neq 0$. Since any nonzero element of $A$ is a nonzero divisor in $A^*$, we have $\det(D_i a_j) \notin P$, which shows that $D_1, \ldots, D_n$ induce linearly independent derivations of $A^*/P$. Therefore $n \leqq \text{rank } \mathrm{Der}_k(A^*/P) \leqq \dim A^*/P \leqq \dim A^* = \dim A = n$, so that we must have $\dim A^*/P = n$, as wanted. A noetherian local ring whose completion is equidimensional is known to be universally catenary (cf. [EGA IV, (7.1.8), (7.1.9), and (7.1.11)]).

***Theorem 4*** (Nomura) *Let $(R, \mathfrak{m})$ be a regular local ring of dimension $n$ containing a field. Let $R^*$ be the completion of $R$ and $k$ be a coefficient field of $R^*$. Let $x_1, \ldots, x_n$ be a regular system of parameters of $R$. Then $R^* = k[[x_1, \ldots, x_n]]$, a formal power series ring over $k$, and* $\mathrm{Der}_k(R^*)$ *is a free module over $R^*$ with the partial derivations $\partial/\partial x_1, \ldots, \partial/\partial x_n$ as a basis. The following conditions are equivalent*:

(1) $\partial/\partial x_i$ $(i = 1, \ldots, n)$ *map $R$ into $R$, i.e.* $\partial/\partial x_i \in \mathrm{Der}_k(R)$;
(2) *there exist* $D_1, \ldots, D_n \in \mathrm{Der}_k(R)$ *and* $a_1, \ldots, a_n \in R$ *such that* $D_i a_j = \delta_{ij}$;
(3) *there exist* $D_1, \ldots, D_n \in \mathrm{Der}_k(R)$ *and* $a_1, \ldots, a_n \in R$ *such that* $\det(D_i a_j) \notin \mathfrak{m}$;
(4) $\mathrm{Der}_k(R)$ *is a free $R$-module of rank $n$*;
(5) rank $\mathrm{Der}_k(R) = n$.

*Proof* Let $K$ and $L$ denote the quotient fields of $R$ and $R^*$ respectively. The implications (1) ⇒ (2) ⇒ (3) and (4) ⇒ (5) are trivial.

(3) ⇒ (4) Clearly $D_1, \ldots, D_n$ are linearly independent over $R$ as well as over $R^*$. Therefore every $D \in \mathrm{Der}_k(R)$ can be written as $D = \sum c_i D_i$ with $c_i$ in $L$. Solving the equations $Da_j = \sum c_i(D_i a_j)$, we get $c_i \in R$.

(5) ⇒ (1) Let $D_1, \ldots, D_n$ be linearly independent over $R$. This means that there exist $a_1, \ldots, a_n \in R$ such that $\det(D_i a_j) \neq 0$. Therefore $D_1, \ldots, D_n$ are linearly independent over $R^*$ also. Hence $\partial/\partial x_i = \sum_j c_{ij} D_j$ with $c_{ij}$ in $L$. Now $\delta_{ik} = \sum_j c_{ij}(D_j x_k)$; therefore the matrix $(c_{ij})$ is the inverse matrix of $(D_j x_k)$ and $c_{ij} \in K$. Then $(\partial/\partial x_i)(R) \subset K \cap R^* = R$. QED

***Theorem 5*** (Classical) *Let $R$ be a regular local ring and $\Delta$ be a subset of* $\mathrm{Der}(R)$. *Let $P$ be a prime ideal of $R$. Then the following two conditions are equivalent*:

(1) rank $J(P; \Delta)(P) = \mathrm{ht}\, P$;

(2) *let $Q$ be a prime ideal contained in $P$. Then $R_P/QR_P$ is regular if and only if* rank $J(Q; \Delta)(P) = \mathrm{ht}\, Q$.

*Proof* Obviously (2) implies (1) (put $Q = P$). On the other hand, if rank $J(Q; \Delta)(P) = \mathrm{ht}\, Q$, then $R_P/QR_P$ is regular by Theorem 1. Suppose that rank $J(P; \Delta)(P) = \mathrm{ht}\, P$ and that $R_P/QR_P$ is regular. Put $\mathrm{ht}\, P = r$ and $\mathrm{ht}\, Q = s$. Then there exist $x_1, \ldots, x_s \in Q$ and $x_{s+1}, \ldots, x_r \in P$ such that $QR_P = \sum_1^s x_i R_P$ and $PR_P = \sum_1^r x_i R_P$. It follows that rank $J(x_1, \ldots, x_r; \Delta)(P) = r$. Hence rank $J(x_1, \ldots, x_s; \Delta)(P) = s$, so that rank $J(Q; \Delta)(P) = s$.

***Definition*** When the conditions of Theorem 5 hold with $\Delta = \mathrm{Der}(R)$, we say that the *weak Jacobian condition* (WJ) holds at $P$. We say that (WJ) holds in $R$ if it holds at every $P \in \mathrm{Spec}(R)$. When we use $\Delta = \mathrm{Der}_k(R)$ with some subring $k$ of $R$ (or $R^*$) we say that (WJ) holds at $P$ over $k$, etc.

Let $A$ be a noetherian ring, $k$ a subring, and $P$ a prime ideal of $A$. We denote the $A_P$-submodule of $\mathrm{Der}_k(A_P)$ generated by the image of the canonical map $\mathrm{Der}_k(A) \to \mathrm{Der}_k(A_p)$ by $\mathrm{Der}_k(A) \cdot A_P$. If $A$ is an integral domain or if $A$ is finitely generated over $k$, then $\mathrm{Der}_k(A) \cdot A_P \simeq \mathrm{Der}_k(A) \otimes_A A_P$.

***Lemma 1*** *Let $(A, \mathfrak{m})$ be a local ring and $D_1, \ldots, D_n \in$* $\mathrm{Der}(A)$ *and $x_1, \ldots, x_n \in \mathfrak{m}$ such that* (1) $\det(D_i x_j) \notin \mathfrak{m}$, (2) $[D_i, D_j] \in \sum_1^n A D_\nu$ *for all $i, j$. Then, putting $(D_i x_j)^{-1} = (c_{ij})$ and $\partial_i = \sum c_{ij} D_j$ $(i = 1, \ldots, n)$, we have $\partial_i x_j = \delta_{ij}$ and* $[\partial_i, \partial_j] = 0$ for all $i, j$.

*Proof* Since $[aD, bD'] = aD(b)D' - bD'(a)D + ab[D, D']$, condition (2) is invariant under invertible linear transformations on the $D_i$'s. Therefore $[\partial_i, \partial_j] = \sum_1^n b_{ijk} \partial_k$, but then $b_{ijk} = [\partial_i, \partial_j](x_k) = 0$.

***Definition*** Let $R$ be a regular ring containing $\mathbb{Q}$ and $P$ be a prime ideal. We say that the *strong Jacobian condition* (SJ) holds at $P$ if

(1) there exist $D_1, \ldots, D_r \in \mathrm{Der}(R) \cdot R_P$ and $x_1, \ldots, x_r \in P$, where $r = \mathrm{ht}\, P$, such that $\det(D_i x_j) \notin PR_P$;

(2) $[D_i, D_j] \in \sum_1^r R_P D_\nu$.

We say that (SJ) holds in $R$ if it holds at every prime of $R$.

Let $A$ be a ring and $I$ be an ideal. A derivation $D$ is said to induce a derivation $\bar{D}$ of $A/I$ if $\phi \circ D = \bar{D} \circ \phi$, where $\phi$: $A \to A/I$ is the natural map.

***Theorem 6*** *Let $(R, \mathfrak{m})$ be a regular local ring containing $\mathbb{Q}$, and $k$ be a coefficient field of the completion $R^*$. Then the following are equivalent:*

(1) (WJ) *holds at $\mathfrak{m}$ over $k$;*
(2) rank $\mathrm{Der}_k(R) = \dim R$;
(3) (SJ) *holds in $R$ over $k$.*

*Moreover, if these conditions are satisfied, then for any prime ideal $P$, all elements of* $\mathrm{Der}_k(R/P)$ *are induced by derivations in* $\mathrm{Der}_k(R)$ *and we have* rank $\mathrm{Der}_k(R/P) = \dim R/P$.

*Proof* (1)$\Rightarrow$(2) By assumption there exists $x_1, \ldots, x_n \in \mathfrak{m}$ and $D_1, \ldots, D_n \in \mathrm{Der}_k(R)$ such that $\det(D_i x_j) \notin \mathfrak{m}$, where $n = \dim R$. Then $D_1, \ldots, D_n$ are linearly independent over $R$. On the other hand, rank $\mathrm{Der}_k(R) \leqq n$ by Theorem 2. Thus rank $\mathrm{Der}_k(R) = n$.

(2)$\Rightarrow$(3) Let $x_1, \ldots, x_n$ be a regular system of parameters of $R$. By Theorem 3 we have $\partial/\partial x_i \in \mathrm{Der}_k(R)$, $i = 1, \ldots, n$. These derivations commute with each other. Let $P$ be a prime ideal of $R$. Since any derivation of $R/P$ can be uniquely extended to a derivation of $(R/P)^* = R^*/PR^*$ and since $R^* = k[[x_1, \ldots, x_n]]$, we see that an element $D'$ of $\mathrm{Der}_k(R/P)$ is determined by its values at $\phi(x_1), \ldots, \phi(x_n)$, where $\phi$: $R \to R/P$ is the natural map. Therefore, if we take a preimage $b_i \in R$ of $D'(\phi(x_i))$ for each $i = 1, \ldots, n$, then the derivation $D = \sum b_i\, \partial/\partial x_i$ induces $D'$.

The $R/P$-module $\mathrm{Der}_k(R, R/P)$ is a free $R/P$-module of rank $n$ with $\phi \circ \partial/\partial x_i$ $(i = 1, \ldots, n)$ as a basis. We can identify $\mathrm{Der}_k(R/P)$ with the submodule

$$N = \{\delta \in \mathrm{Der}_k(R, R/P) \mid \delta(f) = 0 \text{ for all } f \in P\}$$

of $\mathrm{Der}_k(R, R/P)$. Therefore

$$\text{rank } \mathrm{Der}_k(R/P) = \text{rank } N = n - \text{rank } J(P; \partial/\partial x_1, \ldots, \partial/\partial x_n)(P).$$

On one hand, rank $J(P; \partial/\partial x_1, \ldots, \partial/\partial x_n)(P) \leqq \text{ht } P$ by Theorem 1, and on the other hand, rank $\mathrm{Der}_k(R/P) \leqq \dim R/P = n - \text{ht } P$ by Theorem 2. Therefore rank $\mathrm{Der}_k(R/P) = \dim R/P$ and rank $J(P; \partial/\partial x_1, \ldots, \partial/\partial x_n)(P) = \text{ht } P$. Since $[\partial/\partial x_i, \partial/\partial x_j] = 0$, (SJ) holds at $P$. QED

*Remark* The equivalence of (SJ) and (WJ) in this case was pointed out to me by Carla Massaza. Actually she proved the equivalence in the more general case where $k$ is a subfield (of characteristic 0) of $R$ such that the residue field $R/m$ has a finite transcendence degree over it.

***Theorem 7*** *Let $R$ be a regular ring containing $\mathbb{Q}$ in which* (WJ) *or* (SJ) *holds. Then the same condition holds in the polynomial ring $R[X]$.*

*Proof* The ring $R[X]$ is regular, and if $P$ is a prime ideal of $R[X]$ and if $\mathfrak{p} = P \cap R$, then $R[X]_P$ is a localization of $R_\mathfrak{p}[X]$. Moreover, $R_\mathfrak{p}[X]/\mathfrak{p}R_\mathfrak{p}[X] = k(\mathfrak{p})[X]$, where $k(\mathfrak{p}) = R_\mathfrak{p}/\mathfrak{p}R_\mathfrak{p}$. Therefore either

(i) $P = \mathfrak{p}R[X]$, $\text{ht } P = \text{ht } \mathfrak{p}$, or
(ii) $\text{ht } P = \text{ht } \mathfrak{p} + 1$ and $PR_\mathfrak{p}[X]$ is generated by $\mathfrak{p}$ and a monic polynomial $f(X) \in R_\mathfrak{p}[X]$. We have $f'(X) \notin PR_\mathfrak{p}[X]$.

Since any derivation $D \in \text{Der}(R)$ can be extended to a derivation of $R[X]$ by setting $D(X) = 0$, and since this extension commutes with $d/dX$, the theorem follows easily from what we have just seen.

The following lemma may be viewed as a purely algebraic version of a well-known theorem of Frobenius in differential geometry. It was first used by Nagata [9], and then by Zariski [21] and by Lipman [4].

***Lemma 2*** *Let $A$ be a ring containing $\mathbb{Q}$, and $\mathfrak{m}$ be an ideal of $A$. Suppose that $A$ is $\mathfrak{m}$-adically complete and separated. Let $D_1, \ldots, D_r \in \text{Der}(A)$ and $x_1, \ldots, x_r \in \mathfrak{m}$ be such that*

$$[D_i, D_j] \in \sum_{\nu=1}^{r} AD_\nu, \qquad \det(D_i x_j) = \text{unit in } A.$$

*Put $F = \{a \in A \mid D_1 a = \cdots = D_r a = 0\}$. Then $x_1, \ldots, x_r$ are analytically independent over $F$ and $A = F[[x_1, \ldots, x_r]]$. If, moreover, $(A, \mathfrak{m})$ is a regular complete local ring and $r = \dim A$, then $\mathfrak{m} = (x_1, \ldots, x_r)$ and $F$ is a coefficient field of $A$.*

*Proof* By Lemma 1 we may assume that $D_i x_j = \delta_{ij}$ and $[D_i, D_j] = 0$. For the rest of the proof, see [9, 21, or 4]. (For a generalization to characteristic $p$, cf. Matsumura [8].)

***Theorem 8*** *Let $R$ be a regular ring containing $\mathbb{Q}$, and suppose* (SJ) *holds at every maximal ideal of $R$. Then* (SJ) *holds in $R$.*

*Proof* Let $P$ be a prime ideal of $R$ and $\mathfrak{m}$ be a maximal ideal containing $P$. By the hypothesis there exist $D_1, \ldots, D_n \in \text{Der}(R) \cdot R_\mathfrak{m}$, where $n = \text{ht } \mathfrak{m}$, and $x_1, \ldots, x_n \in \mathfrak{m}$, such that $D_i x_j = \delta_{ij}$ and $[D_i, D_j] = 0$. Let $F_\mathfrak{m}$ denote the subring of $(D_1, \ldots, D_n)$-constants in the completion $(R_\mathfrak{m})^*$ of the local ring $R_\mathfrak{m}$. Then $F_\mathfrak{m}$ is a coefficient field of $(R_\mathfrak{m})^*$, and we can apply the proof of Theorem 6 with $k = F_\mathfrak{m}$ to get

$$\text{rank } J(PR_\mathfrak{m}; D_1, \ldots, D_n)(PR_\mathfrak{m}) = \text{ht } P.$$

Thus there exist $y_1, \ldots, y_r \in P$, where $r = \operatorname{ht} P$, such that rank $J(y_1, \ldots, y_r;$ $D_1, \ldots, D_n)(PR_{\mathfrak{m}}) = r$. Since $[D_i, D_j] = 0$ and since we may view $D_i$ as elements of $\operatorname{Der}(R) \cdot R_P$, the condition (SJ) holds at $P$. QED

***Theorem 9*** *Let R be a regular ring containing* $\mathbb{Q}$, *and suppose that* (SJ) *holds in R. Then* (SJ) *holds also in the following rings:*

(a) $R[X]$,
(b) $R[[X]]$,
(c) $S^{-1}R$, *where S is a multiplicative subset of R,*
(d) $R/I$, *where I is an ideal of R such that R/I is regular.*

*Proof* (a) is already proved in Theorem 7.
(b) Since the element $X$ is in the Jacobson radical of $R[[X]]$, any maximal ideal of $R[[X]]$ is of the form $(\mathfrak{m}, X)$, where $\mathfrak{m}$ is a maximal ideal of $R$. It is easy to prove that (SJ) holds at $(\mathfrak{m}, X)$. Therefore (SJ) holds in $R[[X]]$ by the preceding theorem.
(c) is immediate from the definition.
(d) Let $P$ be a prime ideal of $R$ containing $I$, and put $\overline{P} = P/I$, $\overline{R} = R/I$. Since $\overline{R}_{\overline{P}} = R_P/IR_P$ is a regular local ring by hypothesis, the ideal $IR_P$ is prime and there exists a regular system of parameters $x_1, \ldots, x_n$ of $R_P$ such that $(x_1, \ldots, x_r) = IR_P$, where $n = \operatorname{ht} P$ and $r = \operatorname{ht} IR_P$. There exist $D_1, \ldots, D_n \in \operatorname{Der}(R) \cdot R_P$ such that $D_i x_j = \delta_{ij}$, $[D_i, D_j] = 0$. Then $D_{r+1}, \ldots, D_n$ vanish on $IR_P$, therefore they induce derivations $\overline{D}_{r+1}, \ldots, \overline{D}_n$ of $\overline{R}_{\overline{P}}$. Since $D_i \in \operatorname{Der}(R) \cdot R_P$, there exists $a_i \in R - P$ such that $D_i = a_i^{-1} D_i'$, $D_i' \in \operatorname{Der}(R)$. Then $D_i'(I) = 0$ in $R_P$ $(i = r+1, \ldots, n)$, hence there exist $b_{r+1}, \ldots, b_n \in R - P$ such that $b_i D_i'(I) = 0$ in $R$. Thus $b_i D_i'$ induces a derivation of $R/I$, and $D_i = (a_i b_i)^{-1}(b_i D_i')$. Therefore $\overline{D}_i \in \operatorname{Der}(\overline{R}) \cdot \overline{R}_{\overline{P}}$ $(i = r+1, \ldots, n)$, and $\overline{P}\overline{R}_{\overline{P}} = (\overline{x}_{r+1}, \ldots, \overline{x}_n)$, $\overline{D}_i \overline{x}_j = \delta_{ij}$, $[\overline{D}_i, \overline{D}_j] = 0$. QED

***Theorem 10*** (Mizutani) *Let R be a regular ring containing* $\mathbb{Q}$ *in which* (WJ) *holds. Then R is excellent.*

*Proof* First we prove that $R$ is a $G$-ring. Localizing, we can assume that $R$ is a regular local ring, and we have to prove that the formal fibers of $R$ are regular. Let $Q$ be a prime ideal of the completion $R^*$ and put $\mathfrak{q} = R \cap Q$. Let $\operatorname{ht} \mathfrak{q} = r$. Then there exist $D_1, \ldots, D_r \in \operatorname{Der}(R)$ such that rank $J(\mathfrak{q}; D_1, \ldots, D_r)(\mathfrak{q}) = r$. We can extend the derivations $D_i$ to $R^*$ and view the matrix $J(\mathfrak{q}; D_1, \ldots, D_r)(\mathfrak{q})$ as $J(\mathfrak{q}R^*; D_1, \ldots, D_r)(Q)$. On the other hand, we have $\operatorname{ht} \mathfrak{q}R^* = \operatorname{ht} \mathfrak{q} = r$ by [6, (13.B)]. Therefore $R^*_Q/\mathfrak{q}R^*_Q$ is regular by Theorem 1.

Next we prove $R$ is $J$-2. Since $R$ contains $\mathbb{Q}$ it suffices to show that, for any prime ideal $P$ of $R$, the set of regular points $\operatorname{Reg}(R/P)$ is open in $\operatorname{Spec}(R/P)$ [6, p. 246, Th. 73(3)]. This is easy (and classical) by Theorem 5.

Bingener [1] and Scheja and Storch [14] studied the "universal finite module of differentials." Let $A$ be a ring containing a field $k$. Suppose that there exist a finite $A$-module $M_0$ and a derivation $d_0 : A \to M_0$ over $k$ which have the universal mapping property: for any derivation $D$: $A \to N$ over $k$ into a finite $A$-module $N$, there exists a unique $A$-linear map $f : M_0 \to N$ satisfying $D = f \circ d_0$. Then $M_0$ is called the universal finite module of differentials of $A$ over $k$, and is denoted by $D_k(A)$. It commutes with localization by maximal ideals. If $A$ is a noetherian semilocal ring with radical $\mathfrak{m}$, and if $D_k(A)$ exists, then it has the universal mapping property with respect to any derivation of $A$ into $\mathfrak{m}$-adically separated $A$-module, and the completion $D_k(A)^* = D_k(A) \otimes A^*$ is the universal finite module of differentials $D_k(A^*)$ of $A^*$. Bingener and Scheja and Storch proved that, when $k$ is a field of characteristic zero, a noetherian $k$-algebra $A$ with $D_k(A)$ is excellent. Our next theorem shows that, when $A$ is assumed to be regular, this is a special case of Theorem 10. On the other hand, there are rings with universal finite module of differentials which may not be homomorphic images of regular rings.

***Theorem 11*** *Let R be a regular ring containing a field k of characteristic* 0. *Suppose that R has the universal finite module of differentials* $D_k(R)$. *Then* (SJ) *holds in R, and for any maximal ideal* $\mathfrak{m}$ *we have* $\mathrm{Der}_k(R)_\mathfrak{m} = \mathrm{Der}_k(R_\mathfrak{m})$.

*Proof* Take a maximal ideal $\mathfrak{m}$, and denote by $B$ the completion of the local ring $R_\mathfrak{m}$. We have $D_k(R_\mathfrak{m}) = D_k(R)_\mathfrak{m}$ and $D_k(B) = D_k(R_\mathfrak{m}) \otimes_{R_\mathfrak{m}} B = D_k(R) \otimes B$. Therefore

$$\mathrm{Der}_k(R_\mathfrak{m}) = \mathrm{Hom}_{R_\mathfrak{m}}(D_k(R_\mathfrak{m}), R_\mathfrak{m}) = \mathrm{Hom}_R(D_k(R), R)_\mathfrak{m} = \mathrm{Der}_k(R)_\mathfrak{m}.$$

Let $\bar{u}_1, \ldots, \bar{u}_s$ be a transcendence basis of the residue field $R/\mathfrak{m}$ over $k$ (the transcendence degree is finite by [14, (8.1)]; we do not need this fact), and choose a preimage $u_i$ of $\bar{u}_i$ in $R$ for each $i$. Then $R_\mathfrak{m}$ contains the field $k(u_1, \ldots, u_s)$. Put $k' = k(u_1, \ldots, u_s)$ and let $K$ be a coefficient field of $B$ containing $k'$ (actually such $K$ is uniquely determined by $k'$). Then $K$ is algebraic over $k'$. Let $x_1, \ldots, x_n$ be a regular system of parameters of $R_\mathfrak{m}$. Then $B = K[[x_1, \ldots, x_n]]$, and putting $\partial_i = \partial/\partial x_i$ we have $\partial_i \in \mathrm{Der}_{k'}(B)$, $\partial_i x_j = \delta_{ij}$. Moreover, $D_{k'}(R_\mathfrak{m}) = D_k(R_\mathfrak{m})/\sum R_\mathfrak{m}\, du_i$ is a finite $R_\mathfrak{m}$-module, hence

$$\begin{aligned}\mathrm{Der}_{k'}(B) &= \mathrm{Hom}_B(D_{k'}(B), B) = \mathrm{Hom}_B(D_{k'}(R_\mathfrak{m}) \otimes B, B)\\ &= \mathrm{Hom}_{R_\mathfrak{m}}(D_{k'}(R_\mathfrak{m}), R_\mathfrak{m}) \otimes_{R_\mathfrak{m}} B = \mathrm{Der}_{k'}(R_\mathfrak{m}) \otimes_{R_\mathfrak{m}} B\\ &= \text{the } \mathfrak{m}\text{-adic completion of the finite } R_\mathfrak{m}\text{-module } \mathrm{Der}_{k'}(R_\mathfrak{m}).\end{aligned}$$

Therefore there exist $D_1, \ldots, D_n \in \mathrm{Der}_{k'}(R_\mathfrak{m}) = \mathrm{Der}_{k'}(R)_\mathfrak{m}$ such that $D_i \equiv \partial_i \bmod \mathfrak{m}R_\mathfrak{m}$. Then $D_i x_j \equiv \partial_i x_j = \delta_{ij} \bmod \mathfrak{m}R_\mathfrak{m}$, therefore $\det(D_i x_j) \equiv 1 \bmod$

$\mathfrak{m}R_{\mathfrak{m}}$. By Theorem 6 (noting that $\mathrm{Der}_{k'}(R_{\mathfrak{m}}) = \mathrm{Der}_K(R_{\mathfrak{m}})$) we see that (SJ) holds at any prime $P$ contained in $\mathfrak{m}$. QED

We now turn to the case of characteristic $p > 0$, where many of the preceding theorems need modification. Here the most important result is Nagata's Jacobian criterion for formal power series rings, and we must heavily rely upon it. We begin with some preparation.

A subfield $K'$ of a field $K$ will be called *cofinite* if $[K : K'] < \infty$. Let $K$ be a field and $\mathbb{F} = (k_\alpha)_{\alpha \in I}$ be a family of subfields of $K$. We say $\mathbb{F}$ is directed downwards if, for any pair of indices $\alpha, \beta \in I$, there exists $\gamma \in I$ such that $k_\gamma \subset k_\alpha \cap k_\beta$.

***Lemma 3*** *Let $K$ be a field of characteristic $p$ and $k$ be a subfield. Let $\mathbb{F} = (k_\alpha)_{\alpha \in I}$ be a downward directed family of subfields of $K$ containing $k$. Then the following are equivalent*:

(i) $\bigcap_\alpha k_\alpha(K^p) = k(K^p)$;
(ii) *the canonical map* $\Omega_{K/k} \to \varprojlim_\alpha \Omega_{K/k_\alpha}$ *is injective*;
(iii) *for any finite subset* $F = \{u_1, \ldots, u_n\}$ *of $K$ which is $p$-independent over $k$, there exists an index $\alpha$ such that $F$ is $p$-independent over $k_\alpha$*;
(iv) *there exists a $p$-basis $B$ of $K$ over $k$ such that, for any finite subset $F$ of $B$, there exists an index $\alpha$ such that $F$ is $p$-independent over $k_\alpha$.*

*Proof* (i) $\Rightarrow$ (iii) The set $\{u_1, \ldots, u_n\}$ is $p$-independent over $k$ iff the $p^n$ monomials $u_1^{e_1} \cdots u_n^{e_n}$ $(0 \leqq e_i < p)$ are linearly independent over the field $k(K^p)$. Since this last field is the intersection of the $k_\alpha(K^p)$'s it is easy to see that they must be linearly independent over some $k_\alpha(K^p)$.

(ii) $\Leftrightarrow$ (iii) Obvious.

(iii) $\Rightarrow$ (iv) Obvious (any $p$-basis will do).

(iv) $\Rightarrow$ (ii) Take any nonzero element $\omega$ of $\Omega_{K/k}$. We can write uniquely: $\omega = \sum_{i=1}^n a_i d_{K/k} b_i$, where $b_i \in B$ and $a_i \in K, a_i \neq 0$. Take $\alpha$ such that $b_1, \ldots, b_n$ are $p$-independent over $k_\alpha$. Then $\sum a_i d_{K/k_\alpha} b_i \neq 0$.

(ii) $\Rightarrow$ (i) If $a \in K, a \notin k(K^p)$, then $d_{K/k} a \neq 0$ in $\Omega_{K/k}$, whence there exists $\alpha$ such that $d_{K/k_\alpha}(a) \neq 0$. Thus $a \notin k_\alpha(K^p)$.

The following lemma is a slightly weaker form of [EGA $0_{\mathrm{IV}}$, (21.8.5)]. For the sake of completeness we include if here with an elementary proof.

***Lemma 4*** *Let $K$, $k$, $\mathbb{F} = (k_\alpha)$ be as above and suppose that $\bigcap_\alpha k_\alpha(K^p) = k(K^p)$. Then, for any finitely generated extension field (or any separable extension field) $K_1$ of $K$, we have $\bigcap_\alpha k_\alpha(K_1{}^p) = k(K_1{}^p)$.*

*Proof* *Case 1* $K_1$ is *separable* over $K$. Then a $p$-basis of $K_1$ over $k$ can be obtained by adjoining a $p$-basis $C$ of $K_1$ over $K$ to a $p$-basis $B$ of $K$ over $k$ (because $0 \to \Omega_{K/k} \otimes K_1 \to \Omega_{K_1/k} \to \Omega_{K_1/K} \to 0$ is split exact). Let $a_1, \ldots, a_n \in C$

and $b_1, \ldots, b_m \in B$ be mutually distinct elements. By assumption and by the preceding lemma there exists $\alpha$ such that $\{b_1, \ldots, b_m\}$ is $p$-independent over $k_\alpha$. Since $K_1$ is separable over $K$, the set $\{a_1, \ldots, a_n, b_1, \ldots, b_m\}$ is $p$-independent over $k_\alpha$. By the lemma we conclude that $\bigcap k_\alpha(K_1{}^p) = k(K_1{}^p)$.

*Case 2* $K_1$ is *finitely generated* over $K$. By the obvious transitivity of the assertion and by Case 1, we may limit to the case $K_1 = K(x)$ with $x^p = a \in K$, $a \notin K^p$. If $d_{K/k}a \neq 0$, then there exists a $p$-basis $B$ of $K$ over $k$ which contains $a$, $B = \{a\} \amalg \{b_\lambda\}_{\lambda \in \Lambda}$. It is easy to see from the definition that $B' = \{x\} \amalg \{b_\lambda\}$ is a $p$-basis of $K_1$ over $k$. If $\{a, b_1, \ldots, b_n\}$ is $p$-independent in $K$ over $k_\alpha$, then $\{x, b_1, \ldots, b_n\}$ is $p$-independent in $K_1$ over $k_\alpha$. Therefore $\bigcap k_\alpha(K_1{}^p) = k(K_1{}^p)$ by Lemma 3. If $d_{K/k}a = 0$, let $B$ be a $p$-basis of $K$ over $k$. Then $B \amalg \{x\}$ is a $p$-basis of $K_1$ over $k$, as one can easily check. Using this base one can argue as before. Therefore Lemma 4 is proved completely.

The main point of the Jacobian criterion of Nagata is the following

***Theorem 12*** (Nagata) *Let $k$ be a field of characteristic $p > 0$, $R = k[[X_1, \ldots, X_n]]$, and $P$ be a prime ideal of $R$. Let $(k_\alpha)_{\alpha \in I}$ be a downward-directed family of cofinite subfields of $k$ such that $\bigcap k_\alpha = k^p$. Then there exists an index $\alpha$ such that, for any cofinite subfield $k'$ of $k_\alpha$ containing $k^p$, we have*

$$\operatorname{rank} \operatorname{Der}_{k'}(R/P) = \dim(R/P) + \operatorname{rank} \operatorname{Der}_{k'}(k).$$

*Proof* See Nagata [10], but note that Lemma 4 above was implicitly used in his proof.

We are going to apply Nagata's theorem to noncomplete local rings. The difficulty lies in the fact that, if $A$ is a noetherian ring containing $\mathbb{Q}$, all members of $\operatorname{Ass}(A)$ are differential ideals (i.e. stable under any derivation of $A$), while in the case of characteristic $p$ one can only say the following:

***Lemma 5*** *Let $A$ be a ring, $P$ a prime ideal, and $I$ an ideal such that $P \not\supset I$, $P \cap I = 0$. Then $P$ is a differential ideal.*

*Proof* Let $D \in \operatorname{Der}(A)$, $x \in P$. Take an element $y$ of $I$ not contained in $P$. Then $xy = 0$, hence $y\,Dx + x\,Dy = 0$. Multiplying with $y$ we get $y^2\,Dx = 0 \in P$. Since $y^2 \notin P$ we have $Dx \in P$.

It follows from this lemma that, if $A$ is reduced, then all associated primes are differential. We can ask the following question: If $A$ is a noetherian local domain, if $P \in \operatorname{Ass}(A^*)$ and if $D \in \operatorname{Der}(A)$, does the extension of $D$ to $A^*$ map $P$ into itself? But it has easy counterexamples. So we cannot eliminate the hypothesis of reducedness from our next theorem.

***Theorem 13*** *Let $(A, \mathfrak{m}, k_A)$ be a noetherian local domain containing a field $k$ of characteristic $p$ such that the residue field $k_A$ is separable over $k$ and* rank $\operatorname{Der}_k(k_A) < \infty$. *Suppose the completion $A^*$ of $A$ is reduced. Then there exists a cofinite subfield $k'$ of $k$ containing $k^p$ such that, for all cofinite subfields $k''$ of $k'$ we have*

$$\operatorname{rank} \operatorname{Der}_{k''}(A) \leqq \dim A + \operatorname{rank} \operatorname{Der}_{k''}(k_A). \qquad (*)$$

*Proof* Take a minimal prime $P$ of $A^*$. Then by Lemma 5 there is a natural map $\operatorname{Der}(A) \to \operatorname{Der}(A^*/P)$, and, as in the proof of Theorem 3, we have rank $\operatorname{Der}_{k'}(A) \leqq$ rank $\operatorname{Der}_{k'}(A^*/P)$ for any subfield $k'$ of $k$. On the other hand, we have $\dim A^*/P \leqq \dim A^* = \dim A$. Therefore we may replace $A$ be $A^*/P$ and assume that $A$ is complete. Then there is a coefficient field $K$ of $A$ containing $k$, and $A$ is of the form $A = K[[X_1, \ldots, X_n]]/I$, where $I$ is a prime ideal. Let $(k_\alpha)$ be a downward-directed family of subfields of $k$ with $\bigcap k_\alpha = k^p$. (Such a family can be constructed by fixing a $p$-basis $B$ and adjoining complements of finite subsets of $B$ to $k^p$.) Then $\bigcap k_\alpha(K^p) = K^p$ by Lemma 4. Moreover, since rank $\operatorname{Der}_{k_\alpha}(K) =$ rank $\operatorname{Der}_{k_\alpha}(k) +$ rank $\operatorname{Der}_k(K)$, we have $[K : k_\alpha(K^p)] < \infty$. Therefore we can take some $k_\alpha$ as the required subfield $k'$ by Theorem 12.

We keep the definition of (WJ) without change. As for (SJ), we tentatively modify the definition as follows:

***Definition*** Let $R$ be a regular ring of characteristic $p > 0$, and $P$ be a prime ideal of $R$. We say that (SJ) holds at $P$ if

(1) There exist $D_1, \ldots, D_r \in \operatorname{Der}(R) \cdot R_P$ and $x_1, \ldots, x_r \in P$, where $r = \operatorname{ht} P$, such that $\det(D_i x_j) \notin PR_P$;
(2) $[D_i, D_j] \in \sum_1^r R_P D_\nu$, $D_i^p \in \sum_1^r R_P D_\nu$.

Condition (2) is equivalent to saying that $\mathfrak{g} = \sum_1^r R_P D_\nu$ is a $p$-Lie algebra (i.e., $D, D' \in \mathfrak{g} \Rightarrow D^p \in \mathfrak{g}$, $[D, D'] \in \mathfrak{g}$). In fact, the equivalence follows from the formulas

$$[aD, bD'] = ab[D, D'] + aD(b)D' - bD'(a)D,$$

$$(aD)^p = a^p D^p + (aD)^{p-1}(a)D \qquad [3, \text{p. } 191, \text{Ex. } 15],$$

$$(D + D')^p = D^p + D'^p + \sum_{i=1}^{p-1} (p - i)^{-1} g_{i,\, p-1-i}(\operatorname{ad} D, \operatorname{ad} D')D'$$

[2, p. 105, Ex. 19],

where $g_{i,\, p-1-i}(x, y)$ is the sum of the ${}_{p-1}C_i$ formally distinct products $t_1 t_2 \cdots t_{p-1}$ in which $i$ of the $t_\alpha$ are equal to $x$ and the remaining $t_\alpha$ are equal to $y$.

If (1) and (2) hold, we can assume, by performing an invertible linear substitution with coefficients in $R_p$ as in Lemma 1, that

$$D_i x_j = \delta_{ij}, \qquad [D_i, D_j] = 0, \qquad D_i^p = 0.$$

Theorem 10 must be modified as follows.

***Theorem 14*** *Let R be a regular ring. Suppose that, for $n = 0, 1, 2, \ldots$, (WJ) holds in $R[X_1, \ldots, X_n]$. Then R is excellent.*

*Proof* First we prove $R$ is a $G$-ring. Take a maximal ideal $m$ of $R$. We have to prove that the formal fibers of $R_m$ are *geometrically* regular. Replacing $R$ by $R_m$, we can assume $R$ is a regular local ring. Then, by [6, (33.E), Lemma 3], it suffices to prove that, for each finite $R$-algebra $B$ which is a domain and for each prime ideal $Q$ of $B^*$ with $Q \cap B = (0)$, the local ring $B^*_Q$ is regular. Now $B^*$ is the direct product of the completions of the localizations with respect to maximal ideals, and $B$ is of the form $R[X_1, \ldots, X_n]/P$, where $P$ is a prime ideal of $R[X_1, \ldots, X_n]$. Therefore $B^*_Q$ is of the form $C^*_{\mathfrak{q}}/PC^*_{\mathfrak{q}}$, where $C = R[X_1, \ldots, X_n]_{\mathfrak{M}}$ with some maximal ideal $\mathfrak{M}$ and $\mathfrak{q}$ is a prime ideal of the completion $C^*$ such that $\mathfrak{q} \cap C = PC$. By hypothesis there exist $D_1, \ldots, D_r \in \mathrm{Der}(R[X])$ and $f_1, \ldots, f_r \in P$, $r = \mathrm{ht}\, P$, such that $\det(D_i f_j) \notin P$. Therefore the argument of the proof of Theorem 10 shows that $C^*_{\mathfrak{q}}/PC^*_{\mathfrak{q}}$ is regular.

Next we prove that $R$ is $J$-2 (cf. [6, (32.B)]). It suffices to prove that for every finite $R$-algebra $B$, $\mathrm{Reg}(B)$ is open in $\mathrm{Spec}(B)$. Since $B$ is of the form $R[X_1, \ldots, X_n]/I$, the assertion is immediate from Theorem 5.

*Remark* As one can see from the proof, for $R$ to be excellent it is sufficient that (WJ) holds at every prime ideal $P$ of $R[X_1, \ldots, X_n]$ such that $R[X]/P$ is a finite $R$-module.

Theorem 7 becomes false in characteristic $p$. In fact, let $k$ be a field of characteristic $p$ with $[k : k^p] = \infty$ and $R$ be the subring $k^p[[t]][k]$ of the formal power series ring $k[[t]]$ ([11, p. 205, Ex. 3; 6, p. 259]). Then $R$ is a discrete valuation ring, which is not excellent because its completion $R^* = k[[t]]$ is purely inseparable over it. Since $\mathrm{Spec}(R) = \{(t), (0)\}$ and since $d/dt \in \mathrm{Der}_k(R)$, condition (SJ) holds in $R$. But (WJ) does not hold at the prime ideals $(X^p - u^p)$ of $R[X]$, where $u \in k[[t]]$, $u \notin R$. In fact, any derivation $D$ of $R[X]$ can be extended to a derivation of the $t$-adic completion $S = \varprojlim((R/t^\nu R)[X])$ of $R[X]$, and $S$ contains $R^*[X]$, therefore $X^p - u^p = (X - u)^p$ in $S$ and we must have $D(X^p - u^p) = 0$. Incidentally, putting $A = R[X]/(X^p - u^p)$, we get a local domain $A$ with a derivation $D$ induced by $d/dx$, and the minimal prime $(X - u)$ of $A^* = A \otimes_R R^* = R[X]/(X - u)^p$ is not stable under $D$.

We recall that, when $K$ is a ring and $k$ is a subring, the kernel of the natural map $\Omega_k \otimes_k K \to \Omega_K$ is denoted by $\Upsilon_{K/k}$ and is called the module of imperfection, cf. [EGA $O_{IV}$, (20.6.1)]. Let $K$ and $k$ be fields. Then $K$ is separable over $k$ iff $\Upsilon_{K/k} = 0$. If $K$ is finitely generated over $k$, then $\operatorname{rank}_K \Upsilon_{K/k} < \infty$ [EGA $O_{IV}$, (21.7.1)].

***Lemma 6*** *Let $k$ be a field and $A$ be a noetherian local ring containing $k$ with residue field $K$. Suppose that the universal finite module of differentials $D_k(A)$ exists. Then* $\operatorname{rank}_K \Omega_{K/k} < \infty$.

*Proof* Since $K$ is a homomorphic image of $A$, the universal finite module $D_k(K)$ exists (in fact, $D_k(K) = D_k(A)/(Ad\mathfrak{m} + \mathfrak{m}D_k(A))$, hence $D_k(K) = \Omega_{K/k}$ and rank $\Omega_{K/k} < \infty$.

***Lemma 7*** (Projective trick) *Let $A$ be a ring and $P$ be a prime ideal of $A[X_1, \ldots, X_n]$. Let $\mathfrak{m}$ be a maximal ideal of $A$ containing $P \cap A$. Then there exists a suitable "reciprocation" $(X')$ of $(X)$, i.e. $X_i' = X_i^{e_i}$ $(i = 1, \ldots, n)$ with $e_i = \pm 1$, such that if we put $P' = PA[X, X'] \cap A[X']$, then there exists a maximal ideal $\mathfrak{M}'$ of $A[X']$ containing $P'$ and lying over $\mathfrak{m}$.*

*Proof* This is in substance the same as the theorem of extension of specializations [19, Ch. II, Th. 6]. In the present form it follows immediately from [20, Th. 1] or from [22, Ch. 4, §5, Th. 7].

***Theorem 15*** *Let $A$ be a regular ring containing a field $k$ of characteristic $p > 0$. Suppose that*

(1) *$A$ is pseudogeometric*†;
(2) *for every subfield $k'$ of $k$ containing $k^p$ such that $[k : k'] < \infty$, the universal finite module of differentials $D_{k'}(A)$ exists; and*
(3) *for every maximal ideal $\mathfrak{m}$ of $A$, the residue field $A/\mathfrak{m} = \kappa(\mathfrak{m})$ satisfies rank $\Upsilon_{\kappa(\mathfrak{m})/k} < \infty$.*

*Then* (SJ) *holds in $A[X_1, \ldots, X_n]$, and consequently $A$ is excellent.*

*Proof* Let $P$ be any prime ideal of $B = A[X_1, \ldots, X_n]$, let $\mathfrak{M}$ be a maximal ideal containing $P$ and put $\mathfrak{m} = \mathfrak{M} \cap A$.

First we assume that $\mathfrak{m}$ is a maximal ideal in $A$. Put $R = B_{\mathfrak{M}}$. The residue field $B/\mathfrak{M} = \kappa(\mathfrak{M})$ is finitely generated over $A/\mathfrak{m} = \kappa(\mathfrak{m})$, whence $\Omega_{\kappa(\mathfrak{M})/k}$ and $\Upsilon_{\kappa(\mathfrak{M})/k}$ have finite ranks. Thus there exist a $p$-basis $\Sigma = \{u_i\}_{i \in I}$ of $k$ and a finite subset $\Sigma_0$ of $\Sigma$ such that $\kappa(\mathfrak{M})$ is separable over $k_0(\Sigma_1) = k_0'$ ($\Sigma_1 = \Sigma - \Sigma_0$, $k_0$ is the prime field in $k$). Then the completion $R^*$ contains a coefficient field $K \supset k_0'$, and $K \simeq \kappa(\mathfrak{M})$, $[K : K^p k_0'] < \infty$.

† That is, Nagata ring in the terminology of [6].

Since $\Sigma_1$ is $p$-independent in $K$, if $\Lambda$ is the set of all finite subsets of $\Sigma_1$ and if we put for each $\lambda \in \Lambda$

$$K_\lambda = K^p(\Sigma_1 - \lambda), \qquad k_\lambda = k^p(\Sigma_1 - \lambda),$$

then we have $\bigcap K_\lambda = K^p$, $\bigcap k_\lambda = k^p$, and $D_{K_\lambda}(R) = D_{k_\lambda}(R) = (R \otimes D_{k_\lambda}(A)) + R\,dX_1 + \cdots + R\,dX_n$. (For, if $N$ is a finite $R$-module, then $N$ is $\mathfrak{M}$-adically separated, hence $\mathfrak{m}$-adically separated. Therefore any derivation $A_\mathfrak{m} \to N$ over $k$ factors through $D_{k_\lambda}(A_\mathfrak{m}) = D_{k_\lambda}(A) \otimes A_\mathfrak{m}$.)

By (1) we know that $R^*/PR^*$ is reduced, hence there exists a prime ideal $\mathfrak{P}$ of $R^*$ such that $PR^*_\mathfrak{p} = \mathfrak{P}R^*_\mathfrak{p}$. By Nagata's Jacobian criterion there exist $\lambda \in \Lambda$ and $D_1, \ldots, D_s \in \mathrm{Der}_{K_\lambda}(R^*)$ such that rank $J(\mathfrak{P}; D_1, \ldots, D_s)(\mathfrak{P}) = \mathrm{ht}\, \mathfrak{P} = \mathrm{ht}\, P$. Now $\mathrm{Der}_{K_\lambda}(R^*) = \mathrm{Der}_{k_\lambda}(R^*) = \mathrm{Hom}_{R^*}(D_{k_\lambda}(R) \otimes R^*, R^*) = \mathrm{Der}_{k_\lambda}(R) \otimes R^*$ is the $\mathfrak{M}$-adic completion of $\mathrm{Der}_{k_\lambda}(R)$. Therefore there exist $d_i \in \mathrm{Der}_{k_\lambda}(R)$, $i = 1, 2, \ldots, s$, such that $d_i \equiv D_i \mod \mathfrak{M}^\nu R^*$, where $\nu$ is an arbitrarily given positive integer. Then, if ht $P = s$, $f_1, \ldots, f_s \in \mathfrak{P}$ and $\det(D_i f_j) \notin \mathfrak{P}$, then for sufficiently large $\nu$ we have $\det(d_i f_j) \notin \mathfrak{P}$. Since $d_i f_j \in R$ this means $\det(d_i f_j) \notin P$, so we have rank $J(P; \mathrm{Der}_{k_\lambda}(R))(P) = \mathrm{ht}\, P$. Since $\mathrm{Der}_{k_\lambda}(R)$ is a $p$-Lie algebra finitely generated over $R$, and since $\mathrm{Der}_{k_\lambda}(R) = \mathrm{Der}_{k_\lambda}(B_\mathfrak{M}) = \mathrm{Der}_{k_\lambda}(A) \otimes B_\mathfrak{M} + \sum_1^n B_\mathfrak{M}\, \partial/\partial X_i = \mathrm{Der}_{k_\lambda}(B) \otimes B_\mathfrak{M}$, we see that (SJ) holds at $P$.

When $\mathfrak{M}$ does not lie over a maximal ideal of $A$, we use the projective trick.† Take a maximal ideal $\mathfrak{m}$ of $A$ containing $P \cap A$, and let $(X')$, $P'$ and $\mathfrak{M}'$ be as in Lemma 7. By what we have seen there exist a subfield $k'$ of $k$, $f_1, \ldots, f_s \in P'$ ($s = \mathrm{ht}\, P$), and $D_1, \ldots, D_s \in \mathrm{Der}_{k'}(A[X'])$ such that $\det(D_i f_j) \notin P'$. Then $f_i = u_i \phi_i$, $\phi_i \in P$, and $u_i$ is a unit in $B_P$, and we can extend $D_i$ to $A[X, X']$. The restrictions of $D_i$ to $A[X]$ map $A[X]$ into $A[X, X']$, and (since $D_{k'}(A)$ exists) we have $(X_1 \cdots X_n)^\nu D_i \in \mathrm{Der}_{k'}(A[X])$ for sufficiently large $\nu$. Hence rank $J(P; \mathrm{Der}_{k'}(A[X]))(P) = \mathrm{ht}\, P$. QED

*Remark* If A satisfies the conditions of Theorem 15, so does the formal power series ring $A[[Y]]$. In fact, (3) is obvious because every maximal ideal of $A[[Y]]$ contains $Y$. (1) continues to hold for $A[[Y]]$ by a recent theorem of Marot [5] (cf. also [12] for the theorem of Y. Mori used in [5]). As for condition (2) we use the following lemma.

***Lemma 8*** *Let k be a field and A be a k-algebra. If A has the universal finite module of k-differentials, so does* $A[[Y]]$.

*Proof* Put $B = A[[Y]]$. Let $M$ be a finite $B$-module and $D: B \to M$ be a derivation over $k$. Since the $Y$-adic ring $B$ is a Zariski ring, $M$ is $Y$-adically separated and can be embedded in $\prod M/Y^\nu M$. Each $M/Y^\nu M$ is finite over

† We owe this idea to an advice of M. Artin.

$B/Y^\nu B = A[Y]/Y^\nu A[Y]$, hence is finite over $A$. Therefore the derivation $A \to M$ induced by $D$ factors through $D_k(A)$. (This is the same reasoning as in [14, (1.1)]. It follows easily that $D_k(B) = D_k(A) \otimes B + B\,dY$.

*Remark* The lemma does not hold for $A[Y]$. In fact, for such a simple ring as $R = k[t]_{(t)}[Y]$, $k$ a field, $D_k(R)$ does not exist.

***Corollary*** *Let $k$ be a field. Then* (SJ) *holds in* $k[X_1, \ldots, X_n][[Y_1, \ldots, Y_m]][Z_1, \ldots, Z_r]$, *hence this ring is excellent.*

For arbitrary field $k$, the excellence was proved by Valabrega [17] (cf. also [7, 16]). Some of the theorems in the present paper will undoubtedly have generalizations to the unequal characteristic case. For this, cf. [13, 18].

## References


1 Bingener, R., Über Steinsche Algebren und Moduln. Diplomarbeit Bochum SS, 1971.
2. Bourbaki, N., "Groupes et algèbres de Lie," Chapitre 1 (*Actualités Sci. Indnst.* **1285**). Hermann, Paris, 1960.
3. Jacobson, N., "Lectures in Abstract Algebra," Vol. III. Van Nostrand-Reinhold, Princeton, New Jersey, 1964.
4. Lipman, J., Free derivation modules on algebraic varieties. *Amer. J. Math.* **87** (1965), 874–898.
5. Marot, J., Sur les anneaux universellement japonais, *C. R. Acad. Sci. Paris* **277** (1973), Ser. A, 1029–1031.
6. Matsumura, H., "Commutative Algebra." Benjamin, New York, 1970.
7. Matsumura, H., Formal power series rings over polynomial rings I, *in* "Number Theory, Algebraic Geometry and Commutative Algebra" (in honor of Y. Akizuki), pp. 511–520. Kinokuniya, Tokyo 1973.
8. Matsumura, H., Integrable derivations, *Nagoya Math. J.* (to appear).
9. Nagata, M., Remarks on a paper of Zariski on the purity of branch loci, *Proc. Nat. Acad. Sci. U.S.A.* **44** (1958), 796–799.
10. Nagata, M., A Jacobian criterion of simple points, *Illinois J. Math.* **1** (1957), 427–432.
11. Nagata, M., "Local Rings." Wiley (Interscience), New York, 1962.
12. Nishimura, J., Note on Krull domains, *J. Math. Kyoto Univ.* **15** (1975), 397–400.
13. Nomura, M., Formal power series rings over polynomial rings II, *in* "Number Theory, Algebraic Geometry and Commutative Algebra" (in honor of Y. Akizuki), pp. 521–528. Kinokuniya, Tokyo, 1973.
14. Scheja, G., and Storch, U., Differentielle Eigenschaften der Lokalisierungen analytischer Algebren, *Math. Ann.* **197** (1972), 137–170.
15. Seidenberg, A., Differential ideals in rings of finitely generated type, *Amer. J. Math.* **89** (1967), 22–42.
16. Seydi, H., Un critère jacobien des points simples, *C. R. Acad. Sci. Paris* **276** (1973), 475–478.
17. Valabrega, P., On the excellence property for power series rings over polynomial rings, *J. Math. Kyoto Univ.* **15** (1975), 387–395.
18. Valabrega, P., A few theorems on completion of excellent rings, *Nagoya Math. J.* (1976).
19. Weil, A., "Foundations of Algebraic Geometry." Amer. Math. Soc., Providence, Rhode Island, 1946.

20. Weil, A., Arithmetic on algebraic varieties, *Ann. of Math.* **53** (1951), 412–444.
21. Zariski, O., Studies in equisingularity I, *Amer. J. Math.* **87** (1965), 507–536.
22. Zariski, O., and Samuel, P., "Commutative Algebra," Vol. II. Van Nostrand-Reinhold, Princeton, New Jersey, 1960.

EGA $0_{IV}$. Grothendieck, A., and Dieudonné, J., *Inst. Hautes Études Sci. Publ. Math.* **20** (1964).
EGA IV. Grothendieck, A., and Dieudonné, J., *Inst. Hautes Études Sci. Publ. Math.* **24** (1965).

AMS (MOS) 1970 subject classifications: 13B10, 13C15, 13H05

*Note Added in proof* The reasoning of p. 292 line 17 is, more exactly, as follows: Let $x_1, \ldots, x_n$ be a regular system of parameters of $R$ and $x_{n+1}, \ldots, x_m$ be a $p$-basis of $K$ over $k$. Then $R^* = K[[x_1, \ldots, x_n]]$ and there exists an $R^*$-basis $\{\partial_1', \ldots, \partial_m'\}$ of $\mathrm{Der}_{k_\lambda}(R^*)$ such that $\partial_i' x_j = \partial_{ij}$. Choose $y_1, \ldots, y_m \in R$ as follows: $y_i = x_i$ for $i \leq n$, and $y_i$ is close to $x_i$ for $n < i \leq m$. Similarly, choose $\partial_1'', \ldots, \partial_m'' \in \mathrm{Der}_{k_\lambda}(R)$ so that $\partial_i''$ is close to $\partial_i'$ for each $i$. Then $\det(\partial_i'' y_j)$ is a unit in $R$, and by a linear change on $\partial_i''$ we get $\partial_1, \ldots, \partial_m \in \mathrm{Der}_{k_\lambda}(R)$ such that $\partial_i y_j = \partial_{ij}$. Then $\partial_1, \ldots, \partial_m$ form an $R$-basis of $\mathrm{Der}_{k_\lambda}(R)$, which is a $p$-Lie algebra. Hence $[\partial_i, \partial_j] = 0$ and $\partial_i^p = 0$ as in Lemma 1, and we have rank $J(P; \partial_1, \ldots, \partial_m)(P) = s$.

# *Hopf maps and quadratic forms over* $\mathbb{Z}$

*Takashi Ono*

*The Johns Hopkins University*

Not much has been done in the direction of the arithmetic of general quadratic mappings after the model of the theory of sphere bundles in topology. This kind of arithmetic seems to be still in the cradle. It is rather surprising that until recently we did not even try to consider an elementary diophantine problem, a system of three quadratic forms in four variables, related to the well-known map of 3-sphere to 2-sphere initiated by Hopf [4].

The purpose of this paper is to settle the problem at least for the case of three classical Hopf fibrations $S^3 \to S^2$, $S^7 \to S^4$, and $S^{15} \to S^8$. The methods in previous papers [7, 8] did not work for the last case because of the lack of good theory of ideals in the Cayley algebra. In this paper, we shall adopt a more flexible method so that all three cases can be treated almost uniformly. The new method indicates that for a general quadratic map of Hopf type there is associated a family of lattices, and the original diophantine problem is reduced to the diophantine problem of a single quadratic form considered on each lattice of the family.

The symbols $\mathbb{N}$, $\mathbb{Z}$, $\mathbb{Q}$, $\mathbb{R}$ denote the set of natural numbers, the ring of rational integers, the field of rational numbers, the field of real numbers, respectively.

It is my great pleasure to dedicate this work to Professor E. Kolchin, one of the pioneers of the theory of algebraic groups, on the occasion of his 60th birthday.

## 1. Hurwitz triple and the map $f$

Let $X$ be a vector space over $\mathbb{Q}$ of finite dimension and $q_X$ be a positive definite quadratic form on $X$. The associated inner product on $X \times X$ is given by

$$(x|x') = q_X(x + x') - q_X(x) - q_X(x').$$

We have

$$q_X(x) = \tfrac{1}{2}(x|x).$$

For an endomorphism $\alpha$ of $X$, we denote its adjoint by $\alpha^*$,

$$(\alpha x|x') = (x|\alpha^* x').$$

Let $Y$ be another vector space over $\mathbb{Q}$ and $q_Y$ be a positive definite quadratic form on $Y$. We shall use the same notation for the inner product and the adjoint of an endomorphism for the space $(Y, q_Y)$.

By a *Hurwitz triple*, we shall mean a triple $(q_X, q_Y, B)$, where $B$ is a biliner map $X \times Y \to Y$ satisfying the condition

$$q_Y(B(x, y)) = q_X(x)q_Y(y). \tag{1.1}$$

Denote by End $Y$ the ring of endomorphisms of the space $Y$ and define a linear map $\lambda\colon X \to \operatorname{End} Y$ by

$$B(x, y) = \lambda(x)y.$$

One sees easily that

$$\lambda(x)^*\lambda(x) = \lambda(x)\lambda(x)^* = q_X(x) \qquad \text{and} \qquad q_Y(\lambda(x)^* y) = q_X(x)q_Y(y).$$

Let $\mathfrak{o}_X$, $\mathfrak{o}_Y$ be lattices in $X$, $Y$, respectively. A Hurwitz triple $(q_X, q_Y, B)$ is said to be adapted to the pair $(\mathfrak{o}_X, \mathfrak{o}_Y)$ if the following conditions are satisfied:

$$q_X(\mathfrak{o}_X) \subset \mathbb{Z}, \qquad q_Y(\mathfrak{o}_Y) \subset \mathbb{Z}, \qquad \lambda(\mathfrak{o}_X)\mathfrak{o}_Y \subset \mathfrak{o}_Y, \qquad \lambda(\mathfrak{o}_X)^*\mathfrak{o}_Y \subset \mathfrak{o}_Y.$$

To a Hurwitz triple $(q_X, q_Y, B)$ one associates a map

$$f\colon X \times Y \to \mathbb{Q} \times \mathbb{Q} \times Y \tag{1.2}$$

given by $(x, y) \mapsto (q_X(x), q_Y(y), B(x, y))$. By (1.1), the image of $f$ is contained in the quadratic hypersurface

$$\Sigma = \{\sigma = (t, u, v),\ q_Y(v) = tu\}.$$

If, in addition, the triple is adapted to $(\mathfrak{o}_X, \mathfrak{o}_Y)$, then $f$ induces a map

$$f_{\mathbb{Z}}: \mathfrak{o}_X \times \mathfrak{o}_Y \to \Sigma_{\mathbb{Z}} = \Sigma \cap (\mathbb{Z} \times \mathbb{Z} \times \mathfrak{o}_Y).$$

This map is the main object of our study. Note that if $\sigma = (t, u, v)$ is in the image of $f_{\mathbb{Z}}$, then $t \geqq 0$. If $t = 0$, then one has

$$\begin{aligned} f(x, y) = \sigma \Leftrightarrow q_X(x) = 0, \qquad & q_Y(y) = u, \qquad B(x, y) = v \\ \Leftrightarrow x = 0, \qquad & q_Y(y) = u. \end{aligned}$$

Hence, the problem is reduced to solve a single quadratic equation on the lattice $\mathfrak{o}_Y$. For this reason, we shall pay attention mainly to the subset

$$\Sigma_{\mathbb{Z}}{}^* = \{\sigma = (t, u, v) \in \Sigma_{\mathbb{Z}}, t \in \mathbb{N}\}. \tag{1.3}$$

## 2. The map $f$ and the map $h$

The map $f$ introduced above (1.2) is not quite the Hopf map because it does not map a sphere to a sphere. However, there is little difference between the two maps.

The Hopf map $h$ associated with a Hurwitz triple $(q_X, q_Y, B)$ is a map

$$h: X \times Y \to \mathbb{Q} \times Y$$

defined by $(x, y) \mapsto (q_X(x) - q_Y(y), 2B(x, y))$. Hence, if we call $g$ the map $\Sigma \to \mathbb{Q} \times Y$ defined by $\sigma \mapsto (t - u, 2v)$, we have the relation $h = gf$.

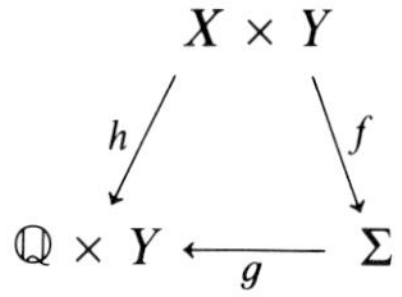

$$\begin{array}{ccc} & X \times Y & \\ {}^{h}\swarrow & & \searrow{}^{f} \\ \mathbb{Q} \times Y & \underset{g}{\longleftarrow} & \Sigma \end{array}$$

Given a quadratic space $(V, q_V)$, lattice $L$ such that $q_V(L) \subset \mathbb{Z}$ and $\rho \in \mathbb{N}$, we put

$$S_V(\rho) = \{x \in V, q_V(x) = \rho\}, \qquad S_L(\rho) = L \cap S_V(\rho).$$

In the above situation, put $q_{X \times Y}(x, y) = q_X(x) + q_Y(y)$, $q_{\mathbb{Q} \times Y}(r, s) = r^2 + q_Y(s)$. Then, we see from (1.1) that $h$ induces a map

$$h_\rho: S_{X \times Y}(\rho) \to S_{\mathbb{Q} \times Y}(\rho^2).$$

If, furthermore, the triple $(q_X, q_Y, B)$ is adapted to the pair $(\mathfrak{o}_X, \mathfrak{o}_Y)$ of lattices, then $h_\rho$ induces

$$h_{\rho, \mathbb{Z}}: S_{\mathfrak{o}_X \times \mathfrak{o}_Y}(\rho) \to S_{\mathbb{Z} \times \mathfrak{o}_Y}(\rho^2)^{\text{even}},$$

where we have put

$$S_{\mathbb{Z}\times\mathfrak{o}_Y}(\rho^2)^{\text{even}} = S_{\mathbb{Z}\times\mathfrak{o}_Y}(\rho^2) \cap (\mathbb{Z} \times 2\mathfrak{o}_Y).$$

Put

$$\Sigma(\rho) = \{\sigma = (t, u, v) \in \Sigma, t + u = \rho\}.$$

Then, $f$ induces

$$f_\rho\colon S_{X\times Y}(\rho) \to \Sigma(\rho),$$

and then

$$f_{\rho,\mathbb{Z}}\colon S_{\mathfrak{o}_X\times\mathfrak{o}_Y}(\rho) \to \Sigma_{\mathbb{Z}}(\rho),$$

where $\Sigma_{\mathbb{Z}}(\rho) = \Sigma(\rho) \cap (\mathbb{Z} \times \mathbb{Z} \times \mathfrak{o}_Y)$. Here the point is that the induced map $g_{\rho,\mathbb{Z}}$ in the commutative diagram

$$\begin{array}{ccc} & S_{\mathfrak{o}_X\times\mathfrak{o}_Y}(\rho) & \\ {}^{h_{\rho,\mathbb{Z}}}\swarrow & & \searrow^{f_{\rho,\mathbb{Z}}} \\ S_{\mathbb{Z}\times\mathfrak{o}_Y}(\rho^2)^{\text{even}} & \underset{g_{\rho,\mathbb{Z}}}{\longleftarrow} & \Sigma_{\mathbb{Z}}(\rho) \end{array} \tag{2.1}$$

is a bijection. The verification is easy and amusing. We only remark that to the point $w = (r, 2v) \in S_{\mathbb{Z}\times\mathfrak{o}_Y}(\rho^2)^{\text{even}}$ corresponds the point

$$\sigma = (\tfrac{1}{2}(\rho + r), \tfrac{1}{2}(\rho - r), v) \in \Sigma_{\mathbb{Z}}(\rho)$$

under this bijection. From this discussion, we see immediately that the fiber $h_{\rho,\mathbb{Z}}^{-1}(w)$ and the fiber $f_{\mathbb{Z}}^{-1}(\sigma)$ coincide, with $\sigma = (t, u, v)$, $t + u = \rho$, and $w = g_{\rho,\mathbb{Z}}(\sigma)$.

## 3. The family $\mathscr{L}$

Notation being as in Section 1, let $(q_X, q_Y, B)$ be a Hurwitz triple adapted to a pair $(\mathfrak{o}_X, \mathfrak{o}_Y)$, $f$ the map $(x, y) \mapsto (q_X(x), q_Y(y), B(x, y))$, and $\Sigma_{\mathbb{Z}}{}^*$ the set (1.3).

To each $\sigma = (t, u, v) \in \Sigma_{\mathbb{Z}}{}^*$, we associate the set

$$\mathfrak{a}_\sigma = \{x \in \mathfrak{o}_X, \lambda(x)^* v \in t\mathfrak{o}_Y\}. \tag{3.1}$$

One sees immediately that $\mathfrak{a}_\sigma$ is a lattice in $X$ such that

$$\mathfrak{o}_X \supset \mathfrak{a}_\sigma \supset t\mathfrak{o}_X.$$

For $(x, y) \in \mathfrak{o}_X \times \mathfrak{o}_Y$ and $\sigma = (t, u, v) \in \Sigma_{\mathbb{Z}}^*$, we have

$$\begin{aligned}(x, y) \in f_{\mathbb{Z}}^{-1}(\sigma) &\Leftrightarrow q_X(x) = t, \quad q_Y(y) = u, \quad \lambda(x)y = v\\ &\Leftrightarrow q_X(x) = t, \quad \lambda(x)^* v = ty\\ &\Leftrightarrow q_X(x) = t, \quad x \in \mathfrak{a}_\sigma\\ &\Leftrightarrow x \in S_{\mathfrak{a}_\sigma}(t).\end{aligned}$$

In other words, the projection $(x, y) \mapsto x$ induces a bijection

$$f_{\mathbb{Z}}^{-1}(\sigma) \approx S_{\mathfrak{a}_\sigma}(t). \tag{3.2}$$

In terms of the original Hopf map, (3.2) yields

(3.3) ***Theorem*** *Let $(q_X, q_Y, B)$ be a Hurwitz triple adapted to a pair $(\mathfrak{o}_X, \mathfrak{o}_Y)$ of lattices and let $h_{\rho, \mathbb{Z}}: S_{\mathfrak{o}_X \times \mathfrak{o}_Y}(\rho) \to S_{\mathbb{Z} \times \mathfrak{o}_Y}(\rho^2)^{\text{even}}$ be the corresponding Hopf map. Let $w = (r, 2v)$ be a point in $S_{\mathbb{Z} \times \mathfrak{o}_Y}(\rho^2)^{\text{even}}$. If $\rho + r = 0$, then the projection $(x, y) \mapsto y$ induces a bijection $h_{\rho, \mathbb{Z}}^{-1}(w) \approx S_{\mathfrak{o}_Y}(\rho)$ and if $\rho + r \neq 0$, then the projection $(x, y) \mapsto x$ induces a bijection $h_{\rho, \mathbb{Z}}^{-1}(w) \approx S_{\mathfrak{a}_\sigma}(\frac{1}{2}(\rho + r))$, where $\sigma = (\frac{1}{2}(\rho + r), \frac{1}{2}(\rho - r), v)$ and $\mathfrak{a}_\sigma$ is the lattice defined by (3.1).*

From now on, we shall denote by $\mathscr{L} = \mathscr{L}(q_X, q_Y, B; \mathfrak{o}_X, \mathfrak{o}_Y)$ the set of all lattices $\mathfrak{a}_\sigma$, $\sigma \in \Sigma_{\mathbb{Z}}^*$. Further study of the map $f$ depends upon the map $\Sigma_{\mathbb{Z}}^* \to \mathscr{L}$ given by $\sigma \mapsto \mathfrak{a}_\sigma$.

First, note that if $\sigma \in \Sigma_{\mathbb{Z}}^*$, then $m\sigma \in \Sigma_{\mathbb{Z}}^*$ for all $m \in \mathbb{N}$ and that $\mathfrak{a}_{m\sigma} = \mathfrak{a}_\sigma$. For a lattice $L = \mathbb{Z}\omega_1 + \cdots + \mathbb{Z}\omega_n$ of a vector space $V$ over $\mathbb{Q}$, and a nonzero vector $x = \Sigma_i x_i \omega_i$, $x_i \in \mathbb{Q}$, we denote by $c_L(x)$ the greatest common divisor of the coefficients $x_1, \ldots, x_n$. One sees immediately that $c_L(x)$ is independent of the choice of a basis. We call $c_L(x)$ the content of $x$ with respect to $L$. We have

$$x \in L \Leftrightarrow c_L(x) \in \mathbb{Z}.$$

We define the subset of the primitive vectors by

$$\Sigma_{\mathbb{Z}}^{(1)} = \{\sigma = (t, u, v) \in \Sigma_{\mathbb{Z}}^*, c_{\mathbb{Z} \times \mathbb{Z} \times \mathfrak{o}_Y}(\sigma) = 1\}.$$

Therefore, any $\sigma \in \Sigma_{\mathbb{Z}}^*$ can be written uniquely as $\sigma = c(\sigma)\sigma_1$ with $c(\sigma) = c_{\mathbb{Z} \times \mathbb{Z} \times \mathfrak{o}_Y}(\sigma)$ and $\sigma_1 = (t_1, u_1, v_1) \in \Sigma_{\mathbb{Z}}^{(1)}$. We have then

$$S_{\mathfrak{a}_\sigma}(t) = S_{\mathfrak{a}_{\sigma_1}}(c(\sigma)t_1). \tag{3.4}$$

We denote by $n(L)$ the norm of a lattice in a quadratic space $(V, q_V)$; by definition, $n(L)$ is the greatest common divisor of the rational numbers $q_V(x) = \frac{1}{2}(x|x)$, $x \in L$. For integers $a$, $b$, $a \neq 0$, we write $a|b$ when $a$ divides $b$ and we denote by $(a, b)$ the greatest common divisor of $a$ and $b$.

(3.5) ***Proposition*** *If* $\sigma = (t, u, v)$ *is primitive, then* $t \mid n(\mathfrak{a}_\sigma)$.

*Proof* Take any $x \in \mathfrak{a}_\sigma$. We have $\lambda(x)^* v = ty$ with $y \in \mathfrak{o}_Y$. Then, we have

$$t^2 q_Y(y) = q_Y(\lambda(x)^* v) = q_X(x) q_Y(v) = q_X(x) tu,$$

and hence $uq_X(x) = tq_Y(y)$. If we put $d = (t, u)$, then we have $td^{-1} \mid q_X(x)$. On the other hand, we have

$$t\lambda(x) y = \lambda(x)\lambda(x)^* v = q_X(x) v.$$

Hence, if we put $\delta = (t, c_{\mathfrak{o}_Y}(v))$, then $t\,\delta^{-1} \mid q_X(x)$. Since $\sigma$ is primitive, we have $(d, \delta) = 1$ and so $t \mid q_X(x)$, for any $x \in \mathfrak{a}_\sigma$, QED

Denote by $S(q_X)_{\mathbb{Q}}$ the group of similarities of the quadratic space $(X, q_X)$ over $\mathbb{Q}$:

$$S(q_X)_{\mathbb{Q}} = \{s \in GL(X),\ q_X(sx) = n(s) q_X(x)\}.$$

The adele group $S(q_X)_{\mathbb{A}}$ acts on the set of lattices on $X$ in a natural way. Denote by $U$ the subgroup of $S(q_X)_{\mathbb{A}}$ formed by adeles leaving the lattice $\mathfrak{o}_X$ invariant. It is well known that the double coset space $S(q_X)_{\mathbb{Q}} \backslash S(q_X)_{\mathbb{A}} / U$ is finite. We denote by $h$ the number of elements in this finite set. Let $\mathfrak{a}$, $\mathfrak{b}$ be lattices in $X$. We say that $\mathfrak{a}$ and $\mathfrak{b}$ are *similar*, written $\mathfrak{a} \sim \mathfrak{b}$, if $\mathfrak{b} = s\mathfrak{a}$ for some $s \in S(q_X)_{\mathbb{Q}}$ and that $\mathfrak{a}$ and $\mathfrak{b}$ are *locally similar*, written $\mathfrak{a} \underset{\text{loc}}{\sim} \mathfrak{b}$, if $\mathfrak{b} = s\mathfrak{a}$ for some $s \in S(q_X)_{\mathbb{A}}$. Thus, the orbit of $\mathfrak{o}_X$ under the action of the group $S(q_X)_{\mathbb{A}}$, i.e., the totality of lattices which are locally similar to $\mathfrak{o}_X$, splits into a finite number of similarity classes. Let $\mathfrak{a}_1, \ldots, \mathfrak{a}_h$ be a set of representatives of these classes.

We shall consider the following properties of the family $\mathscr{L} = \mathscr{L}(q_X, q_Y, B; \mathfrak{o}_X, \mathfrak{o}_Y)$:

(I) $\mathfrak{a}_\sigma \underset{\text{loc}}{\sim} \mathfrak{o}_X$ for all $\mathfrak{a}_\sigma \in \mathscr{L}$.

(I)* $\mathfrak{a}_\sigma \sim \mathfrak{o}_X$ for all $\mathfrak{a}_\sigma \in \mathscr{L}$.

(II) $n(\mathfrak{a}_\sigma) = t$ if $a_\sigma \in \mathscr{L}$ and $\sigma = (t, u, v)$ is primitive.

If (I) holds, we have $\mathfrak{a}_\sigma \sim \mathfrak{a}_i$ for some $i$, $1 \leqq i \leqq h$, and so $\mathfrak{a}_\sigma = s\mathfrak{a}_i$, $s \in S(q_X)_{\mathbb{Q}}$. Write $x \in \mathfrak{a}_\sigma$ as $x = sy$, $y \in \mathfrak{a}_i$. Then, we have

$$q_X(x) = n(s) q_X(y) = n(\mathfrak{a}_\sigma) n(\mathfrak{a}_i)^{-1} q_X(y).$$

Hence, we have a bijection:

$$S_{\mathfrak{a}_\sigma}(t) \approx S_{\mathfrak{a}_i}(tn(\mathfrak{a}_i) n(\mathfrak{a}_\sigma)^{-1}). \tag{3.6}$$

If, in addition to (I), (II) holds, then, writing $\sigma = c(\sigma)\sigma_1$ as before, we have

$$n(\mathfrak{a}_\sigma) = n(\mathfrak{a}_{\sigma_1}) = t_1 = c(\sigma)^{-1} t,$$

and (3.6) becomes

$$S_{\mathfrak{a}_\sigma}(t) \approx S_{\mathfrak{a}_i}(n(\mathfrak{a}_i)c(\sigma)). \tag{3.7}$$

Note that (II) implies $n(\mathfrak{o}_X) = 1$ because $\sigma = (1, 0, 0) \in \Sigma_{\mathbb{Z}}^{(1)}$ and $\mathfrak{a}_\sigma = \mathfrak{o}_X$. Finally, if (I) is replaced by the stronger (I)*, then (3.7) becomes

$$S_{\mathfrak{a}_\sigma}(t) \approx S_{\mathfrak{o}_X}(c(\sigma)) \qquad \text{for all} \quad \sigma \in \Sigma_{\mathbb{Z}}^*. \tag{3.8}$$

In terms of the Hopf map, (3.8) yields the following

(3.9) ***Theorem*** *Assumptions being the same as in* (3.3), *if the family* $\mathcal{L} = \mathcal{L}(q_X, q_Y, B; \mathfrak{o}_X, \mathfrak{o}_Y)$ *satisfies* (I)* *and* (II), *then we have, for* $w = (r, 2v) \in S_{\mathbb{Z} \times \mathfrak{o}_Y}(\rho^2)^{\text{even}}$, *a bijection*:

$$h_{\rho, \mathbb{Z}}^{-1}(w) \approx \begin{cases} S_{\mathfrak{o}_Y}(\rho) & \text{if} \quad \rho + r = 0,\dagger \\ S_{\mathfrak{o}_X}(c(\sigma)) & \text{if} \quad \rho + r \neq 0, \end{cases}$$

*where* $\sigma = (\frac{1}{2}(\rho + r), \frac{1}{2}(\rho - r), v)$ *and* $c(\sigma) = c_{\mathbb{Z} \times \mathbb{Z} \times \mathfrak{o}_Y}(\sigma)$.

It is an interesting problem to examine those properties for a given family $\mathcal{L} = \mathcal{L}(q_X, q_Y, B; \mathfrak{o}_X, \mathfrak{o}_Y)$.

Here is a typical example: Let $X$ be an imaginary quadratic field over $\mathbb{Q}$ or a definite quaternion algebra over $\mathbb{Q}$, $q_X(x) = Nx = \bar{x}x$, the norm of $x$, $B(x, y) = \lambda(x)y = \bar{x}y$, and $\mathfrak{o}_X$ a maximal order in $X$. One checks easily that $\lambda(x)^*y = xy$ and that $(q_X, q_X, B)$ is a Hurwitz triple adapted to $(\mathfrak{o}_X, \mathfrak{o}_X)$. We claim that the family $\mathcal{L}$ for this case has the properties (I), (II). First, each lattice $\mathfrak{a}_\sigma \in \mathcal{L}$ is a left ideal of $\mathfrak{o}_X$. Since the localization of a left ideal is principal and the multiplicative group of $X$ consists of similarities with respect to the norm, (I) follows immediately. (I)* holds if the class number of $X$ is 1. As for (II), note first that, for $\sigma \in \Sigma_{\mathbb{Z}}^*$, we have

$$\mathfrak{a}_\sigma = \mathfrak{o}_X \cap \mathfrak{o}_X t v^{-1}, \tag{3.10}$$

$$\mathfrak{a}_\sigma \supset \mathfrak{o}_X t + \mathfrak{o}_X \bar{v}. \tag{3.11}$$

For a left ideal $\mathfrak{a}$ in $X$ with respect to $\mathfrak{o}_X$, its dual $\mathfrak{a}^\#$ in the sense of quadratic forms is again a left ideal:

$$\mathfrak{a}^\# = \{x \in X, (x \mid y) \in \mathbb{Z} \text{ for all } y \in \mathfrak{a}\}.$$

One checks easily that‡

$$\mathfrak{a}^\# = \mathfrak{o}_X{}^\# \bar{\mathfrak{a}}^{-1} \qquad \text{and} \qquad (\mathfrak{a}\bar{z})^\# = \mathfrak{a}^\# z^{-1}, \qquad x \in X, \quad z \neq 0. \tag{3.12}$$

† Note that $\rho = c(\sigma)$ in this case also, because $\sigma = (0, \rho, 0)$.

‡ As for the ideal theory of associative algebras, see Artin [1].

From (3.10)–(3.12), it follows that

$$\mathfrak{o}_X{}^{\#}\bar{\mathfrak{a}}_\sigma^{-1}t = \mathfrak{a}_\sigma{}^{\#}t = \mathfrak{o}_X{}^{\#}t + \mathfrak{o}_X{}^{\#}\bar{v} = \mathfrak{o}_X{}^{\#}(\mathfrak{o}_X t + \mathfrak{o}_X \bar{v}) \subset \mathfrak{o}_X{}^{\#}\mathfrak{a}_\sigma.$$

Hence, we have $\bar{\mathfrak{a}}_\sigma^{-1}t \subset \mathfrak{a}_\sigma$, i.e. $t \subset \bar{\mathfrak{a}}_\sigma \mathfrak{a}_\sigma$, which implies that $N\mathfrak{a}_\sigma | t^\varepsilon$, where $\varepsilon = 1$ if $X$ is a quadratic field and $\varepsilon = 2$ if $X$ is a quaternion algebra. Since $N\mathfrak{a}_\sigma = n(\mathfrak{a}_\sigma)^\varepsilon$, we have $n(\mathfrak{a}_\sigma)|t$. Property (II) then follows from (3.5). By the way, the above argument shows that

$$\mathfrak{a}_\sigma = \mathfrak{o}_X t + \mathfrak{o}_X \bar{v} \qquad \text{and} \qquad t = \bar{\mathfrak{a}}_\sigma \mathfrak{a}_\sigma \qquad \text{for} \quad \sigma \in \Sigma_{\mathbb{Z}}^{(1)}.$$

If the class number of $X$ is 1, $\mathscr{L}$ has properties (I)* and (II) and (3.9) yields, with the specialization $Y = X$, that

$$h_{\rho,\mathbb{Z}}^{-1}(w) \approx S_{\mathfrak{o}X}(c(\sigma)), \tag{3.13}$$

this being the result proved before by a different approach [7, 8].

## 4. Hopf fibration $S^{15} \to S^8$ over $\mathbb{Z}$

We shall apply our method to the classical Hopf fibration $S^{15} \to S^8$. Let $H$ be the classical quaternion field over $\mathbb{Q}$ with quaternion units $1, i, j, k$ with relations $i^2 = j^2 = -1$, $k = ij = -ji$. Let $\mathfrak{o}_H$ be the unique maximal order of $H$ which contains the standard order $\mathbb{Z} + \mathbb{Z}i + \mathbb{Z}j + \mathbb{Z}k$. As is well known, $\mathfrak{o}_H$ is given by

$$\mathfrak{o}_H = \mathbb{Z}\rho + \mathbb{Z}i + \mathbb{Z}j + \mathbb{Z}k, \qquad \rho = \tfrac{1}{2}(1 + i + j + k).$$

Let $X = H + H\omega$ be the Cayley algebra with units $1, \omega$ over $H$. The multiplication in $X$ is derived from the basic relations

$$\omega^2 = -1, \qquad \omega x = \bar{x}\omega, \qquad x(y\omega) = (yx)\omega,$$

$$(x\omega)y = (x\bar{y})\omega, \qquad (x\omega)(y\omega) = -\bar{y}x, \qquad x, y \in H.$$

For $z = x + y\omega$, the conjugate is defined by $\bar{z} = \bar{x} - y\omega$. The norm and the trace are defined by $Nz = \bar{z}z$, $Tz = \bar{z} + z$. The properties

$$T(xy) = T(yx), \qquad T((xy)z) = T(x(yz)), \qquad x, y, z \in X \tag{4.1}$$

are very useful because $X$ itself is noncommutative and nonassociative.

Following Dickson [2] we consider the lattice in $X$:

$$\mathfrak{o}_X = \mathfrak{o}_H + \mathbb{Z}\omega + \mathbb{Z}\theta + \mathbb{Z}\varphi + \mathbb{Z}\psi,$$

where $\theta = \frac{1}{2}(1 + i + \omega + j\omega)$, $\varphi = \frac{1}{2}(1 + j + \omega + i\omega)$, $\psi = \frac{1}{2}(1 + k + \omega +$

$k\omega$). Put $q_X(x) = Nx$. Then we have $(x|y) = T(\bar{x}y)$. The symmetric matrix of the inner products associated to the basis $\rho$, $i$, $j$, $k$, $\omega$, $\theta$, $\varphi$, $\psi$ of $\mathfrak{o}_X$ is

$$A = \begin{pmatrix} 2 & 1 & 1 & 1 & 0 & 1 & 1 & 1 \\ 1 & 2 & 0 & 0 & 0 & 1 & 0 & 0 \\ 1 & 0 & 2 & 0 & 0 & 0 & 1 & 0 \\ 1 & 0 & 0 & 2 & 0 & 0 & 0 & 1 \\ 0 & 0 & 0 & 0 & 2 & 1 & 1 & 1 \\ 1 & 1 & 0 & 0 & 1 & 2 & 1 & 1 \\ 1 & 0 & 1 & 0 & 1 & 1 & 2 & 1 \\ 1 & 0 & 0 & 1 & 1 & 1 & 1 & 2 \end{pmatrix}.$$

One has $\det A = 1$. Hence, $\mathfrak{o}_X$ is a unimodular lattice, i.e.,

$$\mathfrak{o}_X{}^{\#} = \{x \in X, (x|y) \in \mathbb{Z} \text{ for all } y \in \mathfrak{o}_X\} = \mathfrak{o}_X.$$

If we put $B(x, y) = \lambda(x)y = \bar{x}y$, $x$, $y, \in X$, then we have $\lambda(x)^*y = xy$. One verifies easily that $(q_X, q_X, B)$ is a Hurwitz triple adapted to $(\mathfrak{o}_X, \mathfrak{o}_X)$. The Hopf map $h\colon X \times X \to \mathbb{Q} \times X$ reads $h(x, y) = (Nx - Ny, 2\bar{x}y)$. The map $h$ is the restriction on $\mathbb{Q}^{16}$ of the map $\mathbb{R}^{16} \to \mathbb{R}^9$ which induces the classical fibration $S^{15} \to S^8$ where each fiber is $S^7$ [5].

Before going into more details, we summarize here some facts on lattices which we need.† Let $L$ be a lattice in a quadratic space $(V, q_V)$. The determinant $d(L)$ is $\det((\omega_i|\omega_j))$ when $L = \Sigma_i \mathbb{Z}\omega_i$. The reduced determinant $d_{\text{red}}(L)$ is $n(L)^{-\dim V} d(L)$, this being an integer. We say that $L$ is maximal if there is no lattice $L'$ such that $L \subsetneq L'$ and $n(L) = n(L')$. If $d_{\text{red}}(L)$ has no square factors, $L$ is maximal. We have $d(L^{\#})d(L) = 1$ for the dual $L^{\#}$. For two maximal lattices $L$, $M$, they are locally similar if and only if $d_{\text{red}}(L) = d_{\text{red}}(M)$ and they are in the same genus if and only if $n(L) = n(M)$.

Back to the Cayley algebra $X$, we claim that the family $\mathscr{L} = \mathscr{L}(q_X, q_X, B; \mathfrak{o}_X, \mathfrak{o}_X)$ has properties (I)*, (II). First, for any $\mathfrak{a}_\sigma \in \mathscr{L}$, note that the following relations still hold for $X$:

$$\mathfrak{a}_\sigma = \mathfrak{o}_X \cap \mathfrak{o}_X t v^{-1}, \tag{4.2}$$

$$\mathfrak{a}_\sigma \supset \mathfrak{o}_X t + \mathfrak{o}_X \bar{v}. \tag{4.3}$$

Furthermore, for any lattice $\mathfrak{a}$ in $X$ and $z$ ($\neq 0$) in $X$, using (4.1), we see that

$$(\mathfrak{a}\bar{z})^{\#} = \mathfrak{a}^{\#} z^{-1}. \tag{4.4}$$

† As for details, see Eichler [3, Chapters II and III].

We stated above the property $\mathfrak{o}_X{}^{\#} = \mathfrak{o}_X$ which is peculiar to the Cayley algebra. Therefore, from (4.2)–(4.4), it follows that

$$\mathfrak{a}_\sigma{}^{\#} = \mathfrak{o}_X + \mathfrak{o}_X t^{-1}\bar{v} \qquad \text{and} \qquad \mathfrak{a}_\sigma{}^{\#} t = \mathfrak{o}_X t + \mathfrak{o}_X \bar{v} \subset \mathfrak{a}_\sigma .$$

On the other hand, if $\sigma$ is primitive, (3.5) implies that $t \mid (x \mid y)$ for all $x,\ y \in \mathfrak{a}_\sigma$. Hence, we have $t^{-1}\mathfrak{a}_\sigma \subset \mathfrak{a}_\sigma{}^{\#}$ and so $\mathfrak{a}_\sigma \subset \mathfrak{a}_\sigma{}^{\#} t$. We have thus proved that

$$\mathfrak{a}_\sigma{}^{\#} t = \mathfrak{o}_X t + \mathfrak{o}_X \bar{v} = \mathfrak{a}_\sigma , \qquad \sigma \in \Sigma_{\mathbb{Z}}^{(1)}. \tag{4.5}$$

Taking the determinant of (4.5), we have $t^{16} d(\mathfrak{a}_\sigma{}^{\#}) = d(\mathfrak{a}_\sigma)$, hence $d(\mathfrak{a}_\sigma) = t^8$, which implies that $n(a_\sigma) \mid t$. In view of (3.5), we have $n(\mathfrak{a}_\sigma) = t$, and (II) is settled. From this, it follows that $n(\mathfrak{o}_X) = 1$, a fact which one sees directly also. Anyway, $d_{\text{red}}(\mathfrak{o}_X) = 1$ and $\mathfrak{o}_X$ is maximal. As for property (I), note that $d_{\text{red}}(\mathfrak{a}_\sigma) = 1$ since $d(\mathfrak{a}_\sigma) = t^8 = n(\mathfrak{a}_\sigma)^8$ by (II). Hence $\mathfrak{a}_\sigma$ is maximal and so locally similar to $\mathfrak{o}_X$, i.e. property (I). To prove the stronger property (I)*, it is enough to prove that any lattice $\mathfrak{a}$ in $X$ which is locally similar to $\mathfrak{o}_X$ is actually similar to $\mathfrak{o}_X$. Let $v = n(\mathfrak{a})$. By Lagrange's theorem, there is a quaternion $\gamma \in H$ such that $N\gamma = v$. Put $\mathfrak{b} = \gamma^{-1}\mathfrak{a}$. Then $\mathfrak{b}$ is maximal since $\mathfrak{a}$ is also, and we have $n(\mathfrak{b}) = 1 = n(\mathfrak{o}_X)$. Hence $\mathfrak{b}$ belongs to the genus of $\mathfrak{o}_X$. However, the genus of $\mathfrak{o}_X$ consists of only one class (see Mordell [6]) and so $\mathfrak{b} = \beta\, \mathfrak{o}_X$ with some $\beta \in O(q_X)_{\mathbb{Q}}$, the orthogonal group of $q_X$ over $\mathbb{Q}$. Therefore, we have $\mathfrak{a} = \gamma\mathfrak{b} = \gamma\beta\mathfrak{o}_X = \alpha\mathfrak{o}_X$ with $\alpha \in S(q_X)_{\mathbb{Q}}$, i.e. $\mathfrak{a} \sim \mathfrak{o}_X$, QED

## References


1. E. Artin, Zur Arithmetik hyperkomplexer Zahlen, *Abh. Math. Sem. Univ. Hamburg* **5** (1928), 261–289.
2. L. E. Dickson, A new simple theory of hypercomplex integers, *J. Math. Pures Appl.* **2** (1923), 281–326.
3. M. Eichler, "Quadratische Formen und orthogonale Gruppen." Springer-Verlag, Berlin–Heidelberg–New York, 1952.
4. H. Hopf, Über die Abbildungen der dreidimensionalen Sphäre auf die Kugelfläche, *Math. Ann.* **104** (1931), 637–665.
5. H. Hopf, Über die Abbildungen von Sphären auf Sphären niedrigerer Dimension, *Fund. Math.* **25** (1935), 427–440.
6. L. J. Mordell, The definite quadratic forms in eight variables with determinant unity, *J. Math. Pures Appl.* **17** (1938), 41–46.
7. T. Ono, On the Hopf fibration over $\mathbb{Z}$, *Nagoya Math. J.* **56** (1975), 201–207.
8. T. Ono, On the Hopf fibration $S^7 \to S^4$ over $\mathbb{Z}$, *Nagoya Math. J.* **59** (1975), 59–64.


AMS (MOS) 1970 subject classification: 10B05

# *Families of subgroup schemes of formal groups*

*Frans Oort*

*Mathematisch Instituut*
*Amsterdam*

In this paper we prove that flat families of finite subgroup schemes of a formal group (in characteristic $p \neq 0$) have bounded dimension if at least the family is not constant on positive-dimensional subvarieties. The nice fact about this bound is that it does not depend on the group chosen within the isogeny class. Thus one can compute upper bounds for the dimension of such families, and an application is made to the liftability of abelian varieties of dimension 3.

Some notations: We write AV for abelian variety, $X^t$ for the dual of the AV $X$, and $G^t$ for the dual (in the sense of [15, 2.3]) of a $p$-divisible group $G$. All schemes are supposed to be locally noetherian; we use throughout an algebraically closed field $k$ of characteristic $p \neq 0$. We write $G \sim H$ to mean that $G$ and $H$ are isogenous. For a group scheme $G$ in characteristic $p$, we write

$$a(G) := \dim_L \operatorname{Hom}(\alpha_p \otimes L, G \otimes L),$$

where $L$ is some algebraically closed field.

## 1. Strict families of finite subgroup schemes of formal groups

***Definition 1*** Let $G$ be a $p$-divisible formal group (or an abelian variety) over $k$. Let $T$ be a noetherian prescheme over $k$, and let

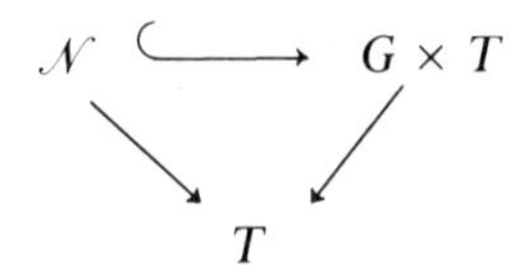

be a flat family of finite subgroup schemes of $G$ parametrized by $T$. We say this is a *strict* family if for every $t \in T$ the set

$$\{t' \in T \mid \mathcal{N}_t = \mathcal{N}_{t'} \subset G\}$$

is finite (i.e., there should not exist curves in $T$ over which the family is constant). We denote by $c(G)$ the *maximum* of $\dim(T)$, where $T$ is irreducible, parametrizing some strict family of finite subgroup schemes of $G$.

*Remark* For any $G$ as above, $c(G)$ exists (i.e., $\dim(T)$ is bounded for strict families); we do not give a proof here.

***Lemma 2*** *Let $G$ and $H$ be isogenous abelian varieties or p-divisable formal groups. Then*

$$c(G) = c(H).$$

*Proof* Let $G \to H$ be an isogeny, and $\mathcal{N} \hookrightarrow H \times T$ a strict family as in the previous definition. Then

$$(G \times T \to H \times T)^{-1}(\mathcal{N}) = M \to G \times T$$

is a strict family parametrized by $T$, hence $c(H) \leq c(G)$. Using an isogeny $H \to G$, we obtain the inequality the other way around, which proves the lemma.

We indicate a procedure to compute $c(G)$ in some simple cases. First note (notation taken from [5, p. 35]):

***Lemma 3*** *Let $n \in \mathbb{Z}$, $n \geq 1$, and suppose*

$$G \sim G_{1,n}.$$

*Then $G$ is isomorphic with $G_{1,n}$.*

*Proof* Let $G_{1,n} \to G$ be an isogeny of degree $p^m$. Then its kernel equals the kernel of

$$F^m\colon G_{1,n}^{(p^m)} \cong G_{1,n} \to G_{1,n},$$

and the lemma is proved.

**Corollary 4** *If $G \sim G_{n,1}$, then $G \cong G_{n,1}$.*

*Proof* The dual $(G_{n,1})^t$ equals $G_{1,n}$, etc.

**Lemma 5** *Let $m, n \in \mathbb{Z}$, $m \geq 1$, $n \geq 1$, and suppose*

$$G \sim G_{1,n} + G_{m,1}.$$

*Then either $G \cong G_{1,n} + G_{m,1}$, or there exists an exact sequence*

$$0 \to \alpha_p \to G_{1,n} + G_{m,1} \to G \to 0.$$

*Proof* The case $n = 1 = m$ follows immediately from [13, Th. 2 and Cor. 7]. Thus we assume $n > 1$ (or, dually, $m > 1$). Let

$$0 \to M \to G_{1,n} + G_{m,1} \to G \to 0;$$

then

$$0 \to M/M' \to (G_{1,n}/M') + G_{m,1} \to G \to 0,$$

with $M' := M \cap G_{1,n}$; note that

$$G_{1,n}/M' \cong G_{1,n}$$

by the previous lemma. Next proceed with $M'' := (M/N') \cap G_{m,1}$. Thus we arrive at an exact sequence

$$0 \to N \to G_{1,n} + G_{m,1} \to G \to 0$$

with

$$N \cap G_{1,n} = 0 = N \cap G_{m,1}.$$

Then either $N = 0$, and we are done, or $N \neq 0$. In that case, $a(N) = 2$ is excluded, because $N \cap G_{1,n} = 0$; thus $a(N) = 1$. Suppose $\alpha_p \subsetneqq N$. Then we construct $M \subset N$ by the exact commutative diagram

$$\begin{array}{ccccccccc}
0 & \longrightarrow & \alpha_p & \longrightarrow & N & \longrightarrow & N/\alpha_p & \longrightarrow & 0 \\
 & & \| & & \uparrow & & \uparrow j & & \\
0 & \longrightarrow & \alpha_p & \longrightarrow & M & \longrightarrow & \alpha_p & \longrightarrow & 0
\end{array}$$

after choosing some $j \neq 0$. The rank of $M$ equals $p^2$; let $f$ and $g$ be defined by

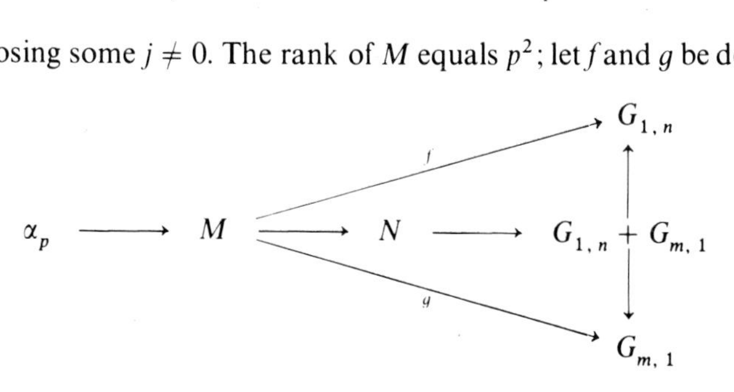

Using $a(N) = 1$ and $N \cap G_{1,n} = 0$, we conclude $g \neq 0$, and analogously $f \neq 0$. However, this leads to a contradiction: use Dieudonné module theory (e.g. cf. [10, 15.5]). This proves Lemma 5.

***Remark 6*** If $G \cong G_{1,n} + G_{m,1}$, then $G \cong (G_{1,n}/\alpha_p) + G_{m,1}$; thus the first case is a special case of the second. In the sequel we use only the second case.

***Remark 7*** Another way of proving the previous lemma is the following: the image of

$$(G_{1,n} \to G_{1,n} + G_{m,1} \twoheadrightarrow G)$$

is isomorphic with $G_{1,n}$ (by Lemma 3), and we obtain an exact sequence

$$0 \to G_{1,n} \to G \to G_{m,1} \to 0$$

(using Corollary 4). Next one computes that for the exact sequence

$$0 \to \alpha_p \to G_{m,1} \xrightarrow{f} G_{m,1} \to 0$$

we obtain the zero map (note that $k$ is algebraically closed, hence perfect)

$$0 = f^*\colon \mathrm{Ext}(G_{m,1}, G_{1,n}) \to \mathrm{Ext}(G_{m,1}, G_{1,n}).$$

Hence we arrive at a commutative, exact diagram:

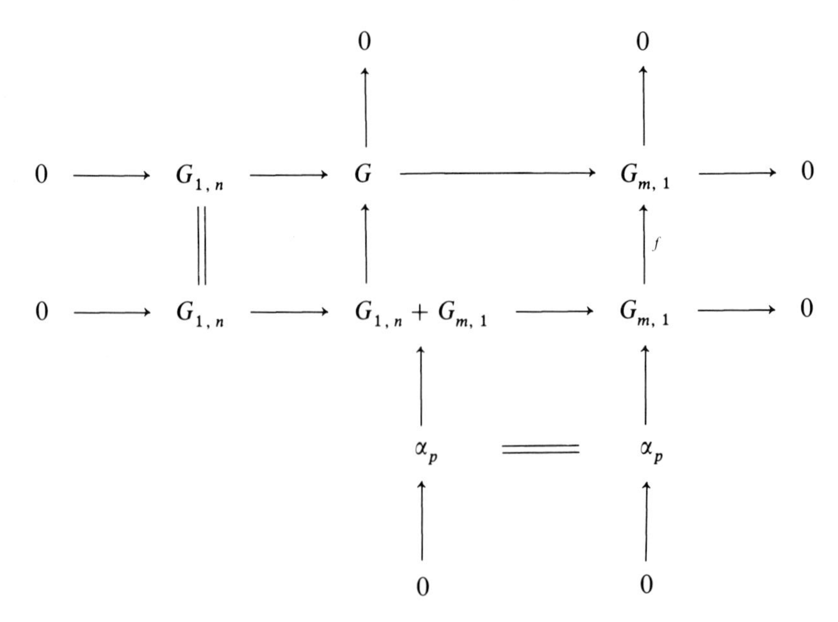

***Lemma 8*** (Rigidity lemma) *Let $S$ be a connected scheme, $G$ and $H$ are $p$-divisible groups over $S$, and let*

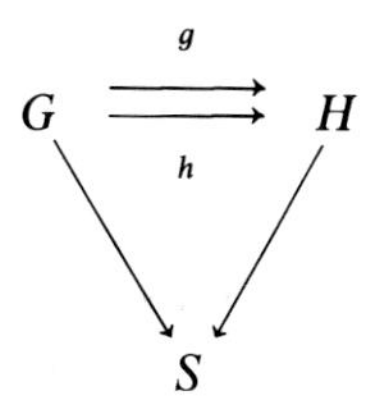


*be homomorphisms such that for at least one point $s \in S$, we have*

$$g_s = h_s\colon G_s \to H_s\,.$$

*Then $g = h$.*

*Proof* It suffices to show that $f := g - h = 0$. Further, using well-known methods, it can be seen that it suffices to study only the following situation: let $R$ be a local, artinian ring with maximal ideal $\mathfrak{m}$, let $I \subset R$ be an ideal with $\mathfrak{m}I = 0$, and $R' := R/I$; let $G$ and $H$ be $p$-divisible groups over $R$, and $f\colon G \to H$ a homomorphism such that

$$0 = f' := f \otimes R'\colon G' := G \otimes_R R' \to H';$$

then $f = 0$. This we prove as follows. Consider the commutative diagram

$$\begin{array}{ccc} G & \xrightarrow{f} & H \\ {\scriptstyle \times p}\downarrow & & \downarrow{\scriptstyle \times p} \\ G & \xrightarrow{f} & H \end{array}$$

In case the groups $G$ and $H$ are étale we are done by [2, Chap. IV$^4$, 18.1.2]. Hence we may suppose $R$ has residue characteristic $p \neq 0$, and we may suppose $G$ and $H$ are connected $p$-divisible groups (use [15, 1.4], and note the fact that a homomorphism from an étale to a connected $p$-divisible group is trivial), and hence work with formal groups (cf. [15, 2.2, Prop. 1]). Let $M_H = M$ be the augmentation ideal of the local ring of $H$, and $\mathfrak{m}$ the maximal ideal of $R$. Note that

$$(\times\, p)^*\colon M \to M^2 + \mathfrak{m}M \subset M.$$

By $f' = 0$, we know

$$f^*\colon M \to I \cdot M_G \subset M_G\,.$$

Thus by $\mathfrak{m}I = 0$ we conclude that

$$(G \xrightarrow{f} H \xrightarrow{\times p} H) = 0.$$

The fact that $G$ is $p$-divisible implies that $\times\, p\colon G \to G$ is epimorphic, e.g., the map on the local rings $\mathcal{O}_G \to \mathcal{O}_G$ is injective. Thus $f = 0$, which proves the rigidity lemma.

***Construction 9*** Let $n \geq 1$, $m \geq 1$, $G = G_{1,n} + G_{m,1}$. We construct a strict family parametrizing a "variable $\alpha_p$ inside $G$." Let $T = \mathbb{P}^1$, and fix $\alpha_p \subset G_{1,n}$ and $\alpha_p \subset G_{m,1}$. For $t = (i : j) \in \mathbb{P}^1$, we define

$$\phi_t\colon \alpha_p \to G \quad \text{by} \quad \alpha_p \begin{array}{l} \xrightarrow{i} \alpha_p \hookrightarrow G_{1,n} \\ \xrightarrow{j} \alpha_p \hookrightarrow G_{m,1} \end{array}$$

The image $\phi_t(\alpha_p) = \mathcal{N}_t \subset G$ only depends on $t$ (and not on the choice for $i$ and $j$), and thus we have constructed

$$\{\mathcal{N}_t\}_{t \in T} =: \mathcal{N} \hookrightarrow G \times T,$$

which is a flat family of finite subgroup schemes of $G$; in fact, this is a strict family.

***Proposition 10*** *Let $n \geq 1$, $m \geq 1$, and $G \sim G_{1,n} + G_{m,1}$; then $c(G) = 1$.*

*Proof* Using Lemma 2 we see that it suffices to consider the case

$$G = G_{1,n} + G_{m,1}.$$

Let $\mathcal{M} \hookrightarrow G \times S$ be a strict, flat family of finite subgroup schemes parametrized by an irreducible, reduced $K$-scheme $S$; we like to show $\dim(S) \leq 1$. Let $s \in S$ be the generic point, $L$ an algebraic closure of $k(s)$,

$$\mathcal{H} := (G \times S)/\mathcal{M}$$

with generic fiber $H := \mathcal{H}_s$. We apply Lemma 5 to $H \otimes L$, and thus we obtain an $L$-homomorphism

$$f\colon \alpha_p \otimes L \to G \otimes L$$

such that

$$(G \otimes L)/f(\alpha_p \otimes L) \cong H \otimes L.$$

Let $K$ be a *finite* extension of $k(s)$ such that $f$ is defined over $K$. We replace $S$ by its normalization in $K$ (again call it $S$), and we pull back $\mathcal{M}$ and $\mathcal{H}$ to this new $S$ (again write $\mathcal{M}$ and $\mathcal{H}$; from now on $H \otimes K$ will be denoted by $H$). Note that

$$H' := (G \otimes K)/f(\alpha_p \otimes K)$$

is isomorphic *over* $L$ with $H$. (In general one cannot conclude in such a situation that $H'$ and $H$ are isomorphic over $K$, or over a finite extension of $K$, but in our situation things are easier.) We choose $q \in \mathbb{Z}$ such that $q \cdot 1_{\mathcal{M}} = 0$, and we construct the following commutative diagram of $K$-formal groups:

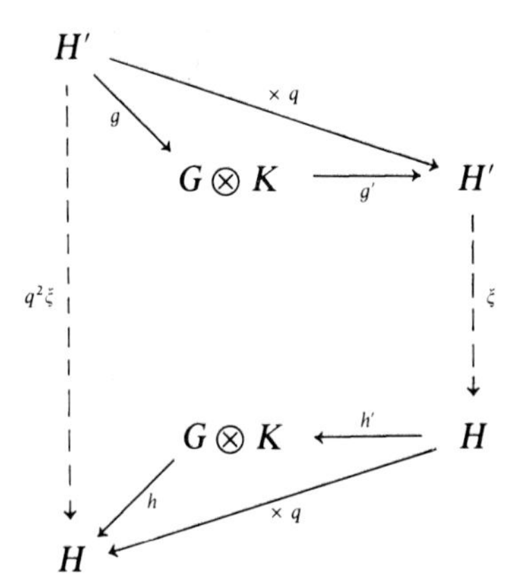


where dashed lines indicate homomorphisms which are defined over $L$ (and $\xi$ is an $L$-isomorphism). Consider

$$h'\xi g' \in \operatorname{End}_L(G = G_{1,n} + G_{m,1});$$

notice that all endomorphisms of $G_{n,m}$ are defined over a field with $p^{n+m}$ elements (cf. [14, Lemma 1.2]), and because $k$ is algebraically closed and $k \subset K$, we conclude

$$(h'\xi g') \in \operatorname{End}_K(G);$$

hence

$$q^2\xi = h(h'\xi g')g \in \operatorname{Hom}_K(H', H);$$

moreover, $\operatorname{Ker}(q^2\colon H' \to H')$ is defined over $K$ and contained in $\operatorname{Ker}(q^2\xi)$; thus $\xi \in \operatorname{Hom}_K(H', H)$, hence $\xi$ is a $K$-isomorphism. We fix, as before, $(\alpha_p)^2 \subset G = G_{1,n} + G_{m,1}$; thus

$$f\colon \alpha_p \otimes K \hookrightarrow (\alpha_p)^2 \otimes K \to G$$

is given by $f = (x, y)$ with $x, y \in K$, and one of these is nonzero. Note that the image of $f$ only depends on the ratio $x : y$. These elements $x, y \in K$ define a rational $k$-map $S \dashrightarrow \mathbb{P}^1 = T$, hence replacing $S$ by a nonempty open subset (again denoted by $S$), we obtain a $k$-morphism

$$b\colon S \to \mathbb{P}^1 = T.$$

We denote by

$$\mathscr{B} := (G \times T)/\mathscr{N}$$

the quotient formal group, where $\mathscr{N} \to G \times T$ is the one constructed in (9). Let $\mathscr{B}' \to S$ be the pullback of $\mathscr{B}$ via $b$. Notice that the $K$-isomorphism $\xi\colon H' \to H$ and the choice of $b$ implies that the generic fibers of $\mathscr{B}' \to S$ and of $\mathscr{H} \to S$ are isomorphic. Hence replacing $S$ by a nonempty open subset, we obtain an $S$-isomorphism

$$\mathscr{B}' \cong \mathscr{H}$$

(probably it is not necessary to replace $S$ by something smaller, but we do not need that here). Take any $t \in T(K)$, and consider

$$C := \overset{-1}{b}(t);$$

clearly $\mathscr{B}' \,|\, C$ is constant over $C$ (suppose $C$ is nonempty):

$$\mathscr{H} \,|\, C \cong \mathscr{B}' \,|\, C = \mathscr{B}_t \times C;$$

using the exact sequence

$$0 \to \mathscr{N} \,|\, C \to G \times C \xrightarrow{\pi} \mathscr{H} \,|\, C \to 0,$$

it follows by the rigidity lemma that $\mathscr{N} \,|\, C$ is constant inside $G \times C$ (take any $d \in C(k)$, and compare $\pi\colon G \times C \to \mathscr{H} \,|\, C \cong (G/\mathscr{N}_d) \times C$ and the quotient morphism $(G \to (G/N_\alpha)) \times C$). Because $\mathscr{N} \hookrightarrow G \times S$ is a strict family, it follows that $C$ is finite, and hence $\dim(S) \leq \dim(T) = 1$, which concludes the proof of Proposition 10.

In Lemma 5 we classified all groups isogenous with $G_{1,n} + G_{m,1}$. Now we write

$$G := G_{1,1} + G_{1,n} + G_{m,1}, \qquad n \geq 1, \quad m \geq 1,$$

and we are going to classify all groups isogenous with this $G$.

**Lemma 11** *Let $H \sim G$. Then there exists finite subgroup schemes*

$$N_2 \subset N_1 \subset G := G_{1,1} + G_{1,n} + G_{m,1}$$

*such that*

$$H \cong G/N_1,$$

$N_1/N_2 \cong \alpha_p$, *and*

$$\begin{array}{ccc} N_2 \cong (\alpha_p)^2 & \longrightarrow & G \\ \big\downarrow \text{id} & & \big\downarrow \\ \alpha_p + \alpha_p & \hookrightarrow & G_{1,n} + G_{m,1}. \end{array}$$

*Proof* An isogeny $G \to H$ defines a nontrivial homomorphism $G_{1,1} \to H$, and hence by Lemma 3 we can find $G_{1,1} \hookrightarrow H$. We define $X$ by the exact sequence

$$0 \to G_{1,1} \to H \to X \to 0.$$

By Lemma 5 we can choose an exact sequence

$$0 \to \alpha_p \to G_{1,n} + G_{m,1} \to X \to 0.$$

Further we notice that for

$$0 \to \alpha_p \to G_{1,n} \xrightarrow{f} G_{1,n} \to 0,$$

$$0 \to \alpha_p \to G_{m,1} \xrightarrow{v} G_{m,1} \to 0,$$

the maps

$$0 = f^*\colon \operatorname{Ext}(G_{1,n}, G_{1,1}) \to \operatorname{Ext}(G_{1,n}, G_{1,1})$$

and

$$0 = v^*\colon \operatorname{Ext}(G_{m,1}, G_{1,1}) \to \operatorname{Ext}(G_{m,1}, G_{1,1})$$

are zero. Thus we arrive at a commutative diagram with exact rows,

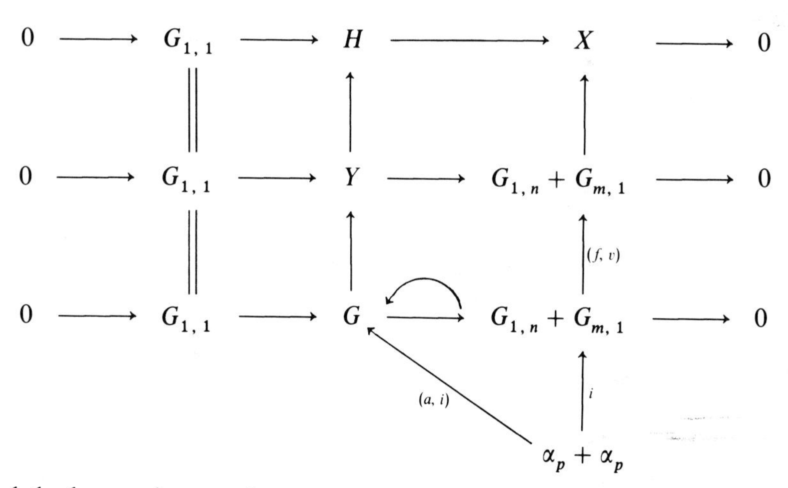


and the lemma is proved.

***Construction 12*** We construct two families of formal groups, such that all fibers are isogenous with $G$ (with $G$ as in Lemma 11). Consider $V = \mathbb{A}^2$, and parametrize a family of homomorphisms

$$a_v\colon \alpha_p + \alpha_p \to \alpha_p \subset G_{1,1}$$

by $V$ in the obvious way. Then we obtain

$$\phi_v := (a_v, i)\colon \alpha_p + \alpha_p \to G_{1,1} + G_{1,n} + G_{m,1},$$

where

$$i\colon \alpha_p + \alpha_p \xrightarrow{\sim} \alpha_p + \alpha_p \subset G_{1,n} + G_{m,1}$$

is fixed, and the images $\phi_v(\alpha_p + \alpha_p)$ define a flat family

$$\{\mathrm{Im}\, \phi_v\}_{v \in V} =: \mathcal{N}_2 \to G \times V$$

of finite subgroup schemes parametrized by $V$. We write $\mathcal{Y} := (G \times V)/\mathcal{N}_2$. For any $v \in V$ the exact sequence

$$0 \to \mathcal{N}_{2,v} \to G \to \mathcal{Y}_v \to 0$$

gives rise to an exact sequence

$$0 \to (\mathcal{N}_{2,v})^D \cong \alpha_p + \alpha_p \to \mathcal{Y}_v{}^t \to G^t \to 0$$

(cf. [10, 19.1]); hence $a(\mathcal{Y}_v{}^t) \geq 2$. It can be shown that $a(Y^t) = a(Y)$ for any $p$-divisible group $Y$; thus $a(\mathcal{Y}_v) \geq 2$. It is easy to see that for general $v \in V$, equality $= 2$ holds. Let

$$U := \{v \mid v \in V,\ a(\mathcal{Y}_v) = 2\};$$

this is an open, nonempty subset of $V$. We write $V' := V \backslash U$; clearly $v \in V'$ implies $a(Y) = 3$. For any $v \in U(k)$

$$\dim_k(\mathrm{Hom}(\alpha_p, \mathcal{Y}_v)) = 2;$$

therefore for each $v \in U(k)$ there is a $\mathbb{P}^1$-family of subgroup schemes $\alpha_p \to Y_v$. All these families with variable $v \in U$ can be glued together to a family: there exists a scheme $W$, a morphism $\pi\colon W \to U$ (all fibers $\cong \mathbb{P}^1$), and a flat family

$$\mathcal{N}_1' \hookrightarrow \overset{-1}{\pi}\mathcal{Y}$$

of subgroup schemes of the pullback of $\mathcal{Y} \mid U$ to $W$; for every $w \in W$, the group scheme $\mathcal{N}'_{1,w}$ is isomorphic with $\alpha_p$; the existence of $W \to U$ can be seen as follows; there exists a unique flat $U$-group scheme

$$\mathcal{A} \to \mathcal{Y} \mid U$$

such that each fiber over $U$ is $(\alpha_p)^2$ (this follows immediately from $a(\mathcal{Y}_v) = 2$ for all $v \in U$); the Lie algebra of the $U$-group scheme $A$ is a rank-2 vector bundle $L$ on $U$ (e.g. cf. [1, II.4.4.8]), any $\alpha_p \to \mathcal{A}_v$ is given by a line in $L_v$, and conversely any line in $L_v$ gives a unique subgroup scheme $\alpha_p \subset \mathcal{A}_v$ (cf. [1, II.7.4.2]); thus $W$ can be constructed as the Grassmann scheme over $U$ consisting of all lines in all fibers of $L$. We can now pull back $\mathcal{N}_2$ to $W$,

$$\mathcal{N}_2' = \overset{-1}{\pi}(\mathcal{N}_2),$$

and arrive at the exact, commutative diagram

$$\begin{array}{ccccccccc} 0 & \longrightarrow & \mathcal{N}_2' & \longrightarrow & G \times W & \longrightarrow & \pi^{-1}Y & \longrightarrow & 0 \\ & & \| & & \uparrow & & \uparrow & & \\ 0 & \longrightarrow & \mathcal{N}_2' & \longrightarrow & \mathcal{N}_1 & \longrightarrow & \mathcal{N}_1' & \longrightarrow & 0 \end{array}$$

thus arriving at a flat family

$$\mathcal{N}_2' \subset \mathcal{N}_1 \subset G \times W$$

parametrized by $W$. Note that all fibers of this situation have properties as mentioned in Lemma 11, and note that any situation such as described in Lemma 11 appears as one of these fibers if $a(G/N_2) = 2$.

In the same way we produce a morphism $W' \to V'$ with fibers $\cong \mathbb{P}^2$ parametrizing all embeddings of $\alpha_p$ into all $\mathcal{Y}_v$, $v \in V'$, and finally we obtain

$$\mathcal{N}_2'' \subset \mathcal{N}_1'' \subset G \times W'$$

with the obvious properties. Any situation as described in Lemma 11 with $a(G/N_2) = 3$ appears as a fiber over $W'$. Note that

$$\dim W = \dim(U) + 1 = 3 \qquad \text{and} \qquad \dim W' = \dim(V') + 2 \leq 3.$$

This ends Construction 12.

***Corollary 13*** *Let $n \geq 1$, $m \geq 1$, and $H \sim G_{1,1} + G_{1,n} + G_{m,1}$; then $c(H) \leq 3$ (actually $= 3$ holds).*

The proof of this corollary goes along the same lines as that of Proposition 10. Suppose $H \cong G = G_{1,1} + G_{1,n} + G_{m,1}$, and let $S$ parametrize a strict, flat family of subgroup schemes $\mathcal{N} \to G \times S$, with $S$ integral. Use Lemma 11 for the generic fiber $Y$ of $(G \times S)/\mathcal{N}$ over some algebraic closure $L$ of $k(s)$, where $s$ is the generic point of $S$; thus

$$N_2 \subset N_1 \subset Z, \qquad Z/N_1 = Y \otimes L;$$

according to $a(Z/N_2) = 2$, respectively $a(Z/N_2) = 3$, we use $W$, respectively $W'$, of Construction 12 as we used $T$ in the proof of (10). By rigidity and because $S$ parametrizes a strict family, we conclude

$$\dim(S) \leq \dim(W) \qquad \text{or} \qquad \dim(S) \leq \dim(W')$$

analogously as we did in the proof of Proposition 10; thus we achieve $c(G) \leq 3$.

***Remark 14*** A situation similar to that in Construction 12 can probably be handled much better as follows: parametrize all extensions $X/G_{1,n} \cong G_{m,1}$ (by a one-dimensional scheme), for any such $X$, parametrize all extensions $G/G_{1,1} \cong X$ (this becomes a two-dimensional scheme for every such $X$), and glue together all these parametrizations to a "universal family" for groups $G$, plus $G_{1,1} \subset G$, plus $G_{1,n} \subset (G/G_{1,1}) = X$ with $X/G_{1,n} \cong G_{m,1}$. This method would need a theory of representability for functors of the form $\mathrm{Ext}(X, Y)$ with $X$ and $Y$ being formal groups.

## 2. Liftability of abelian varieties

Let $X$ be an AV defined over a field $k$ of characteristic $p \neq 0$. We are interested to know whether $X$ can be lifted to characteristic zero (in the week sense, cf. [11]). First note that if $X$ is *ordinary* (i.e., the $p$-rank of $X$ is $g := \dim X$, or: the formal group $\hat{X}$ of $X$ has a dual $\hat{X}^t$ which is étale), then $X$ has a canonical lift to characteristic zero (cf. [4; 6, p. 173, Th. 3.3]): one chooses the canonical lift $Y$ of $X$, and a polarization of $X$ can be lifted also, because

$$\mathrm{End}_{W(k)\text{-gr}}(Y \times Y^t) \xrightarrow{\sim} \mathrm{End}_{k\text{-gr}}(X \times X^t).$$

Now suppose $X$ is not necessarily ordinary, take a polarization $\lambda$ on $X$, let $d^2 = \deg(\lambda)$, hence $(X, \lambda)$ can be viewed as a point of the moduli space

$$(X, \lambda) \in \mathscr{A}_{g,d}$$

(cf. [7, p. 129, Def. 7.3 with $n = 1$, and p. 139, Th. 7.10]); let $V \subset \mathscr{A}_{g,d}$ be an irreducible component of the moduli space which contains the point $(X, \lambda)$. It follows that $(X, \lambda)$ can be lifted to characteristic zero if the generic point of $V$ corresponds to a polarized AV which can be lifted to characteristic zero. Hence, for liftability of $(X, \lambda)$, it suffices to show that the generic point of $V$ corresponds to an *ordinary* AV.

Let $V \subset \mathscr{A}_{g,d}$ be fixed as above, and let $\alpha$ be an isogeny type for formal groups (cf. [5]). We denote by $V_\alpha \subset V$ an irreducible component of the subset of points of $V$ which corresponds to abelian varieties whose formal group (over some algebraically closed field) has the isogeny type $\alpha$; note that $V_\alpha$ is a locally closed subset of $V$.

***Proposition 15*** *If $g \leq 3$, and $\alpha$ is not the isogeny type of an ordinary* AV, *then*

$$V_\alpha \subsetneqq V.$$

By what is said above, from (15) we immediately deduce:

***Theorem 16*** *Let $X$ be an* AV *of dimension $g \leq 3$ and let $\lambda$ be any polarization on $X$. Then $(X, \lambda)$ can be lifted to characteristic zero.*

We are going to prove Proposition 15 in the case $g = 3$ (the case $g = 1$ is obvious, and the case $g = 2$ goes along the same line as the proof for $g = 3$).

***Table 1***

| | Isogeny type | pr | $c(G)$ | $\dim(V_\alpha) \leq$ |
|---|---|---|---|---|
| I | $3 \times (1, 0)$ | 3 | 0 | 6 |
| II | $2 \times (1, 0) + (1, 1)$ | 2 | 0 | 5 |
| III | $1 \times (1, 0) + 2 \times (1, 1)$ | 1 | 1 | 5 |
| IV | $(1, 2) + (2, 1)$ | 0 | 1 | 4 |
| V | $3 \times (1, 1)$ | 0 | 3 | 3 |

In the case $g = 3$ we have the isogeny types shown in Table 1. That is, e.g. by $1 \times (1, 0) + 2 \times (1, 1)$, we understand the formal group $G_{1,0} + (G_{1,1})^2$. We denote by pr the $p$-rank, and in the right hand column we give an upper bound for $\dim(V_\alpha)$. All we have to do is to show that these entries in the last column are correct, because by [11, 2.2.1 and 2.3.3] we know that

$$\dim(V) \geq g^2 - \tfrac{1}{2}g(g-1) = \tfrac{1}{2}g(g+1).$$

We start with $\alpha = \text{II, III}$, or IV. Let $v$ be the generic point of $V_\alpha$, and $(X, \lambda)$ be the AV plus polarization corresponding to $v$, defined over an algebraic closure $L$ of $K(v)$. By [9, Cor. 1, p. 234] we can find a principally polarized AV $(Y, \mu)$ and an isogeny

$$f\colon X \to Y \qquad \text{with} \quad f^*(\mu) = \lambda.$$

Let

$$T \subset A_{3,1} \times A_{3,d}$$

be the isogeny correspondence, and let $T_\alpha$ be an irreducible component of $T \cap (A_{3,1} \times V_\alpha)$ such that $T_\alpha$ contains $f$ (note that $T_\alpha$ projects into $V_\alpha$). Let $S_\alpha$ be the first projection of $T_\alpha$,

$$p_1(T_\alpha) = S_\alpha \subset A_{3,1}.$$

Notice that all points of $S_\alpha$ correspond to AV having isogeny type $\alpha$, and hence by [3, IV.1, Th. 7] we know that

$$\dim(S_\alpha) \leq 5, 4, \qquad \text{resp. } 3 \text{ if} \quad \alpha = \text{II, III}, \quad \text{resp. IV}.$$

Let $v \in S_\alpha(k)$, and let

$$\overset{-1}{p_1}(v) \cap S_\alpha = T_v \subset \mathscr{A}_{3,1} \times \mathscr{A}_{3,d}$$

be the set of isogenies $Z \to Y_v$ with $Y_v$ fixed and $Z$ arbitrary, and such that this isogeny is contained in $S_\alpha$. Over a scheme $T \to T_v$ which is finite over $T_v$ this defines a family of finite subgroup schemes $\mathscr{N} \to Y_v^t \times T$, by taking all kernels of the isogenies $Y_v^t \to Z^t$. By rigidity (cf. [7, p. 115, Prop. 6.1]; or, use Lemma 8) this is a strict family, and hence

$$\dim(T_v) \leq c(G) = 0, \quad 1, \quad \text{resp. } 1$$

by Lemma 3, Proposition 10, respectively again Proposition 10. Note that $(X, \lambda)$ is in the image of the projection of $T_\alpha$ onto $V_\alpha$: hence this projection is generically surjective; thus by the dimension formula (cf. [8, p. 93, Th. 3]), we obtain

$$\dim(V_\alpha) \leq 5 + 0, \quad 4 + 1, \quad \text{resp. } 3 + 1,$$

if $\alpha = \text{II, III}$, resp. IV.

Now consider $\alpha = \text{V}$. Let $(X, \lambda)$ be the AV plus polarization corresponding to the generic point $v$ of $V_\alpha$. There exists a finite extention $K$ of $k(v)$ such that $X$ is defined over $K$, and such that there exists a supersingular elliptic curve $E$ and a $K$-isogeny

$$g\colon E^3 \to X$$

(use [12, Th. 4.2]). Notice that $E$ is defined over $k$. Let $T$ be the normalization of $V_\alpha$ in $K$. The kernel of $g$ extends to a flat family $\mathcal{N} \hookrightarrow E^3 \times T$ of finite subgroup schemes of $E^3$, such that the generic fiber of $E^3 \times T/\mathcal{N}$ is isomorphic with $X$ over $K$. By Corollary 13 we conclude $\dim(V_\alpha) = \dim(T) \leq 3$. This ends the proof of Proposition 15, and hence the proof of Theorem 16.

*Remark* Liftability of an AV of arbitrary dimension from characteristic $p$ to characteristic zero was proved by Mumford (unpublished, cf. D. Mumford, Bi-extensions of formal groups, in "Algebraic Geometry" (papers presented at the Bombay Colloq., 1968), pp. 307–322, Oxford Univ. Press, London and New York, 1969); the method to derive Theorem 16 from results such as Proposition 15 stems from Mumford. We do not know whether the methods expressed in this paper can be generalized so that they can apply in the case of arbitrary $g$.

*Note added in proof*: Liftability of polarized AVs from characteristic $p$ to characteristic zero has been proved (cf. P. Norman and F. Oort, Moduli of abelian varieties (to appear)).

## References

1. M. Demazure and P. Gabriel, "Groupes algébriques," Vol. I. Masson, Paris, North-Holland Publ., Amsterdam, 1970.
2. A. Grothendieck and J. Dieudonné, "Éléments de géométrie algébrique," Chap. IV[4] (Publ. Math. **32**). Inst. Hautes Études Sci., Paris, 1967.
3. N. I. Koblitz, $p$-adic variation of the zeta-function over families of varieties defined over finite fields, *Compos. Math.* **31** (1975), 119–218.

4. J. Lubin, J.-P. Serre, and J. Tate, Elliptic curves and formal groups, Woods Hole Summer Institute, 1964 (mimeographed notes).
5. Yu. I. Manin, The theory of commutative formal groups over fields of finite characteristic, *Russian Math. Surveys* **18** (1963), 1–80.
6. W. Messing, "The Crystals Associated to Barsotti–Tate Groups: With Applications to Abelian Schemes" (Lect. Notes in Math. **264**). Springer-Verlag, Berlin–Heidelberg–New York, 1972.
7. D. Mumford, "Geometric Invariant Theory" (Ergeb. Math., Neue Folge **34**). Springer-Verlag, Berlin–Heidelberg–New York, 1965.
8. D. Mumford, "Introduction to algebraic geometry." Preliminary version of first 3 chapters.
9. D. Mumford, "Abelian Varieties" (Tata Inst. F. R. Studies in Math. **5**). Oxford Univ. Press, London and New York, 1970.
10. F. Oort, "Commutative Group Schemes" (Lect. Notes in Math. **15**). Springer-Verlag, Berlin–Heidelberg–New York, 1966.
11. F. Oort, Finite group schemes, local moduli for abelian varieties and lifting problems, *Compos. Math.* **23** (1971), 265–296 (also "Algebraic Geometry," Oslo 1970, Wolters-Noordhoff, 1972).
12. F. Oort, Subvarieties of moduli space, *Invent. Math.* **24** (1974), 95–119.
13. F. Oort, Which abelian surfaces are products of elliptic curves? *Math. Ann.* **214** (1975), 35–47.
14. F. Oort, Isogenics of formal groups, *Indag. Math.* **37** (1975), 391–400.
15. J. T. Tate, *p*-Divisible groups, *Proc. Conf. Local Fields, Driebergen, 1966.* Springer-Verlag, Berlin–Heidelberg–New York, 1967.

We thank IHES (Institut des Hautes Études Scientifiques, Bures-sur-Yvette, France) for hospitality; part of the research for this paper was done at that institute. I thank N. Katz and M. Raynaud for discussions on this subject.

AMS (MOS) 1970 subject classifications: 14K99, 14L05, 14D15

# *An effective lower bound on the "diophantine" approximation of algebraic functions by rational functions (II)*

*Charles F. Osgood*
*Naval Research Laboratory*

In an article in the *Proceedings* in 1959 [2] Professor Kolchin showed an analogue of Liouville's theorem (on the diophantine approximation of algebraic numbers by rational numbers) which dealt with the approximation of solutions of algebraic differential equations. (For technical reasons Kolchin's theorem is a good deal harder to prove than Liouville's theorem, however.) Perhaps the "most concrete" application of Kolchin's theorem is to the approximation by rational functions of formal power series solutions of a nontrivial algebraic differential equation. Assume that $K$ is a field of characteristic 0, $z$ is transcendental over $K$, and that $y_1$ is a formal power series in $z^{-1}$ (plus a finite number of terms involving positive integral powers of $z$) having coefficients in $K$. Suppose $y = y_1$ is not a rational function of $z$ but is a zero of a not identically zero polynomial, $\alpha$, say, in $z, y, dy/dz, \ldots$ with coefficients in $K$. Let $r, s \not\equiv 0$ denote polynomials in $z$ over $K$. Let ord denote the valuation "order of vanishing at $z = \infty$."

Then Kolchin's theorem says that there exist a natural number $n$ and a real number $c$ such that for every $r, s \not\equiv 0$ as above:

$$\operatorname{ord}(y_1 - r/s) < n(\deg s) + c. \tag{1}$$

(If one changes to a multiplicative valuation, the analogy with various diophantine inequalities becomes even more striking.)

An integer $n$ which will suffice in (1) may be read off from any not identically zero differential polynomial $\alpha$ of which $y_1$ is a nonsingular zero. Thus an $n$ for (1) may be "effectively computed." (The $c$ is merely known to exist and is *not* "effectively computed" by Kolchin.) The smallest integer $n$ which Kolchin's theorem permits us to put in (1), using the differential polynomial $\alpha$, is the "denomination of $\alpha$," $d(\alpha)$.

***Definition of*** $d(\alpha)$ Replace $y$ by $uv^{-1}$ in $\alpha$. Set $d(\alpha)$ equal to the least natural number $t$ such that $v^t\alpha$ equals a polynomial in $z$ and the derivatives of $u$ and $v$.

Clearly $d(\alpha + \beta) \leq \max\{d(\alpha), d(\beta)\}$. That the inequality can also hold is of crucial importance for the theorem of this paper. (For a "differential monomial" $\prod_j (y^{(j)})^{e_j}$ the denomination is $\sum_j (j + 1)e_j$.)

One may easily prove, by essentially Liouville's original method, that for any irrational algebraic function $y_1$ over any field *of any characteristic*, (1) holds with $n$ equal to the degree of the irreducible polynomial for $y_1$ over $K[z]$. (This result has been known at least since the turn of the century, see [4].) If $K$ equals the rational field, it involves a certain amount of work, but it is not too difficult to see that $c$ may be effectively computed. Outside the context of differential algebra this is (apparently) the only analogue for algebraic functions of Liouville's theorem possible. Since we *are* using differential algebra, I shall refer to this result as the "simple" Liouville-type theorem for algebraic functions. In [5] I proved what one might refer to as a "less simple" Liouville-type theorem for algebraic functions, and in the present paper I shall prove a "yet less simple" Liouville-type theorem for algebraic functions over certain fields.

Because of Roth's theorem [8] one would expect to be able to replace $n$ in (1) by $2 + \varepsilon$ (for each $\varepsilon > 0$ where $c = c(\varepsilon, y_1)$) for each irrational algebraic function $y_1$ over a field of characteristic 0. This was shown to be true by Uchiyama (see [9]) using Roth's method. The result is *not* effective.

Except for isolated examples, all of the known bounds on the diophantine approximation of algebraic numbers or of algebraic functions were, until recently, not effective. This state of affairs was changed by the methods of Alan Baker, which have proven of great importance to number theory; they lead to effective algorithms for finding all solutions of certain diophantine equations. Fel'dman [1], building on Baker's methods, obtained an effective bound of the present type in which Liouville's exponent has been decreased, but only by a small positive amount.

In [5] I showed how one can adapt Kolchin's argument in order to obtain, if $y_1$ is algebraic and $\deg y_1 > 2$, a (usually large) effective improvement of the bound from the "simple" Liouville-type theorem mentioned above. Part of my idea was that, from the standpoint of using Kolchin's theorem, *the algebraic equation satisfied by* $y_1$ is one of the poorest algebraic differential equations with which to work. Instead, consider the following: Choose $N(y_1)$ to be the least natural number such that there are at least $(\deg y_1) + 1$ distinct differential monomials of denomination $\leq N(y_1)$ [$N(y_1) \leq (\log_e(\deg y_1))^2$, if $\deg y_1 \geq 8$, cf. [5]]. By a very simple vector space argument there must exist a not identically zero algebraic differential equation of denomination $\leq N(y_1)$ which is satisfied by $y_1$; also, there exists such an equation for which $y_1$ is a nonsingular solution. By restricting $K$ to be the rational field (or more generally, requiring the generators of the subfield of $K$ with which we actually deal to be "effectively known," see [5]) we can show, after some work, inequality (1) with $n$ replaced by $N(y_1)$, *where also c is effectively computable*. In comparison with effective results obtained previously, the statements appear surprisingly strong. If $\deg y = 3$, then the $n$ in (1) is 2 and if $\deg y_1 \geq 8$, then $n < (\log_e(\deg y_1))^2$. The result for $n = 3$ is very gratifying since it is stronger than even the best non-effective result known (i.e., Uchiyama's result, where $n$ in (1) is replaced by $2 + \varepsilon$). The bound for $n = 3$ is fully as good as that which holds for quadratic irrational algebraic functions. (The analagous statement for algebraic numbers is presumably false, although no one has been able to show this. In any event, it does not seem that a close analogue of any present purely number theoretic argument could yield such a strong result.) Further, using very special properties of the generalized Ricatti differential equation, it is possible to take $K$ equal to *any* field of characteristic 0, drop the "effectively known" concept, and have an effective upper bound for $c$, in terms of upper bounds on the degrees (in $z$) of the coefficient polynomials of the irreducible cubic equation for $y_1$. This result is almost a model of what one would like to obtain for all irrational algebraic functions $y_1$, regardless of degree.

In a recent paper [6] I studied the more complicated situation where $y_1$ is an irrational algebraic function, but $K$ has positive characteristic. (An example due to Artin and Mahler [3] shows $n = \deg(y_1)$ can be best possible if the characteristic is positive!) Another paper of mine [7] deals with an attempt to replace $n$ in (1) by $2 + \varepsilon$ in the general "most concrete" case of Kolchin's theorem, i.e., where $K$ has characteristic 0 and $y_1 \notin K(z)$ is a formal power series solution of a nontrivial algebraic differential equation, $\alpha = 0$, over $K(z)$. (A bound of $2 + \varepsilon$ would be best possible, aside—possibly—from the $\varepsilon > 0$ and the choice of $c = c(\varepsilon, y_1)$.) *The exact result obtained is that for each* $\varepsilon > 0$ *there exists a nontrivial algebraic differential equation* $P_\varepsilon = 0$ *such that if, for a given* $r/s$, (1) *is violated, with* $n$ *replaced by* $2 + \varepsilon$, *then* $r/s$ *is a solution of* $P_\varepsilon = 0$. If $\alpha$ has "effectively known" constant coefficients, then $P_\varepsilon$ can be effectively computed. (Extensions of this result to simultaneous approximation over a common denominator have also been obtained by me and have been submitted for publication.)

There are two features of the approach in [7] which make it much different from the approach in [5]. The first feature is that I construct in $P_\varepsilon = 0$ a (usually) much higher order algebraic differential equation (than $\alpha = 0$) of which $y_1$ is a highly *singular* solution. In fact, all the mixed partial derivatives of $P_\varepsilon$ with respect to $y, dy/dz, \ldots$ of (total) order less than (almost) $\frac{1}{2}$ the denomination of $P_\varepsilon$ vanish at $y = y_1$. [That $(\frac{1}{2})^{-1} = 2$ is a crucial point here in obtaining $2 + \varepsilon$.] The other new feature is that in order to achieve such a ratio I must somehow utilize the fact that for some algebraic differential equations the denomination is much less than the largest denomination among its constituent monomials, e.g., $(y')^2 - \frac{1}{2}yy''$ has denomination $\leq 3$. (The fact that $y_1$ is a singular solution of $P_\varepsilon$ leads directly to my inability to show that it cannot be a limit point of rational solutions of $P_\varepsilon = 0$. Consider $yy' + xyy'' - x(y')^2 = 0$. Now $y = 0$ is a [singular] solution, but so is $y = z^{-n}$, for $n = 0, 1, 2, \ldots$. Hopefully, an analogue of this "bad case" would happen only rarely in my equations $P_\varepsilon = 0$, but the explicit construction of one of these equations involves an extremely large number of calculations.)

Professor Wolfgang Schmidt suggested to me that I might see if I could improve my "nonsimple" Liouville-type theorem using the second new feature of the proof in [7]. The result is an "even less simple" Liouville-type theorem with a better (and more easily described) bound. The remainder of this paper will be devoted to the proof of the following result. We assume the hypotheses of Theorem I of [5], which are: Suppose that $y_1 \in \overline{K}(z)$ is an algebraic function over $K(z)$, where $K$ has characteristic 0. Suppose that we are given an irreducible polynomial equation, having coefficients in $K[z]$ and effectively known constant coefficients, which is satisfied by $y_1$. (Brackets will now denote the greatest integer function.)

***Theorem*** *There exists an effectively computable constant* $c$ *such that*

$$\operatorname{ord}(y_1 - r/s) < [\log_2(\deg y_1) + 1] \deg s + c, \tag{2}$$

*for all polynomials* $r, s \not\equiv 0$ *in* $\overline{K}[z]$.

Set $\Delta = [\log_2(\deg y_1) + 1]$. Notice $\Delta = 2$ if $\deg y_1 = 2$ or 3 and $\Delta \leq 3$ if $\deg y_1 \leq 7$.

*Proof* We shall effectively compute a nonzero differential polynomial having denomination $\leq \Delta$ which has $y_1$ as a nonsingular zero. The argument proving Theorem I of [5] will then apply. As in [7] we deal first with a differential polynomial over $K[z]$ in two quantities $u$ and $v$ whose derivatives are to be algebraically independent over $K(z)$. Consider, for $\theta = 1, 2, \ldots,$ the set $S(\theta)$ of all linear combinations over $K(z)$ of differential monomials of the form $\prod_{i,j} (u^{(i)})^{e_i}(v^{(j)})^{f_j}$ where $\sum_i e_i + \sum_j f_j \leq \Delta$, $0 < i < \theta - 1$ and $0 \leq j \leq \theta - 1$. We shall show that for all sufficiently large $\theta$, $S(\theta)$ contains a nonzero element $p(u, v)$ which vanishes identically at $(u, v) = (vy_1, v)$. We have, therefore, a problem of finding a nontrivial solution to a system of linear homogeneous equations; the unknowns are the coefficients of the different $\prod_{i,j} (u^{(i)})^{e_i}(v^{(j)})^{f_j}$ potentially occurring in $p(u, v)$ and the equations are the statements that, at $(u, v) = (vy_1, v)$, the coefficients of the differing $y_1{}^t \prod_j (v^{(j)})^{g_j}$ in $p(u, v)$ are zero $(0 \leq t \leq \deg y_1 - 1$, $\sum_j g_j \leq \Delta$, and $0 \leq j \leq \theta - 1)$. By Lemma I of [7], for $\gamma = 1, 2, \ldots,$ the number of solutions in nonnegative integers $\delta_j$ of the inequality $\sum_{j=0}^{\gamma-1} \delta_j \leq \Delta$ is $\binom{\gamma+\Delta}{\gamma}$. Thus we need

$$\binom{\Delta + 2\theta}{2\theta} > (\deg y_1)\binom{\Delta + \theta}{\theta}$$

in order to see the existence of a nontrivial solution. Now

$$\begin{aligned}\binom{\Delta + 2\theta}{2\theta}\bigg/\binom{\Delta + \theta}{\theta} &= \frac{\theta!}{(2\theta)!}\left(\frac{(\Delta + 2\theta)\cdots(\Delta + 1)}{(\Delta + \theta)\cdots(\Delta + 1)}\right)\\ &= \frac{(\Delta + 2\theta)\cdots(\Delta + \theta + 1)}{(2\theta)\cdots(\theta + 1)}\\ &= \frac{(\Delta + 2\theta)\cdots(2\theta + 1)}{(\Delta + \theta)\cdots(\theta + 1)},\\ &\text{if}\quad \theta + 1 < \Delta + \theta + 1 < 2\theta < \Delta + 2\theta.\end{aligned}$$

Thus we need to show that for all sufficiently large $\theta$,

$$2^{\Delta}\left(\prod_{j=0}^{\Delta-1}\left(\frac{\theta + \frac{1}{2}(\Delta - j)}{\theta + \Delta - j}\right)\right) > \deg y_1. \tag{3}$$

Since $2^{\Delta} > \deg y_1$, (3) *does* hold if $\theta$ is larger than some effectively computable bound. As was mentioned in [5], we can effectively put a system of linear homogeneous equations in triangular form and effectively compute a solution, $p(u, v)$.

It will be trivial that the argument which follows is effective. Notice that, where $x$, $u$, and $v$ and their derivatives are algebraically independent over

$K(z)$, $p((xv)y_1, xv) \equiv 0$. Thus writing $p(xu, xv)$ as a linear combination of differential monomials in $x$, we see that each nonzero coefficient, say $\rho$, is a differential monomial in $u$ and $v$ which satisfies $\rho(vy_1, v) \equiv 0$. Further, $\rho$ is homogeneous, i.e., the total degree of each constituent differential monomial in the derivatives of $u$ and $v$ is the same. Substitute some nonzero $\rho$ for $p$ above (so that without loss of generality $p$ is homogeneous) and go through the above construction again. We distinguish two cases:

*Case* (*i*) $p(xu, xv) \equiv x^b \rho(u, v) \equiv x^b p(u, v)$, for some nonzero coefficient $\rho$ and some integer $b \leq \Delta$.

*Case* (*ii*) There exists a nonzero coefficient of a differential polynomial in $x$ which involves derivatives of $x$ (beyond the zeroth order).

In Case (i) set $x = v^{-1}$. Then $p(uv^{-1}, 1) \equiv v^{-b}p(u, v) \not\equiv 0$. Setting $uv^{-1} = y$ we have a differential polynomial in $y$ of denomination $\leq \Delta$ which has $y_1$ as a [possibly singular] zero. We shall now see that Case (ii) eventually leads to Case (i).

Let $\Gamma(p) = max\{\sum_i e_i + \sum_j f_j | (u^{(i)})^{e_i}(v^{(j)})^{f_j}$ is a constituent monomial in $p\}$. Then if $\rho$ is as in Case (ii), $\Gamma(\rho) \leq \Gamma(p) - 1$. Substituting $\rho$ for $p$ repeatedly, if we never end up in Case (i) we eventually obtain a homogeneous polynomial in $u$ and $v$ for $p$. But such a polynomial puts us in Case (i). We must next deal with the possibility that $y_1$ is a singular solution of $p(y, 1) = 0$. Suppose $p(y, 1)$ has denomination $b \leq \Delta$, and that $y_1$ is a singular solution. Now if $m$ is the order of $p(y, 1)$, we see that

$$v^b \frac{\partial}{\partial y^{(m)}} p(y, 1) \equiv \frac{\partial}{\partial y^{(m)}} p(vy, v) \equiv \left( \sum_{k \geq 0} \binom{k}{m} v^{(k-m)} \frac{\partial}{\partial u^{(k)}} \right) p(u, v), \tag{4}$$

using $y = u/v$. The right-hand side of (4), which we shall denote by $q$, is a differential polynomial of total degree $\leq \Delta$ in the derivatives of $u$ and $v$ which vanishes at $(u, v) = (vy_1, v)$. Let $\varphi(p) = \max\{\sum_i e_i | (u^{(i)})^{e_i}(v^{(j)})^{f_j}$ is a constituent monomial of $p\}$. Then $\varphi(q) \leq \varphi(p) - 1$. Working with $q$ instead of $p$, etc., we either obtain eventually an algebraic differential equation of denomination $\leq \Delta$ of which $y_1$ is a nonsingular solution, or we arrive at a nonzero differential polynomial in $v$ which vanishes at $u = vy_1$, a contradiction.

## References


1. Fel'dman, An effective refinement of the exponent in Liouville's theorem, *Izv. Akad. Nauk SSSR* **35** (1971), 973–990 [English transl.: *Math. USSR-Izv.* **5** (1971), 985–1002].
2. Kolchin, E. R., Rational approximation to solutions of algebraic differential equations, *Proc. Amer. Math. Soc.* **10** (1959), 238–244.

3. Mahler, K., On a theorem of Liouville in fields of positive characteristic, *Canad. J. Math.* **1** (1949), 397–400.
4. Maillet, E., "Introduction à la Théorie des Nombres Transcendants." Gauthier-Villars, Paris, 1906.
5. Osgood, C. F., An effective lower bound on the "diophantine" approximation of algebraic functions by rational functions, *Mathematika* **20** (1973), 4–15.
6. Osgood, C. F., On the effective "diophantine approximation" of algebraic functions over fields of arbitrary characteristic and applications to differential equations, *Proc. Konink. Nederl. Akad. Van Wetens.* (Amsterdam) Ser. A, **78**, No. 2 (1975) 105–119. Errata, *Ibid.* No. 5 (1975).
7. Osgood, C. F., Concerning a possible "Thue–Siegel–Roth theorem" for algebraic differential equations, *in* "Number Theory and Algebra" (H. Zassenhaus, ed.), pp. 223–234. Academic Press, New York, 1977.
8. Roth, K. F., Rational approximations to algebraic numbers, *Mathematika* **2** (1955a), 1–20; Corrigendum, *Ibid.* (1955b), 168.
9. Uchiyama, S., Rational approximations to algebraic functions, *J. Fac. Sci. Hokkaido Univ.* Ser. 15 (1961), 173–192.

AMS (MOS) 1970 subject classifications: 10F99, 12A90, 12H05

# *On elementary, generalized elementary, and liouvillian extension fields*

*Maxwell Rosenlicht*
*University of California Berkeley*

*Michael Singer*
*North Carolina State University*

This paper contains a new proof of Singer's result that a relatively algebraically closed differential subfield of a generalized elementary extension field is itself generalized elementary, together with a number of new results and extensions of old results on the three types of field in question.

By a *differential field* is meant a field together with an indexed family of derivations (the *given derivations*) of the field. The given derivations of a differential extension field have the same index set and extend the original derivations.

The first proposition brings together a number of known results, e.g. [4, Prop. 1].

***Proposition 1*** *Let $k$ be a differential field of characteristic zero and for each given derivation $D$ let $\alpha_D \in k$. Then there exists a differential extension field $k(t)$ of $k$ which has the same constants as $k$ and in which $Dt = \alpha_D$ for each given $D$. This $k(t)$ is unique to within differential $k$-isomorphism and $t$ is either transcendental over $k$ or in $k$.*

To prove this, first let $k(t)$ be any differential extension field of $k$ such that $Dt = \alpha_D$ for each given derivation $D$. If $t$ is transcendental over $k$, then any element of $k(t)$ can be written as $f(t)/g(t)$, with $f(t)$, $g(t)$ relatively prime elements of $k[t]$ and $g(t)$ monic. For each $D$, $Df(t)$ and $Dg(t)$ are also in $k[t]$ and have degrees respectively at most the degree of $f(t)$ and less than the degree of $g(t)$. Suppose that $f(t)/g(t)$ is constant. Then for each $D$ we have $g(t)\,Df(t) = f(t)\,Dg(t)$, relative primeness gives $g(t) \mid Dg(t)$, $f(t) \mid Df(t)$, and using degrees we get $Dg(t) = 0 = Df(t)$. Hence if $t$ is transcendental over $k$ and $k(t)$ has constants other than the constants of $k$, then there exists $f(t) \in k[t]$, $f(t) \notin k$, such that $Df(t) = 0$ for each given $D$. Thus in either the case where $t$ is transcendental over $k$ and $k(t)$ has constants not in $k$ or the case where $t$ is algebraic over $k$, there exist $n > 0$ and $a_0, a_1, \ldots, a_n \in k$, $a_0 \neq 0$, such that $D(a_0 t^n + a_1 t^{n-1} + \cdots + a_n) = 0$ for all given $D$. But

$$D(a_0 t^n + a_1 t^{n-1} + \cdots) = (Da_0)t^n + (Da_1 + na_0\alpha_D)t^{n-1} + \cdots,$$

so that $Da_0 = Da_1 + na_0\alpha_D = 0$. Hence $\alpha_D = D(-a_1/na_0)$ for each given derivation $D$ and $t + a_1/na_0$ is constant. If now $k(t)$ is any differential extension field of $k$ with the same constants as $k$ and such that $Dt = \alpha_D$ for each $D$, then either $t$ is transcendental over $k$ and there is no $a \in k$ such that $Da = \alpha_D$ for all $D$, or $t \in k$. In case there is no $a \in k$ such that $Da = \alpha_D$ for all $D$, we get a differential extension field $k(t)$ of $k$ as desired by taking $t$ transcendental over $k$ and defining $Dt = \alpha_D$ for all $D$, and this proves the result.

***Proposition 2*** *Let $k$ be a differential field of characteristic zero and for each given derivation $D$ of $k$ let $\alpha_D \in k$. Then there exists a differential extension field $k(t)$ of $k$ with $t \neq 0$ which has the same constants as $k$ and in which $Dt = \alpha_D t$ for each given $D$. Such a differential extension field $k(t)$ is algebraic over $k$ if and only if some positive power $t^n$ of $t$ is in $k$, which is true if and only if there exists a nonzero $a \in k$ such that $n\alpha_D = Da/a$ for each $D$, and in this case the least such $n$ is $[k(t):k]$.*

First let $k(t)$ be a differential extension field of $k$ having the same constants as $k$ and with $t \neq 0$ and $Dt = \alpha_D t$ for all $D$. Suppose that $t$ is algebraic over $k$, with $[k(t):k] = m$, and let $(a_0, \ldots, a_m)$ be a nonzero $(m+1)$-tuple of elements of $k$ (unique to within multiplication by a nonzero factor in $k$) such that $\sum_{i=0}^m a_i t^i = 0$. For each $D$ we have $D(a_i t^i) = (Da_i/a_i + i\alpha_D)a_i t^i$ if $a_i \neq 0$, so that

$$0 = D\left(\sum_{i=0}^{m} a_i t^i\right) = \sum (Da_i/a_i + i\alpha_D)a_i t^i,$$

the latter sum ranging over all $i = 0, \ldots, m$ such that $a_i \neq 0$. Since $t \neq 0$ for at least two elements $i, j = 0, \ldots, m$ we have $a_i, a_j \neq 0$, and therefore $Da_i/a_i + i\alpha_D = Da_j/a_j + j\alpha_D$, or $(i-j)\alpha_D = D(a_j/a_i)/(a_j/a_i)$. Assuming, as we may, that $i > j$, we set $i - j = n$, $a_j/a_i = a$ to get $n\alpha_D = Da/a$ for any given derivation $D$. Conversely, if $a$ is a nonzero element of $k$ such that $n\alpha_D = Da/a$ for each $D$, then $D(t^n/a) = 0$, so $t^n/a$ is a constant, hence in $k$, and so $t$ is algebraic over $k$. Going back to the polynomial equation $\sum_{i=0}^m a_i t^i = 0$, we have $n = i - j \leq m \leq n$, so that the least value of $n$ is $[k(t) : k]$. This proves the last part of the proposition. We proceed to prove the first part, first assuming that there is no nonzero $a \in k$ and no positive integer $n$ such that $n\alpha_D = Da/a$ for each $D$. In this case we take $t$ transcendental over $k$ and make $k(t)$ a differential extension field of $k$ by setting $Dt = \alpha_D t$ for each given $D$. We have to show that any constant of $k(t)$ is in $k$. As before, any nonzero element of $k(t)$ can be written as $f(t)/g(t)$, with $f(t)$, $g(t)$ relatively prime elements of $k[t]$, and if $f(t)/g(t)$ is constant, then for each $D$ we have $g(t)\,Df(t) = f(t)\,Dg(t)$. Noting that $Df(t)$, $Dg(t)$ are elements of $k[t]$ of degrees at most those of $f(t)$, $g(t)$ respectively, we get that $Df(t)$, $Dg(t)$ are multiples by an element of $k$ of $f(t)$, $g(t)$, respectively. If either $f(t)$ or $g(t)$ has more than one term, the argument above will give us an integer $n > 0$ and an element $a \in k$ such that $n\alpha_D = Da/a$ for all $D$, contrary to our assumption. Therefore $f(t)$, $g(t)$ each have one term, so $f(t)/g(t) = t^n/a$, for some $n \in \mathbb{Z}$ and some nonzero $a \in k$. The constancy of $f(t)/g(t)$ then gives $n\alpha_D = Da/a$ for each $D$, again contrary to assumption. This shows that the constants of $k(t)$ are those of $k$. It remains finally to prove the existence of a $k(t)$ with the desired properties under the assumption that there exists a positive integer $n$ and a nonzero $a \in k$ such that $n\alpha_D = Da/a$ for each given derivation $D$ of $k$. Suppose the pair $(n, a)$ is chosen so that $n$ is minimal. Let $k(t)$ be an algebraic extension of $k$ such that $t^n = a$. Then $k(t)$ has a unique structure of differential extension field of $k$, and in this structure $Dt/t = (1/n)\,Da/a = \alpha_D$ for each $D$. We shall complete the proof of Proposition 2 by showing that $k(t)$ and $k$ have the same constants. Let $[k(t) : k] = m \leq n$. Then each element of $k(t)$ can be uniquely written in the form $\sum_{i=0}^{m-1} a_i t^i$, with each $a_i \in k$. For each given derivation $D$ we have $D(\sum_{i=0}^{m-1} a_i t^i) = \sum (Da_i/a_i + i\alpha_D)a_i t^i$, the last sum ranging over all $i = 0, \ldots, m-1$ for which $a_i \neq 0$, and we have a constant only if each $Da_i/a_i + i\alpha_D = 0$, or $i\alpha_D = D(1/a_i)/(1/a_i)$. Since $i \leq m - 1 < n$, the minimality property of $n$ implies that $i = 0$. Thus the constants of $k(t)$ are in $k$, as claimed.

In the differential extension field $k(t)$ of $k$ of Proposition 2, the element $t$ of $k(t)$ is determined to within multiplication by a nonzero element of the subfield of constants $C$ of $k$. If $t$ is transcendental over $k$, the differential field $k(t)$ is unique to within differential $k$-isomorphism. If $t$ is algebraic over $k$ and $n = [k(t) : k]$, the element $t^n$ of $k$ is determined to within multiplication by the $n$th power of a nonzero element of $C$ and there is a one–one correspondence between differential $k$-isomorphism classes of differential extension fields $k(t)$ of $k$ and elements of the multiplicative group of $C$ modulo its subgroup of $n$th powers; $k(t)$ is unique to within differential $k$-isomorphism if and only if each element of $C$ is an $n$th power.

We recall that if $k$ is a differential field and $x$ an element of a differential extension field of $k$, then $x$ is called *primitive* over $k$ (respectively *exponential* over $k$) if $Dx \in k$ (respectively $Dx/x \in k$) for each given derivation $D$; in either case, $k(x)$ is a differential extension field of $k$. If $x$, $y$ are elements of a differential field and $y \neq 0$, then $x$ is called a *logarithm* of $y$ and $y$ an *exponential* of $x$ if $Dx = Dy/y$ for each given derivation $D$. An element $x$ of a differential extension field of $k$ is called an *elementary primitive* over $k$ (respectively an *elementary exponential* over $k$) if there exist elements $u_1, \ldots, u_n, v$ of $k$, all except possibly the last being nonzero, and constants $c_1, \ldots, c_n \in k$, such that for each given derivation $D$ we have $Dx$ (respectively $Dx/x$) equal to $\sum_{i=1}^{n} c_i\, Du_i/u_i + Dv$. Clearly a logarithm (respectively an exponential) of an element of $k$ is an elementary primitive over $k$ (respectively an elementary exponential over $k$), and an elementary primitive over $k$ (respectively elementary exponential over $k$) is primitive over $k$ (respectively exponential over $k$). A differential extension field of $k$ is called a *liouvillian extension* of $k$ (respectively an *elementary extension* of $k$, respectively a *generalized elementary extension* of $k$) if it is of the form $k(t_1, \ldots, t_N)$, where for each $i = 1, \ldots, N$, the quantity $t_i$ is primitive, or exponential, or algebraic, over the field $k(t_1, \ldots, t_{i-1})$ (respectively a logarithm or an exponential of an element of, or algebraic over, $k(t_1, \ldots, t_{i-1})$, respectively an elementary primitive, or elementary exponential, or algebraic, over $k(t_1, \ldots, t_{i-1})$); in each case each field $k(t_1, \ldots, t_i)$ is a differential extension field of $k$ of the appropriate type. An elementary extension of $k$ is always a generalized elementary extension of $k$, and the latter always a liouvillian extension of $k$. By Liouville's theorem (cf. [3, Th. 3] or the first part of the present Theorem 2) and Propositions 1 and 2, an element $x$ of a differential extension field of $k$ having the same constants as $k$ that is primitive (respectively exponential) over $k$ is an elementary primitive over $k$ (respectively elementary exponential over $k$) if and only if there exists an elementary differential extension field of $k$ having the same constants and containing an element $x_1$ such that for each given derivation $D$ we have $Dx = Dx_1$ (respectively $Dx/x = Dx_1/x_1$). Repeated use of Propositions 1 and 2 shows that any generalized elementary extension of a differential field $k$ that has the same constants is a differential subfield of an elementary extension field with the same constants.

The main tools in the proofs given below are Theorems 1 and 2 of [3], whose statements we reproduce here for the convenience of the reader.

***Theorem A*** *Let $k$ be a differential field of characteristic zero, $K$ a differential extension field of $k$ with the same constants $C$. For each $i = 1, \ldots, n$ and $j = 1, \ldots, v$ let $c_{ij} \in C$ and let $v_i$ be an element of $K$, $u_j$ a nonzero element of $K$.*

*Suppose that for each $i = 1, \ldots, n$ and each given derivation $D$ of $K$, we have*

$$\sum_{j=1}^{\nu} c_{ij} \frac{Du_j}{u_j} + Dv_i \in k.$$

*Then either* deg tr $k(u_1, \ldots, u_\nu, v_1, \ldots, v_n)/k \geq n$ *or the $n$ elements of $\Omega_{K/k}$ given by $\sum_{j=1}^{\nu} c_{ij}\, du_j/u_j + dv_i$, $i = 1, \ldots, n$, are linearly dependent over $C$.*

***Theorem B*** *Let $k$ be a differential field of characteristic zero, $K$ a differential extension field of $k$ with the same constants, with $K$ algebraic over $k(t)$ for some given $t \in K$. Suppose that $c_1, \ldots, c_n$ are constants of $k$ that are linearly independent over $\mathbb{Q}$, that $u_1, \ldots, u_n, v$ are elements of $K$, with $u_1, \ldots, u_n$ nonzero, and that for each given derivation $D$ of $K$ we have $\sum_{i=1}^{n} c_i\, Du_i/u_i + Dv \in k$. If for each given derivation $D$ of $K$ we have $Dt \in k$, then $u_1, \ldots, u_n$ are algebraic over $k$ and there exists a constant $c$ of $k$ such that $v - ct$ is algebraic over $k$. If for each given derivation $D$ of $K$ we have $Dt/t \in k$, then $v$ is algebraic over $k$ and there are integers $\nu_0, \nu_1, \ldots, \nu_n$, with $\nu_0 \neq 0$, such that each $u_i^{\nu_0}/t^{\nu_i}$ is algebraic over $k$.*

The following Theorem 1 is the main result of [5].

***Theorem 1*** *Let $k \subset E \subset K$ be differential fields of characteristic zero with the same subfield of constants and with $E$ algebraically closed in $K$. If $K$ is a generalized elementary extension of $k$, then so is $E$.*

For the proof, we first note that if an element $x$ of a differential extension field of $k$ is either an elementary primitive or an elementary exponential over $k$, then there exist elements $u_1, \ldots, u_m, v \in k$, all except possibly the last being nonzero, and nonzero constants $c_1, \ldots, c_m$ of $k$ such that $x$ occurs exactly once in the sequence $u_1, \ldots, u_m, v$, all other terms of the sequence being elements of $k$, and such that for each given derivation $D$ we have $\sum_{i=1}^{m} c_i\, Du_i/u_i + Dv = 0$. Let $t_1, \ldots, t_N \in K$ be such that $K = k(t_1, \ldots, t_N)$ and each $t_i$ is an elementary primitive, or an elementary exponential, over $k(t_1, \ldots, t_{i-1})$, or algebraic over this latter field. Let $x_1, \ldots, x_n$ be the subset of $t_1, \ldots, t_N$ consisting of those $t_i$, $i = 1, \ldots, N$, that are transcendental over $k(t_1, \ldots, t_{i-1})$, taken in the same order. For $i = 0, \ldots, n$ let $k_i$ be the algebraic closure in $K$ of $k(x_1, \ldots, x_i)$. Then $k_0$ is the algebraic closure of $k$ in $K$, $k_n = K$, $x_1, \ldots, x_n$ is a transcendence basis for $K$ over $k$, and for each $i = 1, \ldots, n$, $x_i$ is an elementary primitive or an elementary exponential over the differential field $k_{i-1}$ and a transcendence basis for $k_i$ over $k_{i-1}$. For each $i = 1, \ldots, n$, we can find nonzero elements $u_{i1}, \ldots, u_{im(i)} \in K$, an element $v_i \in K$, and nonzero constants $c_{i1}, \ldots, c_{im(i)}$ of $k$, such that all the terms except one of the sequence $u_{i1}, \ldots, u_{im(i)}, v_i$ are in $k_{i-1}$, that one term being equal to $x_i$, and such that for each given derivation $D$ we have

$$\sum_{j=1}^{m(i)} c_{ij} \frac{Du_{ij}}{u_{ij}} + Dv_i = 0.$$

The $k$-differentials of $K$ given by

$$\left\{ \sum_{j=1}^{m(i)} c_{ij} \frac{du_{ij}}{u_{ij}} + dv_i \right\}_{i=1, \ldots, n}$$

are clearly a basis for the $K$-vector space $\Omega_{K/k}$ of all $k$-differentials of $K$. We may assume that deg tr $E/k > 0$, for otherwise the theorem is trivial. Then $x_1, \ldots, x_n$ are algebraically dependent over $E$, so there is a least number $s$, $1 \leq s \leq n$, such that $x_1, \ldots, x_s$ are algebraically dependent over $E$, and then $x_1, \ldots, x_{s-1}$ are algebraically independent over $E$. Applying Theorem A to the pair of differential fields $E \subset E(k_s)$ and using the natural injection $\Omega_{E(k_s)/E} \to \Omega_{K/E}$, we get that the differentials

$$\left\{ \sum_{j=1}^{m(i)} c_{ij} \frac{du_{ij}}{u_{ij}} + dv_i \right\}_{i=1, \ldots, s}$$

are elements of $\Omega_{K/E}$ that are linearly dependent over the field of constants $C$ of $k$. Let $c_1, \ldots, c_s$ be elements of $C$, not all zero, such that

$$\sum_{i=1}^{s} c_i \left( \sum_{j=1}^{m(i)} c_{ij} \frac{du_{ij}}{u_{ij}} + dv_i \right) = 0$$

in $\Omega_{K/E}$. Let $\gamma_1, \ldots, \gamma_r \in C$ be a basis for the $\mathbb{Q}$-vector space spanned by $\{c_i c_{ij}\}_{i=1, \ldots, n;\, j=1, \ldots, m(i)}$, so that we can write $c_i c_{ij} = \sum_{\rho=1}^{r} n_{ij\rho} \gamma_\rho$, with each $n_{ij\rho} \in \mathbb{Q}$. Replacing $\gamma_1, \ldots, \gamma_r$ by $\gamma_1/A, \ldots, \gamma_r/A$ for a suitable $A \in \mathbb{Z}$ if necessary, we can assume that each $n_{ij\rho} \in \mathbb{Z}$. Then

$$\sum_{i=1}^{s} c_i \left( \sum_{j=1}^{m(i)} c_{ij} \frac{du_{ij}}{u_{ij}} + dv_i \right) = \sum_{\rho=1}^{r} \gamma_\rho \frac{d \prod_{i,j} u_{ij}^{n_{ij\rho}}}{\prod_{i,j} u_{ij}^{n_{ij\rho}}} + d(c_1 v_1 + \cdots + c_s v_s) = 0$$

in $\Omega_{K/E}$, each product $\prod_{i,j}$ ranging over $i = 1, \ldots, s; j = 1, \ldots, m(i)$. Since $\gamma_1, \ldots, \gamma_r$ are linearly independent over $\mathbb{Q}$, by a general property of differentials (cf. [3, Prop. 4]) each $\prod_{i,j} u_{ij}^{n_{ij\rho}}$ and $c_1 v_1 + \cdots + c_s v_s$ are algebraic over $E$, hence in $E$. If $c_s = 0$, then since $u_{ij}$ and $v_i$ are in $k_{s-1}$ whenever $i < s$, we have each $\prod_{i,j} u_{ij}^{n_{ij\rho}}$ and $c_1 v_1 + \cdots + c_s v_s$ in $k_{s-1}$, therefore in $k_{s-1} \cap E = k_0$, the latter intersection being $k_0$ because of the algebraic independence of $x_1, \ldots, x_{s-1}$ over $E$. This would imply that

$$\sum_{i=1}^{s} c_i \left( \sum_{j=1}^{m(i)} c_i \frac{du_{ij}}{u_{ij}} + dv_i \right) = 0$$

in $\Omega_{K/k}$, which is known to be false, so we deduce that $c_s \neq 0$. If $x_s$ is an elementary primitive over $k_{s-1}$, then $v_s = x_s$ and for each $i = 1, \ldots, s$ and $j = 1, \ldots, m(i)$ we have $u_{ij} \in k_{s-1}$, so that each $\prod_{i,j} u_{ij}^{n_{ij\rho}} \in k_{s-1} \cap E = k_0$. But $k_{s-1}(c_1 v_1 + \cdots + c_s v_s) = k_{s-1}(x_s)$, a pure transcendental extension of $k_{s-1}$, so that $c_1 v_1 + \cdots + c_s v_s$ is transcendental over $k_{s-1}$, therefore over $k$. For any given derivation $D$ we have

$$0 = \sum_{i=1}^{s} c_i \left( \sum_{j=1}^{m(i)} c_{ij} \frac{Du_{ij}}{u_{ij}} + Dv_i \right)$$

$$= \sum_{\rho=1}^{r} \gamma_\rho \frac{D \prod_{i,j} u_{ij}^{n_{ij\rho}}}{\prod_{i,j} u_{ij}^{n_{ij\rho}}} + D(c_1 v_1 + \cdots + c_s v_s),$$

so that $c_1 v_1 + \cdots + c_s v_s \in E$ is an elementary primitive over $k_0$. In the case where $x_s$ is an elementary exponential over $k_{s-1}$, we could have taken $u_{s1} = x_1, u_{s2}, \ldots, u_{sm(s)}, v_s \in k_{s-1}$, and also $c_s c_{s1} = 1$ and $\gamma_1 \in \mathbb{Q}$, in which case the equation $c_s c_{s1} = \sum_{\rho=1}^{r} n_{s1\rho} \gamma_\rho$ implies that $n_{s1\rho} = 0$ for $\rho > 1$, so that for each $\rho > 1$ we have $\prod_{i,j} u_{ij}^{n_{ij\rho}} \in k_{s-1}$; also $c_1 v_1 + \cdots + c_s v_s \in k_{s-1}$, so all of these latter quantities are in $k_{s-1} \cap E = k_0$. We again have the equation displayed above, showing this time that $\prod_{i,j} u_{ij}^{n_{ij1}}$ is an elementary exponential over $k_0$. Since $n_{s11} \neq 0$ we have $\prod_{i,j} u_{ij}^{n_{ij1}}$ transcendental over $k_{s-1}$, therefore over $k$. Thus in either case, whether $x_s$ is an elementary primitive or an elementary exponential over $k_{s-1}$, we get the existence in $E$ of an element $t$ that is an elementary primitive or an elementary exponential over $k_0$ and transcendental over $k$. Thus $k_0(t)$ is a generalized elementary extension of $k$ which is contained in $E$. Replacing $k$ by $k_0(t)$ reduces us to proving the theorem in the case of a lower value for deg tr $E/k$, so that induction on this latter number will complete the proof.

*Remark 1* Theorem 1 may fail if the assumption that $E$ is algebraically closed in $K$ is dropped. For example, let $k$ be the ordinary differential field $\mathbb{C}(x)$ of rational functions of one complex variable $x$ with the usual derivation $x' = 1$, let $K = k\langle \sin^{-1} x \rangle = k((1-x^2)^{1/2}, \sin^{-1} x)$, and let $E = k\langle y \rangle$, where $y = (1-x^2)^{1/2} \sin^{-1} x$. We have $y' = 1 - xy/(1-x^2)$, so that $E = k(y)$, a pure transcendental extension of $k$, and $[K : E] = 2$. Suppose we have a relation $\sum_{i=1}^{n} c_i u_i'/u_i + v' \in k$, where $c_1, \ldots, c_n$ are elements of $\mathbb{C}$ that are linearly independent over $\mathbb{Q}$ and $u_1, \ldots, u_n, v \in K$. Theorem B implies that $u_1, \ldots, u_n \in k((1-x^2)^{1/2})$ and that $v - c \sin^{-1} x \in k((1-x^2)^{1/2})$ for some $c \in \mathbb{C}$. Thus any such relation $\sum_{i=1}^{n} c_i u_i'/u_i + v' \in k$ with $u_1, \ldots, u_n, v \in E$ would necessarily have $u_1, \ldots, u_n, v \in k$. This shows that $E$ is not a generalized elementary extension of $k$.

Another counterexample, less instructive since it also violates the condition that $k$ and $K$ have the same constants, is obtained from $k = \mathbb{R}(x) \subset$

$K = \mathbb{C}(x, \tan^{-1} x) = \mathbb{C}(x, \log((x - i)/(x + i)))$ and the intermediate field $E = k(\tan^{-1} x)$.

A related matter is Singer's example [4, Prop. 2] of a generalized elementary extension no algebraic extension of which is elementary, namely $k(y) \supset k = \mathbb{C}(x)$, where $x' = 1$ and $y' = \pi/x + 1/(x + 1)$. Since $y'\,dx = \pi\,dx/x + dx/(x + 1)$, we get $y$ transcendental over $k$, for otherwise the last differential would be exact, an impossibility by residue considerations. If an algebraic extension of $k(y)$ were elementary over $k$ it would have as a transcendence base over $k$ an element $t$ which is a logarithm or an exponential of an element algebraic over $k$. Theorem B would then imply that $t$ is a logarithm and that there exists $c \in \mathbb{C}$ such that $y - ct$ is algebraic over $k$. If $t' = u'/u$, for some $u$ algebraic over $k$, we get $y' - ct' = \pi/x + 1/(x + 1) - cu'/u$, so $\pi\,dx/x + dx/(x + 1) - c\,du/u$ is an exact differential of the algebraic function field $k(u)$, which again is impossible by residue considerations.

*Remark 2* The analogue of Theorem 1 for liouvillian extensions, in which the words "generalized elementary" are replaced by "liouvillian," does not hold. For example, let $k = \mathbb{C}(x) \subset K = k(e^{x^2}, \int e^{x^2})$ and $E = k\langle y \rangle$, where $y = (\int e^{x^2})/e^{x^2}$. The quantities $e^{x^2}$ and $\int e^{x^2}$ are algebraically independent over $k$, the latter function being nonelementary. We have $y' = 1 - 2xy$, so that $E = k(y)$. Since $K = E(e^{x^2})$, a pure transcendental extension of $E$, $E$ is algebraically closed in $K$. If we had a relation $\sum_{i=1}^{n} c_i u_i'/u_i + v' \in k$, where $c_1, \ldots, c_n \in \mathbb{C}$ are linearly independent over $\mathbb{Q}$ and $u_1, \ldots, u_n, v \in K$, then Theorem B applied to the pair $k(e^{x^2}) \subset K$ would give $u_1, \ldots, u_n \in k(e^{x^2})$ and $v - c \int e^{x^2} \in k(e^{x^2})$ for some $c \in \mathbb{C}$. Thus for such a relation as above with $u_1, \ldots, u_n, v \in E$, we must have $u_1, \ldots, u_n, v \in k$, and this shows that $E$ is not liouvillian over $k$. (Similarly we can show that if $k = \mathbb{C}(a, b)$, where $a$ and $b$ are differential indeterminates, and we set $K = k(e^{\int a}, \int be^{\int a})$ and $y = (\int be^{\int a})/e^{\int a}$, then $y' + ay = b$ and the differential extension field $K$ of $k$ is liouvillian, whereas the intermediate field $k\langle y \rangle = k(y)$ is not.)

***Proposition 3*** *Let $k \subset K$ be differential fields of characteristic zero, with the constants of $K$ algebraic over the subfield of constants of $k$, and let $K_1, K_2$ be differential fields between $k$ and $K$ such that $K_1$ is obtained from $k$ by repeated algebraic extensions and extensions by primitives, while $K_2$ is obtained from $k$ by repeated algebraic and exponential extensions. Then $K_1$ and $K_2$ are free over $k$.*

Assuming, as we may, that $K_1$ and $K_2$ are finite extensions of $k$, we have to prove that $\deg \operatorname{tr} K_1(K_2)/k = \deg \operatorname{tr} K_1/k + \deg \operatorname{tr} K_2/k$. Choose $x_1, \ldots, x_m \in K_1$ such that if $k_i$ is the algebraic closure in $K_1(K_2)$ of $k(x_1, \ldots, x_i)$, $i = 0, \ldots, m$, then each $x_i$ is primitive and transcendental over $k_{i-1}$ and $K_1 \subset k_m$. Similarly choose $y_1, \ldots, y_n \in K_2$ such that if $\kappa_j$ is the algebraic closure in $K_1(K_2)$ of $k(y_1, \ldots, y_j)$, $j = 0, \ldots, n$, then each $y_j$ is exponential and transcendental over $\kappa_{j-1}$ and $K_2 \subset \kappa_n$. Then $x_1, \ldots, x_m$ and $y_1, \ldots, y_n$ are transcendence bases over $k$ for $K_1$ and $K_2$ respectively. We have to show that the assumption that $x_1, \ldots, x_m, y_1, \ldots, y_n$ are algebraically dependent over $k$ leads to a contradiction. Making this assumption, we have $m, n \geq 1$. If $m > 1$ and $x_1, y_1, \ldots, y_n$ are algebraically dependent over $k$, it will be enough to produce a contradiction when $K_1$ is replaced by $k_1$. If $x_1, y_1, \ldots, y_n$ are algebraically independent over $k$, then $x_2, \ldots, x_m, y_1, \ldots, y_n$ are algebraically dependent over $k_1$ and it suffices to produce a contradiction when $k, K_1, K_2$ are replaced by $k_1, K_1, k_1(K_2)$ respectively, a change which reduces $m$ to $m - 1$ and leaves $n$ unchanged. Continuing this process reduces us to the case $m = 1$, with the same $n$ as before. In the same way we can reduce ourselves to the case $n = 1$, leaving $m$ unchanged. Thus we can assume that $m = n = 1$, with $x_1$ and $y_1$ algebraically dependent over $k_0 = \kappa_0$. The differential field $k_0(x_1, y_1)$ is then algebraic over the field $k_0(x_1)$, which is transcendental over $k_0$. Since for each given derivation $D$ we have $Dx_1 \in k_0$ and $Dy_1/y_1 \in k_0$, Theorem B implies that $y_1$ is algebraic over $k_0$, our desired contradiction.

***Proposition 4*** *Let $K$ be a liouvillian extension of the differential field $k$ of characteristic zero, with the constants of $K$ algebraic over the subfield of constants of $k$ and with* $\deg \operatorname{tr} K/k = n$. *Then there exists a transcendence basis $x_1, \ldots, x_n$ for $K$ over $k$ such that if for $i = 0, \ldots, n$ we let $k_i$ be the algebraic closure in $K$ of $k(x_1, \ldots, x_i)$, then each $k_i$ is a differential field and each $x_i$ is either primitive or exponential over $k_{i-1}$. Furthermore, the number of $x_i$'s that are primitive over $k_{i-1}$ and the number of $x_i$'s that are exponential over $k_{i-1}$ depend only on $k$ and $K$ and not on the particular $x_1, \ldots, x_n$ chosen.*

If $k \subset k(t_1) \subset k(t_1, t_2) \subset \cdots \subset k(t_1, \ldots, t_N) = K$ is a chain of differential fields such that each $t_i$ is primitive, or exponential, or algebraic over $k(t_1, \ldots, t_{i-1})$, we may choose $x_1, \ldots, x_n$ to be those $t_i$'s which are transcendental over their predecessor fields $k(t_1, \ldots, t_{i-1})$, in the same order. Since $n = \deg \operatorname{tr} K/k$ depends only on $k$ and $K$, it remains only to show that the number of $x_i$'s which are primitive over $k_{i-1}$ also depends only on $k$ and $K$, and not on the sequence $x_1, \ldots, x_n$. If for a given pair of fields $k$, $K$ this number is indeed independent of the choice of $x_1, \ldots, x_n$, we denote it by $\pi(k, K)$. Our problem is to prove that $\pi(k, K)$ exists, and we do this by induction on $n = \deg \operatorname{tr} K/k$, the case $n = 0$ being trivial and the case $n = 1$ an easy consequence of Proposition 3. Therefore assume that $n > 1$ and that the result holds for all smaller $n$. Let $x_1, \ldots, x_n, k_0, \ldots, k_n$ be as in the statement, and let $\xi_1, \ldots, \xi_n$ be another transcendence basis for $K$ over $k$

such that if for $i = 0, \ldots, n$ we let $\kappa_i$ be the algebraic closure in $K$ of $k(\xi_1, \ldots, \xi_{i-1})$, then each $\kappa_i$ is a differential field and each $\xi_i$ primitive or exponential over $\kappa_{i-1}$. We have to show that

$$\pi(k, k_1) + \pi(k_1, K) = \pi(k, \kappa_1) + \pi(\kappa_1, K).$$

But

$$\pi(k_1, K) = \pi(k_1, k_1(\kappa_1)) + \pi(k_1(\kappa_1), K)$$

and

$$\pi(\kappa_1, K) = \pi(\kappa_1, k_1(\kappa_1)) + \pi(k_1(\kappa_1), K),$$

so it remains to show that

$$\pi(k, k_1) + \pi(k_1, k_1(\kappa_1)) = \pi(k, \kappa_1) + \pi(\kappa_1, k_1(\kappa_1)).$$

Since deg tr $k_1(\kappa_1)/k \leq 2$, we are reduced to proving the existence of $\pi(k, K)$ in the case $n =$ deg tr $K/k = 2$. If both $x_1$ and $x_2$ are primitive over $k_0$ and $k_1$ respectively, a double application of Proposition 3 shows that $\xi_1$ and $\xi_2$ are primitive over $\kappa_0$ and $\kappa_1$ respectively. Similarly, if $x_1$ and $x_2$ are exponential over $k_0$ and $k_1$ respectively, then so are $\xi_1$ and $\xi_2$ over $\kappa_0$ and $\kappa_1$ respectively. By symmetry, if $\xi_1$ and $\xi_2$ are both primitive, or both exponential, over $\kappa_0$ and $\kappa_1$ respectively, then $x_1$ and $x_2$ are of the same type over $k_0$ and $k_1$ respectively. The only remaining possibilities are for either $x_1$ to be primitive over $k_0$ and $x_2$ exponential over $k_1$, or vice versa, and similarly for $\xi_1, \xi_2$ relative to $\kappa_0, \kappa_1$, and so we are done.

***Lemma*** *Let $k \subset K$ be differential fields of characteristic zero with the same field of constants $C$. Suppose that $k$ is a liouvillian extension of $C$ and that $K$ is algebraic over $k$. Suppose that $c_1, \ldots, c_n \in C$ are linearly independent over $\mathbb{Q}$, that $u_1, \ldots, u_n, v \in K$, with $u_1, \ldots, u_n$ nonzero, and that for each given derivation $D$ we have $\sum_{i=1}^n c_i\, Du_i/u_i + Dv \in k$. Then $v \in k$ and there is a nonzero integer $N$ such that $u_i^N \in k$, $i = 1, \ldots, n$.*

The proof proceeds by induction on deg tr $k/C$. When this number is zero, we have $C = k = K$ and the result is trivial. Therefore assume that this number is not zero and that the result is true for all smaller values of deg tr $k/C$. First assume that $\sum_{i=1}^n c_i\, Du_i/u_i + Dv = 0$ for each given derivation $D$. We choose a liouvillian extension $k_0$ of $C$ contained in $k$ and an element $t \in k$ that is transcendental over $k_0$ and either primitive or exponential over $k_0$ and such that $k$ is algebraic over $k_0(t)$. If $t$ is primitive over $k_0$ we can conclude from Theorem B that $u_1, \ldots, u_n$ are algebraic over $k_0$ and that there exists $c \in C$ such that $v + ct$ is algebraic over $k_0$. In this case, $\sum_{i=1}^n c_i\, Du_i/u_i + D(v + ct) = c\, Dt \in k_0$ for each $D$ and we can apply the induction hypothesis to conclude that $v + ct \in k_0$, so that $v \in k$, and that there is a nonzero integer $N$ such that $u_i^N \in k_0 \subset k$ for $i = 1, \ldots, n$. If $t$ is exponential over $k_0$ we conclude from Theorem B that $v$ is algebraic over $k_0$ and that there are integers $\nu_0, \nu_1, \ldots, \nu_n$, with $\nu_0 \neq 0$, such that each $u_i^{\nu_0} t^{\nu_i}$ is algebraic over $k_0$. For each given derivation $D$ we have

$$\sum_{i=1}^{n} c_i \, D(u_i^{\nu_0} t^{\nu_i})/(u_i^{\nu_0} t^{\nu_i}) + D\nu_0 v = \left(\sum_{i=1}^{n} c_i \nu_i\right) Dt/t \in k_0$$

and we again apply the induction hypothesis, this time getting $v \in k_0 \subset k$ and the existence of a nonzero integer $M$ such that each $(u_i^{\nu_0} t_i^{\nu_i})^M \in k_0$. Letting $N = \nu_0 M$, we get $u_i^N \in k_0(t) \subset k$. This proves the lemma in our special case.

For the general case, let $\sum_{i=1}^{n} c_i \, Du_i/u_i + Dv = \alpha_D \in k$ for each given derivation $D$. Taking traces with respect to the algebraic field extension $k(u_1, \ldots, u_n, v)$ of $k$, we get

$$\sum_{i=1}^{n} c_i \, D\mathrm{N}(u_i)/\mathrm{N}(u_i) + D \,\mathrm{Tr}(v) = m\alpha_D,$$

where $m$ is the extension field degree. Therefore

$$\sum_{i=1}^{n} c_i \, D(\mathrm{N}(u_i)u_i^{-m})/(\mathrm{N}(u_i)u_i^{-m}) + D(\mathrm{Tr}(v) - mv) = 0.$$

Applying what was shown in the preceding paragraph, we get $\mathrm{Tr}(v) - mv \in k$, so that $v \in k$, and that there is a nonzero integer $N$ such that each $(\mathrm{N}(u_i)u_i^{-m})^N \in k$, and hence $u_i^{mN} \in k$. This concludes the proof.

*Remark* The lemma holds only with some restriction on the field $k$. A counterexample (due to *Ax*) when $k$ is not liouvillian over $C$ takes $k$ to be the formal power series field $C((x))$ with the usual derivation $x' = 1$. We let $K$ be the field $C((x^{1/2}))$. Then $u'/u = v'$, where $u$ is the series $\exp(x^{1/2})$ and $v = x^{1/2}$. Here neither $v$ nor any nonzero power of $u$ is in $k$.

***Theorem 2*** *Let $k$ be a differential field of characteristic zero, $K$ an elementary extension field of $k$ with the same subfield of constants $C$, and $y \in K$ such that $Dy \in k$ for each given derivation $D$. Then there exist $c_1, \ldots, c_n \in C$ and elements $u_1, \ldots, u_n, v \in k$, with $u_1, \ldots, u_n$ nonzero, such that $Dy = \sum_{i=1}^{n} c_i \, Du_i/u_i + Dv$ for each given $D$. Furthermore, if $k$ is a liouvillian extension of $C$, then $c_1, \ldots, c_n, u_1, \ldots, u_n, v$ may be chosen such that there are elements $w_1, \ldots, w_n \in K$ with $Dw_i = Du_i/u_i$ for $i = 1, \ldots, n$ and each given $D$.*

The first part of the theorem is simply Liouville's theorem, of which there are many proofs. We shall nevertheless give yet another proof of it, deriving from the ideas of [2], since this proof will have to be reexamined in detail to obtain the second assertion. Set $m = \deg \operatorname{tr} K/k$. Since $K$ is elementary over

$k$, there exist $s_1, \ldots, s_m, t_1, \ldots, t_m \in K$ such that $Ds_i/s_i = Dt_i$ for $i = 1, \ldots, m$ and each given derivation $D$ and such that for each $i = 1, \ldots, m$ precisely one member of each pair $\{s_i, t_i\}$ is algebraic over $k(s_1, \ldots, s_{i-1}, t_1, \ldots, t_{i-1})$. It then follows that $K$ is algebraic over $k(s_1, \ldots, s_m, t_1, \ldots, t_m)$ and that $\{ds_i/s_i - dt_i\}_{i=1,\ldots,m}$ is a $K$-basis of $\Omega_{K/k}$. Applying Theorem A to the equations $Ds_i/s_i - Dt_i = 0$ and $Dy \in k$, we conclude the existence of $\gamma_1, \ldots, \gamma_m \in C$ such that $dy = \sum_{i=1}^m \gamma_i(ds_i/s_i - dt_i)$ in $\Omega_{K/k}$. Let $c_1, \ldots, c_n$ be a $\mathbb{Q}$-basis for the $\mathbb{Q}$-span of $\gamma_1, \ldots, \gamma_m$, so that we can write each $\gamma_i = \sum_{j=1}^n n_{ij} c_j$, with each $n_{ij} \in \mathbb{Q}$. Replacing each $c_j$ by $c_j/\mathrm{LCD}(n_{ij}, \ldots, n_{mj})$ if necessary, we can assume each $n_{ij} \in \mathbb{Z}$. Then

$$dy = \sum_{j=1}^n c_j(d(s_1^{n_{1j}} \cdots s_m^{n_{mj}})/(s_1^{n_{1j}} \cdots s_m^{n_{mj}}) - d(n_{1j}t_1 + \cdots + n_{mj}t_m)).$$

Letting $z_j = s_1^{n_{1j}} \cdots s_m^{n_{mj}}$ and $y_j = n_{1j}t_1 + \cdots + n_{mj}t_m$ for $j = 1, \ldots, n$, we see that $Dz_j/z_j = Dy_j$ for all $j = 1, \ldots, n$ and each given derivation $D$ and that $\sum_{j=1}^n c_j\, dz_j/z_j - d(y + \sum_{j=1}^n c_j y_j) = 0$. Since $c_1, \ldots, c_n$ are linearly independent over $\mathbb{Q}$, the quantities $z_1, \ldots, z_n$ and $z_0 = y + \sum_{j=1}^n c_j y_j$ are all algebraic over $k$. For each given derivation $D$ we have

$$Dy = Dz_0 - \sum_{j=1}^n c_j\, Dy_j = -\sum_{j=1}^n c_j\, Dz_j/z_j + Dz_0\,.$$

Taking traces with respect to the algebraic field extension $k_1 = k(z_0, z_1, \ldots, z_n)$ of $k$ and dividing by $[k_1 : k]$ gives

$$Dy = \sum_{j=1}^n (-c_j/[k_1 : k])\, D\mathrm{N}(z_j)/\mathrm{N}(z_j) + D(\mathrm{Tr}(z_0)/[k_1 : k]),$$

which proves the first part of the theorem.

To prove the second part, assume that $k$ is a liouvillian extension of $C$. The elements $z_0, z_1, \ldots, z_n$ of $K$ which appear above are all algebraic over $k$ and for each given derivation $D$ satisfy $-\sum_{j=1}^n c_j\, Dz_j/z_j + Dz_0 \in k$. Applying the lemma, we get $z_0 \in k$ and the existence of an integer $N > 0$ such that $z_j^N \in k$, $j = 1, \ldots, n$. For each $D$ we have

$$Dy = \sum_{j=1}^n (-c_j/N)\, Dz_j{}^N/z_j{}^N + Dz_0$$

and $D(Ny_j) = Dz_j{}^N/z_j{}^N$ for $j = 1, \ldots, n$. Since $Ny_j \in K$, $j = 1, \ldots, n$, we are done.

*Remark* The second part of Theorem 2 does not hold without some special condition on the base field $k$. For an example where this result fails, let $k$ be the ordinary differential field $\mathbb{C}(x, t)$, where $x' = 1$ and $t$ is to be determined, and let $K = \mathbb{C}(x, t, \alpha, \log(1 + \alpha), \log(2 + \alpha), \ldots, \log(n + \alpha))$, where $\alpha^2 = x$. Let $c_1, \ldots, c_n \in \mathbb{C}$ be linearly independent over $\mathbb{Q}$. We have

$$\sum_{i=1}^{n} c_i \frac{(i+\alpha)'}{i+\alpha} + (t\alpha)' = \sum_{i=1}^{n} \frac{-c_i}{2(i^2 - x)} + \alpha\left(\sum_{i=1}^{n} \frac{ic_i}{2x(i^2 - x)} + t' + \frac{t}{2x}\right).$$

If we choose $t$ such that $t' = -t/2x - \sum_{i=1}^{n} ic_i/2x(i^2 - x)$, then $k = \mathbb{C}(x, t)$ will be a differential field, as desired. In fact,

$$0 = \sum_{i=1}^{n} \frac{ic_i\alpha}{2x(i^2 - x)} + \alpha t' + \frac{t\alpha}{2x} = \sum_{i=1}^{n} \frac{c_i}{4\alpha}\left(\frac{1}{i+\alpha} + \frac{1}{i-\alpha}\right) + (t\alpha)',$$

so that we may take

$$t\alpha = -\sum_{i=1}^{n} \frac{c_i}{2} (\log(i+\alpha) - \log(i-\alpha)).$$

Thus $K$ is a subfield of $\mathbb{C}(\alpha, \log(1+\alpha), \ldots, \log(n+\alpha), \log(1-\alpha), \ldots, \log(n-\alpha))$, which can be taken to be a field of meromorphic functions on the upper half plane of the complex variable $x$. If we set

$$y = \sum_{i=1}^{n} c_i \log(i+\alpha) + t\alpha,$$

then $y \in K$ and

$$y' = \sum_{i=1}^{n} c_i(i+\alpha)'/(i+\alpha) + (t\alpha)' = -\sum_{i=1}^{n} c_i/2(i^2 - x) \in k.$$

It is clear that $K$ is an elementary extension of $k$. That $y \notin k$ comes from the algebraic independence of $\log(1+\alpha), \ldots, \log(n+\alpha), \log(1-\alpha), \ldots, \log(n-\alpha)$ over $\mathbb{C}(\alpha)$, an easy consequence of Ostrowski's theorem (cf. [1, p. 1155]). We claim that the second part of the theorem fails for the present $k, K, y$ if $n > 1$. We shall prove this by showing that if $w \in K$, $u \in k$, and $w' = u'/u$, then $u \in \mathbb{C}$.

So suppose that $w, u \in K$, $w' = u'/u$. Looking at the chain of differential fields

$$\begin{aligned}\mathbb{C}(\alpha) &\subset \mathbb{C}\left(\alpha, \sum_{i=1}^{n} c_i \log \frac{i-\alpha}{i+\alpha}\right) \\ &\subset \mathbb{C}\left(\alpha, \sum_{i=1}^{n} c_i \log \frac{i-\alpha}{i+\alpha}, \log(1+\alpha)\right) \subset \cdots \\ &\subset \mathbb{C}\left(\alpha, \sum_{i=1}^{n} c_i \log \frac{i-\alpha}{i+\alpha}, \log(1+\alpha), \ldots, \log(n+\alpha)\right) = K\end{aligned}$$

and applying Theorem B to the two top fields, we get $w = \gamma_n \log(n+\alpha) + w_1$, for some $\gamma_n \in \mathbb{C}$, with $w_1$ and $u$ in the field just below

$K$. Since $\gamma_n/2\alpha(n+\alpha) + w_1' = u'/u$, we can apply Theorem B again, etc., finally obtaining

$$w = \gamma_n \log(n+\alpha) + \cdots + \gamma_1 \log(1+\alpha) + \gamma \sum_{i=1}^{n} c_i \log \frac{i-\alpha}{i+\alpha} + \bar{w},$$

with $\gamma_n, \ldots, \gamma_1, \gamma \in \mathbb{C}$ and $\bar{w}, u \in \mathbb{C}(\alpha)$. Thus

$$\frac{\gamma_n}{n+\alpha} + \cdots + \frac{\gamma_1}{1+\alpha} - \gamma \sum_{i=1}^{n} c_i \left( \frac{1}{i+\alpha} + \frac{1}{i-\alpha} \right) + \frac{d\bar{w}}{d\alpha} = \frac{du/d\alpha}{u}.$$

Since the right hand side is a finite sum $\sum$ (integer)/($\alpha$-(element of $\mathbb{C}$)), we get $\bar{w} \in \mathbb{C}$ and each $\gamma c_i \in \mathbb{Z}$. Therefore $\gamma = 0$ if $n > 1$, and each $\gamma_i \in \mathbb{Z}$, so that $u/\prod_{i=1}^{n} (i+\alpha)^{\gamma_i} \in \mathbb{C}$. If $u \in k$, using the fact that $\mathbb{C}(x)$ is relatively algebraically closed in $k = \mathbb{C}(x, t)$ we see that $u \in \mathbb{C}(x)$, and hence $u \in \mathbb{C}$.

## References

1. E. Kolchin, Algebraic groups and algebraic dependence, *Amer. J. Math.* **90** (1968), 1151–1164.
2. R. Risch, Implicitly elementary integrals, *Proc. Amer. Math. Soc.* **57** (1976), 1–7.
3. M. Rosenlicht, On Liouville's theory of elementary functions, *Pacific J. Math.* **65** (1976), 485–492.
4. M. Singer, Functions satisfying elementary relations, *Trans. Amer. Math. Soc.* (to appear).
5. M. Singer, Elementary solutions of differential equations, *Pacific J. Math.* **59**, No. 2 (1975), 535–547.

Research supported by National Science Foundation grant number MPS-73-08528.

AMS (MOS) 1970 subject classification: 12H05

# *Derivations and valuation rings*

*A. Seidenberg*

*University of California*
*Berkeley*

Roughly, derivations are related to contact, and so are valuations, so one may ask for a study connecting derivations and valuations. Here, a first existence theorem is established: it is shown that if a derivation of a local ring of an algebraic variety sends the ring into itself, then it also sends some dominating valuation ring into itself. If the ring is 2-dimensional and regular and the maximal ideal is not differential, then the valuation ring is unique. The ground field is assumed to be of characteristic 0.

***Theorem 1*** *Let $\mathcal{O} = k[x_1, \ldots, x_n]$ be a finite integral domain over a base field $k$ of characteristic zero, let $m$ be a prime ideal therein, and let $D$ be an integral derivation of the local ring $\mathcal{O}_m$ (i.e., $D\mathcal{O}_m \subset \mathcal{O}_m$). Then there exists a valuation ring centered at $m$ that is also sent into itself by $D$.*

*Proof* One may assume $\mathcal{O}$ to be integrally closed. In fact, by [5, Vol. 1, p. 267, Th. 9] the integral closure $\overline{\mathcal{O}}$ of $\mathcal{O}$ is also a finite integral domain. By the Lying-Over Theorem, let $\bar{m}$ be a prime ideal in $\overline{\mathcal{O}}$ with $\bar{m} \cap \mathcal{O} = m$. Then $\overline{\mathcal{O}}_{\bar{m}}$ is also a localization of a finite integral domain. Moreover, $D\overline{\mathcal{O}}_{\bar{m}} \subset \overline{\mathcal{O}}_{\bar{m}}$. In fact, $\overline{\mathcal{O}}_{\bar{m}} \supset \mathcal{O}_m$, hence $\overline{\mathcal{O}}_{\bar{m}}$ also contains the integral closure $\overline{\mathcal{O}}_m$ of $\mathcal{O}_m$. Let $\overline{\mathcal{O}}_{\bar{m}} \cdot \bar{m} \cap \overline{\mathcal{O}}_m = m_1$, whence $m_1 \cap \overline{\mathcal{O}} = \bar{m}$. Then $\overline{\mathcal{O}}_{\bar{m}} \supset (\overline{\mathcal{O}}_{\bar{m}})_{m_1} \supset \overline{\mathcal{O}}_{\bar{m}}$, so $\overline{\mathcal{O}}_{\bar{m}} = (\overline{\mathcal{O}}_m)_{m_1}$. By [1], if a derivation maps a Noetherian domain containing the rationals into itself, then it also maps the integral closure into itself. Hence $D\overline{\mathcal{O}}_m \subset \overline{\mathcal{O}}_m$; and $D(\overline{\mathcal{O}}_m)_{m_1} \subset (\overline{\mathcal{O}}_m)_{m_1}$. Hence we may assume that $\mathcal{O}$ is integrally closed.†

There are two cases: (i) $D\mathcal{O}_m m \not\subset \mathcal{O}_m m$ and (ii) $D\mathcal{O}_m m \subset \mathcal{O}_m m$.

*Case* (*i*) We pass to the completion $\hat{\mathcal{O}}_m$ of $\mathcal{O}_m$; by [5, Vol. 2, p. 320, Th. 32], $\hat{\mathcal{O}}_m$ is an integral domain (since $\mathcal{O}_m$ is integrally closed). The derivation $D$ obviously extends to a continuous integral derivation of $\hat{\mathcal{O}}_m$, still to be denoted $D$. By [4, p. 526, Lemma 4], $\hat{\mathcal{O}}_m = \hat{\mathcal{O}}_1[[x]]$, where $x$ is analytically independent over $\hat{\mathcal{O}}_1$, $D\hat{\mathcal{O}}_1 = 0$; and, after changing $D$ by a unit in $\mathcal{O}_m$, $D = \partial/\partial x$. $\hat{\mathcal{O}}_1$ is Noetherian, in fact, is isomorphic to $\hat{\mathcal{O}}_m/(x)$. Let $v$ be a valuation centered at the maximal ideal of $\hat{\mathcal{O}}_1$. Extend $v$ to $\hat{\mathcal{O}}_1[[x]]$ as follows. Place $v^a(\alpha) = v^a(a_0 + a_1 x + \cdots) = \min\{v(a_i)\}$; we can do this since $\hat{\mathcal{O}}_1$ is Noetherian. Define: $\exp \alpha = \min\{n \mid v(a_n) = v^a(\alpha)\}$. Now place: $v(\alpha) = (v^a(\alpha), \exp \alpha)$. Defining addition in $\{v(\alpha)\}$ componentwise, and order, lexicographically, one finds that $v(\alpha)$ varies in an ordered group. Moreover, as one checks, $v$ is a valuation; let $\mathfrak{L}_v$ be the corresponding valuation ring. One has $\mathfrak{L}_v \supset \hat{\mathcal{O}}_m$ and $v(\alpha) = (0, 0)$ if and only if $\alpha$ is a unit in $\hat{\mathcal{O}}_1[[x]]$, so $\mathfrak{L}_v$ is correctly centered. We still have to check that $D\mathfrak{L}_v \subset \mathfrak{L}_v$. Note that $v(a') = v(Da) \geq v(a) - v(x)$ for $a \in \hat{\mathcal{O}}_m$. Now let $a, b \in \hat{\mathcal{O}}_m$ with $v(a) > v(b)$. Then $v(a') \geq v(b)$, since $v(a) \geq v(b) + v(x)$ and $v(a') \geq v(a) - v(x)$. Hence if $v(a/b) > 0$, then $v((a/b)') = v((a'b - ab')/b^2) \geq 0$. Finally, if $v(a/b) = 0$, then

$$a = \cdots + a_n x^n + \cdots, \qquad n = \exp a,$$

$$b = \cdots + b_n x^n + \cdots, \qquad n = \exp b, \quad v(a_n) = v(b_n).$$

Checking the case $n = 0$ separately, we may suppose $n > 0$. Then $v^a(a'b) = v^a(ab)$, $v^a(ab') = v^a(ab) = v^a(b^2)$, so $v^a(a'b - ab') \geq v^a(b^2)$. If greater holds, then $v(a'b - ab') > v(b^2)$, and if equal holds, then the $\exp(a'b - ab')$ is $\geq 2n - 1$, and actually $> 2n - 1$, since the contribution $na_n b_n$ from $a'b$ cancels with a like contribution from $ab'$. So also in the case $v(a/b) = 0$, we have $v(D(a/b)) \geq 0$. Hence $D\mathfrak{L}_v \subset \mathfrak{L}_v$ and case (i) is complete (after contracting $v$ to the quotient field of $\mathcal{O}$).

*Case* (*ii*) For a moment, drop the assumption that $\mathcal{O}$ is integrally closed. Let $\mathcal{O}_m \cdot m = (x_1, \ldots, x_n)$. Then $Dx_1 = x_1' = a_{11}x_1 + \cdots + a_{1n}x_n$, $x_2' = a_{21}x_1 + \cdots + a_{2n}x_n$, where the coefficients are in $\mathcal{O}_m$. $(x_2/x_1)' = x_2'/x_1 - x_1'x_2/x_1^2$. Therefore $D$ sends $\mathcal{O}_m[x_2/x_1, \ldots, x_n/x_1]$ into itself. $\mathcal{O}_m[x_2/x_1, \ldots, x_n/x_1] \cdot m = (x_1)$. This $\neq (1)$ if notation is correctly chosen; in fact, let $v$ be a valuation centered at $m$ and take the notation so that $v(x_1) = \min\{v(x_i)\}$. By [2, p. 24, Th. 1] the prime ideals of a $D$-invariant ideal in a Noetherian ring containing the rational numbers, and hence in particular

† If $\mathcal{O}$ is 1-dimensional, every valuation ring centered at $m$ is sent into itself by $D$, for in this case every such ring is of the form $\overline{\mathcal{O}}_{\bar{m}}$.

the primes of $(x_1)$, are themselves $D$-invariant. Let $p$ be a minimal prime of $(x_1)$; then $p$ is minimal in the ring, by the Principal Ideal Theorem. By elementary properties of quotient rings, $\mathcal{O}_m[x_2/x_1, \ldots, x_n/x_1]_p$ is the quotient ring of a minimal prime of $\mathcal{O}[x_2/x_1, \ldots, x_n/x_1]$. Therefore we may assume $m$ minimal. As in the first part of the proof, let $\overline{\mathcal{O}}$ be the integral closure of $\mathcal{O}$ and let $\overline{m}$ lie over $m$. Then $\overline{m}$ is minimal in $\overline{\mathcal{O}}$ (cf. [5, Vol. 1, p. 259, (1)]). Hence we may assume $\mathcal{O}$ is integrally closed *and* $m$ is minimal. In this case, $\mathcal{O}_m$ is itself a valuation ring, and the theorem follows.

The sought valuation $v$ is not necessarily unique, though in the case $\mathcal{O} = k[X, Y]$, a polynomial ring in two variables, $m = (X, Y)$, and $D = \partial/\partial X$, it is, as we showed in [1, p. 168]: it is the valuation in which $v(X)$ is infinitely small with respect to $v(Y)$. The above proof must, of course, yield this valuation, so the example serves as a mild check on the proof. The example generalizes completely, as follows.

***Theorem 2*** *Let $\mathcal{O}$ be a 2-dimensional regular local ring containing the rational numbers and with $m$ as maximal ideal. Let $D$ be an integral derivation of $\mathcal{O}$ with $Dm \not\subset m$. Then there is only one valuation of the quotient field of $\mathcal{O}$ centered at $m$ whose valuation ring is sent into itself by $D$.*

For the proof we need a lemma.

***Lemma*** *Let $\mathcal{O}$ be a 2-dimensional regular local ring with maximal ideal $m$ and let $\hat{\mathcal{O}}$ be its completion. Let $v$ be a valuation centered at $m$. Then $v$ has an extension to $\hat{\mathcal{O}}$.*

*Proof* Let $m > q_2 > q_3 > \cdots$ be the initial sequence of $v$-ideals in $\mathcal{O}$; it is known that $\bigcap q_i$ is prime (cf. [3, p. 156]). Since $\hat{\mathcal{O}}q_i \cap \mathcal{O} = q_i$, we have the sequence $\hat{\mathcal{O}}m > \hat{\mathcal{O}}q_2 > \hat{\mathcal{O}}q_3 > \cdots$. We assert that also $\bigcap \hat{\mathcal{O}}q_i$ is a prime ideal. In fact, let $\alpha \in \hat{\mathcal{O}}q_i - \hat{\mathcal{O}}q_{i+1}$. One can approximate elements in $\hat{\mathcal{O}}$ with elements in $\mathcal{O}$ so that the differences are in $\hat{\mathcal{O}}m^k$, for any $k$, and can take $k$ so that $\hat{\mathcal{O}}m^k \subset \hat{\mathcal{O}}q_{i+1}$. Therefore we can write $\alpha = a + \alpha'$ with $a \in q_i - q_{i+1}$ and $\alpha' \in \hat{\mathcal{O}}q_{i+1}$. Similarly, let $\beta \in \hat{\mathcal{O}}q_j - \hat{\mathcal{O}}q_{j+1}$ and write $\beta = b + \beta'$ with $b \in q_j - q_{j+1}$ and $\beta' \in \hat{\mathcal{O}}q_{j+1}$. Since $q_i q_j \not\subset \bigcap q_i$, there exists a $q_k$ with $v(q_i q_j) = v(q_k)$. Then $\alpha\beta \notin \hat{\mathcal{O}}q_k - \hat{\mathcal{O}}q_{k+1}$. So $\bigcap \hat{\mathcal{O}}q_i$ is prime.

If $\bigcap \hat{\mathcal{O}}q_i = 0$, then $v(\alpha) = v(a)$ gives the desired extension; here one uses the simple sequence $\{\hat{\mathcal{O}}q_i\}$ to check the valuation axioms. If $\bigcap \hat{\mathcal{O}}q_i \neq 0$, then $\bigcap \hat{\mathcal{O}}q_i = (f) \neq 0$. In this case, if $\alpha = \alpha' f^\rho$ with $\alpha' \not\equiv 0(f)$ and $\alpha' \in \hat{\mathcal{O}}q_i - \hat{\mathcal{O}}q_{i+1}$, we place $v(\alpha) = (\rho, v(q_i))$. Again we get the desired valuation. Here we use the well-ordered $\omega^2$-sequence of ideals $\{f^\rho \hat{\mathcal{O}}q_i\}$ to check the required conditions.†

† The extension $v$ of $v$ to $\hat{\mathcal{O}}$ is *unique*, for $v$ is *the* valuation whose initial sequence of $v$-ideals is $\{\hat{\mathcal{O}}q_i\}$.

*Proof of the theorem* Let $x \notin m$ such that $Dx \notin m$. We may replace $D$ by $(1/Dx)D$ and so suppose $Dx = 1$. Then we extend $D$ to $\hat{\mathcal{O}}$ and write $\hat{\mathcal{O}} = \hat{\mathcal{O}}_1[[x]]$. We have the endomorphism $\alpha \mapsto \alpha - xD\alpha + x^2D^2\alpha/2! - \cdots$, which maps $\hat{\mathcal{O}}$ onto $\hat{\mathcal{O}}_1$ [4, p. 256, Lemma 4]. Since $x \in m - m^2$, we may take it as a parameter in $\mathcal{O}$: let $m = (x, y)$. Then $x \mapsto 0$ and $y \mapsto y^* = y - xy' + x^2y''/2! - \cdots$ in the above endomorphism, so $y^*$ generates the maximal ideal in $\hat{\mathcal{O}}_1$; so $\hat{\mathcal{O}}_1$ is regular (and 1-dimensional).

Let $y_n{}^* = y - xy' + x^2y''/2! - \cdots \pm x^ny^{(n)}/n!$. We say that $v(y_n{}^*) > v(x^n)$. The proof is by induction; the case $n = 0$ holds since $v(y) > 0$. Suppose $v(x^n) \geq v(y_n{}^*)$, i.e., $v(x^n/y_n{}^*) \geq 0$. Then $v(nx^{n-1}/y_n{}^* \pm x^{2n}y^{(n+1)}/n!y_n^{*2}) \geq 0$. The first term in the parentheses has negative value, by induction, and the second term has nonnegative value, so we have a contradiction, and the induction is complete.

We have $\hat{\mathcal{O}}m = (x, y^*)$. Since $y^*$ is a parameter, $(y^*)$ is prime, as is $(x)$, and the extension of $v$ to $\hat{\mathcal{O}}$ given by the lemma can be described by saying $v(y^*) > 0$, $v(x) > 0$, and $v(y^*)$ is infinitely great with respect to $v(x)$. Thus the extension of $v$ to $\hat{\mathcal{O}}$ is unique, and a fortiori $v$ itself is unique.

***Corollary 1*** *The extension $v$ of $v$ to $\hat{\mathcal{O}}$ is unique and the (continuous) extension of $D$ sends the valuation ring of $v$ into itself.*

*Proof* We have $\hat{\mathcal{O}} = \hat{\mathcal{O}}_1[[x]] = k[[y^*]][[x]]$ and $v(x)$ is infinitely small with respect to $v(y^*)$, so, as already noted, the extension of $v$ to $\hat{\mathcal{O}}$ is unique. That $D$ sends the valuation ring into itself follows as in Case (i) of Theorem 1.

***Corollary 2*** *The residue field of $v$ is $k$.*

*Proof* Let $\alpha, \beta \in \hat{\mathcal{O}} = k[[y^*]][[x]]$ with $v(\alpha) = v(\beta)$. Then any two terms of $\alpha$ have distinct values and $v(\alpha)$ is the value of the term $by^{*i}x^j$ $(b \in k - 0)$ of minimum value, whence $v(\alpha - c\beta) > v(\alpha)$ for some $c \in k$.

*Remark* Let $v$ be a valuation nonnegative on a complete local domain $\hat{\mathcal{O}}$, let $\alpha_1 + \alpha_2 + \cdots$ be a convergent series in $\hat{\mathcal{O}}$, and assume $v(\alpha_i) > \gamma$ for some $\gamma$ in the value group and for $i \geq 1$. Then $v(\alpha_1 + \alpha_2 + \cdots) > \gamma$.

*Proof* Let $B$ be the ideal of elements with value $> \gamma$. Then $\alpha_i \in B$ for $i \geq 1$, and $\alpha_1 + \alpha_2 + \cdots \in B$, since $B$, as every ideal, is closed.

***Corollary 3*** *$m$ has a basis $m = (x, y)$ with $v(y) > v(x)$.*

*Proof* This follows directly from Corollary 2, or it may be seen as follows. Take $x \in m$ such that $Dx \notin m$ (and assume $Dx = 1$). Since $x \in m - m^2$ there is a basis of the form $(x, y)$ for $m$. Replace $y$ by $y - xy'$. We have seen that $v(y - xy') > v(x)$.

Let $m = (x, y)$. Then $x$, $y$ are analytically independent over $k$, $\hat{\mathcal{O}} = k[[x, y]]$, and every element ($\neq 0$) of $\hat{\mathcal{O}}$, and in particular every element ($\neq 0$) of $\mathcal{O}$, has, in an obvious sense, a leading form: it is an element of $k[x, y]$, a polynomial ring in two letters over $k$. If $v(y) > v(x)$, then every element of order 1 and value $> v(x)$ has $y$ or a nonzero constant multiple of it as leading form. Here $y$ is the so-called *directional form* of $v$ (cf. [5, Vol. 2, p. 364]). Taking $v$ to be the valuation associated with $D$, one sees that $D$ canonically determines a direction at $m$.

***Theorem 3*** *Let $\mathcal{O}$ be a regular r-dimensional local ring containing the rational numbers and having $m = (x_1, \ldots, x_r)$ as maximal ideal. Let D be an integral derivation of $\mathcal{O}$ with $Dm \not\subset m$ and let v be a valuation centered at m whose valuation ring is sent into itself by D. Then the initial sequence $m > q_2 > q_3 > \cdots$ of v-ideals depends only on D (and not on v). The intersection $\bigcap \hat{\mathcal{O}} q_i$ is a 1-dimensional prime ideal p having at $\hat{\mathcal{O}} m$ a simple point. Defining "leading form" in an obvious way, the leading forms of elements in $\mathcal{O}$ of order 1 and value $> v(m)$ generate a 1-dimensional prime ideal in $\mathcal{O}/m[x_1, \ldots, x_r]$, which depends only on D (and not on v). This prime gives the tangent to p at m. (Thus, also for $r > 2$, D canonically determines a direction at m.)*

*Proof* Let $Dx_1 \notin m$ and assume, without loss of generality, that $Dx_1 = 1$. Let $x_2^* = x_2 - x_1 x_2', \ldots, x_r^* = x_r - x_1 x_r'$. Then $v(x_i^*) > v(x_1)$, so obviously $q_2 = (x_2^*, \ldots, x_r^*, x_1^2)$. Now let $x_i^* = x_i - x_1 x_i' + x_1^2 x_1''/2!$, $i > 1$. Then obviously $q_3 = (x_2^*, \ldots, x_r^*, x_1^3)$. Similarly for all the $q_i$. Let now $x_i^* = x_i - x_1 x_i' + x_1^2 x_i''/2! - \cdots$. Then $\bigcap \hat{\mathcal{O}} q_i = (x_2^*, \ldots, x_r^*)$, as one easily checks, and thus is a 1-dimensional prime having $\hat{\mathcal{O}} m$ as simple point. Replacing $x_i$ by $x_i - x_1 x_i'$, let us assume that $v(x_i) > v(x_1)$, $i = 1, \ldots, r$. Then $x_i' \in m$ and $x_i$ is the leading form of $x_i$. One sees then that $(x_2, \ldots, x_r)$ is the ideal generated in $k[x_1, \ldots, x_r]$ by the leading forms of order 1 and value $> v(x_1)$.

## References

1. A. Seidenberg, Derivations and integral closure, *Pacific J. Math.* **16** (1966), 167–173.
2. A. Seidenberg, Differential ideals in rings of finitely generated type, *Amer. J. Math.* **89** (1967), 22–42.
3. O. Zariski, Polynomial ideals defined by infinitely near base points, *Amer. J. Math.* **60** (1938), 151–204.
4. O. Zariski, Studies in equisingularity I, *Amer. J. Math.* **87** (1965), 507–536.
5. O. Zariski and P. Samuel, "Commutative Algebra," Vols. I and II. Van Nostrand-Reinhold, Princeton, New Jersey, 1958 and 1960.

Supported in part by a National Science Foundation grant.

AMS (MOS) 1970 subject classifications: 14B99, 13B10

# *On theorems of Lie–Kolchin, Borel, and Lang*

*Robert Steinberg*

*University of California*
*Los Angeles*

This somewhat expository note is concerned with two topics. The first one comes from the observation that Kolchin's proof of the Lie–Kolchin theorem (see Corollary C) as given in [5, §7, Th. 1] can be adapted to yield an extension (see Theorem A) which has the Borel subgroup conjugacy theorem and the Borel fixed point theorem as quick consequences. (See [1, §§15, 16] for the orginal proofs of these theorems.) The development is quite elementary. The notion of quotient space is not needed, and of completeness not until the very end. It is our understanding that M. Sweedler has also found an "easy proof" of the conjugacy theorem (see [4, p. 138, Notes]).

Our second concern is the theorem, apparently new, that if $G$ is a connected algebraic group and $\sigma$ an endomorphism such that $G_\sigma$ is finite, then $1 - \sigma \colon G \to G$ is a finite (and dominant) morphism. This implies Lang's theorem (see [6], or [2, §16] for a different proof) that $1 - \sigma$ is surjective in case $\sigma$ is the Frobenius map for some rational structure on $G$, and our development yields a new, especially simple, proof of this theorem in case $G$ is affine.

As general references for the theory of algebraic groups we cite [2–4]. Here is the key to our first development.

**A. *Theorem*** *Let $G$ be a connected solvable algebraic group acting linearly on a vector space $V$ of finite dimension and stabilizing there a nonzero homogeneous closed cone $C$. Then $G$ stabilizes some line in $C$.*

*Remark* This is a special case of Borel's fixed point theorem (Corollary E), but this will not emerge from our development since the completeness of projective space will not be used.

*Proof* We assume first that $G$ is Abelian, and then using induction on $\dim V$, that $\dim V > 1$. Since $G$ is Abelian, its elements have a common eigenvector, hence fix some line $V_1$ of $V$. If $\dim V = 2$ then either $C$ is all of $V$ and hence contains $V_1$ or else is a finite union of lines each of which is fixed since $G$ is connected. To reduce to this case in general we may assume $V_1 \nsubseteq C$. Consider the natural projection $p: V \to V/V_1$. We have $p(C)$ nonzero since $V_1 \nsubseteq C$. We claim that $p(C)$ is closed. To see this without using the completeness of projective space observe that $p$ has the comorphism $k[X_2, X_3, \ldots, X_n] \to k[X_1, X_2, \ldots, X_n]$ if $V_1$ is taken as the first coordinate axis of $V$. Since $V_1 \nsubseteq C$, there exists a homogeneous polynomial $f$ on $V$ such that $f(V_1) \neq 0$ and $f(C) = 0$, whence $X_1$ is integral over $k[X_2, \ldots, X_n]$ on $C$. Thus $p: C \to \overline{p(C)}$ is a finite morphism and $p(C)$ is closed. Now by induction $G$ fixes a line $V_2/V_1$ of $p(C)$. We have $V_2$ fixed by $G$ and $C \cap V_2 \neq 0$, hence the required reduction to dimension 2.

Assume now that $G$ is not Abelian. By induction on $\dim G$ there exists a line $V_1 \subseteq C$ fixed by $G'$. The sum of all such lines yields a subspace $W$ such that $C \cap W \neq 0$. Write $W = \sum W_i$ according to the various characters with which $G'$ acts on the lines that it fixes. Since $G$ normalizes $G'$, it permutes the $W_i$'s, hence fixes each of them since it is connected. On each $W_i$ we thus have $G'$ acting via scalars of determinant 1, hence trivially since $G'$ is connected. Thus $G$ is Abelian on $W$, and we are done.

**B. *Corollary*** *If $G$ in Theorem A stabilizes a nonempty closed set of flags on $V$, then $G$ fixes one of them.*

*Proof* We imbed the flag variety in the projective space $\mathbb{P}(V \otimes \wedge^2 V \otimes \wedge^3 V \otimes \cdots)$ in the usual way (see, e.g., [2, §10.3] for details) and apply Theorem A to the action of $G$ on this space.

**C. *Corollary*** (Lie–Kolchin) *A connected solvable algebraic group acting linearly on a finite-dimensional vector space $V$ fixes some flag there, i.e., can be put in upper triangular form.*

*Proof* We apply Corollary B to the set of all flags on $V$, or else we apply Theorem A and use induction on $\dim V$.

**D.** ***Corollary*** (Borel) *Let $G$ be a connected linear algebraic group. Then any two Borel subgroups of $G$ are conjugate.*

*Proof* (Standard, adapted from [2, §11.1]) Let $B$ be a Borel subgroup of $G$ of maximum dimension. There exists a representation space $V$ for $G$ and a line $L$ of $V$ whose stabilizer is $B$ (see [2, §5.1]), and by Corollary C a flag $F$ extending $L$ and fixed by $B$. Now $G$ acts on the flag variety and the orbit through $F$ has minimum dimension (each orbit has as dimension the codimension in $G$ of the stabilizer of any of its elements, a solvable subgroup of $G$), hence is closed. If $B'$ is another Borel subgroup, then $B'$ has a fixed point on $GF$ by Corollary B: $B'xF = xF$ for some $x \in G$. Then $x^{-1}B'x$ fixes $F$ and is contained in $B$, as required.

**E.** ***Corollary*** (Borel fixed-point theorem) *A connected solvable linear algebraic group acting on a nonempty complete variety has a fixed point there.*

*Proof* Let $G$ be the group, $X$ the variety. By replacing $X$ by an orbit which has minimum dimension and hence is closed and hence complete, we may assume that there is a single orbit. Fix $x \in X$ and choose a representation space $V$ for $G$ containing a line $L$ whose stabilizer is $G_x$, the stabilizer of $x$ in $G$. Let $y$ be the point of $\mathbb{P}(V)$ corresponding to $L$, $Y$ the orbit of $y$ under $G$, and $Z$ the orbit of $(x, y) \in X \times \mathbb{P}(V)$ under $G$. The natural projections yield bijective $G$-morphisms from $Z$ to $X$ and to $Y$. Now $Z$ is closed in $X \times \mathbb{P}(V)$ since any $G$-orbit there projects onto $X$, hence has dimension $\geq \dim X = \dim Z$. Since $X$ is complete, the projection of $Z$ on the second factor, $Y$, is also closed. By Theorem A applied to $Y$, or rather to the cone in $V$ over $Y$, we get a fixed point for $G$ on $Y$ (which thus reduces to a point), hence also for $G$ on $X$ by the above-mentioned bijectivity.

We continue with our second theme.

**F.** ***Theorem*** *Let $G$ be a connected algebraic group and $\sigma$ an endomorphism such that $G_\sigma$ is finite. Then $1 - \sigma\colon x \to x\sigma(x^{-1})$ is a finite morphism.*

Throughout this part, finite morphisms are required to be dominant. Here $1 - \sigma$ is so since its nonempty fibers are the elements of $G/G_\sigma$ and hence have dimension 0. The rest of the proof will be given in several steps, the first of which yields the simple proof of Lang's theorem mentioned in the introduction.

**G.** ***Lemma*** *If $G$ is affine in Theorem F and $\sigma$ is the Frobenius map for some rational structure on $G$, then $1 - \sigma$ is finite, and hence surjective.*

*Proof* Let $k$ be the base field, $|k| = q$, and let $\sigma_0$ denote the comorphism of $\sigma$. Then $\sigma_0 f = f^q$ for every $f$ in $A = k[G]$. For such an $f$ and $y$, $z$ in $G$

we have $f(yz) = \sum e_i(y)f_i(z)$ with $\{e_i\}$ (resp. $\{f_i\}$) a basis for the space of right (resp. left) translates of $f$, whence $f_j(yz) = \sum e_{ji}(y)f_i(z)$ with each $e_{ji}$ in the space of left-right translates of $f$. With $y = (\sigma - 1)z$ this yields $f_j^q = \sigma_0 f_j = \sum (\sigma_0 - 1)e_{ji} \cdot f_i$. Thus $(\sigma_0 - 1)A[\{f_j\}]$ is finitely generated as a module over $(\sigma_0 - 1)A$, and it contains $f$. Hence $A$ is integral over $(\sigma_0 - 1)A$ and $\sigma - 1$ is a finite morphism. Thus $1 - \sigma$ is also.

**H. *Lemma*** *If G is semisimple in Theorem F, then* $1 - \sigma$ *is finite.*

*Proof* If $\sigma$ is surjective, then it is shown in [8, §10.5] that some $\sigma^n$ $(n \geq 1)$ is the Frobenius map for a suitable rational structure on $G$. Then $1 - \sigma^n$ is finite by Lemma G and so are both of its factors in $1 - \sigma^n = (1 + \sigma + \sigma^2 + \cdots + \sigma^{n-1})(1 - \sigma)$. Assume next that $G$ is simply connected, and also, as is permissible, that $\sigma$ is not surjective. Let $K = \ker \sigma$ and let $H$ be the product of the simple components of $G$ that are not in $K$, so that $G = HK$ (direct). For $y \in H$, $z \in K$, we have $\sigma(yz) = \alpha(y)\beta(y)$ with $\alpha, \beta$ homomorphisms from $H$ into $H$, $K$, and $(1 - \sigma)(yz) = (1 - \alpha)y \cdot z\beta(y^{-1})$. Since $G_\sigma$ is finite, $H_\alpha$ is also, whence $1 - \alpha$ is a finite morphism, by induction on dim $G$. Then so is $1 - \sigma$, the composition of $(1 - \alpha) \times \mathrm{id}_K$ and the automorphism (of varieties) $yz \to yz\beta(y^{-1})$. Finally, in the general case let $p\colon \overline{G} \to G$ be the universal covering, $\overline{A}$ and $A$ the affine algebras of $\overline{G}$ and $G$, and $\bar{\sigma}$ the lifting of $\sigma$ to $\overline{G}$ (for this see, e.g., [8, §9.16]). We have $p\overline{G}_{\bar{\sigma}} \subseteq G_\sigma$, whence $\overline{G}_{\bar{\sigma}}$ is finite and $1 - \bar{\sigma}$ is a finite morphism since $\overline{G}$ is simply connected. Thus

$$\overline{A} \text{ is integral over } (1 - \bar{\sigma}_0)\overline{A}. \qquad (*)$$

But $\overline{A}$ is also integral over $p_0 A$. This follows since $\overline{A}$ is algebraic over $p_0 A$ and the latter consists of the elements of $\overline{A}$ that are fixed by $\ker p$ and annihilated by $\ker dp$ (differential of $p$). It also follows from Corollary M. Thus $(1 - \bar{\sigma}_0)\overline{A}$ is integral over $(1 - \bar{\sigma}_0)p_0 A = p_0(1 - \sigma_0)A$, and so is $\overline{A}$ by $(*)$, whence $p_0 A$ is also. Thus $A$ is integral over $(1 - \sigma_0)A$, as required. We have used here the injectivity of $1 - \bar{\sigma}_0$ and of $p_0$.

**I. *Corollary*** *In Lemmas G and H if n is any positive integer, then* $1 + \sigma + \sigma^2 + \cdots + \sigma^{n-1}$ *is finite.*

For $G_{\sigma^n}$ is finite in these cases (but not in general) and the above factorization of $1 - \sigma^n$ may be used.

**J. *Theorem*** *In Theorem F the map* $1 - \sigma$ *is surjective.*

*Proof* If $G$ is Abelian, then $1 - \sigma$ is a homomorphism, whence $(1 - \sigma)G$ is closed and $1 - \sigma$ is surjective, while if $G$ is semisimple, then $1 - \sigma$ is surjective by Lemma H. From this the general case readily follows since $G$ has a normal series such that each term is fixed by $\sigma$ and each factor group is either Abelian or semisimple (see, e.g., [8, §§10.3, 10.4] for the patching procedure).

**K.** ***Lemma*** *Let $f: X \to Y$ be a morphism of irreducible varieties which is dominant and has finite fibers. Then there exist affine open subsets $U$ of $X$, $V$ of $Y$ such that $f^{-1}(V) = U$ and $f: U \to V$ is a finite morphism.*

*Proof* This well-known result is easily established by two localizations, the first yielding a reduction to the affine case (see, e.g., the reasoning of [7, pp. 99, 94, 95]).

**L.** ***Corollary*** *Let $G$ be a connected algebraic group and $f: X \to Y$ a $G$-morphism of homogeneous $G$-spaces with finite fibers. Then $f$ is finite.*

*Proof* Choose $V$ and $U = f^{-1}V$ as in Lemma K. Then apply the conclusion of Lemma K to all pairs of sets $gU$, $gV$ $(g \in G)$ to conclude that $f$ is finite.

**M.** ***Corollary*** *A surjective homomorphism of connected algebraic groups with finite kernel is finite.*

**N.** *Completion of proof of Theorem F* We apply Corollary L with $X = G$, $Y = G$, $f = 1 - \sigma$, and $G$ acting on $X$ by $g \cdot x = gx$ and on $Y$ by $g \cdot y = gy\sigma(g^{-1})$, which is permissible since $Y$ is homogeneous by Theorem J.

**O.** ***Corollary*** *If $1 - \sigma$ in Theorem F is separable, then $G/G_\sigma$ as a variety is isomorphic to $G$ itself. This holds in the cases of Lemmas G and H.*

*Proof* The map $1 - \sigma: G \to G$ is surjective by Theorem J, has as its fibers the elements of $G/G_\sigma$, and is separable. Thus it defines a quotient by [2, §6.7]. In Lemma G $d\sigma$ is 0 and in Lemma H nilpotent (see the proof of Lemma H). Thus $d(1 - \sigma)$ is an isomorphism in both cases and $1 - \sigma$ is separable.

**P.** ***Corollary*** *The map $1 - \sigma$ of Theorem F is closed and open, as is the map $1 + \sigma + \sigma^2 + \cdots + \sigma^{n-1}$ of Corollary I.*

*Proof* The map $1 - \sigma$ is closed since it is finite (and dominant), hence, via complements, is open on sets consisting of complete fibers. Thus if $U$ is any open subset of $G$, then $(1 - \sigma)U = (1 - \sigma)(UG_\sigma)$ is open. For $1 + \sigma + \sigma^2 + \cdots + \sigma^{n-1}$, the proof is similar.

## References

1. A. Borel, Groupes algébriques linéaires, *Ann. of Math.* (2) **64** (1956), 20–82.
2. A. Borel, "Linear Algebraic Groups." Benjamin, New York, 1969.
3. C. Chevalley, *Sém. classification groupes Lie algébriques, Paris, 1956–1968.*

4. J. E. Humphreys, "Linear Algebraic Groups." Springer-Verlag, New York–Heidelberg–Berlin, 1975.
5. E. R. Kolchin, Algebraic matric groups and the Picard–Vessiot theory of homogeneous linear differential equations, *Ann. of Math.* (2) **49** (1948), 1–42.
6. S. Lang, Algebraic groups over finite fields, *Amer. J. Math.* **78** (1956), 555–563.
7. D. Mumford, Introduction to algebraic geometry, Harvard lecture notes.
8. R. Steinberg, Endomorphisms of linear algebraic groups, *Amer. Math. Soc. Mem.* **80** (1968).

AMS (MOS) 1970 subject classification: 20G15

# *A differential-algebraic study of the intrusion of logarithms into asymptotic expansions*

*Walter Strodt*

*St. Lawrence University*

## A. Introduction

### A1 An intuitive glance at one of the rank-rise problems to be considered

The Riccati equation

$$\frac{dy}{dx} = f(x) + g(x)y + h(x)y^2, \tag{1}$$

with rational functions $f$, $g$, $h$ as coefficients, obviously can, for special choices of $f$, $g$, $h$, have logarithms involved in its solutions (e.g., if $f(x) = g(x) = 0$ and $h(x) = -x^{-1}$, then it has the solution $(\log x)^{-1}$), but (what is less obvious) it *cannot* have $\log \log x$ appearing. If (1) is generalized slightly by allowing a term in $y^3$ to appear,

$$\frac{dy}{dx} = f(x) + g(x)y + h(x)y^2 + k(x)y^3, \tag{2}$$

with $f$, $g$, $h$, and $k$ rational functions, then $\log \log x$ *can* be involved in the solutions (e.g., if $f(x) = g(x) = 0$ and $h(x) = k(x) = -x^{-1}$, then (2) has a solution with an asymptotic expansion $(\log x)^{-1} - (\log x)^{-2} \log \log x + \cdots$).

But even if (2) is generalized further by allowing the degree in $y$ to be unrestricted,

$$\frac{dy}{dx} = \sum_{j=0}^{n} f_j(x) y^j \tag{3}$$

(with the $f_j$ rational functions), no additional logarithmic complications can occur; that is, no solution of (3) can have an asymptotic expansion in which log log log $x$ appears, or $\log_4 x$, etc. [Note 1 (see Appendix)].

The nonappearances of *high-ranking* logarithms ($\log_q x$ will be said to have *logarithmic rank* equal to $q$), as asserted in the preceding paragraph, are elementary instances of results which are obtained in the present paper. For example, the nonappearance of log log log $x$, $\log_4 x, \ldots$, asserted for (3) follows from a general theorem (F14) which has as a corollary the result that, for first-order algebraic differential equations with rational functions for coefficients, solutions cannot involve $\log_q x$ if $q$ is more than twice the highest power of $dy/dx$ appearing in the equation.

**A2 Abstract formulation of two rank-rise problems** Although the discussion in the preceding section is phrased in the language of analysis, this paper's treatment of the various rank-rise problems will be carried out in an abstract setting. The results will immediately carry over to corresponding problems in analysis.

The domain of operations is a *graduated logarithmic field* [3, §2.4] which, roughly speaking, is an abstract differential field having extra structure designed to provide a setting for asymptotic expansions. More precisely, it is a differential field $(K, D)$ provided with logarithms $x_0, x_1, \ldots, x_p, \ldots$ (i.e. elements of $K$ satisfying $Dx_0 = 1$, $x_0 \; Dx_1 = 1, \ldots, x_0 x_1 \cdots x_{p-1} \; Dx_p = 1, \ldots$) and with a relation $\prec$ of *asymptotic dominance* suitably responsive to the algebraic-differential structure. (For example, if $f \prec h$ and $g \prec h$, then $f + g \prec h$; if $f \prec m$ and $m$ is a logarithm, then $Df \prec Dm$.) [Note 2.]

A relation $\sim$ of *asymptotic equivalence* is defined by the formula: $f \sim k$ iff $f - k \prec k$.

We say that $m$ is a *logarithmic monomial* if $m = c x_0{}^a x_1{}^b \cdots x_s{}^w$ with $c$ constant and $\{a, b, \ldots, w\} \subset \mathbb{Q}$. If $w \neq 0$, then $s$ is called the *logarithmic rank* of $m$; if $a = b = \cdots = w = 0$, then the number $-1$ is assigned to $m$ as its logarithmic rank. The set of all logarithmic monomials of logarithmic rank $\leq r$ is denoted by $M_r$. The set of all logarithmic monomials is denoted by $M_\infty$. $A_r$, the *asymptotic envelope* of $M_r$, is defined by the formula: $f \in A_r$ iff there exists $g \in M_r$ such that $f \sim g$. $A_\infty$ is the asymptotic envelope of $M_\infty$.

A differential subfield $F$ of $K$ is called *r-constrained* if $M_r \subset F \subset A_r \cup \{0\}$. A differential subfield $H$ of $K$ is called *monomic* if $M_\infty \subset H \subset A_\infty \cup \{0\}$

[Note 3]. A differential polynomial whose coefficients belong to an $r$-constrained field is called *r-inscribed*. (For instance, differential polynomials with rational functions of $x_0$ as coefficients are easily seen to be 0-inscribed.)

We write $y_0 \sim \sum u_n$, and we say that $y_0$ has an *asymptotic expansion in terms of the sequence* $(u_0, u_1, \ldots, u_n, \ldots)$ if $y_0 \sim u_0$, $y_0 - u_0 \sim u_1, \ldots,$ $y_0 - (u_0 + u_1 + \cdots + u_n) \sim u_{n+1}, \ldots$. In this paper every asymptotic expansion under discussion will be in terms of a sequence of logarithmic monomials [Note 4].

The *rank-rise problem for asymptotic expansions* is to determine, for an $r$-inscribed first-order differential polynomial $P$, how large the logarithmic rank of $m_n$ can be if $m_n$ is a term in an asymptotic expansion of a solution of $P$ [Note 5].

In the succeeding sections, before attacking the rank-rise problem for asymptotic expansons we shall take up a closely related problem which will be referred to as the *adjunction rank-rise problem*: Suppose $P$ has its coefficients in an $r$-constrained field $F$. Suppose that $P(y_0) = 0$, and that $y_0$ lies in an algebraically closed monomic field $H$ containing $F$. The problem is to determine whether $y_0$ belongs, for some $q \geq r$, to some $q$-constrained field $G$ such that $F \subset G \subset H$, and if so, how small a number $q$ may be used. (One answer to this problem is given by Theorem F12.) [Note 6.]

## B. Some basic notations, procedures, and lemmas for the adjunction rank-rise problem

**B1 Notation and remarks** We shall call $(F, r, P, y_0, H)$ an *adjunction challenge* if $F$ is $r$-constrained, $P$ is a first-order differential polynomial with coefficients in $F$, $H$ is an algebraically closed monomic field containing $F$, and $y_0$ is an element of $H$ such that $P(y_0) = 0$.

Given an adjunction challenge $(F, r, P, y_0, H)$ we try to find a $q$-constrained field $G$ such that $y_0$ lies in $G$ and $F \subset G \subset H$.

Let us first search for such a $G$ with $q = r$. Obviously, for the existence of such a $G$ it is necessary and sufficient that the *differential* field $F\{y_0\}$ generated by $F$ and $y_0$ be contained in $A_r \cup \{0\}$. What is less obvious is that it suffices for the *algebraic* field $F(y_0)$ to be contained in $A_r \cup \{0\}$. This sufficiency comes about because on the one hand the relation $F(y_0) \subset A_r \cup \{0\}$ implies that $[F(y_0)]^\gamma \subset A_r \cup \{0\}$ ($\gamma$ stands for algebraic closure [Note 7]) and on the other hand, the fact that $P$ is an *algebraic* differential equation of *first order* implies that $F\{y_0\} \subset [F(y_0)]^\gamma$.

Thus the first step in a search for $G$ becomes the determination of whether $F(y_0) \subset A_r \cup \{0\}$. If it is not, the smallest $q$ that can be hoped for is $q = r + 1$. Considerations for the case $q = r + 1$, similar to those mentioned for the case $q = r$, lead to the examination of whether

$F_1(y_0) \subset A_{r+1} \cup \{0\}$ (here $F_1$ is the $(r+1)$-constrained field $F(M_{r+1})$ generated by $F \cup M_{r+1}$), and so on. In some adjunction rank-rise problems it is desirable to obtain a $G$ which has an additional property of *symmetry* (see Section B5), and this calls for the testing of inclusion relations of the more general form $I_j(y_0) \subset A_{r+j} \cup \{0\}$ (where $I_j$ is a given $(r+j)$-constrained field). All in all, considerations like these lead to the concepts of *adjunction trial*, and *adjunction failure*, defined in the next section.

**B2** ***Definition*** Let $\lambda = (F, r, P, y_0, H)$ be an adjunction challenge. Let $n$ be a nonnegative integer. Then the ordered $(n+1)$-tuple $s = (I_0, I_1, \ldots, I_n)$ will be called an *adjunction trial*, of *length* $n+1$, for $\lambda$, if $F \subset I_0 \subset I_1 \subset \cdots \subset I_n \subset H$, and $I_j$ is $(r+j)$-constrained for each $j$ in $\{0, 1, \ldots, n\}$.

If $s = (I_0, I_1, \ldots, I_n)$ is an adjunction trial for $\lambda$, and for each $j$ in $\{0, 1, \ldots, n\}$ the relation $I_j(y_0) \not\subset A_{r+j} \cup \{0\}$ is valid, then $s$ is called an *adjunction failure* of *magnitude* $n+1$, for $\lambda$

**B3** ***Definition*** Let $\lambda$ be an adjunction challenge. Let $S_\lambda = \{k :$ there exists an adjunction failure, of magnitude $k$, for $\lambda\}$. We define the *adjunction gap*, for $\lambda$, to be 0 if $S_\lambda$ is empty, to be $\infty$ if $S_\lambda$ is an infinite set, to be max $S_\lambda$ if $S_\lambda$ is a nonempty finite set.

*Note* Evidently, if $h$ is a positive integer less than or equal to the adjunction gap for $\lambda$, then there exists an adjunction failure, of magnitude $h$, for $\lambda$.

**B4** ***Fundamental Lemma I*** *Let $\lambda = (F, r, P, y_0, H)$ be an adjunction challenge. Let $h$ be a nonnegative integer such that the adjunction gap for $\lambda$ is $\leq h$. Then there exists an $(r+h)$-constrained $G$ such that $F(y_0) \subset G \subset H$.*

*Proof* Let $F_j = F(M_{r+j})$ $(j = 0, 1, \ldots, h)$. Then $s = (F_0, F_1, \ldots, F_h)$ is an adjunction trial, of length $h+1$, for $\lambda$. But $s$ cannot be an adjunction failure for $\lambda$, since the length of $s$ exceeds $h$. Therefore, there exists $k \in \{0, 1, \ldots, h\}$ such that $F_k(y_0) \subset A_{r+k} \cup \{0\}$. Let $J = [F_k(y_0)]^\gamma$. Then $J$ is $(r+k)$-constrained. We may take $J(M_{r+h})$ to serve as $G$.

**B5 Preliminary remarks on symmetry** In that important analytic case where the universal field is constructed from functions meromorphic in sectors bisected by the real axis, an apparently indispensable tool in the existence theory is the *Schwarzian reflection* $f^*$ of the functon $f$ ($f^*$ is defined by the formula $f^*(x) = \overline{f(\bar{x})}$, where the bar indicates complex conjugate). To make the abstract theory applicable to this analytic situation, the operation of Schwarzian reflection needs a model in the abstract structure. This model is provided in the form of an involution * which is suitably responsive to the algebraic, differential, and asymptotic dominance structures [Note 8]. The rank-rise adjunction problem undergoes a corresponding shift, in which the given $F$ and $H$ are given to be symmetric (e.g., $f \in F \Rightarrow f^* \in F$) and the intermediate field $G$ is required to be symmetric. This shift motivates the next section.

**B6** ***Fundamental Lemma II*** *Let $\lambda = (F, r, P, y_0, H)$ be an adjunction challenge with $F$ and $H$ symmetric. Let $h$ be a nonnegative integer such that the adjunction gap for $\lambda$ is $\leqq h$. Then there exists a symmetric $(r + h)$-constrained $G$ such that $F(y_0) \subset G \subset H$.*

*Proof* Let $F_j = F(M_{r+j})$ $(j = 0, 1, \ldots, h)$. Then $s = (F_0, F_1, \ldots, F_h)$ is an adjunction trial of length $h + 1$, for $\lambda$. But $s$ cannot be an adjunction failure for $\lambda$, since the length of $s$ exceeds $h$. Hence there exists $k \in \{0, 1, \ldots, h\}$ such that

$$F_k(y_0) \subset A_{r+k} \cup \{0\}. \tag{1}$$

We choose $k$ as small as possible for (1) to hold. Then for each $j$ in $\{0, 1, \ldots, k - 1\}$ we have

$$F_j(y_0) \not\subset A_{r+j} \cup \{0\}. \tag{2}$$

From (1) and the symmetry of $F_k$ it follows that $F_k(y_0^*) \subset A_{r+k} \cup \{0\}$. Let $I_k = [F_k(y_0^*)]^\gamma$. Then $I_k$ is $(r + k)$-constrained [Note 9]. We define

$$I_j = F_j \qquad (j < k), \tag{3}$$

and

$$I_j = I_k(M_{r+j}) \quad (j > k). \tag{4}$$

Then $t = (I_0, I_1, \ldots, I_h)$ is an adjunction trial, of length $h + 1$, for $\lambda$. But $t$ cannot be an adjunction failure, for $\lambda$, because the length of $t$ exceeds $h$. Hence there exists $b \in \{0, 1, \ldots, h\}$ such that

$$I_b(y_0) \subset A_{r+b} \cup \{0\}. \tag{5}$$

Because of (2) and (3), relation (5) implies that $b \geqq k$. Let $J = [I_b(y_0)]^\gamma$. Then $J$ is $(r + b)$-constrained. Since $b \geqq k$, we have $I_b = I_k(M_{r+b})$. Hence $I_b = [F_k(y_0^*)]^\gamma(M_{r+b})$. Hence $J = (\{[F_k(y_0^*)]^\gamma(M_{r+b})\}(y_0))^\gamma$. Hence $J = [F_k(y_0, y_0^*, M_{r+b})]^\gamma$, which is symmetric. We may take $J(M_{r+h})$ to serve as $G$.

**B7 Preliminary remarks on instability ladders** We recall several definitions [3, §2.19, 2.79]: $P$ is *stable* at $u$ if $P(y) \sim P(u)$ whenever $y \sim u$; *unstable* at $u$ otherwise; if $m$ is a logarithmic monomial at which $P$ is unstable, then $P$ is called a *critical monomial* for $P$, and we write $m \in \text{crit}\ P$ [Note 10].

Lemma B9 makes it possible to replace the study of adjunction failures by the study of a different kind of rank-rise sequence (an *instability ladder*, as defined in Definition B8) which is defined in terms of instability properties of $P$, instead of in terms of a solution of $P$. This change in point of view gives a decisive added flexibility to the calculations.

**B8 *Definition*** Let $Q$ be an $r$-inscribed first-order differential polynomial. Let $f = (f_1, f_2, \ldots, f_s)$ be a finite sequence of elements of $K$. We shall say that $f$ is an *instability ladder*, of *length* $s$, for $(Q, r)$, if all the following conditions are satisfied:

(1) $f_1 \succ f_2 \succ \cdots \succ f_s$;
(2) $Q$ is unstable at $f_1$, $Q(f_1 + Y)$ is unstable at $f_2, \ldots, Q(f_1 + f_2 + \cdots + f_{s-1} + Y)$ is unstable at $f_s$;
(3) there exists an ordered $(s+1)$-tuple $(F_0, F_1, \ldots, F_s)$ such that $F_k$ is $(r+k)$-constrained ($k = 0, 1, \ldots, s$), $F_0 \subset F_1 \subset \cdots \subset F_s$, the coefficients of $Q$ belong to $F_0$, and $f_i \in F_i$ ($i = 1, 2, \ldots, s$); and
(4) logarithmic rank $f_i = r + i$ [Note 11].

When (1)–(4) hold, $(F_0, F_1, \ldots, F_s)$ is called an *inclusion sequence* for $(Q, r, f_1, \ldots, f_s)$.

**B9** *Let $\lambda = (F, r, P, y_0, H)$ be an adjunction challenge. Let $(I_0, I_1, \ldots, I_n)$ be an adjunction failure for $\lambda$. Then if $J_k = I_k^{\gamma}$ ($k = 0, 1, \ldots, n$) and $J_{n+1} = J_n(M_{n+1})$, there exists an element $f_0$ of $J_0$ such that either $f_0 \prec y_0$ or $f_0 \sim y_0$, and such that $(P(f_0 + Y), r)$ has an instability ladder $(f_1, f_2, \ldots, f_{n+1})$, with $f_1 \sim y_0 - f_0$, and with inclusion sequence $(J_0, J_1, \ldots, J_{n+1})$ for $(P(f_0 + Y), r, f_1, f_2, \ldots, f_{n+1})$.*

*Proof* It is easy to see that since $(I_0, I_1, \ldots, I_n)$ is an adjunction failure for $\lambda$, so is $(J_0, J_1, \ldots, J_n)$. Hence $J_0(y_0) \not\subset A_r \cup \{0\}$. Since $J_0$ is algebraically closed, it follows that there exists an element $f_0$ of $J_0$ such that

$$y_0 - f_0 \notin A_r \cup \{0\}. \tag{1}$$

Since $f_0 \in A_r \cup \{0\}$ it follows from (1) that either $f_0 \prec y_0$ or $f_0 \sim y_0$. Let $Q(Y) = P(f_0 + Y)$, and let $z_0 = y_0 - f_0$. Then coeff $Q \subset J_0$ and $Q(z_0) = 0$. Since $z_0 \in H$ and $z_0 \neq 0$, $z_0$ is $\sim$ to a critical monomial $m_1$ for $Q$; by [3, §3.16(1)], $m_1$ must belong to $M_{r+1}$ [Note 12].

Therefore $z_0$, which is not in $A_r$, must be of logarithmic rank equal to $r + 1$.

Since $J_1(y_0) \not\subset A_{r+1} \cup \{0\}$, it follows that $J_1(z_0) \not\subset A_{r+1} \cup \{0\}$, so there exists an element $f_1$ of $J_1$ such that $z_0 - f_1 \notin A_{r+1} \cup \{0\}$. Let $R(Y) = Q(f_1 + Y)$ and let $w_0 = z_0 - f_1$. Then coeff $R \subset J_1$ and $R(w_0) = 0$. Since $w_0 \in H - \{0\}$, $w_0$ is $\sim$ to a critical monomial $m_2$ for $R$; $m_2$ [3, §3.16(1)], must belong to $M_{r+2}$. Therefore $w_0$, which is not in $A_{r+1}$, must be of logarithmic rank equal to $r+2$. Since $z_0$ and $f_1$ are of logarithmic rank equal to $r+1$, while their difference $w_0$ is of logarithmic rank equal to $r+2$, we must have $f_1 \sim z_0$, whence $w_0 \prec f_1$. Since $f_1 \sim z_0$, it follows that $Q$ is unstable at $f_1$. An application of these considerations to the relation $J_2(w_0) \not\subset A_{r+2} \cup \{0\}$ shows that there exists an element $f_2$ of $J_2$ such that $w_0 - f_2 \notin A_{r+2} \cup \{0\}$. Then $f_2 \sim w_0$, so $f_2 \prec f_1$, and $R$ is unstable at $f_2$. Continuing in this way we determine $(f_1, f_2, \ldots, f_n)$ as an instability ladder, of length $n$, for $(Q, r)$, with inclusion sequence $(J_0, J_1, \ldots, J_n)$. The construction is carried out in such a way that $z_0 - (f_1 + f_2 + \cdots + f_j)$ is of logarithmic rank equal to $r + j + 1$ $(j = 1, 2, \ldots, n)$. If $u_0 = z_0 - (f_1 + \cdots + f_{n-1})$ and $t_0 = u_0 - f_n$, then (logarithmic rank $u_0$) $= r + n$, (logarithmic rank $f_n$) $= r + n$, and (logarithmic rank $t_0$) $= r + n + 1$, so $f_n \sim u_0$. Therefore $t_0 \prec f_n$. If $S(Y) = Q(f_1 + \cdots + f_n + Y)$, then $S(t_0) = 0$, so there exists $m_{n+1} \in \text{crit } S$ such that $t_0 \sim m_{n+1}$. We may take $f_{n+1}$ equal to $m_{n+1}$ to complete the required instability ladder $(f_1, f_2, \ldots, f_{n+1})$.

## C. Instability ladders for differential polynomials of Class($V$, $r$)

**C1** *Remark* The most familiar instance of rank-rise, from coefficients of rank $r$ to a solution of rank $r+1$, is the case of the equation

$$x_0 x_1 \cdots x_r \, DY - 1 = 0, \tag{1}$$

with solution $y_0 = x_{r+1}$. This section is devoted to the study of instability ladders for certain differential polynomials (called of *Class*($V$, $r$)), which may be thought of as perturbations of the left-hand member of (1).

**C2** ***Definition*** Let F be $r$-constrained: Let $P(Y) = \sum p_{ij} Y^i (D_r Y)^j$ [Note 13], with $p_{ij} \in$ F. Let $p_{ij} \prec 1$ whenever $i + j \geq 2$, let $p_{10} \prec 1$, let $p_{01} \sim 1$, and let $p_{00} \sim -1$. Then $P$ is said to be of *Class*($V$, $r$).

**C3** ***Lemma*** *Let F be q-constrained. Let e be an element of F such that $e \prec 1$. Then the relation*

$$e + D_q e \approx x_q^{-1} \qquad \text{[Note 14]} \tag{1}$$

*is impossible.*

*Proof* Suppose to the contrary that (1) is satisfied. Then obviously $e \neq 0$. Hence $e \in A_q$. Therefore $e \sim k x_s^{\ a} x_{s+1}^b \cdots x_q^{\ w}$ with $k$ constant, $s \leq q$, and $\{a, b, \ldots, w\} \subset \mathbb{Q}$. Since $e \prec 1$, it follows that $\{a, b, \ldots, w\} \neq \{0\}$. Hence we may assume $a \neq 0$. Then

$$D_q e \sim k a x_s^{\ a} x_{s+1}^{b+1} \cdots x_q^{w+1}. \tag{2}$$

It follows that if $s < q$, then $D_q e \succ e$, whence

$$e + D_q e \sim D_q e \sim kax_s^{\,a} x_{s+1}^{b+1} \cdots x_q^{w+1},$$

which contradicts (1). Therefore $s = q$, and

$$e \sim kx_q^{\,a}, \qquad D_q e \sim kax_q^{\,a}. \tag{3}$$

Then if $a \neq -1$, it follows from (3) that $e + D_q e \sim k(1 + a)x_q^{\,a}$, which contradicts (1).

Thus we must have $a = -1$. Then we have

$$e \sim kx_q^{-1}, \qquad D_q e \sim -kx_q^{-1}. \tag{4}$$

From (4) we have $e + D_q e \prec kx_q^{-1}$, which again contradicts (1).

**C4** ***Definition*** By $E$ is meant $\{f : f \prec 1\}$. By $E_r$ is meant $\{f : \exists g \in M_r \cap E$ such that $f \prec g\}$.

*Remark* It follows that $E_{-1} = \{0\}$. Also, if $f \in A_r \cap E$, then $f \in E_r$.

**C5** ***Lemma*** *Let $P$ be of* Class$(V, r)$. *Let $(f_1, f_2, \ldots, f_n)$ be an instability ladder for $(P, r)$. Then if $f_1 \sim x_{r+1}$, we have $n = 1$.*

*Proof* Assume to the contrary that $n \geqq 2$. Let $(F_0, F_1, \ldots, F_n)$ be an inclusion sequence for $(P, r, f_1, \ldots, f_n)$. Let $Q(Y) = P(f_1 + Y)$. Then coeff $Q \subset F_1$, and $Q$ is unstable at $f_2 \in F_2$. Let $f_2 \sim m \in M_{r+2} - M_{r+1}$. Then $m = ux_{r+1}^h x_{r+2}^k$ with $u \in M_r$ and $k \neq 0$. Since $f_2 \prec f_1$, we have $m \prec x_{r+1}$, and therefore $u \lessapprox 1$.

*Case 1* $u \prec 1$. Then $D_r u \gtrapprox u$ [Note 15]. Therefore, since

$$D_r m = (D_r u)x_{r+1}^h x_{r+2}^k + uhx_{r+1}^{h-1} x_{r+2}^k + ukx_{r+1}^{h-1} x_{r+2}^{k-1},$$

we have $D_r m \sim (D_r u)x_{r+1}^h x_{r+2}^k$, and from this it follows that $D_r m \gtrapprox m$. Therefore

$$m^{-1} D_r m \succ (\text{each element of } E_r), \qquad \text{in Case 1.} \tag{1}$$

*Case 2* $u \approx 1$, $h \neq 0$. Then $D_r m \sim uhx_{r+1}^{h-1} x_{r+2}^k$, so $m^{-1} D_r m \sim hx_{r+1}^{-1}$. Therefore

$$m^{-1} D_r m \succ (\text{each element of } E_r), \qquad \text{in Case 2.} \tag{2}$$

*Case 3* $u \approx 1$, $h = 0$. Then $D_r m = ukx_{r+1}^{-1} x_{r+2}^{k-1}$, so $m^{-1} D_r m = kx_{r+1}^{-1} x_{r+2}^{-1}$. Therefore

$$m^{-1} D_r m \succ (\text{each element of } E_r), \qquad \text{in Case 3.} \tag{3}$$

Let $P(Y) = \sum p_{ij} Y^i (D_r Y)^j$, and let $Q(Y) = \sum q_{ab} Y^a (D_r Y)^b$. Then

$$q_{ab} = \sum p_{ij} \binom{i}{a} \binom{j}{a} f_1^{i-a} (D_r f_1)^{j-b}. \tag{4}$$

Since $p_{10}$, $p_{01} - 1$, and $p_{ij}$ (for $i + j \geq 2$) are all in $E_r$, while $f_1 \sim x_{r+1}$, and $D_r f_1 \sim 1$, it is easy to see from (4) that

$$q_{10}, \quad q_{01} - 1, \quad \text{and} \quad q_{ij} \qquad (\text{for } i + j \geq 2) \qquad \text{are all in } E_r. \tag{5}$$

Now $f_1 = x_{r+1}(k + e)$ with $e \in F_1 \cap E$, and

$$\begin{aligned} q_{00} &= p_{00} + p_{10} f_1 + p_{01} D_r f_1 + \sum\{p_{ij} f_1{}^i (D_r f_1)^j : i + j \geq 2\} \\ &= -1 + 0 f_1 + D_r f_1 + s, \end{aligned}$$

with $s \in E_r$. Obviously $D_r f_1 = 1 + e + D_{r+1} e$. Therefore $q_{00} = e + D_{r+1} e + s$.

Let $Q_1(Y) = Q(Y) - q_{00}$, $Q_0(Y) = q_{00}$. Then because of (1)–(3) and (5) we have $Q_1(y) \sim D_r m$ whenever $y \sim m$. Thus $Q_1$ is stable at $m$. Therefore, since $Q_1(Y) + q_{00}$ is not stable at $m$, we must have $q_{00} \neq 0$. Hence $Q_0$ is stable at $m$. Therefore the instability of $Q$ at $m$ implies that

$$D_r m \sim -q_{00} \qquad \text{[Note 16].} \tag{6}$$

Since $q_{00} \in F_1$, while $D_r m$ involves $x_{r+2}$ to a nonzero degree in Cases 1 and 2, it follows from relation (6) that Cases 1 and 2 are impossible. Therefore we must have Case 3. If $k \neq 1$, $D_r m$ involves $x_{r+2}$ to a nonzero degree, which is impossible because of (6). Therefore we must have Case 3 with $k = 1$. Then (6) implies that $x_{r+1}^{-1} \approx q_{00}$, whence $x_{r+1}^{-1} \approx e + D_{r+1} e + s$. Since $s \prec x_{r+1}^{-1}$ it follows that $x_{r+1}^{-1} \approx e + D_{r+1} e$. This contradicts Lemma C3 (with $q = r + 1$).

## D. Instability ladders for $r$-normal differential polynomials

**D1** *Remark* The class of *r-normal* differential polynomials considered in this section is a class of quasilinear polynomials defined in a technical manner [3, §3.10] and possessed of certain technically agreeable properties (such as preservation under each transformation $Y \to Y + e$, with $e \prec 1$ [3, §3.24]). This class serves as an easily reached steppingstone in passing from *asymptotically nonsingular* polynomials (which are easily seen to be reasonable and important [cf. Remark E1]) to polynomials of Class($V$, $r$), for which the instability ladders were examined in Section C.

**D2** ***Lemma*** *If $g \in A_r \cap E$, then $g^a (D_r g)^b \prec g$ whenever $a$ and $b$ are non-negative integers such that $a + b \geq 2$* [proof given in Note 17].

**D3** ***Lemma*** *Let $A$ be first order, r-inscribed, and r-normal. Let $m \in (\text{crit } A) \cap E - M_r$. Then*

$$m \in M_{r+1}, \tag{1}$$

$$m = gx_{r+1} \quad \textit{for some} \quad g \in M_r \cap E, \tag{2}$$

$$A(0) \neq 0, \tag{3}$$

$$\textit{if} \quad C(Y) = -A(gY)/A(0), \qquad \textit{then} \quad C \textit{ is of } \text{Class}(V, r). \tag{4}$$

*Proof* (1) is implied by [3, §3.16(1)]. Therefore $m \in M_{r+1} - M_r$. Hence $m = gx_{r+1}^k$ for some $k \in \mathbb{Q} - \{0\}$, some $g \in M_r$. Let $A(Y) = \sum a_{ij} Y^i (D_r Y)^j$. Then $a_{00} \prec 1$, $a_{10} \sim 1$, $a_{ij} \lesssim 1$ for all $(i, j)$, $a_{01} \neq 0$, and $a_{ij} \lesssim a_{01}$ whenever $j \geq 1$.

*Case 1* $g$ is a nonzero constant. Then $m = cx_{r+1}^k$, so since $m \prec 1$ we have $k < 0$. Let $y \sim m$. Then $D_r y \sim D_r m = ckx_{r+1}^{k-1} \prec m$. Therefore $a_{10} y \sim m$, and $a_{01} D_r y \prec m$ (since $a_{01} \lesssim 1$). For each $(i, j)$ with $i + j \geq 2$ we have $a_{ij} y^i (D_r y)^j \lesssim m^2 \prec m$. Also, $a_{00} \prec m$ since $a_{00} \in E_r$. Hence $A(y) \sim m$ wherever $y \sim m$. Thus $A$ is stable at $m$. This contradiction shows that Case 1 is impossible. Therefore we must have

*Case 2* $g$ is not constant. Then since $m \prec 1$ we must have $g \prec 1$. Therefore, since $g \in M_r$, we must have $D_r g \gtrsim g$ [Note 15].

Let $B(Y) = A(gY)$. Then $B$ is unstable at $x_{r+1}^k$ [3, §2.20]. Let $B(Y) = \sum b_{st} Y^s (D_r Y)^t$. Then $b_{00} = a_{00}$, $b_{10} = a_{10} g + a_{01} D_r g$, $b_{01} = a_{01} g$, and in general

$$b_{st} = \sum \left\{ a_{ij} \binom{j}{t} g^{i+t} (D_r g)^{j-t} : j \geq t;\ i + j = s + t \right\}.$$

We write $B(Y) = B_0(Y) + B_1(Y)$, where

$$B_0(Y) = \sum b_{s0} Y^s.$$

$$B_1(Y) = (D_r Y)[b_{01} + \sum \{b_{st} Y^s (D_r Y)^{t-1} : t \geq 1,\ s + t \geq 2\}].$$

Now if $t \geq 1$, and $s + t \geq 2$, and $j \geq t$, and $i + j = s + t$, then using Lemma D2 we have

$$a_{ij} \binom{j}{t} g^{i+t} (D_r g)^{j-t} \prec a_{01} g = b_{01}.$$

Hence

$$b_{st} \prec b_{01} \tag{5}$$

whenever $t \geq 1$ and $s + t \geq 2$. Since $B$ is $r$-inscribed it follows from (5) that $b_{st}/b_{01} \in E_r$ whenever $t \geq 1$ and $s + t \geq 2$. Therefore $b_{st}\, y^s(D_r\, y)^{t-1} \prec b_{01}$ whenever $t \geq 1$, and $s + t \geq 2$, and $y \sim x_{r+1}^k$. Therefore if $y \sim x_{r+1}^k$, then $B_1(y) \sim b_{01} D_r(x_{r+1}^k) = b_{01} k x_{r+1}^{k-1}$. Thus $B_1$ is stable at $x_{r+1}^k$.

Evidently, if $b_{s0}$ were zero for every $s$, then $B_0(y)$ would be zero for all $y$, and in particular would be zero whenever $y \sim x_{r+1}^k$, and therefore $B(y)$ would be $\sim b_{01} k x_{r+1}^{k-1}$ whenever $y \sim x_{r+1}^k$. Then $B$ would be stable at $x_{r+1}^k$. This contradiction shows that there is at least one $s$ such that $b_{s0} \neq 0$. Let $v = \max\{]b_{s0}[ : b_{s0} \neq 0\}$ [Note 18]. Then $v \in U_r$. Let $E = \{s : b_{s0} \approx v\}$. Let $h = \min\ E$ if $k < 0$, $h = \max\ E$ if $k > 0$. Then $B_0(y) \sim b_{h0}\, x_{r+1}^{hk}$ whenever $y \sim x_{r+1}^k$. Thus $B_0$ is stable at $x_{r+1}^k$. Therefore the instability of $B$ at $x_{r+1}^k$ implies that $B_1(x_{r+1}) \sim -B_0(x_{r+1})$. Hence $b_{01} k x_{r+1}^{k-1} \sim -b_{h0}\, x_{r+1}^{hk}$. Since $b_{01}$ and $b_{h0}$ are in $A_r$, this implies that $hk = k - 1$, and that

$$b_{01} k \sim -b_{h0}. \tag{6}$$

*Case 2A* $k < 0$. Then $h > 1$. Since $h = \min\ E$, we have $1 \notin E$. Hence $b_{10} \prec b_{h0} \approx b_{01}$. Hence $a_{10} g + a_{01}\, D_r g \prec a_{01} g \lessapprox a_{01}\, D_r g$, so

$$a_{10} g \sim -a_{01}\, D_r g. \tag{7}$$

Now $b_{h0} = a_{h0}\, g^h + a_{h-1,\,1} g^{h-1}\, D_r g + \cdots + a_{0h}(D_r g)^h$. Since $a_{hj} \lessapprox a_{01}$ whenever $j \geq 1$, and since it follows from Lemma D2 that $g^{h-j}(D_r g)^j \prec g$ whenever $0 \leq j \leq h$, we have $a_{h-j,\,j}\, g^{h-j}(D_r g)^j \prec a_{01} g = b_{01}$ whenever $j \geq 1$. Therefore the relation $b_{h0} \approx b_{01}$ implies that $b_{01} \approx a_{h0}\, g^h$. Thus $a_{01} g \approx a_{h0}\, g^h$. Therefore, since $a_{h0} \lessapprox 1$, we have $a_{01} g \lessapprox g^h$, so $a_{01} \lessapprox g^{h-1}$. Hence $a_{01} D_r g \lessapprox g^{h-1} D_r g \prec g \sim a_{10}\, g$. But this implies $a_{01} D_r g \prec a_{10}\, g$ which contradicts (7). This contradiction shows that Case 2A is impossible.

Hence we must have Case 2B: $k > 0$. Then $h < 1$. Therefore $h = 0$. Hence $k = 1$. This gives (2).

Since $h = \max\ E$, we have $E = \{0\}$. Therefore $b_{00} \approx v$, and

$$b_{s0} \prec v \qquad \text{for all} \quad s > 0. \tag{8}$$

Since $b_{00} \approx v$, we have $B(0) \neq 0$. That is, $A(0) \neq 0$. This gives (3). Using (6), we get

$$b_{01} \sim -b_{00} \approx v. \tag{9}$$

Using (5) and (9), we get

$$b_{st} \prec v \qquad \text{wherever} \quad t \geq 1 \quad \text{and} \quad s + t \geq 2. \tag{10}$$

Now $C(Y) = -B(Y)/B(0)$. Therefore, using (8)–(10) we see that if we set $C(Y) = \sum c_{ij}\, Y^i(D_r\, Y)^j$, then we have $c_{00} \sim -1$, $c_{01} \sim 1$, $c_{10} \prec 1$, and $c_{st} \prec 1$ whenever $s + t \geq 2$. That is, $C$ is of Class$(V, r)$.

**D4** ***Lemma*** *Let $A$ be first order, r-inscribed, and r-normal. Let $(f_1, f_2, \ldots, f_n)$ be an instability ladder for $(A, r)$ with $f_1 \prec 1$. Then $n = 1$.*

*Proof* Let $f_1 \sim m \in M_{r+1} - M_r$. It follows from Lemma D3 that $m = gx_{r+1}$ for some $g \in M_r \cap E$, that $A(0) \neq 0$, and that if $C(Y) = -A(gY)/A(0)$, then $C$ is of Class$(V, r)$. It is easy to see that $(f_1/g, f_2/g, \ldots, f_n/g)$ is an instability ladder for $(C, r)$, with $f_1/g \sim x_{r+1}$. Therefore it follows from Lemma C5 that $n = 1$.

## E. Instability ladders for asymptotically nonsingular differential polynomials

**E1** *Remark* For a differential polynomial $P$ and a logarithmic monomial $m$, the statement that $m$ is a critical monomial for $P$ can be considered to be a statement that $m$ is a particular kind of *approximate solution* of $P$. As an approximate solution of $P$, $m$ can be assigned, in various ways, an index of *multiplicity*. One index of multiplicity, in the first-order case, is the *singularity grade* for $P$ at $m$; this is, roughly, how many derivatives of $P(Y, DY)$ with respect to $DY$ have $m$ as a critical monomial. The asymptotically nonsingular case is the "usual" case, where the singularity grade is zero, that is where the *separant* $\partial P/\partial(DY)$ of $P$ is stable at $m$ [Note 19].

**E2** ***Lemma*** *Let $P$ be a first-order r-inscribed differential polynomial. Let $f_1 \sim m \in$ (crit $P$) $- M_r$. Let $P$ be asymptotically nonsingular at $m$. Then* [Note 20] $(\Omega_0 P)(m) \neq 0$, *and if $A(Y) = P(f_1 + mY)/[\text{Proj}\ \Omega_0 P(m)]$, then $A$ is $(r + 1)$-inscribed and $(r + 1)$-normal.*

*Proof* By [3, §3.16(4)] we have inst$(P, m) = 1$. Therefore it follows from [3, §3.11] (with $k = 1$, $h = f_1$, and $r$ replaced by $r + 1$) that $A$ is $[r + 1, 1]$-normal (i.e. $(r + 1)$-normal).

Let $F$ be an $r$-constrained field containing the coefficients of $P$. Let $F_1 = F(M_{r+1})$. Then $F_1$ is $(r + 1)$-constrained and contains the coefficients of $A$. Hence $A$ is $(r + 1)$-inscribed.

**E3** ***Lemma*** *Let $P$ be a first-order r-inscribed differential polynomial, and let $(f_1, f_2, \ldots, f_n)$ be an instability ladder for $(P, r)$. Let $P$ be asymptotically nonsingular at $f_1$. Then $n \leq 2$.*

*Proof* Assume $n \geq 3$. Let $f_1 \sim m \in M_{r+1} - M_r$. Then the construction described in Lemma E2 carries $P$ into the $(r + 1)$-normal differential polynomial $A$ of that section. It is easy to see that $(m^{-1}f_2, m^{-1}f_3, \ldots, m^{-1}f_n)$ is an instability ladder for $(A, r + 1)$ with $m^{-1}f_2 \prec 1$. It follows from Lemma D4 that this last instability ladder has one term only. That is, $n = 2$. This contradiction with our assumption $n \geq 3$ completes the proof.

## F. Rank-rise results for the general first-order equation

**F1** *Remark* We shall use the notations $\Gamma P = \partial P/\partial(DY)$ [3, §3,3], and (for $f \neq 0$), $SG(P, f) = \min\{k : k \in N;$ either $\Gamma^{k+1}P = 0$, or $\Gamma^{k+1}P$ is stable at $f\}$. (This second definition generalizes the definition given in [3, §6.12] for the case $f \in \text{crit } P$.)

**F2** ***Lemma*** *Let P be a first-order differential polynomial with coefficients in K. Let $z \in K$. Then*

(1) *if $w \prec z$ and $v \sim z$, and $Q(Y) = P(w + Y)$, we have $SG(Q, v) \leq SG(P, z)$.*
(2) *If $w \sim z$ and $v \prec z$ and $Q(Y) = P(w + Y)$, we have $SG(Q, v) \leq SG(P, z)$.*

*Proof* See Note 21.

**F3** ***Lemma*** *Let $(f_1, f_2, f_3)$ be an instability ladder for $(P_0, r)$. Let $P_2(Y) = P_0(f_1 + f_2 + Y)$. Then $SG(P_0, f_1) > SG(P_2, f_3)$.*

*Proof* Assume to the contrary that

$$SG(P_0, f_1) \leq SG(P_2, f_3). \tag{1}$$

Let $P_1(Y) = P_0(f_1 + Y)$. Now it follows from Lemma F2(2), with $P = P_0$, $z = f_1$, $w = f_1$, $v = f_2$ that $SG(P_1, f_2) \leq SG(P_0, f_1)$. Similarly it follows from Lemma F2(2) with $P = P_1$, $z = f_2$, $w = f_2$, $v = f_3$ that $SG(P_2, f_3) \leq SG(P_1, f_2)$. Thus we have

$$SG(P_2, f_3) \leq SG(P_1, f_2) \leq SG(P_0, f_1). \tag{2}$$

It follows from (1) and (2) that $SG(P_0, f_1) = SG(P_1, f_2) = SG(P_2, f_3)$. Let $k = SG(P_0, f_1)$. Let $Q_j = \Gamma^k P_j$ $(j = 0, 1, 2)$. If $k = 0$, obviously $Q_j$ is unstable at $f_j (j = 0, 1, 2)$. If $k > 0$, it follows from the definition of $SG$ that $Q_j$ is unstable at $f_j$ $(j = 0, 1, 2)$. Also $Q_1(Y) = Q_0(f_1 + Y)$ and $Q_2(Y) = Q_0(f_1 + f_2 + Y)$. It is easy to see, therefore, that $(f_1, f_2, f_3)$ is an instability ladder for $(Q_0\ r)$. Again from the definition of SG, either $\Gamma Q_0$ is identically zero, or $\Gamma Q_0$ is stable at $f_1$. If $\Gamma Q_0 = 0$, then $Q_0$ is of order zero and cannot be unstable at $f_1$ since the logarithmic rank of $f_1$ is $r + 1$, while the coefficients of $Q_0$ belong to an $r$-constrained field. Thus $\Gamma Q_0$ cannot be zero. Therefore $\Gamma Q_0$ is stable at $f_1$. That is, $Q_0$ is asymptotically non-singular at $f_1$. But $(f_1, f_2, f_3)$ is an instability ladder, of length three, for $(Q_0, r)$. This contradiction with Lemma E3 completes the proof.

**F4** ***Lemma*** *Let $(f_1, f_2, \ldots, f_n)$ be an instability ladder for $(P, r)$. Let $h = SG(P, f_1)$. Then $n \leq 2 + 2h$.*

*Proof* Assume to the contrary that $n > 2 + 2h$. Let

$$P_0 = P, \qquad P_j(Y) = P(f_1 + f_2 + \cdots + f_j + Y) \qquad (j = 1, 2, \ldots, n-1).$$

Then $(f_1, f_2, f_3)$ is an instability ladder for $(P_0, r)$ so by Lemma F3 we have $SG(P_0, f_1) > SG(P_2, f_3)$. Also, $(f_3, f_4, f_5)$ is an instability ladder for $(P_2, r+2)$, so $SG(P_2, f_3) > SG(P_4, f_5)$. Continuing in this way, we get

$$\begin{aligned} SG(P_0, f_1) > SG(P_2, f_3) > SG(P_4, f_5) > \cdots \\ > SG(P_{2h}, f_{2h+1}) > SG(P_{2h+2}, f_{2h+3}). \end{aligned} \tag{1}$$

Since $SG(P_0, f_1) = h$, it follows from (1) that $SG(P_{2h+2}, f_{2h+3}) < 0$, which is absurd.

**F5** ***Lemma*** *Let $\lambda = (F, r, P, y_0, H)$ be an adjunction challenge. Let $SG(P, y_0) = h$. Then the adjunction gap for $\lambda$ is $\leq 2 + 2h$.*

*Proof* Assume to the contrary that the adjunction gap for $\lambda$ is $\geq 3 + 2h$. Let $(I_0, I_1, \ldots, I_{2+2h})$ be an adjunction failure of magnitude $3 + 2h$, for $\lambda$. Let $J_0 = I_0^{\gamma}$. Then it follows from Lemma B9 that there exists an element $f_0$ of $J_0$, with $f_0 \prec y_0$ or $f_0 \sim y_0$, and an instability ladder $(f_1, f_2, \ldots, f_{3+2h})$ for $(P(f_0 + Y), r)$, with $f_1 \sim y_0 - f_0$. But it follows from Lemma F2 with $w = f_0$, $z = y_0$, and $v = f_1$ that $SG(P(f_0 + Y), f_1) \leq h$, and therefore it results from Lemma F4 that $(P(f_0 + Y), r)$ cannot have an instability ladder with more than $2 + 2h$ terms. This contradiction completes the proof.

**F6** ***Theorem*** *Let $P$ be a first-order differential polynomial with coefficients in an $r$-constrained field $F$. Let $y_0$ be a solution of $P$ lying in a monomic extension $H$ of $F$. Let $h$ be the singularity grade for $P$ at $y_0$. Then there exists an $(r + 2 + 2h)$-constrained field $G$ such that $F(y_0) \subset G \subset H$. Moreover, if $F$ and $H$ are symmetric, then $G$ may be chosen to be symmetric.*

*Proof* Corollary of Lemmas F5, B4, and B6.

**F7** ***Definition*** Let $P$ be a first-order differential polynomial with coefficients in an $r$-constrained field. Let $(m_0, m_1, \ldots, m_n, \ldots)$ be a sequence of logarithmic monomials such that $P$ is unstable at $m_0$, $P(m_0 + Y)$ is unstable at $m_1$, $\ldots$, $P(m_0 + m_1 + \cdots + m_{n-1} + Y)$ is unstable at $m_n$, $\ldots$, and $m_0 \succ m_1 \succ \cdots \succ m_n \succ \cdots$. Then $(m_0, m_1, \ldots, m_n, \ldots)$ is called a *critical chain* for $P$.

*Remark* It is easy to see that if $P(y_0) = 0$ and $y_0 \sim \sum m_n$ (where the $m_n$ are logarithmic monomials), then $(m_0, m_1, \ldots, m_n, \ldots)$ is a critical chain for $P$. Thus each bound on logarithmic rank-rise for critical chains is automatically a bound on logarithmic rank-rise for asymptotic expansions in logarithmic monomials [Note 22].

**F8** ***Lemma*** *Let $P$ be a first-order $r$-inscribed differential polynomial. Let $(m_0, m_1, \ldots, m_n, \ldots)$ be a critical chain for $P$. Let $h = SG(P, m_0)$. Then the logarithmic rank of $m_n$ is $\leq r + 2 + 2h$.*

*Proof* This is essentially a special case of Lemma F4. (In detail: Suppose there exists $k$ such that $m_k \notin M_{r+2+2h}$. Let $F$ be an $r$-constrained field containing the coefficients of $P$. Let $F_j = F(M_{r+j})$ $(j = 0, 1, \ldots, 3 + 2h)$.

Let $a = \min\{n : m_n \notin M_r\}$. Let $f_0 = m_0 + m_1 + \cdots + m_{a-1}$ (conceivably $f_0 = 0$). Let $Q(Y) = P(f_0 + Y)$. Since $m_n \in M_r$ for $n < a$, it follows that coeff $Q \subset F_0$. Now $(m_a, m_{a+1}, \ldots, m_n, \ldots)$ is a critical chain for $Q$. Since $m_a \in$ crit $Q$ and $Q$ is $r$-inscribed, it follows from [3, §3.16(1)] that $m_a \in M_{r+1}$. Thus $m_a \in M_{r+1} - M_r$.

Let $b = \min\{n : m_n \notin M_{r+1}\}$. Then $b > a$. Let $f_1 = m_a + m_{a+1} + \cdots + m_{b-1}$. Then $f_1 \in F_1$, $f_1$ has logarithmic rank $r + 1$, and $Q$ is unstable at $f_1$. Let $c = \min\{n : m_n \notin M_{r+2}\}$. Then $c > b$. Let $f_2 = m_b + m_{b+1} + \cdots + m_{c-1}$. Then $f_2 \in F_2$, $f_2$ has logarithmic rank $r + 2$, and $Q(f_1 + Y)$ is unstable at $f_2$. Continuing in this manner we construct an instability ladder $(f_1, f_2, \ldots, f_{3+2h})$ for $(Q, r)$, with inclusion sequence $(F_0, F_1, \ldots, F_{3+2h})$. But this contradicts Lemma F4 since (by Lemma F2, with $z = m_0$, $w = f_0$, $v = f_1$) we have $SG(Q, f_1) \leq h$).

**F9** ***Theorem*** *Let $P$ be a first-order differential polynomial with coefficients in an $r$-constrained field. Let $P(y_0) = 0$ and let $y_0 \sim \sum m_n$ (where the $m_n$ are logarithmic monomials). Let $h$ be the singularity grade for $P$ at $m_0$. Then the logarithmic rank of $m_n$ is $\leq r + 2 + 2h$ for all $n$.*

*Proof* This follows from Lemma F8 and the Remark in Definition F7.

**F10** ***Lemma*** *Let $Q$ be a first-order differential polynomial and let $(f_1, f_2, \ldots, f_n)$ be an instability ladder for $(Q, r)$. Let $d$ be the degree of $Q$ in $DY$. Then $n \leq 2d$.*

*Proof* Let $h = SG(Q, f_1)$. Then $\Gamma^h Q$ is unstable at $f_1$, and is not identically zero. This implies, since $\Gamma^h Q$ has coefficients in an r-constrained field and $f_1$ has logarithmic rank equal to $r + 1$, that $\Gamma^h Q$ is effectively of order one. Hence $h \leq d - 1$. Hence it follows from Lemma F4 that $n \leq 2 + 2(d - 1)$.

**F11** ***Lemma*** *Let $\lambda = (F, r, P, y_0, H)$ be an adjunction challenge. Let $d$ be the degree of $P$ in $DY$. Then the adjunction gap for $\lambda$ is $\leq 2d$.*

*Proof* This is similar to the proof of Lemma F5, using Lemma F10 instead of Lemma F4.

**F12** ***Theorem*** *Let $P$ be a first-order differential polynomial with coefficients in an $r$-constrained field $F$. Let $d$ be the degree of $P$ in $DY$. Let $y_0$ be a solution of $P$ lying in a monomic extension $H$ of $F$. Then there exists an*

*$(r + 2d)$-constrained field $G$ such that $F(y_0) \subset G \subset H$. Moreover, if $F$ and $H$ are symmetric, then $G$ may be chosen to be symmetric.*

*Proof* Corollary of Lemmas F11 and B4 and Section B5.

**F13** ***Lemma*** *Let $P$ be first-order and $r$-inscribed. Let $P$ be of degree $d$ in $DY$. Let $(m_0, m_1, \ldots, m_n, \ldots)$ be a critical chain for $P$. Then the logarithmic rank of $m_n$ is $\leq r + 2d$ for all $n$.*

*Proof* Similar to the proof of Lemma F8, using Lemma F10 instead of Lemma F4.

**F14** ***Theorem*** *Let $P$ be first order, with coefficients in an $r$-constrained field, and let $d$ be the degree of $P$ with respect to $DY$. Let $P(y_0) = 0$, and let $y_0 \sim \sum m_n$ (with the $m_n$ equal to logarithmic monomials). Then the logarithmic rank of $m_n$ is $\leq r + 2d$ for all $n$.*

*Proof* This follows from Lemma F13 and the Remark in Definition F7.

## G. Other rank-rise results [Note 23]

For brevity, let us give the name Scheme F to that style of argument which was used in the sequence of results (F4, F5, F6, F8, F9), and used again in the sequence of results (F10–F14) [namely, (control over length of instability ladders) ⇒ (control over adjunction gap) ⇒ (control over adjunction rank-rise (with or without symmetry)), and (control over length of instability ladders) ⇒ (control over rank-rise in critical chains) ⇒ (control over rank-rise in asymptotic expansions)].

Scheme F can be followed in the case of equations of Class($V$, $r$) (beginning with Lemma C5) or in the case of $r$-normal equations (beginning with Lemma D4). In either of these two cases we easily get an upper bound equal to 1 for each version of rank-rise. This bound, which is best-possible, is not implied by the results in Section F.

Similarly, Scheme F can be followed in the case of asymptotically non-singular equations, beginning with Lemma E3, but in this case gives the upper bound 2 for all the rank rises, an upper bound which is obtainable by specializing the results of Section F to the case $SG = 0$. The example $xdy/dx = -y^2 - y^3$ mentioned in connection with equation (2) of Section A1 shows that this result is best-possible.

For Riccati polynomials $P(Y) = DY + a + bY + cY^2$, with $a$, $b$, $c$ in an $r$-constrained field, it can be proved by an argument similar to, but shorter than, the proof of Lemma D3, that if $m$ is a critical monomial for $P$ and $m \notin M_r$, then either

(A) $m = gx_{r+1}$ for some $g$ in $M_r$, or
(B) $m = gx_{r+1}^{-1}$ for some $g$ in $M_r$,

and that if $C(Y) = g^{-1}P(gY)$ in Case (A), or $C(Y) = -g^{-1}Y^2P(gY^{-1})$ in Case (B), then in either case $C$ is of Class($V$, $r$). This makes it possible, using an argument similar to the proof of Lemma D4, to show that for the Riccati equation the number of terms in an instability ladder is $\leq 1$. Then following Scheme F we immediately get an upper bound equal to 1 for all versions of rank-rise. (This result justifies the assertion in Section A1 that the solutions of the Riccati equation with rational functions for coefficients cannot involve log log $x$.)

Linear equations, as a special case of Riccati, are subject to the same rank-rise bounds. (In the case of linear, the critical monomial $m$ must take the first form (A): $m = gx_{r+1}$ mentioned in the preceding paragraph.)

For quadratic differential polynomials of total degree 2 in $Y$ and $DY$ a somewhat lengthy argument similar to the proof of Lemmas D3 and D4 shows that the maximum length of an instability ladder is 2, from which Scheme F leads at once to the result that for quadratic differential polynomials each version of rank-rise is bounded by the number 2. This result, which does not follow from the results in Section F, is best-possible, as is shown by the differential polynomial $P(y) = 4y - 4x\ dy/dx - x^2(dy/dx)^2$, which has a solution $y_0 \sim (\log x)^2 - 2 \log x \log \log x + \cdots$.

*Remark* There is some evidence for the conjecture that for every first-order inscribed differential polynomial, each version of rank-rise is bounded by the number 2. The author has very recently verified this conjecture so far as it applies to the case of rank-rise for asymptotic expansions of solutions of equations with constant coefficients.

**Appendix**

*Note 1* Asymptotic expansions are defined, in Section A2, in terms of a strong sense of asymptotic equivalence introduced in that section (cf. also Note 4).

*Note 2* All nonstandard terms used, but not defined, in this paper are defined in [3] (which has a terminological index).

*Note 3* I. The adjective *constrained* [3, §2.104] is applied to any subfield $G$ of $K$ such that $G$ is $q$-constrained for some $q$.

II. The concept *monomic* generalizes the concept *finitely constrained*, which is defined in [2, §14], and is referred to in Note 6. (A differential subfield $I$ of $K$ is called *finitely constrained* if $M_\infty \subset I$ and each finite subset of $I$ is contained in a constrained field.)

*Note 4* The concept of asymptotic expansion is strong because $\prec$ and $\sim$ are strong. (For example, for the case of asymptotic expansions in terms of logarithmic monomials the relation $y_0 \sim \sum m_n$ implies $Dy_0 \sim \sum Dm_n$.)

*Note 5* I. As remarked above, one answer to this problem is given by Theorem F14. A better but more technical answer is given by Theorem F9.

II. In this paragraph we make the assumption that the underlying graduated logarithmic field $((K, \prec, U, C), D, \mathbf{x})$ has $C$ algebraically closed. Let $P$ be a first order d.p. with coefficients in an $r$-constrained field $F$, and let $P$ be *nonhomogeneous* (cf. Note 10). Then $P$ has a *critical chain* (as defined in Definition F7). If $(m_0, m_1, \ldots, m_n, \ldots)$ is such a critical chain, then the existence for $P$ of a solution $y_0$ satisfying $y_0 \sim \sum m_n$ is guaranteed by results of J. Turcheck [4, especially *First order differential closure theorem*, p. 101, together with 3.3.1, p. 102], and moreover, $y_0$ may be chosen to belong to an algebraically closed monomic extension of $F$. ($K$, in general, is not finitely constrained (cf. Note 3), as required in Turcheck's theory, but it may be replaced by the finitely constrained field generated by $F \cup M_\infty$ (cf. [2, §18, Cor. II]).

*Note 6* I. A better but more technical answer is given by Theorem F6. Throughout the remainder of this Note, $F$ is a fixed $r$-constrained field and $P$ is a fixed first order d.p. with coefficients in $F$.

II. We remark first that it follows from Theorem F12 or F6 that if $y_0$ is a solution of $P$ lying in an algebraically closed monomic extension $H$ of $F$, then there does exist a constrained field $G$ such that $F(y_0) \subset G \subset H$. Hence we might express the adjunction rank-rise problem as the calculation of $b(y_0, H) = \min\{q : \exists q\text{-constrained } G \text{ such that } F(y_0) \subset G \subset H\}$. An easy argument, using the fact that the intersection of constrained fields is constrained [3, §2.108] shows that $b(y_0, H)$ is independent of $H$. Let $c(y_0)$ be the common value which $b(y_0, H)$ assumes for all revelant $H$. $c(y_0)$ is defined whenever $y_0$ is a solution of $P$ and is *monomically adjoinable*, i.e., belongs to an algebraically closed monomic extension of $F$. Theorems F6 and F12 give upper bounds on $c(y_0)$.

III. It happens commonly (cf. e.g. [3, §6.12)] that when $P$ has its coefficients in an $r$-constrained field $F$, then $P$ has a solution $y_0$ which is *constrainedly adjoinable* to $F$, i.e., belongs to a constrained extension of $F$. A natural question is, for each constrainedly adjoinable $y_0$, to calculate $a(y_0) = \min\{q : \exists q\text{-constrained } G \text{ such that } F(y_0) \subset G\}$. As the discussion in II indicated, it follows from Theorem F12 or F6 that monomically adjoinable $\Rightarrow$ constrainedly adjoinable. Then it is obvious that, for each monomically adjoinable $y_0$, we have $a(y_0) \le c(y_0)$.

If $Z$ is algebraically closed [3, §2.113] then it is easy to show that constrainedly adjoinable $\Rightarrow$ monomically adjoinable, and that $c(y_0) \le a(y_0)$. Thus in the case where $Z$ is algebraically closed, the concepts of monomically adjoinable and constrainedly adjoinable turn out to coincide, and

$a(y_0) = c(y_0)$, and Theorems F12 and F6 provide upper bounds on $a(y_0)$ for all constrainedly adjoinable $y_0$.

If $Z = ((K, \prec, U, C), D, \mathbf{x})$ is not necessarily algebraically closed, but the subfield $C$ is algebraically closed, Turcheck's theory (see Note 5) on extensions of graduated logarithmic fields may be used to show that Theorems F12 and F6 still give upper bounds on $a(y_0)$ for constrainedly adjoinable $y_0$. (If $y_0$ lies in a constrained extension $G$ of $F$, then $K$ is to be replaced, in the application of Turcheck's theory, by the finitely constrained field generated by $G \cup M_\infty$.)

IV. If $y_0$ is a constrainedly adjoinable solution, it is easy to see that $y_0 \sim \sum m_n$ with (logarithmic rank $m_n$) $\leq a(y_0)$ (for all $n$). Examples exist in which (logarithmic rank $m_n$) is strictly less than $a(y_0)$ (for all $n$).

*Note 7* Since $F(y_0) \subset H$, and $H$ is algebraically closed, we have $[F(y_0)]^\gamma \subset H$, so $[F(y_0)]^\gamma \subset A_\infty \cup \{0\}$. Thus if $z \in [F(y_0)]^\gamma$, and $z \neq 0$, then $z \sim m \in M_\infty$. Therefore $m$ is a critical monomial (cf. Section B7) for an algebraic polynomial with coefficients in $A_r \cup \{0\}$. It is easy to see that therefore $m \in M_r$. Thus $[F(y_0)]^\gamma \subset A_r \cup \{0\}$.

*Note 8* For details see [3, §2.5].

*Note 9* If $P(Y) = \sum p_{ij} Y^i(DY)^j$ with $p_{ij} \in F$, and if $P^*(Y) = \sum p_{ij}^* Y^i(DY)^j$, then $P^*(y_0^*) = 0$. Thus $Dy_0^*$ is algebraic over $F(y_0^*)$, so $Dy_0^* \subset I_k$, from which it follows easily that $DI_k \subset I_k$.

*Note 10* If $P$ is a first-order d.p. with coefficients in a constrained field, $F$, an algorithm due to Bank [1], formulated abstractly by Strodt and Wright [3, §2.85], determines the set of all critical monomials for $P$. If $P$ is *nonhomogeneous* (i.e. effectively involves at least two terms of unlike total degree in $Y$ and $DY$), and the underlying graduated logarithmic field $((K, \prec, U, C), D, \mathbf{x})$ has $C$ algebraically closed, then the set of critical monomials for $P$ is nonempty.

*Note 11* If $f \sim m$, and $m$ is a logarithmic monomial, we define (logarithmic rank $f$) to be equal to (logarithmic rank $m$).

*Note 12* Every $r$-inscribed d.p. is $r$-modulated. [3, §§2.105, 2.77]

*Note 13* $D_r$ is the differential operator $x_0 x_1 \cdots x_r D$.

*Note 14* $f \approx g$ (read $f$ matched with $g$) means $\exists a$ nonzero constant $c$ such that $f \sim cg$.

*Note 15* (Proof that if $u \prec 1$ and $u \in M_r$, then $D_r u \gtrapprox u$). Let $u = cx_s^a x_{s+1}^b \cdots x_r^w$, with $s \leq r$, $a \neq 0$. Then $D_r u \sim ca\, x_s^a x_{s+1}^{b+1} \cdots x_r^{w+1}$, so if $s < r$, then $D_r u \succ u$, while if $s = r$, then $D_r u \approx u$.

*Note 16* It is almost obvious that if $A$ and $B$ are stable at $f$, and $\{A(f), B(f)\} \subset A_\infty$, then $A + B$ is stable at $f$ unless $A(f) \sim -B(f)$.

*Note 17* Let $g \approx x_s^{-\delta} x_{s+1}^i x_{s+2}^j \cdots x_r^v$, with $\delta > 0$. It is easy to see that $g^a(D_r g)^b \approx x_s^{-(a+b)\delta} x_{s+1}^w \cdots x_r^z$ for some rational numbers $w, \ldots, z$, from which the required inequality follows at once.

*Note 18* If $f \approx x_0^a x_1^b \cdots x_r^w$, then $x_0^a x_1^b \cdots x_r^w$ is called the gauge of $f$, and is denoted by $]f[$.

*Note 19* The condition of asymptotic nonsingularity is easily tested by Bank's algorithm [Note 10].

*Note 20* $\Omega_0 P = Y\, \partial P/\partial Y + (DY)\, \partial P/\partial(DY)$. If $f \sim c x_0^a x_1^b \cdots x_r^w$, with $c$ constant, then $c x_0^a x_1^b \cdots x_r^w$ is called the *Projection* of $f$, and is denoted by Proj $f$.

*Note 21* I. *Proof of Lemma* F2(1). Let $h = SG(P, z)$. Let $P_1 = \Gamma^{h+1}P$, $Q_1 = \Gamma^{h+1}Q$. Then either (A) $P_1 = 0$, or (B) $P_1$ is stable at $z$. In Case (A), the degree of $P$ in $DY$ is $\leq h$, so the degree of $Q$ in $DY$ is $\leq h$, which implies that $Q_1 = 0$, and therefore that $SG(Q, v) \leq h$. In Case (B) we note that $Q_1(Y) = P_1(w + Y)$, and that $w \prec v$, and therefore that whenever $y \sim v$ we have $w + y \sim z$, from which $Q_1(y) = P_1(w + y) \sim P_1(z) \sim P_1(w + v) = Q_1(v)$, so that $Q_1$ is stable at $v$, which implies that $SG(Q, v) \leq h$.

II. *Proof of Lemma* F2(2). This is similar to the proof of Lemma F2(1).

*Note 22* In a wide variety of circumstances it can [Note 5.II] be proved that for each critical chain there is a solution with an asymptotic expansion in terms of that chain. Under those circumstances the concepts of rank-rise for critical chains and rank-rise for asymptotic expansions coincide.

*Note 23* In Section G we shall loosely use the terms *rank-rise* to refer to the difference between the logarithmic rank $r$ of the coefficient domain, and the maximum logarithmic rank appearing in the asymptotic expansion of a solution, or between $r$ and the minimum $q$ such that the solution lies in a $q$-constrained extension of the coefficient domain. (For instance, in this language Theorem F12 would say that the adjunction rank-rise is less than or equal to twice the degree in $DY$.)

## References


1. S. Bank, On the instability theory of differential polynomials, *Ann. Mat. Pura Appl.* (4) **74** (1966), 83–111. MR 34#4623.
2. W. Strodt, On the Briot and Bouquet theory of singular points of ordinary differential equations, Univ. of Wisconsin, U.S. Army MRC Tech. Summary Rep. No. 508 (September 1964).

3. W. Strodt and R. K. Wright, Asymptotic behavior of solutions and adjunction fields for nonlinear first order differential equations, *Mem. Amer. Math. Soc.* No. 109 (1971), 284 pp. MR 44 # 1884.
4. J. Turcheck, First order differential closures of certain partially ordered fields, *Trans. Amer. Math. Soc.* **195** (1974), 97–114. MR 49 # 4984.

AMS (MOS) 1970 subject classifications: 12H05, 34E05

# *A "theorem of Lie–Kolchin" for trees*

*J. Tits*

*Collège de France*
*Paris*

## Introduction

It is known that the automorphism group of a (sufficiently homogeneous) tree $T$ behaves in some respects as $SL_2$ over a field with a non-archimedean valuation. In that analogy, the end space of $T$ plays the role of the projective line. The Lie–Kolchin theorem [2, 7], as generalized à la Rosenlicht [9], asserts that a connected $k$-split solvable subgroup of $SL_n(k)$—for any field $k$—has a fixed point in the projective space $\mathbf{P}_{n-1}(k)$. Thus, our central—and quite easy—Corollary 2, according to which

*a solvable fixed point free automorphism group of a tree $T$ leaves invariant an end or a pair of ends of $T$*

can be regarded as an analogue of the theorem of Lie–Kolchin for trees. Our main application of the above result (or rather of the somewhat stronger Corollary 1, dealing with nilpotent groups) aims, through the general Proposition 3, at showing that the group $G$ of rational points of an algebraic almost simple group of relative rank $\geq 2$ over a field $k$ cannot operate without fixed point or fixed end on a tree; actually, for lack of complete knowledge of the structure of $G$, we can only state the above property in full generality for a "big" normal subgroup of $G$ (cf. Corollary 4, but also Remark 4.3a). In the rank 1 case, treated in Section 5, it turns out, roughly speaking, that the only way for a group of the above type to act on a tree without fixed point or fixed end is the known one, that is, through a valuation of the "standard root datum" of the group in question (cf. [6, §6]). These results on the action of algebraic simple groups on trees are, in some sense, the substitutes of the main theorem of [5] on "abstract" homomorphisms of algebraic simple groups when the receiving group is replaced by the automorphism group of a tree. Note that our notion of tree is somewhat more general than the usual one, and devised so as to include the buildings of groups of rank 1 over fields with nondiscrete valuations.

The study of groups which cannot operate on trees without fixed point was initiated by J.-P. Serre [10]. The viewpoint we adopt in the presentation of our results is inspired by his. There is also some analogy between our methods of proofs and the techniques used in [10]. For trees in the usual sense, our "theorem of Lie–Kolchin" is implicitly contained in a paper of H. Bass [1]. Finally, it should be mentioned that much deeper results on the existence of fixed points for actions of arithmetic groups on trees have recently been obtained by G. A. Margulis (unpublished).

## 1. Trees

**1.1** We call *segment* (resp. *half-line*; resp. *line*) of a metric space, the image of a closed interval $[0, d]$ (resp. of the half-line $[0, \infty)$; resp. of $\mathbf{R}$) by an isometric embedding $\alpha$ in that space. The *extremities* of the segment $\alpha([0, d])$ are the points $\alpha(0)$ and $\alpha(d)$; the *extremity* of the half-line $\alpha([0, \infty))$ is $\alpha(0)$. We define a *tree* as a nonempty, complete metric space containing no homeomorphic image of a circle and such that any two points are the extremities of a segment. That segment is then unique; indeed, if $\alpha$ and $\alpha'$ are two distinct isometric embeddings of $[0, d]$ in a metric space, with $\alpha(0) = \alpha'(0)$ and $\alpha(d) = \alpha'(d)$, if $t \in [0, d]$ is such that $\alpha(t) \neq \alpha'(t)$, and if $t'$, $t''$ denote respectively the largest element of $[0, t]$ and the smallest element of $[t, d]$ on which $\alpha$ and $\alpha'$ coincide, then $\alpha([t', t'']) \cup \alpha'([t', t''])$ is homeomorphic to a circle. A subset of a tree which is itself a tree when endowed with the induced metric is called a *subtree*. Clearly, the intersection of any number of subtrees is empty or is a subtree. In the set of all half-lines of a tree, the relation "$A \cap B$ is a half-line" is an equivalence relation whose equivalence classes are called the *ends* of the tree. An end is said to belong to a subtree if the latter contains a representative of it.

In the sequel, $T$ always denotes a tree, $d$ its distance function, and $E = E(T)$ the set of its ends. The segment joining two points $p$, $q$ of a tree is denoted by $[p, q]$.

**1.2** *The intersection of any number of segments or half-lines with a common extremity $p$ is a segment or a half-line having $p$ as an extremity.*

This is obvious.

**1.3** *If $p, q, r \in T$ are such that $[p, q] \cap [q, r] = \{q\}$, then $[p, r] = [p, q] \cup [q, r]$.*

Indeed, set $[p, q] \cap [p, r] = [p, x]$ and $[q, r] \cap [p, r] = [y, r]$. If one had $x \neq q$ or $y \neq q$, the set $[x, q] \cup [q, y] \cup [x, y]$ would be homeomorphic to a circle.

**1.4** *Let $A$, $B$ be two subtrees of $T$ whose intersection consists of at most one point. Then, there exist unique points $p \in A$ and $q \in B$ such that $A \cap [p, q] = \{p\}$ and $[p, q] \cap B = \{q\}$. The segment $[p, q]$ is contained in every subtree which has a nonempty intersection with both $A$ and $B$.*

Let $C$ be any subtree having a nonempty intersection with $A$ and $B$ (such a subtree obviously exists: take for instance any segment joining a point of $A$ and a point of $B$). Let $x \in A \cap C$ and $y \in B \cap C$ and set $[x, y] \cap A = [x, p]$ and $[x, y] \cap B = [q, y]$. It is clear that $p, q$ have the property stated. Furthermore, if $p', q'$ is any pair of points with that property, one has, by Section 1.3, $[p, p'] \cup [p', q'] = [p, q'] = [p, q] \cup [q, q']$, hence $p = p'$ and $q = q'$. Since $[p, q] \subset C$, the last assertion also holds.

**1.5** From Sections 1.3 and 1.4, it readily follows that if $p \in T$ and $e \in E$, there is a unique representative of the end $e$ with extremity $p$; we denote it by $[p, e)$. Similarly, two distinct ends $e, f$ belong to a unique line, denoted by $(e, f)$.

**1.6** ***Lemma*** *Let $\{A_i | i \in I\}$ be a set of subtrees of $T$ such that any two of them have a nonempty intersection. Then, $\bigcap_I A_i \neq \varnothing$ or all $A_i$'s have a unique common end.*

(Compare with the last lemma of [10].)

Choose a point $p \in T$. For $i \in I$, let $q_i \in A_i$ be the unique point such that $[p, q_i] \cap A_i = \{q_i\}$. For $i, j \in I$ and $x \in A_i \cap A_j$, one has, by Section 1.4, $[p, q_i] \cup [p, q_j] \subset [p, x]$; hence either $[p, q_i] \subset [p, q_j]$ or $[p, q_j] \subset [p, q_i]$. It readily follows that the closure $L$ of $\bigcup_I [p, q_i]$ is a segment or a half-line, and that $L \cap A_i \cap A_j \neq \varnothing$ for every $i, j \in I$. If $L \cap A_k$ is a segment for some $k \in I$, the sets $L \cap A_k \cap A_i$ are pairwise nondisjoint subsegments; therefore $\bigcap_I A_i \neq \varnothing$. Otherwise, all $L \cap A_i$ are half-lines representing the same end. Finally, if the $A_i$'s have two distinct common ends $e$, $f$, one has $\bigcap_I A_i \supset (e, f) \neq \varnothing$.

## 2. Various fixed-point properties

**2.1** By *automorphism* of a tree $T$, we mean an isometry of $T$ onto itself. If $X$ is a set of automorphisms of $T$ or a subset of a group acting on $T$, we denote by $T^X$ and $E^X$ the fixed-point sets of $X$ in $T$ and $E = E(T)$. An end $e \in E^X$ is said to be *neutral* (resp. *attracting*; resp. *repulsing*) for $X$ if, for every $x \in X$, there exists a half-line $L$ representing $e$ and such that $x(L) = L$ (resp. $\subsetneq L$; resp. $\supsetneq L$). (N.B. This terminology differs from that of [13], where we reserved the expression "fixed by $X$" to designate the neutral ends. Here, talking of a single end, we use the words "fixed," "invariant," and "stable" as synonymous; as usual, "fixed," applied to a set, means "pointwise fixed.")

**2.2** Let $G$ be a group acting on a tree $T$ by isometries. We shall be concerned with the following mutually exclusive possibilities.

(P1) $T^G \neq \varnothing$ (in which case $T^G$ is a tree);
(P2) $T^G = \varnothing$ and $G$ has a neutral fixed end (in which case, $E^G$ is reduced to that end);
(P3) $T^G = \varnothing$ and card $E^G \geq 2$ (in which case $E^G$ consists of exactly two ends which are not neutral for $G$);
(P4) $T^G = E^G = \varnothing$ and there is a pair of ends invariant by $G$ (that pair is then unique);
(P5) $T^G = \varnothing$ and $E^G$ consists of a single end which is not neutral for $G$.

When there is no doubt as to which action of $G$ on a tree $T$ is being considered, we shall sometimes, by abuse of language, say that a subgroup $H$ of $G$ has property (P$n$), or satisfies (P$n$), if that property holds for the restriction of the given action to $H$.

The following assertions are obvious.

**2.2.1** If $G$ acting on $T$ has property (P$n$) ($1 \leq n \leq 5$), then, every subgroup $H$ of $G$ has property (P$m$) for some $m \leq n$; more precisely, one has $(n, m) = (n, 1)$, $(n, n)$, $(5, 2)$, $(4, 3)$ or $(5, 3)$. If $m = 1$, every end fixed by $G$ belongs to the tree $T^H$.

**2.2.2** Let $H$ be a normal subgroup of $G$. If $H$ has one of the properties (P2) and (P5) (resp. (P3) and (P4)), so does $G$; indeed, more generally, if $H$ has a single invariant end (resp. pair of ends), so does $G$. If $H$ has property (P1), then $G$ satisfies (P$n$) if and only if $G/H$ acting on the tree $T^H$ does.

**2.2.3** If $G$ has property (P2), every finitely generated subgroup of $G$ has property (P1).

**2.2.4** If (P3) or (P5) holds, $G$ has a nontrivial homomorphism in $\mathbf{R}$. If (P4) holds, $G$ has a homomorphic image which is the semidirect product of $\{\pm 1\}$

and a nontrivial subgroup of **R** on which $\{\pm 1\}$ acts by multiplication. In particular, $G$ cannot be perfect in any of those cases.

**2.3** For $n \in \{1, 2, 3, 4, 5\}$, we shall say that a group $G$ is $F_n$, or is an *$F_n$-group*, if every action of $G$ on a tree satisfies one of the conditions (P$m$), with $1 \leq m \leq n$. Except for the difference in the definition of trees, properties $F_1$ and $F_2$ are respectively properties $(FA)$ and $(FA')$ of [10] and [1]. Clearly, $F_1 \Rightarrow F_2 \Rightarrow F_3 \Rightarrow F_4 \Rightarrow F_5$.

**2.3.1** *Every finite group is* $F_1$, by [6, Lemme (3.2.3)].

**2.3.2** *Every finitely generated* $F_2$*-group is* $F_1$, by Section 2.2.3.

**2.3.3** *Every perfect* $F_5$*-group is* $F_2$, by Section 2.2.4.

**2.3.4** *Let* $(m, n) = (1, 1), (2, 2), (2, 4)$, *or* $(5, 5)$. *Let G be a group and H a subgroup containing a normal subgroup N such that G/N is* $F_m$. *If, for a given action of G on a tree, H has one of the properties* (P1) *to* (P$n$), *then so does G.*

We shall only consider the case $m = 2$, the other ones being similar but easier. By Section 2.2.1, if $H$ has one of the properties (P1) to (P$n$), $N$ has property (P$n'$) for some $n' \leq n$. If $n' \neq 2$, our assertion readily follows from Section 2.2.2. If $n' = 2$, the unique fixed end of $N$ is fixed by $G$ and, being neutral for $N$, it must be neutral for $G$ because $G/N$ has no nontrivial homomorphism in **R** (otherwise, $G/N$ would have a fixed point free action on **R**, in contradiction with the assumption that it is $F_2$).

From Section 2.3.4, we deduce:

**2.3.5** *If* $n = 1, 2$, *or* 5, *every extension of an* $F_n$*-group by an* $F_n$*-group is* $F_n$.

**2.3.6** *If* $n = 1, 2, 4$, *or* 5, *every group having an* $F_n$*-subgroup of finite index is* $F_n$.

## 3. Solvable groups

**3.1** ***Proposition 1*** *Let f be an automorphism of a tree T. If* $T^f \neq \varnothing$, *every subtree of T invariant by f has a nonempty intersection with* $T^f$ *and every end fixed by f belongs to* $T^f$. *If* $T^f = \varnothing$, *then* $E^f$ *consists of exactly two ends, an attracting one a and a repulsing one r; the line* $(a, r)$ *is contained in every subtree of T invariant by f.*

(Compare [13, Prop. 3.2].)

Suppose $T^f \neq \varnothing$. If $T'$ was a subtree invariant by $f$ and disjoint from $T^f$, the segment $[p, q]$ defined by $[p, q] \cap T^f = \{p\}$ and $[p, q] \cap T' = \{q\}$ (cf. Section 1.4) would be fixed by $f$, hence contained in $T^f$, a contradiction.

Further, if $e$ is an end fixed by $f$, the half-line $[p, e)$ is fixed by $f$ for any $p \in T^f$, therefore $e$ belongs to $T^f$.

From now on, we assume that $T^f$ is empty. Let $p \in T$ and set $[p, f^{-1}p] \cap [p, fp] = [p, q]$. Thus, $[q, f^{-1}p] \cap [q, fp] = \{q\}$. Since $q \in [f^{-1}p, p]$, we have $fq \in [p, fp]$. Similarly, $f^{-1}q \in [p, f^{-1}p]$. Suppose that $fq \in [p, q]$. Then $q \in [fp, fq]$, hence $f^{-1}q \in [p, q]$, and since $d(q, fq) = d(q, f^{-1}q)$, it follows that $fq = f^{-1}q$. Therefore, the segment $[q, fq]$ is invariant by $f$ and so is its middle point, a contradiction. This shows that $fq \notin [p, q]$, hence $fq \in [q, fp]$. Similarly, $f^{-1}q \in [q, f^{-1}p]$. Consequently, one has $[f^{-1}q, q] \cap [q, fq] = \{q\}$ and, transforming by $f^n$, $[f^{n-1}q, f^n q] \cap [f^n q, f^{n+1}q] = \{f^n q\}$. From this relation, assertion 1.3, and the nonexistence of "circles" in $T$, it readily follows that the set $\bigcup_{n \in \mathbf{Z}} [f^n q, f^{n+1} q]$ is a line whose two ends are invariant by $f$. Now, if $e, e'$ are any two distinct ends invariant by $f$, one of them must be attracting and the other repulsing, because the line $(e, e')$ is invariant and not fixed by $f$. Therefore, $f$ can have at most two fixed ends. Finally, applying the result just proved to the restriction of $f$ to any invariant subtree $T'$, we see that the two ends fixed by $f$ must belong to $T'$. The proof is complete.

**3.2 *Proposition 2*** *If the quotient of a group $G$ by its center is $F_3$, then $G$ itself is $F_3$.*

Let $G$ operate on a tree $T$. We want to prove that one of the properties (P1), (P2), (P3) of Section 2.2 holds.

We first consider the case where $G$ is commutative. If there exists $g \in G$ such that $T^g$ is empty, then, by Proposition 1, $E^g$ consists of one attracting and one repulsing end; it follows that $G$, which centralizes $g$, fixes $E^g$, and (P3) holds. Assume therefore that $T^g \neq \varnothing$ for every $g \in G$. For any $g, g' \in G$, the tree $T^{g'}$ is invariant by $g$ (because $g$ centralizes $g'$); therefore $T^g \cap T^{g'} \neq \varnothing$, by Proposition 1. We are now in a position to use Lemma 1.6: either $T^G = \bigcap_{g \in G} T^g \neq \varnothing$, which means that (P1) holds, or $T^G = \varnothing$ and the subtrees $T^g$ have a common end, which is obviously neutral for $G$, and this is property (P2). The proposition is thus proved for a commutative group $G$.

Going over to the general case, we denote by $C$ the center of $G$. If $T^C$ is not empty, it is a tree on which $G/C$ operates and our assertion follows from the hypothesis made on $G/C$. If $T^C$ is empty, the special case of the proposition already proved implies that $E^C$ consists either of two ends, or of a single one $e$ which is neutral for $C$. All we have to show is that $E^C$ is fixed by every element $g$ of $G$ and that, in the second case, $e$ is neutral for $g$. But for this, it suffices to apply again the proposition already proved in the commutative case to the group generated by $C$ and $g$.

**3.3 *Corollary 1*** *Every nilpotent group is $F_3$.*

**3.4** ***Corollary 2*** *Every solvable group is* $F_5$.

(Use Corollary 1 and Section 2.3.5.)

**3.5** ***Corollary 3*** *Every solvable torsion group is* $F_2$.

(Use Corollary 2 and Section 2.2.4.)

## 4. Algebraic simple groups of relative rank $\geq 2$

**4.1** ***Proposition 3*** *Let* $G$ *be a group and* $X$ *a subset generating a subgroup of finite index of* $G$. *For* $g \in G$, *let* $N(g)$ *denote the union of all normalizers of nilpotent subgroups containing* $g$. *Suppose that, for any* $x, y \in X$, *either* $y \in N(x)$ *or* $N(x) \cap N(y)$ *generates a subgroup of finite index of* $G$. *Then* $G$ *is* $F_4$.

(N.B. The following proof shows that the proposition remains valid if one replaces "nilpotent" by "$F_3$" and "of finite index" by "containing a normal subgroup $H$ of $G$ such that $G/H$ is $F_2$." Here, $F_2$ can be replaced by $F_5$ at the cost of changing the conclusion in: $G$ is $F_5$.)

Let $G$ operate on a tree $T$. We claim that one of the properties (P1)–(P4) holds for some subgroup of finite index of $G$. By Section 2.3.6, this will prove our proposition.

Suppose first that $T^c$ is empty for some $c \in G$. Then, the unique line invariant by $c$ (Section 3.1) is also the unique line invariant by every nilpotent subgroup containing $c$ (Section 3.3). Therefore, it is also invariant by $N(c)$ and we are through since the hypotheses made on $X$ imply, whether $N(c)$ contains $X$ or not, that the group generated by $N(c)$ is of finite index in $G$.

Thus we may and shall assume that $T^x \neq \varnothing$ for all $x \in X$. Our next step is to show that, for $g, h \in G$,

$$\text{if} \quad T^g \neq \varnothing, \quad T^h \neq \varnothing, \quad \text{and} \quad h \in N(g), \qquad \text{then} \quad T^g \cap T^h \neq \varnothing. \quad (1)$$

Indeed, let $H$ be a nilpotent subgroup containing $g$ and normalized by $h$. By Section 3.3, either $T^H \neq \varnothing$, or $H$ leaves invariant a unique line $L$ or it fixes a unique end $e$. Since $h$ normalizes $H$, it leaves invariant $T^H$, $L$, or $e$. But then, it follows from Proposition 1 that, according to the case, $T^g \cap T^h \supset T^H \cap T^h \neq \varnothing$, or $L \subset T^g \cap T^h$, or $e$ belongs to both $T^g$ and $T^h$, and (1) is proved.

If $T^x \cap T^y \neq \varnothing$ for every $x, y \in X$, it follows from Section 1.6 that either $T^X \neq \varnothing$, in which case $T^X$ is a subtree fixed by the subgroup of finite index $G'$ of $G$ generated by $X$, or the subtrees $T^x$ $(x \in X)$ have a common end, which is obviously neutral for $G'$; in both cases, our assertion is proved.

Suppose therefore that $T^b \cap T^c = \emptyset$ for some $b, c \in X$. Let $p, q \in T$ be defined by $[p, q] \cap T^b = \{p\}$ and $[p, q] \cap T^c = \{q\}$. From (1), it follows that $c \notin N(b)$ and that, for every $x \in N(b) \cap N(c)$, $T^x$ has a nonempty intersection with both $T^b$ and $T^c$, hence contains $p$ and $q$, by Section 1.4. As a result, $p$ and $q$ are fixed by the subgroup generated by $N(b) \cap N(c)$, which is of finite index in $G$ by hypothesis. This completes the proof.

**4.2** *Corollary 4* *If $\mathcal{G}$ is an algebraic almost simple group of relative rank $\geq 2$ over a field k, then the normal subgroup G of $\mathcal{G}(k)$ generated by all rational unipotent elements contained in k-split unipotent subgroups of $\mathcal{G}$ is $F_2$.*

(N.B. If card $k \geq 4$, $G$ can also be described as the unique minimal non-commutative normal subgroup of $\mathcal{G}(k)$: cf. [12] and [4].)

Let $\mathcal{S}$ be a maximal $k$-split torus of $\mathcal{G}$ and $\Phi$ the system of nondivisible roots of $\mathcal{G}$ with respect to $\mathcal{S}$. For $a \in \Phi$, let $\mathcal{U}_{(a)}$ denote the corresponding unipotent "root group" (cf. [3, 5.2]) and set $U_{(a)} = \mathcal{U}_{(a)}(k)$. Then, the set $X = \bigcup_{a \in \Phi} U_{(a)}$ fulfills the conditions of Section 4.1. Indeed, it generates $G$ (cf. [5, 6.2(v)]). Furthermore, let $x \in U_{(a)}$ and $y \in U_{(b)}$ $(a, b \in \Phi)$. If $b \neq -a$, $x$ and $y$ generate a nilpotent group and one has $y \in N(x)$. If $b = -a$, the same argument shows that $N(x) \cap N(y)$ contains all $U_{(c)}$ for $c \neq \pm a$, and hence generates $G$. The assertion now follows from the above proposition and Sections 2.3.3 and 2.3.1, since $G$ is perfect or finite.

**4.3** *Remarks* (a) In many cases, possibly always, $\mathcal{G}(k)/G$ is an abelian torsion group. Then, by Sections 2.3.5 and 3.5, $\mathcal{G}(k)$ itself is $F_2$.

(b) The argument proving Corollary 4 applies without change to the classical groups "of rank $\geq 2$" over arbitrary division rings (cf. e.g. [6, §10]), the Ree groups of type ${}^2F_4$, and, using Corollary 3 of [11, p. 115], the groups $\mathcal{G}(R)$ where $\mathcal{G}$ is an almost simple Chevalley group-scheme of rank $\geq 2$ and $R$ a Euclidean ring. In particular, by Section 2.3.2, $\mathcal{G}(\mathbf{Z})$ is $F_1$ (compare [10, no. 5, remarque 2]). But Proposition 3 also applies to other arithmetic groups. It can for instance be used to prove that

*if $\mathcal{G}$ is an almost simple Chevalley group scheme of rank $\geq 2$, every subgroup of finite index of $\mathcal{G}(\mathbf{Z})$ is $F_1$,*

which partially answers a question of J.–P. Serre ([10, no. 6]; a complete answer, based on methods similar to those of [8], has been announced by G. A. Margulis). Here, one takes for $X$ the set of all unipotent elements of the subgroup in question which are central in a maximal unipotent subgroup of $\mathcal{G}(\mathbf{C})$; to prove that this $X$ meets the requirements of Proposition 3 is beyond the scope of the present paper and will be done elsewhere†

## 5. The rank 1 case

**5.1** Let $G$ be a group, $U_+$ and $U_-$ two proper subgroups generating $G$, $S$ the intersection of their normalizers, and $M$ the set of all $m \in G$ such that ${}^mU_+ = U_-$ and ${}^mU_- = U_+$. We assume that

(R) for every $u \in U_+ - \{1\}$, there exist unique elements $u', u'' \in U_-$ such that $m(u) = u'uu'' \in M$.

A typical example of that situation is the following: $G = SL_2(K)$ for some division ring $K$,

$$U_+ = \left\{ \begin{pmatrix} 1 & t \\ 0 & 1 \end{pmatrix} \middle| \, t \in K \right\}, \qquad U_- = \left\{ \begin{pmatrix} 1 & 0 \\ t & 1 \end{pmatrix} \middle| \, t \in K \right\},$$

$$S = \left\{ \begin{pmatrix} t & 0 \\ 0 & t' \end{pmatrix} \right\} \cap G, \qquad M = \left\{ \begin{pmatrix} 0 & t \\ t' & 0 \end{pmatrix} \right\} \cap G.$$

Another example (including the previous one if $K$ is finite-dimensional over its center) is obtained by taking for $U_+$ and $U_-$ the groups of $k$-rational points of the unipotent radicals of two distinct proper $k$-parabolic subgroups in an almost simple algebraic group of relative rank 1 over some field $k$.

**5.2** Let $\eta: S \to \mathbf{R}$ be a nontrivial homomorphism such that $\eta(msm) = -\eta(s)$ for all $s \in S$ and $m \in M$, choose arbitrarily $m_0$ in $M$ and define the function $\varphi: U_+ \to \mathbf{R} \cup \{\infty\}$ by

$$\varphi(u) = \begin{cases} \frac{1}{2}\eta(m(u) \cdot m_0) & \text{if } u \neq 1, \\ \infty & \text{if } u = 1. \end{cases}$$

We shall say that $\eta$ is a *valuation* of the system $(U_+, U_-)$ if for every real number $r$, $\varphi^{-1}([r, \infty])$ is a group, a condition which is clearly independent of the choice of $m_0$. (N.B. The terminology adopted here for the convenience of the exposition somewhat deviates from that of [6]; in particular, our valuations also include the "quasi-valuations" of [6, 6.2.3e].) In the case of $SL_2(K)$, with $U_+$ and $U_-$ defined as above, the valuations are nothing else but the homomorphisms of the form

$$\begin{pmatrix} t & 0 \\ 0 & t' \end{pmatrix} \mapsto 2\omega(t) = -2\omega(t'),$$

---

† *Added in proof*: In view of results obtained after this paper was written (cf. *C. R. Acad. Sci. Paris* **283** (1976), 693–695), our statement concerning subgroups of finite index of $\mathscr{G}(\mathbf{Z})$ can now be deduced from Proposition 3 (whose full force is no longer necessary) in exactly the same way as Corollary 4. Also Serre's method [10] can be applied.

where $\omega$ denotes a non-archimedean valuation of $K$ (cf. [6, 10.2.10]), and one has

$$\varphi\left(\begin{pmatrix} 1 & t \\ 0 & 1 \end{pmatrix}\right) = \omega(t) \qquad \left(\text{for} \quad m_0 = \begin{pmatrix} 0 & 1 \\ -1 & 0 \end{pmatrix}\right).$$

**5.3** To a valuation $\eta$ of $(U_+, U_-)$, one associates, exactly as in [6, §§7, 8], a *complete building* $T_\eta$, which is a tree on which $G$ operates by isometries. We briefly recall the construction of $T_\eta$, without going into details. For $r \in \mathbf{R}$, let $G_r$ denote the group generated by $\varphi^{-1}([r, \infty])$, $m_0 \cdot \varphi^{-1}([-r, \infty]) \cdot m_0^{-1}$, and Ker $\eta$. In $G \times \mathbf{R}$, let us introduce the equivalence relation

$$(g, r) \sim (g', r') \quad \Leftrightarrow \quad r = r' \quad \text{and} \quad g^{-1}g' \in G_r.$$

Then, $G$ operates on the left on the quotient $T_\eta^0 = (G \times \mathbf{R})/\sim$, and there is a unique metric in $T_\eta^0$ which is invariant by $G$ and such that $r \mapsto (1, r) \bmod \sim$ is an isometry of $\mathbf{R}$ into $T_\eta^0$. Finally, the tree $T_\eta$ is the completion of the metric space $T_\eta^0$.

**5.4** ***Proposition 4*** *Let* $G, U_+, U_-$ *be as in Section* 5.1, *suppose* $U_+$ *and* $U_-$ *are nilpotent, and let* $G$ *act on a tree* $T$ *without fixed point and without fixed end. Then, there exist a unique valuation* $\eta$ *of* $(U_+, U_-)$ *and a unique isometric embedding of the complete building* $T_\eta$ *into* $T$ *compatible with the actions of* $G$ *onto* $T_\eta$ *and* $T$.

(N.B. In that statement, "nilpotent" can be replaced by "$F_3$" as will be obvious from the proof.)

Since $U_+$ and $U_-$ are nilpotent and conjugate, it follows from Section 3.3 that one of the following cases occurs:

(i) $U_+$ (resp. $U_-$) leaves invariant a unique line $L_+$ (resp. $L_-$), on which it induces a nontrivial group of translations;

(ii) the fixed-point sets $T_+$ and $T_-$ of $U_+$ and $U_-$ are not empty;

(iii) there is a unique end $e_+$ (resp. $e_-$) fixed by $U_+$ (resp. $U_-$) and neutral for it.

We shall treat them successively.

*Case* (*i*) Since $G$ is generated by $U_+$ and $U_-$ and has no fixed end, the lines $L_+$ and $L_-$ have no common end; consequently, their intersection is empty or is a segment. Let $u$ be an element of $U_+$ which does not fix $L_+$ and such that $L_+ \cap L_- \cap uL_- = \varnothing$ (since $L_+ \cap L_-$ is bounded, a sufficiently high power of any element of $U_+$ which does not fix $L_+$ has that property) and let $u'$, $u''$, $m(u)$ be as in Section 5.1(R).

Suppose first that $L_+ \cap L_-$ is empty and let $p_+ \in L_+$ and $p_- \in L_-$ be defined by $[p_+, p_-] \cap L_\pm = \{p_\pm\}$. By Section 1.3, one has $[p_-, up_-] = [p_-, p_+] \cup [p_+, up_+] \cup [up_+, up_-]$, and

$$[p_-, up_-] \cap L_- = \{p_-\}, \qquad [p_-, up_-] \cap uL_- = \{up_-\}$$

(the reader is advised to draw a picture), hence, denoting by $d(X, Y)$ the distance of the sets $X$ and $Y$ and using Section 1.4, $d(L_-, uL_-) = d(p_-, up_-) > d(p_-, p_+)$. On the other hand,

$$d(L_-, uL_-) = d(u'L_-, u'uL_-) = d(L_-, m(u)L_-) = d(L_-, L_+) = d(p_-, p_+),$$

a contradiction.

Thus, $L_+ \cap L_-$ is a segment $[p, q]$. Consequently, one has $L_- \cap uL_- = L_- \cap u'^{-1}L_+ = u'^{-1}(L_- \cap L_+) \neq \varnothing$. It follows that the segment $[p, up]$, which joins a point of $L_-$ and a point of $uL_-$, meets their intersection. Since $[p, up] \subset L_+$, this contradicts the hypothesis made on $u$, and we conclude that the first case cannot occur.

*Case* (*ii*) Since the intersection of the trees $T_+$ and $T_-$ is fixed by $G$, it must be empty. Let $p_+ \in T_+$ and $p_- \in T_-$ be defined by $[p_+, p_-] \cap T_\pm = \{p_\pm\}$. The trees $T_+$, $T_-$ and hence the points $p_+$, $p_-$ are permuted by every element of the set $M$ defined in Section 5.1. Let $u$ be an element of $U_+$ which does not fix the middle point of $[p_+, p_-]$ and set $[p_+, p_-] \cap [p_+, up_-] = [p_+, q]$. Then, $d(p_-, q) > \frac{1}{2}d(p_-, p_+)$ and, by Section 1.3, $d(p_-, up_-) = d(p_-, q) + d(q, up_-) = 2d(p_-, q) > d(p_-, p_+)$. On the other hand, if $u'$, $u''$, $m(u)$ are as in Section 5.1(R), one has $d(p_-, up_-) = d(u'p_-, u'up_-) = d(p_-, m(u)p_-) = d(p_-, p_+)$, a contradiction which shows that also the second case cannot occur.

*Case* (*iii*) Since $G$ fixes no end, one has $e_+ \neq e_-$. The group $S$ and the set $M$ of Section 5.1 clearly preserve the line $A = [e_+, e_-]$: the ends $e_+$, $e_-$ are permuted by $M$ and fixed by $S$. Let $u \in U_+ - \{1\}$ and let $u'$, $u''$, $m(u)$ be as in Section 5.1(R). The element $u$ does not fix $e_-$ (otherwise $m(u)$ would), therefore $T^u \cap A$ is a half-line $[q, e_+)$. We shall show that

$$q \text{ is fixed by } u' \text{ and } u'', \text{ hence by } m(u). \tag{1}$$

Suppose the contrary. Upon replacing $u'$, $u$, $u''$ by $u''^{-1}$, $u^{-1}$, $u'^{-1}$, if necessary, we may assume that $q$ is not fixed by $u''$. Setting $T^{u''} \cap A = [q', e_-)$, we then have $q' \notin [q, e_+)$, hence $[q, uq'] \cap A = \{q\}$, therefore $q \in [uq', e_-)$ and, transforming by $u'$, $u'q \in [u'uq', e_-) = [m(u)q', e_-) \subset A$. This implies that $u'q = q$. But then, one has $[u''^{-1}q, e_-) = [m(u)^{-1}q, e_-) \subset A$, which clearly implies that $q$ is fixed by $u''$, a contradiction establishing our assertion (1).

The elements $m(u)$ of $M$ cannot all have the same fixed point in $A$, otherwise, as follows from (1), this point would be fixed by $U_+$, hence by $U_-$ and finally by $G$. Therefore $S$ $(= M \cdot M)$ does not fix $A$. Let us now choose an element $m_0$ in $M$ and identify $A$ with $\mathbf{R}$ isometrically in such a way that 0 is fixed by $m_0$ and that $e_+$ corresponds to $+\infty$. Let $\eta$: $S \to \mathbf{R}$ be the homomorphism defined by $st = t + \eta(s)$ for $s \in S$ and $t \in A = \mathbf{R}$. Then, it readily follows from (1) that the function $\varphi$: $U_+ \to \mathbf{R} \cup \{\infty\}$ of Section 5.2 is given by $\varphi(u) = \inf\{t \in \mathbf{R} \mid ut = t\}$. Therefore, $\eta$ is a valuation and it is easy to see that the mapping $(g, r) \mapsto gr$ of $G \times \mathbf{R}$ into $T$ factorizes through an isometry $\alpha$: $T_\eta \to T$. The unicity of $\eta$ and $\alpha$ are clear from the way they have been obtained.

**5.5** Restated in loose form in the case of $SL_2(K)$, Proposition 4 essentially means that

*Every fixed point free and fixed end free action of $SL_2(K)$ on a tree is through a well-defined non-archimedean valuation of $K$.*

## References


1. H. Bass, Some remarks on group actions on trees, *Commun. Algebra* **4** (1976), 1091–1126.
2. A. Borel, "Linear Algebraic Groups." Benjamin, New York, 1969.
3. A. Borel and J. Tits, Groupes réductifs, *Inst. Hautes Études Sci. Publ. Math.* **27** (1965), 55–151.
4. A. Borel and J. Tits, Eléments unipotents et sous-groupes paraboliques de groupes réductifs, I, *Invent. Math.* **12** (1971), 95–104.
5. A. Borel and J. Tits, Homomorphismes "abstraits" de groupes algébriques simples, *Ann. of Math.* **97** (1973), 499–571.
6. F. Bruhat and J. Tits, Groupes réductifs sur un corps local. I: Données radicielles valuées, *Inst. Hautes Études Sci. Publ. Math.* **41** (1972), 5–251.
7. E. Kolchin, Algebraic matric groups and the Picard–Vessiot theory of homogeneous linear ordinary differential equations, *Ann. of Math.* (2) **49** (1948), 1–42.
8. G. A. Margulis, Diskretnye Gruppy dviženii mnogoobrazii položitel'noi krivizny, *Proc. Internat. Congr. Math., Vancouver, 1974* **2** (1975), 21–34.
9. M. Rosenlicht, Some rationality questions on algebraic groups, *Ann. di Mat.* (IV) **43** (1957), 23–50.
10. J.-P. Serre, Amalgames et points fixes, *Proc. Internat. Conf. Group Theory, Canberra, 1973* (*Lect. Notes in Math.* **372**), pp. 633–640. Springer-Verlag, Berlin–Heidelberg–New York, 1974.
11. R. Steinberg, Lectures on Chevalley groups, Notes by J. Faulkner and R. Wilson, Yale Univ., 1967.
12. J. Tits, Algebraic and abstract simple groups, *Ann. of Math.* (2) **80** (1964), 313–329.
13. J. Tits, Sur le groupe des automorphismes d'un arbre, *in* "Essays on Topology" (Mémoires dédiés à G. de Rham), pp. 188–211. Springer-Verlag, Berlin–Heidelberg–New York, 1970.


AMS (MOS) 1970 subject classifications: 05C05, 20E99, 20G15, 20G25, 20H25, 54H25

# *Regular elements in anisotropic tori*

*F. D. Veldkamp*

*University of Utrecht*

## Introduction

For the representation theory of finite forms of quasisimple algebraic groups, one needs information about the existence of regular elements in anisotropic maximal tori, and of regular characters as well as characters in general position on such tori (cf. [3, 9]). It is known [7] that they certainly exist if the ground field is large enough, but for small fields there are exceptions. The analogous problem for Lie algebras is treated in [6].

The problem about characters is, in a sense, dual to that for elements in tori. In the present paper we shall deal with regular elements and elements in general position; see further on in the introduction for precise definitions. The characters will be treated in a subsequent paper [12], since besides duality there are a few more interesting connections between tori and their character groups that play a role here. The classical groups $A_l$, $B_l$, $C_l$, $D_l$ and their twisted forms will be dealt with in a more or less general way. Exceptional groups of type ${}^2C_2$, ${}^3D_4$, $F_4$, and $G_2$ are treated using a case by case inspection. For groups of type $E_{6,7,8}$ this would lead, however, to unwieldy calculations; we use a method based on results of Springer [7, 8] to handle at least some anisotropic tori in these groups.

Before we state the results of this paper, we have to present some definitions, notations, and known results; for more details the reader is referred to [3, 7, 9, 10]. $G$ is a connected quasisimple linear algebraic group,

$\sigma$ an endomorphism of $G$ such that the group of fixed points $G_\sigma$ is finite, $T$ a maximal torus in $G$ fixed by $\sigma$, $X$ the character groups of $T$, $V = X \otimes \mathbb{R}$, and $\sigma^*$ the action of $\sigma$ on $X$ and $V$. Then $\sigma^* = q\tau$, $\tau$ being an isometry with respect to the Killing form on $V$. The torus $T$ is called *anisotropic* if it is contained in no proper parabolic subgroup of $G$ fixed by $\sigma$; this is the case if and only if $\tau$ does not have 1 as an eigenvalue. By abuse of language, we shall often say that $T_\sigma = T \cap G_\sigma$ is anisotropic.

An element $x$ of $G$ is called *regular* if its centralizer $Z_G(x)$ in $G$ has minimal dimension, which equals the rank $l$ of $G$. An element $x$ of a maximal torus $T$ is regular if and only if $Z_G(x)^0 = T$, that is, if $\alpha(x) \neq 1$ for all roots $\alpha$ relative to $T$. An element of $T$ is said to be *in general position* if its stabilizer in the Weyl group $W$ of $T$ is 1 or, equivalently, if its stabilizer in $W_\sigma$ is 1, where $W_\sigma$ denotes the centralizer of $\sigma$ in $W$. Elements in general position are regular; the converse is true in simply connected groups, but not in general. We look for elements that are regular or in general position in *anisotropic maximal tori* $T_\sigma$ in $G_\sigma$. Our methods equally well apply to other maximal tori, but for the sake of brevity we have confined ourselves to anisotropic ones since these seem to be the most interesting for the representation theory of $G_\sigma$.

The classes of maximal tori fixed by $\sigma$ under conjugation by $G_\sigma$ are in 1–1 correspondence with the classes of $W\tau$ under conjugation by $W$, the Weyl group of a fixed maximal torus $T$ in $G$. If $\sigma$ centralizes $W$, then these classes correspond to the conjugacy classes of $W$, which are listed in [1, 2]; we shall use the notation of these papers to indicate classes of $W$ and the corresponding maximal tori. The class of $W\tau$ corresponding to a maximal torus $T$ is called the *type* of $T$; for a somewhat more detailed description see Section 2.

Given $\sigma$ and a $\sigma$-invariant maximal torus $T$, there exist a permutation $\pi$ of the roots relative to $T$, and powers $q_\alpha$ of the characteristic $p$ of the ground field of $G$ such that $\sigma^*\alpha = q_\alpha \pi\alpha$ for each root $\alpha$. For each quasisimple type of $G$ (up to isogeny) we make particular choices of $\sigma$, to be referred to as $\sigma_0$, with contragredient $\sigma_0^* = q\tau_0$. We may assume $\sigma_0$ to fix a Borel subgroup $B_0$ and a maximal torus $T_0$ therein. If we order the roots correspondingly to $B_0$, the permutation $\pi$ corresponding to $\sigma_0$ as above permutes the positive roots. The following list gives a complete enumeration of the possible $\sigma_0$, up to isomorphism of the corresponding groups $G_{\sigma 0}$. In I and II, $G$ and $T_0$ are supposed to be defined and split over the field $\mathbb{F}_q$, in III and IV over $\mathbb{F}_2$, and in V over $\mathbb{F}_3$.

I. $\pi =$ identity, $q_\alpha = q$ for all roots $\alpha$, and $G_{\sigma 0}$ is a Chevalley group parametrized by $\mathbb{F}_q$, denoted as ${}^1A_l(q)$, ${}^1B_l(q)$, etc.

II. $\pi \neq$ identity, $q_\alpha = q$ for all $\alpha$, and $G_{\sigma 0}$ is a Steinberg group ${}^2A_l(q^2)$, ${}^2D_l(q^2)$, ${}^3D_4(q^3)$, or ${}^2E_6(q^2)$.

III. $G$ is of type $C_2$, with simple roots $\alpha_1$ (long) and $\alpha_2$ (short). $\pi$ interchanges $\alpha_1$ and $\alpha_2$; $q_{\alpha_1} = 2^{n+1}$, $q_{\alpha_2} = 2^n$, $q = 2^{n+\frac{1}{2}}$. For $G_{\sigma 0}$ we get a Suzuki group ${}^2C_2(q^2)$.

IV. $G$ is of type $F_4$, with simple roots $\alpha_1, \alpha_2$ (long) and $\alpha_3, \alpha_4$ (short). $\pi$ interchanges $\alpha_1$ and $\alpha_4$, and $\alpha_2$ and $\alpha_3$; $q_{\alpha_1} = q_{\alpha_2} = 2^{n+1}$, $q_{\alpha_3} = q_{\alpha_4} = 2^n$, $q = 2^{n+\frac{1}{2}}$. In this case, $G_{\sigma 0}$ is a Ree group ${}^2F_4(q^2)$.

V. $G$ is of type $G_2$, with simple roots $\alpha_1$ (short) and $\alpha_2$ (long). $\pi$ interchanges $\alpha_1$ and $\alpha_2$, $q_{\alpha_1} = 3^n$, $q_{\alpha_2} = 3^{n+1}$, $q = 3^{n+\frac{1}{2}}$. This time, $G_{\sigma 0}$ is a Ree group ${}^2G_2(q^2)$.

In case I (Chevalley groups) a particularly interesting type of maximal tori are the Coxeter tori, corresponding to the Coxeter class in $W$. Similarly, in case II (Steinberg groups) one has the twisted Coxeter tori, corresponding to the twisted Coxeter class in $W\tau_0$ (see [8, §8]). Both types are anisotropic. In almost all cases they contain elements in general position. In fact, we have the following general results.

***Theorem 1*** *Coxeter and twisted Coxeter tori contain elements in general position except when $G_{\sigma 0} = {}^1G_2(2)$ or ${}^2A_2(2^2)$ (simply connected or adjoint). In the adjoint ${}^2A_2(2^2)$ the twisted Coxeter tori do contain regular elements, however.*

***Theorem 2*** *Given any quasisimple algebraic group $G$ and any $\tau_0$ in one of the cases I–V, there exist elements $\tau$ in $W\tau_0$ with the properties: for any possible $q$ in the given case I, ..., V, a maximal torus $T$ corresponding to the class of $\tau$ is anisotropic with respect to $\sigma_0$, and $T_{\sigma 0}$ contains elements in general position, except that the adjoint group ${}^2A_2(2^2)$ contains no elements in general position. However, the same result for regular elements holds without restriction.*

The proofs of these theorems are contained in the case by case treatment that will follow; more detailed results will be described in the separate cases (see Section 5, Tables 1 and 2, and Propositions 1–6, 8–11).

Some notations to be used in the sequel: If $q$ is a fixed power of a prime, $\Gamma_{s/t}$ denotes the Galois group of the field $\mathbb{F}_{q^s}$ over $\mathbb{F}_{q^t}$, if $t$ divides $s$. Instead of $\Gamma_{s/1}$ we just write $\Gamma_s$. By $\rho$ we always denote the automorphism $x \to x^q$ of some finite extension field $\mathbb{F}_{q^s}$ of $\mathbb{F}_q$; it is a generator of $\Gamma_s$, and $\rho^t$ one of $\Gamma_{s/t}$. $N_{s/t}$ is the Galois norm in $\mathbb{F}_{q^s}$ over $\mathbb{F}_{q^t}$, $N_s = N_{s/1}$; the kernel of $N_{s/t}$ is $N^1_{s/t}$, that of $N_s$ is $N_s{}^1$.

## 1.

Let $G$ be a semisimple algebraic group and $\pi$: $G' \to G$ its universal covering. If $\sigma$ is an endomorphism of $G$ such that $G_\sigma$ is finite, then there exists a unique endomorphism $\sigma'$ of $G'$ such that $\pi\sigma' = \sigma\pi$. Instead of $\sigma'$ we will write $\sigma$ from now on. $G'_\sigma$ is finite, and $\pi G'_\sigma = G_{\sigma u}$, the subgroup generated by the unipotent elements of $G_\sigma$. On the other hand, $G_\sigma/G_{\sigma u} \cong F/(1-\sigma)F$, where $F$ is the kernel of $\pi$, and $G_{\sigma u} \cong G'_\sigma/F_\sigma$. See [10, 9.16 and 12.6]. If $\sigma'$ is an endomorphism of $G'$ with finite $G'_{\sigma'}$, then there exists an endomorphism $\sigma$ of $G$ with $\pi\sigma' = \sigma\pi$ if and only if $\sigma' F \subseteq F$; then $G_\sigma$ is also finite.

In the above situation let $T$ be a maximal torus of $G$ invariant under $\sigma$, and $T'$ a maximal torus of $G'$ with $\pi T' = T$. Then $\sigma T' = T'$, $\pi T'_\sigma = T_\sigma \cap G_{\sigma u}$, and $T_\sigma/T_\sigma \cap G_{\sigma u} \cong F/(1-\sigma)F$. Since the value of a root of $T'$ in an element $x \in T'$ is the same as the value of the corresponding root of $T$ in $\pi x$, $x$ is regular if and only if $\pi x$ is so. Hence, if $T'_\sigma$ contains regular elements, so does $T_\sigma$. On the other hand, if, $\pi x \in T_\sigma$ is in general position, then so is $x \in T'_\sigma$. For $wx = x$ for some $w \in W$, $w \neq 1$, implies $w\pi x = \pi x$. The converse is not true, of course. To find the elements in general position in $T_\sigma$, we can proceed as follows. If $t \in T'$, then $\pi t \in T_\sigma$ if and only if $\sigma t = ct$ for some $c \in F = \ker(\pi)$. Then $\pi t$ is in general position if and only if $wt \neq c_1 t$ for all $c_1 \in F_\sigma$, $w \in W_\sigma$, $w \neq 1$. In several cases we have been able to find elements in general position in nonsimply connected groups in this way. But if $F$ is too large, which happens notably in case of type $A_l$, it seems impossible to perform the necessary computations. In that situation the following lemma is useful, although rather crude.

**Lemma 1** *Let $f$ be the least common multiple of the orders of all elements in $F_\sigma$. If $T'_\sigma$ contains an element $y$ such that $y^f$ is in general position, then $T_\sigma$ contains an element in general position, viz., $\pi y$.*

*Proof* Assume $\pi y$ is not in general position, so there is an element $w \neq 1$ of the Weyl group with $w\pi y = \pi y$. Then $wy = cy$ with $c \in F$, hence $wy^f = c^f y^f = y^f$, since $c$ lies in the center of $G'$ and $c^f = 1$, so $y^f$ is not in general position.

A similar result holds for going up with regular elements from $T_\sigma$ to $T'_\sigma$, but we will not need this.

## 2.

Let $G$ be quasisimple, $\sigma_0$, $B_0$, and $T_0$ as in the introduction. If $T$ is any maximal torus of $G$ invariant under $\sigma_0$, then $T$ is conjugate to $T_0$ under an inner automorphism Int $g$. If we identify $T$ with $T_0$ by Int $g$, the action of $\sigma_0$ on $T$ has to be replaced by the action of $\sigma = w\sigma_0$ on $T_0$ for some $w \in W$. In fact, if $T_0 = gTg^{-1}$, then we have to take $w = g\sigma_0(g)^{-1}T_0$ in $W = N_0/T_0$, where $N_0$ is the normalizer of $T_0$; $g\sigma_0(g)^{-1} \in N_0$ follows from $\sigma_0 T = T$. Replacing $T$ by a conjugate torus $T' = xTx^{-1}$ with $x \in G_{\sigma_0}$ which is also invariant under $\sigma_0$ amounts to replacing $\sigma = w\sigma_0$ by $w_1 \sigma w_1^{-1}$ for some $w_1 \in W$. That is, the maximal tori $T$ under conjugation by $G_{\sigma_0}$ are classified by the conjugacy classes in $W\sigma_0$ under the action of $W$ defined by $\sigma \mapsto w_1 \sigma w_1^{-1}$ for $\sigma \in W\sigma_0$, $w_1 \in W$ (cf. [9, II.1.3]).

## 3. Classical groups

If $G$ is a quasisimple linear algebraic group, we denote its simply connected covering group by $G_{un}$ and the adjoint group by $G_{\mathrm{ad}}$. We will first give the root systems, $\sigma_0$, Weyl groups, etc., of the classical groups.

*Type* $^1A_l$ Take $G_{un} = SL_{l+1}$. Let $\sigma_0$ be the Frobenius endomorphism $(\alpha_{ij})_{1\le i,\, j\le l+1} \mapsto (\alpha_{ij}^q)_{1\le i,\, j\le l+1}$, $q$ being a power of the characteristic $p$ of the ground field. A $\sigma_0$-invariant maximal torus $T_0$ is the torus of all diagonal matrices in $SL_{l+1}$, which we denote by $t = (t_1, t_2, \ldots, t_{l+1})$ with $\prod_{i=1}^{l+1} t_i = 1$. The character group of $T_0$ is $X = \langle \omega_1, \omega_2, \ldots, \omega_{l+1} \mid \sum_{i=1}^{l+1} \omega_i = 0\rangle$, where $\omega_i\colon t \mapsto t_i$. The action of $\sigma_0$ on $V = X \otimes \mathbb{R}$ is $\sigma_0{}^* = q \cdot 1$. The Weyl group $W$ consists of all permutations of $\omega_1, \omega_2, \ldots, \omega_{l+1}$. The element of $W$ defined by a permutation $\pi$ of $1, 2, \ldots, l+1$ will also be denoted by $\pi : \pi\omega_i = \omega_{\pi i}$ for $i = 1, \ldots, l+1$. The roots are $\omega_i - \omega_j$, $i \ne j$, $1 \le i$, $j \le l+1$.

*Type* $^1B_l$, $l \ge 2$ Here we take $G_{\mathrm{ad}} = SO_{2l+1}$, the rotation group of the quadratic form $Q(x) = \xi_1\xi_2 + \xi_3\xi_4 + \cdots + \xi_{2l-1}\xi_{2l} + \xi_{2l+1}^2$ in the $(2l+1)$-dimensional linear space $U$. Let $T_0$ be the maximal torus of diagonal matrices $t = \{t_1, t_1^{-1}, t_2, t_2^{-1}, \ldots, t_l, t_l^{-1}, 1\}$, to be denoted by $t = (t_1, t_2, \ldots, t_l)$. The character group of $T_0$ is $X = \langle \omega_1, \ldots, \omega_l\rangle$, where $\omega_i\colon t \mapsto t_i$. For $\sigma_0$ we take the Frobenius endomorphism $(\alpha_{ij}) \mapsto (\alpha_{ij}^q)$, $q$ a power of the characteristic $p$ of the ground field; then $\sigma_0{}^* = q \cdot \mathrm{id}$. The Weyl group consists of the transformations

$$w\colon \omega_i \mapsto e_i\omega_{\pi i}, \qquad \pi \in S_l, \quad e_i = \pm 1.$$

If $\pi = (i_1, \ldots, i_s)(j_1, \ldots, j_t) \cdots (l_1, \ldots, l_v)$ is the cycle decomposition in $S_l$, we denote $w$ by

$$w = (i_1, \ldots, i_s)^{e_{i_1}, \ldots, e_{i_s}}(j_1, \ldots, j_t)^{e_{j_1}, \ldots, e_{j_t}} \cdots (l_1, \ldots, l_v)^{e_{l_1}, \ldots, e_{l_v}}$$

where, of course, cycles of length 1 may be included. Using the abbreviation $(m_1, \ldots, m_u)^{\pm}$ for $(m_1, \ldots, m_u)^{1, \ldots, 1, \pm 1}$ (called *positive* or *negative cycles*, respectively), the conjugacy classes of $W$ are given by the signed cycle type, that is, every element of $W$ is conjugate to precisely one element of the form

$$w = (1, \ldots, r_1)^{\pm}(r_1 + 1, \ldots, r_1 + r_2)^{\pm} \cdots (r_1 + \cdots + r_{u-1} + 1, \ldots, l)^{\pm}.$$

The roots are $\pm\omega_i$, $\pm\omega_i \pm \omega_j$, $i \ne j$, $1 \le i, j \le l$.

If the ground field has characteristic $\ne 2$, we also need the simply connected group of type $B_l$, the spin group $\mathrm{Spin}_{2l+1} = G_{un}$. Consider the basis

$e_1 = (1, 0, \ldots, 0)$, $e_2 = (0, 1, 0, \ldots)$, $\ldots$, $e_{2l+1} = (0, \ldots, 0, 1)$ in $U$. For the symmetric bilinear form corresponding to $Q$ we have

$$(e_{2i-1}, e_{2i}) = 1 \qquad \text{for} \quad i = 1, 2, \ldots, l,$$
$$(e_{2l+1}, e_{2l+1}) = 1,$$
$$\text{all other} \quad (e_i, e_j) = 0 \qquad \text{for} \quad i \leq j.$$

In $\mathrm{Spin}_{2l+1}$ the maximal torus $T_0'$ which covers $T_0$ consists of the elements

$$u = [u_1, \ldots, u_l] = \prod_{i=1}^{l} (u_i^{-1} e_{2i-1} + u_i e_{2i})(e_{2i-1} + e_{2i}).$$

These elements are multiplied coordinatewise. Notice that $[u_1, \ldots, u_l] = [v_1, \ldots, v_l]$ if and only if for all $i$: $u_i = e_i v_i$ with $e_i = \pm 1$, $\prod_{i=1}^{l} e_i = 1$. The inner automorphism $x \to uxu^{-1}$ of the Clifford algebra of $(U, Q)$ induces by restriction to $U$ the rotation $t = (u_1^2, \ldots, u_l^2) \in T_0$, that is, the projection $\pi\colon T_0' \to T_0$ is given by $\pi[u_1, \ldots, u_l] = (u_1^2, \ldots, u_l^2)$. The action of the Frobenius endomorphism $\sigma_0$ on $T_0'$ is $[u_1, \ldots, u_l] \mapsto [u_1^q, \ldots, u_l^q]$. The Weyl group acts on $T_0'$ and its character group as on $T_0$ with signed permutations.

*Type* ${}^1C_l$, $l \geq 3$ Consider the simply connected group $G_{un} = \mathrm{Sp}_{2l}$, the symplectic group with respect to the form $\xi_1 \eta_2 - \xi_2 \eta_1 + \cdots + \xi_{2l-1} \eta_{2l} - \xi_{2l} \eta_{2l-1}$. The maximal torus $T_0$ consists of the diagonal matrices $t = \{t_1, t_1^{-1}, t_2, t_2^{-1}, \ldots, t_l, t_l^{-1}\}$, which element we denote by $t = (t_1, t_2, \ldots, t_l)$. For $\sigma_0$ we take again the Frobenius endomorphism $(\alpha_{ij}) \mapsto (\alpha_{ij}^q)$. The Weyl group $W$ and character group are the same as for $B_l$. The roots are $\pm 2\omega_i$, $\pm \omega_i \pm \omega_j$. If $\sigma_1$ is the endomorphism of the adjoint group of type $B_l$ corresponding to some $\tau \in W$, so $\sigma_1^* = q\tau$, and $\sigma_2$ the endomorphism of the simply connected group of type $C_l$ with $\sigma_2^* = q\tau$, then obviously $T_{\sigma 1}$ and $T_{\sigma 2}$ are $W_\tau$-isomorphic, hence they contain the same number of regular elements.

*Type* ${}^1D_l$, $l \geq 4$ Take $G = SO_{2l}$, the rotation group of the quadratic form $\xi_1 \xi_2 + \xi_3 \xi_4 + \cdots + \xi_{2l-1} \xi_{2l}$. For $\sigma_0$ we take again the Frobenius endomorphism, for $T_0$ the maximal torus of the diagonal matrices $t = \{t_1, t_1^{-1}, t_2, t_2^{-1}, \ldots, t_l, t_l^{-1}\}$, again denoted as $t = (t_1, t_2, \ldots, t_l)$. The character group is $X = \langle \omega_1, \ldots, \omega_l \rangle$, with $\omega_i\colon t \mapsto t_i$. The Weyl group consists of the transformations

$$w\colon \omega_i \mapsto e_i \omega_{\pi i}, \qquad \pi \in S_l, \quad e_i = \pm 1, \quad \prod_{i=1}^{l} e_i = 1.$$

Here we will use the same notations as for the Weyl group of $B_l$. The conjugacy classes are determined by the signed cycle type except when all cycles are positive and of even length, in which case there are two classes. The roots are $\pm \omega_i \pm \omega_j$.

In case of characteristic $\neq 2$ we also need the simply connected group $G_{un} = \mathrm{Spin}_{2l}$. The maximal torus $T_0'$ therein which covers $T_0$ is described in exactly the same way as in the case of type ${}^1B_l$.

For some purposes it is useful to consider $SO_{2l}$ in the obvious way as a subgroup of $SO_{2l+1}$, with the same maximal torus $T_0$. Since the root system and the Weyl group of $SO_{2l}$ are contained in those of $SO_{2l+1}$, an element of $T_0$ which is regular or in general position in $SO_{2l+1}$ certainly is so in $SO_{2l}$. Similarly for the spin groups.

*Type* ${}^2A_l, l \geq 2$  Notations as for ${}^1A_l$. Now $\sigma_0$ will be the endomorphism with $\sigma_0{}^* = -q = q(-1)$. Notice that $\sigma_0$ commutes with the elements of the Weyl group, so the conjugacy classes of maximal tori correspond to the conjugacy classes in $W$ multiplied by $-1$.

*Type* ${}^2D_l$, $l \geq 4$  With the notations as for ${}^1D_l$, we choose $\sigma_0$ such that $\sigma_0{}^* = q(l)^-$, i.e.,

$$\sigma_0{}^*\omega_i = q\omega_i \qquad \text{for} \quad i = 1, \ldots, l-1, \qquad \sigma_0{}^*\omega_l = -q\omega_l\,.$$

This is the diagram automorphism which acts on the simple roots $\alpha_1 = \omega_1 - \omega_2$, $\alpha_2 = \omega_2 - \omega_3$, $\ldots$, $\alpha_{l-1} = \omega_{l-1} - \omega_l$, $\alpha_l = \omega_{l-1} + \omega_l$ as follows:

$$\sigma_0{}^*\alpha_i = q\alpha_i \qquad \text{for} \quad i = 1, \ldots, l-2, \qquad \sigma_0{}^*\alpha_{l-1} = q\alpha_l\,,$$

$$\sigma_0{}^*\alpha_l = q\alpha_{l-1}.$$

## 4. Anisotropic tori in classical groups

If $\sigma^* = q\tau$, then the maximal torus $T_\sigma$ is anisotropic if and only if $\tau$ has no eigenvalue 1. With notation as in the previous section we find that this is the case for the following $\tau$, up to conjugacy under $W$.

*Type* ${}^1A_l$  $\tau = (1, 2, \ldots, l+1)$, the Coxeter element of the Weyl group.

*Type* ${}^1B_l$, $l \geq 2$, *or* ${}^1C_l$, $l \geq 3$  $\tau = w = (1, \ldots, r_1)^-(r_1+1, \ldots, r_1+r_2)^- \cdots (r_1 + \cdots + r_{u-1} + 1, \ldots, l)^-$, a product of negative cycles only. Coxeter element: $(1, 2, \ldots, l)^-$.

*Type* ${}^1D_l$, $l \geq 4$  As for ${}^1B_l$, with *u even*, that is, $\tau = w$ is a product of an even number of negative cycles only. Coxeter element: $(1, 2, \ldots, l-1)^-(l)^-$.

*Type* ${}^2A_l$, $l \geq 2$  $\tau = -w = -(1, 2, \ldots, r_1)(r_1+1, \ldots, r_1+r_2) \cdots (r_1 + \cdots + r_{u-1} + 1, \ldots, l+1)$ with all $r_i$ odd (where $r_u = l + 1 - \sum_{j=1}^{u-1} r_j$). Twisted Coxeter element: $-(1, 2, \ldots, l+1)$ if $l$ is even, $-(1, 2, \ldots, l)$ if $l$ is odd.

*Type* ${}^2D_l$, $l \geq 4$ $\tau = (1, \ldots, r_1)^-(r_1 + 1, \ldots, r_1 + r_2)^- \cdots (r_1 + \cdots + r_{u-1} + 1, \ldots, l)^-$ with $u$ *odd*, that is, $\tau$ is a product of an odd number of negative cycles only. Twisted Coxeter element: $(1, 2, \ldots, l)^-$.

## 5. Results for the classical groups

We have collected the results for the classical groups in Tables 1 (regular elements) and 2 (elements in general position).

In column (3) "Exist in all $T_\sigma$ provided" means: regular elements (or elements in general position, respectively) exist in all anisotropic maximal tori $T_\sigma$ provided $q$ satisfies the conditions mentioned in this column.

Column (4), "$T_\sigma$ containing regular elements for all $q$," lists anisotropic maximal tori $T_\sigma$ with $\sigma^* = q\tau$ which, given $\tau$, contain regular elements for all possible values of $q$, and similarly for elements in general position. This list is in general *not necessarily exhaustive.* Notice that the conditions already imposed on $\tau$ in order that $T_\sigma$ be anisotropic are not mentioned again in this column. Thus, e.g., for ${}^1D_l$ it is to be understood that the total number of negative cycles is even, and for ${}^2A_l$ that all $r_i$ are odd.

Column (5) lists some anisotropic maximal tori that contain no regular elements (no elements in general position, respectively). Again, this is not an exhaustive account of all possibilities.

For real $x$, $[x]$ denotes the largest integer $\leq x$.

$C_d$ denotes the cyclic subgroup of order $d$ in the center of the simply connected group $G_{un}$ (not of type $D_l$, $l$ even).

The proofs for the results in Tables 1 and 2 are contained in Sections 6–11. We will use the notations fixed above.

## 6. Type ${}^1A_l$

For $\sigma^* = q(1, 2, \ldots, l + 1)$ we find that $t = (t_1, \ldots, t_{l+1})$ belongs to $T_\sigma$ if $t = (s^{q^l}, s^{q^{l-1}}, \ldots, s^q, s)$ with $s \in \mathbb{F}_{q^l}$ satisfying $s^{q^{l+1}} = s$, and $\prod_{i=0}^{l} s^{q^i} = 1$. Obviously, the latter condition on $s$ implies the former. Let $\rho$ denote the automorphism $x \mapsto x^q$ generating $\Gamma_{l+1}$, the Galois group of $\mathbb{F}_{q^{l+1}}$ over $\mathbb{F}_q$. Then $t = (\rho^l(s), \rho^{l-1}(s), \ldots, \rho(s), s)$, with $s \in N_{l+1}^1 = \ker(N_{l+1})$, $N_{l+1}$: $\mathbb{F}_{q^{l+1}}^* \to \mathbb{F}_q^*$ being the Galois norm. It is easily seen that $|N_{l+1}^1| = (q-1)^{-1} \times (q^{l+1} - 1)$. The roots relative to $T_0$ are $\omega_i - \omega_j$, $i \neq j$, so $t$ is irregular if and only if $\rho^i(s) = \rho^j(s)$ for some $i \neq j$, that is, if $\rho^i(s) = s$ for some $i$, $1 \leq i \leq l$. This means $s \in \mathbb{F}_{q^m}^*$ for some $m < l + 1$, $m \mid l + 1$. Hence the number $R_\sigma$ of regular elements in $T_\sigma$ satisfies

$$R_\sigma \geq (q-1)^{-1}(q^{l+1} - 1) - \sum_{m \mid l+1,\, m < l+1} (q^m - 1).$$

**Table 1**

*Regular Elements in Anisotropic Maximal Tori $T_\sigma$ in Classical Groups*

| (1)<br>Type | (2)<br>Connectedness type of $G$ | (3)<br>Exist in all $T_\sigma$ provided | (4)<br>$T_\sigma$ containing regular elements for all $q$ | (5)<br>$T_\sigma$ containing no regular elements |
|---|---|---|---|---|
| $^1A_l$ | All | All $q$ | All | Nonexistent |
| $B_l(q)$,<br>$l \geq 2$ | Simply connected | $q \geq 2l - 1$ if $l \equiv 0$ or $3 \bmod 4$<br>$q \geq 2l$ if $l \equiv 1$ or $2 \bmod 4$ | $r_i \neq r_j$ for $i \neq j$ | $\tau = -1$ for<br>$q < 2l - 1$ if $l \equiv 0$ or $3 \bmod 4$<br>$q < 2l$ if $l \equiv 1$ or $2 \bmod 4$ |
| | Adjoint | $q \geq 2l - 1$ | | $\tau = -1$ for $q < 2l - 1$ |
| $^1C_l(q)$,<br>$l \geq 3$ | Simply connected | $q \geq 2l$ | $r_i \neq r_j$ for $i \neq j$ | $\tau = -1$, $q < 2l$ |
| | Adjoint | $q \geq 2l - 1$ | | $\tau = -1$, $q < 2l - 1$ |
| $^1D_l(q)$,<br>*even* $l \geq 4$ | Simply connected or half-spin groups | $q \geq 2l - 3$ if $l \equiv 0 \bmod 4$<br>$q \geq 2l - 2$ if $l \equiv 2 \bmod 4$ | $r_i \neq r_j$ for $i \neq j$ | $\tau = -1$, $q < 2l - 3$ if $l \equiv 0 \bmod 4$<br>$q < 2l - 2$ if $l \equiv 2 \bmod 4$ |
| | $SO_{2l}$ or adjoint | $q \geq 2l - 3$ | | $\tau = -1$, $q < 2l - 3$ |
| $^1D_l(q)$,<br>*odd* $l > 4$ | Simply connected | $q \geq 2l - 6$ if $l \equiv 1 \bmod 4$<br>$q \geq 2l - 7$ if $l \equiv 3 \bmod 4$ | $r_i \neq r_j$ for $i \neq j$ | $\tau = (1)^-(2)^- \cdots (l-2)^-(l-1, l)^-$,<br>$q < 2l - 6$ if $l \equiv 1 \bmod 4$<br>$q < 2l - 7$ if $l \equiv 3 \bmod 4$ |
| | $SO_{2l}$ or adjoint | $q \geq 2l - 7$ | | $\tau = (-1)^-(-2)^- \cdots (l-2)^-(l-1, l)^-$,<br>$q < 2l - 7$ |
| $^2A_l(q^2)$,<br>$l \geq 2$ | All | $q \geq 2[\frac{1}{2}(l+1)]$, except when $q = 2$ and some $r_i = 3$ | $r_i \neq r_j$ for $i \neq j$, provided no $r_i = 3$ if $q = 2$.<br>$l = 2$, $\tau = -(1, 2, 3)$ if $G$ *adjoint* | $\tau = -1$, $q < 2[\frac{1}{2}(l+1)]$ for *simply connected G.*<br>$\tau = -(1, 2, 3)$ if $l = q = 2$ and $G$ *simply connected* |
| $^2D_l(q^2)$,<br>*even* $l \geq 4$ | Simply connected | $q \geq 2l - 6$ if $l \equiv 0 \bmod 4$<br>$q \geq 2l - 7$ if $l \equiv 2 \bmod 4$ | $r_i \neq r_j$ for $i \neq j$ | $\tau = (1)^-(2)^- \cdots (l-2)^-(l-1, l)^-$,<br>$q < 2l - 6$ if $l \equiv 0 \bmod 4$<br>$q < 2l - 7$ if $l \equiv 2 \bmod 4$ |
| | $SO_{2l}$ or adjoint | $q \geq 2l - 7$ | | $\tau = (1)^-(2)^- \cdots (l-2)^-(l-1, l)^-$,<br>$q < 2l - 7$ |
| $^2D_l(q^2)$,<br>*odd* $l > 4$ | Simply connected | $q \geq 2l - 3$ if $l \equiv 1 \bmod 4$<br>$q \geq 2l - 2$ if $l \equiv 3 \bmod 4$ | $r_i \neq r_j$ for $i \neq j$ | $\tau = -1$, $q < 2l - 3$ if $l \equiv 1 \bmod 4$<br>$q < 2l - 2$ if $l \equiv 3 \bmod 4$ |
| | $SO_{2l}$ or adjoint | $q \geq 2l - 3$ | | $\tau = -1$, $q < 2l - 3$ |

***Table 2***

*Elements in General Position in Anisotropic Maximal Tori $T_\sigma$ in Classical Groups.*

| (1)<br>Type | (2)<br>Connectedness type of $G$ | (3)<br>Exist in all $T_\sigma$ provided | (4)<br>$T_\sigma$ containing elements in general position for all $q$ | (5)<br>$T_\sigma$ containing no elements in general position |
|---|---|---|---|---|
| ${}^1A_l(q)$ | All | All $q$ | All | Nonexistent |
| $B_l(q)$, $l \geq 2$ | Simply connected | $q \geq 2l - 1$ if $l \equiv 0$ or $3 \bmod 4$<br>$q \geq 2l$ if $l \equiv 1$ or $2 \bmod 4$ | $r_i \neq r_j$ for $i \neq j$ | $\tau = -1$ for<br>$q < 2l - 1$ if $l \equiv 0$ or $3 \bmod 4$<br>$q < 2l$ if $l \equiv 1$ or $2 \bmod 4$ |
| | Adjoint | $q \geq 2l$ | | $\tau = -1$ for $q < 2l$ |
| ${}^1C_l(q)$, $l \geq 3$ | All | $q \geq 2l$ | $r_i \neq r_j$ for $i \neq j$ | $\tau = -1$ for $q < 2l$ |
| ${}^1D_l(q)$, *even* $l \geq 4$ | Simply connected | $q \geq 2l - 3$ if $l \equiv 0 \bmod 4$<br>$q \geq 2l - 2$ if $l \equiv 2 \bmod 4$ | $r_i \neq r_j$ for $i \neq j$ | $\tau = -1$, $q < 2l - 3$ if $l \equiv 0 \bmod 4$<br>$q < 2l - 2$ if $l \equiv 2 \bmod 4$ |
| | Not simply connected | $q \geq 2l - 2$ | | $\tau = -1$, $q < 2l - 2$ |
| ${}^1D_l(q)$, *odd* $l > 4$ | Simply connected | $q \geq 2l - 7$ if $l \equiv 3 \bmod 4$<br>$q \geq 2l - 6$ if $l \equiv 1 \bmod 4$ | $r_i \neq r_j$ for $i \neq j$ | $\tau = (1)^-(2)^- \cdots (l-2)^-(l-1, l)^-$,<br>$q < 2l - 7$ if $l \equiv 3 \bmod 4$<br>$q < 2l - 6$ if $l \equiv 1 \bmod 4$ |
| | Not simply connected | $q \geq 2l - 6$ | | $\tau = (1)^-(2)^- \cdots (l-2)^-(l-1, l)^-$,<br>$q < 2l - 6$ |
| ${}^2A_l(q^2)$, $l \geq 2$ | $G_{un}/C_d$, where $d \mid (q + 1, l + 1)$ | $q \geq 2d[\frac{1}{2}(l + 3)] - 1$ if $q$ odd<br>$q \geq d(2[\frac{1}{2}(l + 1)] + 1) - 1$ if $q$ even, except when some $r_i = 3$ and $q = 2$ | $r_i \neq r_j$ for $i \neq j$, provided no $r_i = 3$ if $q = 2$ | $\tau = -1$, $q < 2[\frac{1}{2}(l + 1)]$, $d = 1$<br>$\tau = -(1, 2, 3)$ if $l = q = 2$, (all $d$)<br>$\tau = -1$ if $l = q = 2$, (all $d$) |
| ${}^2D_l(q^2)$, *even* $l \geq 4$ | Simply connected | $q \geq 2l - 7$ if $l \equiv 2 \bmod 4$<br>$q \geq 2l - 6$ if $l \equiv 0 \bmod 4$ | $r_i \neq r_j$ for $i \neq j$ | $\tau = (1)^-(2)^- \cdots (l-2)^-(l-1, l)^-$,<br>$q < 2l - 7$ if $l \equiv 2 \bmod 4$<br>$q < 2l - 6$ if $l \equiv 0 \bmod 4$ |
| | Not simply connected | $q \geq 2l - 6$ | | $\tau = (1)^-(2)^- \cdots (l-2)^-(l-1, l)^-$,<br>$q < 2l - 6$ |
| ${}^2D_l(q^2)$, *odd* $l > 4$ | Simply connected | $q \geq 2l - 3$ if $l \equiv 1 \bmod 4$<br>$q \geq 2l - 2$ if $l \equiv 3 \bmod 4$ | $r_i \neq r_j$ for $i \neq j$ | $\tau = -1$, $q < 2l - 3$ if $l \equiv 1 \bmod 4$<br>$q < 2l - 2$ if $l \equiv 3 \bmod 4$ |
| | Not simply connected | $q \geq 2l - 2$ | | $\tau = -1$, $q < 2l - 2$ |

For $l > 1$ we find, since $(q-1)^{-1}(q^{l+1}-1) = q^l + q^{l-1} + \cdots + q + 1$,

$$R_\sigma \geq \sum_{m \nmid l+1} q^m + |\{m \mid l+1, m < l+1\}| \geq q^l + 1.$$

For $l = 1$ we would get $R_\sigma \geq 2$ in this way, but it is easy to get a sharper result in this case. For $|N_2{}^1| = q + 1$, $N_2{}^1 \cap \mathbb{F}_q = \{\pm 1\}$, so $R_\sigma = |N_2{}^1 \backslash N_2{}^1 \cap \mathbb{F}_q| = q - 1$ if $q$ is odd, $R_\sigma = q$ if $q$ is even. This takes care of regular elements in the simply connected group $G_{un,\sigma}$, hence in all groups $G_\sigma$ of type ${}^1A_l$. For elements in general position we have to proceed somewhat more carefully. Let $G = G_{un}/C_d$, $C_d$ the cyclic subgroup of order $d$ of the center of $G_{un}$, $d \mid (q-1, l+1)$, and let $\pi$ denote the projection: $G_{un} \to G$. By Lemma 1, $\pi T_\sigma$ contains elements in general position if $T_\sigma$ contains elements $t$ such that $t$ and $t^d$ are regular. This means that we have to count the regular elements $(\rho^l(s^d), \ldots, \rho(s^d), s^d)$ with $s$ as above, $s^d \notin \mathbb{F}_{q^m}^*$. Hence we find that the number $P_\sigma$ of elements in general position in $\pi T_\sigma$ satisfies

$$P_\sigma \geq d^{-1}(q-1)^{-1}(q^{l+1}-1) - \sum_{m \mid l+1,\, m < l+1} (q^m - 1).$$

For $l > 1$ we find, since $d \leq q - 1$,

$$P_\sigma \geq d^{-1}\left\{(q-1)^{-1}(q^{l+1}-1) - (q-1) \sum_{m \mid l+1,\, m < l+1} (q^m - 1)\right\},$$

and the right hand side is easily seen to be positive. For $l = 1$ and $q$ even, we get $d = 1$, so there is no problem; for odd $q$, we have $d = 2$ and $P_\sigma \geq \frac{1}{2}(q-3)$, so $P_\sigma > 0$ if $q > 3$. If $q = 3$, then $\pi T_\sigma$ contains no elements in general position. But in $PGL_2(\mathbb{F}_3)$ the Coxeter torus (which properly contains $\pi T_\sigma$) can easily be shown directly to contain elements in general position.

## 7. Type ${}^1B_l$

We first consider the adjoint group $G_{\mathrm{ad}} = SO_{2l+1}$. For

$$\sigma^* = q(1, \ldots, r_1)^-(r_1 + 1, \ldots, r_1 + r_2)^- \cdots (r_1 + \cdots + r_{u-1} + 1, \ldots, l)^-,$$

the elements of $T_\sigma$ are of the form

$$t = (s_1^{q^{r_1-1}}, s_1^{q^{r_1-2}}, \ldots, s_1{}^q, s_1, s_2^{q^{r_2-1}}, \ldots, s_2{}^q, s_2, \ldots, s_u^{q^{r_u-1}}, \ldots, s_u{}^q, s_u)$$

with $s_i^{q^{r_i}+1} = 1$ for $i = 1, 2, \ldots, u$. Denote by $\rho$ the automorphism $x \mapsto x^q$ of $\bar{\mathbb{F}}_q$ over $\mathbb{F}_q$, or its restriction to any finite extension of $\mathbb{F}_q$. The Galois group of $\mathbb{F}_{q^{2r}}$ over $\mathbb{F}_{q^r}$ is generated by $\rho^r$; the corresponding Galois norm

$$N_{2r/r} = N_{\mathbb{F}_{q^{2r}}/\mathbb{F}_q}: \mathbb{F}_{q^{2r}}^* \to \mathbb{F}_{q^r}^*$$

is $N_{2r/r}(s) = s^{q^r+1}$. The kernel of $N_{2r/r}$, denoted by $N^1_{2r/r}$, is easily seen to have order $q^r + 1$. We can write the elements of $T_\sigma$ as

$$t = (\rho^{r_1-1}(s_1), \rho^{r_1-2}(s_1), \ldots, \rho(s_1), s_1, \rho^{r_2-1}(s_2), \ldots, s_2, \ldots, s_u),$$

with $s_i \in N^1_{2r_i/r_i}$. Since the roots of $G$ with respect to $T_0$ are $\pm\omega_i$, $\pm\omega_i \pm \omega_j$, $t$ is irregular if and only if $\rho^j(s_i) = 1$ or $\rho^j(s_i) = \rho^{j'}(s_{i'})^{\pm 1}$ for some $i, j, i', j'$. So $t$ is regular if and only if it satisfies the following conditions: $\rho^j(s_i) \neq s_i$ for $j = 1, 2, \ldots, 2r_i - 1$ if $r_i > 1$, $s_i \neq 1$ if $r_i = 1$, and furthermore $\rho^j(s_i) \neq \rho^{j'}(s_{i'})$ for all $i \neq i'$, $0 \le j \le 2r_i - 1$, $0 \le j' \le 2r_{i'} - 1$. The condition $\rho^j(s_i) \neq s_i$ for $s_i \in \mathbb{F}_{q^{2r_i}}$ is equivalent to

$$s_i \notin \mathbb{F}_{q^t} \qquad \text{for} \quad t\,|\,2r_i\,, \qquad t \neq 2r_i\,.$$

Thus we find the following conditions on the $s_i$ in order that $t$ be a regular element of $T_\sigma$:

(a) $s_i \in N^1_{2r_i/r_i}$, i.e., $s_i \in \mathbb{F}_{q^{2r_i}}$, $N_{2r_i/r_i}(s_i) = 1$;
(b) $s_i \notin \mathbb{F}_{q^t}$ for $t\,|\,2r_i$, $t \neq 2r_i$, i.e., $\mathbb{F}_q(s_i) = \mathbb{F}_{q^{2r_i}}$, if $r_i > 1$, $s_i \neq 1$ if $r_i = 1$;
(c) the conjugates of $s_i$ under $\Gamma_{2r_i}$ are distinct from all conjugates of $s_j$ under $\Gamma_{2r_j}$, for $i \neq j$.

We first consider conditions (a) and (b) for some $s = s_i$ and $r = r_i$. We have already seen that there are $q^r + 1$ elements $s$ in $\mathbb{F}_{q^{2r}}$ satisfying (a). Assume $s \in \mathbb{F}_{q^t}$ for $t\,|\,2r$, $t \neq 2r$. If $t\,|\,r$, then $s \in \mathbb{F}_{q^r}$, so $N_{2r/r}(s) = 1$ means $s^2 = 1$, that is, $s = \pm 1$. If $t \nmid r$, then $t = 2t_1$ with $t_1\,|\,r$, $t_1 \neq r$, $2t_1 \nmid r$. In that case,

$$|\mathbb{F}_{q^t} \backslash \mathbb{F}_{q^t} \cap \mathbb{F}_{q^r}| = q^{2t_1} - q^{t_1} = q^{t_1}(q^{t_1} - 1),$$

hence

$$\begin{aligned} \Bigl| \bigcup_{t|2r,\, t \neq 2r,\, t \nmid r} \mathbb{F}_{q^t} \backslash \mathbb{F}_{q^t} \cap \mathbb{F}_{q^r} \Bigr| &\le \sum_{t_1|r,\, t_1 \neq r,\, 2t_1 \nmid r} q^{t_1}(q^{t_1} - 1) \\ &< q^{r-1} + q^{r-2} + \cdots + q + 1 \\ &= (q^r - 1)/q - 1. \end{aligned}$$

Altogether we find that

$$\Bigl| N^1_{2r/r} \bigcup_{t|2r,\, t \neq 2r} \mathbb{F}_{q^t} \cap N^1_{2r/r} \Bigr| > (1 - (q-1)^{-1})(q^r - 1),$$

which is certainly $\geq 0$.

Now we must consider condition (c) in addition to (a) and (b). If $s_i$ and $s_j$ satisfy (a) and (b), and $r_i \neq r_j$, then $\mathbb{F}_q(s_i) = \mathbb{F}_{q^{2r_i}} \neq \mathbb{F}_{q^{2r_j}} = \mathbb{F}_q(s_j)$, so the conjugates over $\mathbb{F}_q$ of $s_i$ are certainly distinct from those of $s_j$. But if $r_i = r_j$, we have to find $s_i$ and $s_j$ in $\mathbb{F}_{q^{2r_i}}$ as above lying in distinct orbits under the

Galois group $\Gamma_{2r_i}$ of $\mathbb{F}_{q^{2r_i}}$ over $\mathbb{F}_q$. So consider all $r_i$ equal to, say, $r$. First let $r = 1$. Since $N_2{}^1 \cap \mathbb{F}_q = \{\pm 1\}$, the number of elements $s \in N_2{}^1$, $s \notin \mathbb{F}_q$, is $q - 1$ if $q$ is odd, and $q$ if $q$ is even. The $\Gamma_2$-orbits of such elements are of the form $\{s, s^{-1}\}$, hence we find $[q/2]$ possible orbits. The number of $r_i$ equal to 1 is at most $l$. The $s_i$ satisfying (a)–(c) with $r_i = 1$ can be chosen as follows: one $s_i = -1$ if $q$ is odd, the other $s_i$ in distinct $\Gamma_2$-orbits outside $\mathbb{F}_q$. This can be done provided $q \geq 2l - 1$. On the other hand, it is clear from these conditions that if all $r_i = 1$, i.e., if $w = -1$, then (c) cannot be satisfied for $q < 2l - 1$. Now let $r > 1$. Each $\Gamma_{2r}$-orbit of elements satisfying (a) and (b) consists of $2r$ elements, so the number of $r_i$ equal to $r$ must not exceed

$$\frac{1}{2r}\left| N^1_{2r/r} \setminus \bigcup_{t|2r,\, t \neq 2r} \mathbb{F}_{q^t} \cap N^1_{2r/r} \right|,$$

which number is seen above to be $> (1/2r)(1 - (q-1)^{-1})(q^r - 1)$. On the other hand, the number of cycles of length $r$ in the decomposition of $w$ is $\leq l/r$, so the appropriate number of $\Gamma_{2r}$-orbits of elements satisfying (a) and (b) can certainly be found if

$$\frac{l}{r} \leq \frac{1}{2r}(1 - (q-1)^{-1})(q^r - 1).$$

If we take $q \geq 2l - 1$ (cf. the case $r = 1$), then this is easily verified to hold. So for $q \geq 2l - 1$ every anisotropic $T_\sigma$ contains regular elements. But we recall that if all cycles in $w$ are of different length, then all $a$ satisfy. So, in particular, if $w = (1, 2, \ldots, l)^-$, a Coxeter element of the Weyt group, then $T_\sigma$ always contains regular elements.

For elements in general position in anisotropic maximal tori in the adjoint group of type $B_l$ we use the $W$-isomorphism of these tori with the corresponding ones in the simply connected group of type $C_l$ (cf. Section 3, *Type* ${}^1C_l$). We shall see in Section 8 that all anisotropic maximal tori in the simply connected ${}^1C_l$ contain elements in general position if $q \geq 2l$, no restriction on $q$ being needed if $\tau$ is a product of negative cycles of distinct lengths $r_i$.

Finally, we turn to the simply connected group $G' = \mathrm{Spin}_{2l+1}$ of type $B_l$. Let $G$ be the adjoint group and $\pi\colon G' \to G$ the projection. We have to consider the case of characteristic $\neq 2$ only. $T_\sigma{}'$ consists of the elements

$$t' = [\pm s_1^{q^{r_1-1}}, \ldots, \pm s_1^{q^2}, \pm s_1{}^q, s_1, \pm s_2^{q^{r_2-1}}, \ldots, \pm s_2{}^q, s_2, \ldots, s_u]$$

with $s^{q^{r_i}+1} = g_i = \pm 1$, $\prod_{i=1}^u g_i = 1$. This element is regular if and only if its projection $\pi t' = t$ is so. Now

$$t = (t_1^{q^{r_1-1}}, \ldots, t_1^{q^2}, t_1{}^q, t_1, t_2^{q^{r_2-1}}, \ldots, t_2{}^q, t_2, \ldots, t_u)$$

with $t_i = s_i^2$ for $i = 1, \ldots, u$. This is a regular element of $\pi T_\sigma' \subset T_\sigma$ if and only if the $t_i$ satisfy

(a′) $t_i \in N^1_{2r_i/r_i}$, and $t_i \in (N^1_{2r_i/r_i})^2$ for all but an even number of indices $i$,

and, moreover, conditions (b) and (c) mentioned earlier in this section. To simplify computations we strengthen condition (a′) to

(a″) $t_i \in N_2{}^1$ if $r_i = 1$, $\prod_{r_i=1} t_i \in (N_2{}^1)^2$, $t_i \in (N^1_{2r_i/r_i})^2$ if $r_i > 1$.

Now it is a straightforward computation that elements $t_i$ for $r_i = 1$ satisfying conditions (a″), (b), (c) can be found if $q \geq 2l - 1$ for $l \equiv 0$ or 3 mod 4, $q \geq 2l$ if $l \equiv 1$ or 2 mod 4; these are the precise limits if $\sigma^* = -q$, i.e., if all $r_i = 1$. For $r_i = r > 1$, the argument is similar to that given above in the adjoint case. The order of $(N^1_{2r/r})^2$ is $\frac{1}{2}(q^r + 1)$; subtracting from this an estimate for the number of elements in $\bigcup_{t|2r,\, t \neq 2r} \mathbb{F}_{q^t} \cap N^1_{2r/r}$, one finds

$$\left| (N^1_{2r/r})^2 \Big\backslash \bigcup_{t|2r,\, t \neq 2r} \mathbb{F}_{q^t} \cap N^1_{2r/r} \right| > \left(\tfrac{1}{2} - (q-1)^{-1}\right)(q^r - 3).$$

The right hand side is certainly $\geq 0$ for $q \geq 2l - 1 \geq 3$. If there is more than one $r_i = r$, we get the condition

$$2l \leq \left(\tfrac{1}{2} - (q-1)^{-1}\right)(q^r - 3),$$

which is certainly satisfied for $q \geq 2l - 1$, $l \geq 3$, for then

$$2l \leq \tfrac{1}{4}((2l-1)^2 - 3) \leq \left(\tfrac{1}{2} - (q-1)^{-1}\right)(q^r - 3).$$

## 8. Type ${}^1C_l$

The proof for simply connected groups of type ${}^1C_l$ is almost the same as for the adjoint ${}^1B_l$, except that now condition (b) has to read: $\mathbb{F}_q(s_i) = \mathbb{F}_{q^{2r_i}}$ for *all* $r_i$ (since the roots $\omega_i$ of $B_l$ are replaced by $2\omega_i$ of $C_l$). Thus one gets that all anisotropic maximal tori $T_\sigma$ contain regular elements provided $q \geq 2l$. Actually, $q \geq 2l$ is needed only for $T_\sigma$ with $\sigma^* = -q$; for all other $T_\sigma$, $q \geq 2l - 1$ suffices. For regular elements in the adjoint group, $q \geq 2l - 1$ ensures their existence in all anisotropic $T_\sigma$. If $\sigma^* \neq -q$, this follows from what we just remarked in the simply connected case. If $\sigma^* = -q$, one gets this bound by direct computation in the projective group $PSp_{2l}$.

Finally, to find elements in general position in the adjoint group $G$ we use the simply connected group $G'$ and the projection $\pi\colon G' \to G$ and its restriction $T' \to T$. If $t = (t_1, \ldots, t_l)$ is an element of $T'$, then $\pi t \in T_\sigma$ if and only if

$\sigma t = \pm t$, since $\ker(\pi) = \{\pm 1\}$. For $\sigma t = t$ we find the usual form as before, whereas $\sigma t = -t$ means that

$$t = ((-1)^{r_1 - 1} s_1^{q^{r_1 - 1}}, \ldots, s_1^{q^2}, -s_1{}^q, s_1, (-1)^{r_2 - 1} s_2^{q^{r_2 - 1}}, \ldots, -s_2{}^q, s_2, \ldots, s_u)$$

with $s_i^{q^{r_i}+1} = (-1)^{r_i}$. An element $\pi t$ with $t = (t_1, \ldots, t_l)$ is in general position if and only if $wt \neq \pm t$ for all $w \neq 1$, $w \in W$. With counting arguments like those used before, one finds that such a $\pi t \in T_\sigma$ exists if $q \geq 2l$, which bound is sharp for $\sigma^* = -q$. If all $r_i$ are distinct from each other, no condition on $q$ is necessary.

## 9. Type $^1D_l$

For $G = SO_{2l}$ and $\sigma^* = q\tau$,

$$\tau = (1, \ldots, r_1)^- (r_1 + 1, \ldots, r_1 + r_2)^- \cdots (r_1 + \cdots + r_{u-1} + 1, \ldots, l)^-$$

a product of an *even* number $u$ of negative cycles we proceed again as in the case of $B_l$ (see Section 7). Condition (b) now has to read: "$\mathbb{F}_q(s_i) = \mathbb{F}_{q^{2r_i}}$ for $r_i > 1$" (so there is no condition for $r_i = 1$). If $l$ is even, $\tau$ can be a product of $l$ negative cycles of length 1; but if $l$ is odd, there can be at most $l - 2$ negative cycles of length 1 in $\tau$. This causes a difference between the cases $l$ even and $l$ odd, respectively. The rest of the argument is similar as for $B_l$. This takes care of regular elements in $SO_{2l}$. For elements in general position in anisotropic $T_\sigma$ in $SO_{2l}$, we use the inclusion $SO_{2l} \subseteq SO_{2l+1}$ (cf. Section 3, last paragraph of *Type* $^1D_l$). If $l$ is even and $\sigma^* \neq -q$, this yields that $q \geq 2l - 2$ is certainly sufficient; for $\sigma^* = -q$ a direct verification in $SO_{2l}$ shows that $t \in T_\sigma$ with $wt \neq t$ for all $w \in W$, $w \neq 1$, exist precisely if $q \geq 2l - 2$. In the same way one gets $q \geq 2l - 6$ for odd $l$.

The simply connected groups $\mathrm{Spin}_{2l}$ of type $D_l$ are dealt with in the same way as the groups $\mathrm{Spin}_{2l+1}$ of type $B_l$; only for $l = 4$ does some extra work have to be done.

The results about elements in general position in the adjoint groups of type $^1D_l$ are derived by projecting down from $SO_{2l}$ in the same way as we did for the adjoint $^1C_l$. The existence of regular elements in adjoint groups is also derived by projection down from $SO_{2l}$.

Finally, for even $l$, we have to deal with the half-spin groups. Let $G$ be such a group, $G'$ the corresponding spin group $\mathrm{Spin}_{2l}$, $\pi$: $G' \to G$ the projection; $\ker(\pi) = \langle\sqrt{-1}\rangle$. Regular elements in $G'$ project on regular elements in $G$. That the bounds obtained in this way are sharp in the case of $\sigma^* = -q$ is verified by direct computation; then $T_\sigma$ consists of the elements $\pi t$ with $\sigma t = t$ or $\sqrt{-1}\,t$. For elements in general position in $T_\sigma$, we have to take $\pi t$ with $t \in T'$, $\sigma t = t$ or $\sqrt{-1}\,t$, and $wt \neq t$ or $\sqrt{-1}\,t$ for $w \neq 1$, $w \in W$.

## 10. Type $^2A_l$

Consider $\sigma$ with $\sigma^* = q\tau$,

$$\tau = -(1, 2, \ldots, r_1)(r_1 + 1, \ldots, r_1 + r_2) \cdots (r_1 + \cdots + r_{u-1} + 1, \ldots, l + 1),$$

where all $r_i$ are odd ($r_u = l + 1 - \sum_{j=1}^{u-1} r_j$). The elements of $T_\sigma$ are of the form

$$t = (s_1^{q^{r_1-1}}, s_1^{-q^{r_1-2}}, \ldots, s_1^{q^2}, s_1^{-q}, s_1, s_2^{q^{r_2-1}}, \ldots, s_2^{-q}, s_2, \ldots, s^{q^{r_u-1}}, \ldots, s_u^{-q}, s_u),$$

with

(i) $s^{q^{ri}+1} = 1$ for $i = 1, \ldots, u$,
(ii) $\prod_{i=1}^{u} s_i^{1-q+q^2-\cdots+q^{ri-1}} = 1$.

Let $\Gamma_{2r_i} = \langle \rho_i \rangle$ be the Galois group of $\mathbb{F}_{q^{2r_i}}/\mathbb{F}_q$, with $\rho_i$: $x \mapsto x^q$ for $x \in \mathbb{F}_{q^{2r_i}}$. By (i), $\rho_i^{r_i}(s_i) = s_i^{-1}$, so

$$t = (\rho_1^{r_1-1}(s_1), \rho_1^{2r_1-2}(s_1), \ldots, \rho_1{}^2(s_1), \rho_1^{r_1+1}(s_1), s_1, \ldots, s_u),$$

i.e., the coordinates of $t$ are the $\rho_i^{2j}(s_i)$ with $j = 0, 1, \ldots, r_i - 1$, since all $r_i$ are odd. The condition that $t^{\omega_i - \omega_j} \neq 1$ for all roots $\omega_i - \omega_j$ in order that $t$ be regular then leads to the following set of conditions on the $s_i$:

(a) (i) $s_i \in \mathbb{E}_{q^{2ri}}$, $s_i^{q^{ri}+1} = 1$; (ii) $\prod_{i=1}^{u} s_i^{(q^{ri}+1)/q+1} = 1$;
(b) $\mathbb{F}_{q^2}(s_i) = \mathbb{F}_{q^{2r_i}}$ for $i = 1, \ldots, u$;
(c) the conjugates of $s_i$ over $\mathbb{F}_{q^2}$ are distinct from those of $s_j$ for $i \neq j$.

Let $v$ be the number of $r_i$ which are equal to 1. If $v$ is odd, we take for the corresponding $s_i$: 1 and $\frac{1}{2}(v - 1)$ pairs $\{s, s^{-1}\}$ with $s^{q+1} = 1$, $s \neq \pm 1$; if $v$ is even, we drop the 1. Then these $s_i$ satisfy (a)(i), (b), (c), and the condition

$$\prod_{r_i = 1} s_i = 1,$$

which will partly take care of (a)(ii). For $r_i > 1$ we choose for the $s_i$ primitive $[(q^{r_i} + 1)/q + 1]$th roots of unity in $\mathbb{F}_{q^{2r_i}}$. If certain $r_i$ are equal for distinct $r_i$ we have to do this so as to satisfy condition (c). This leads again to certain conditions on $q$ as in previous cases. In the course of the computations for condition (c) one needs for the Euler function $\phi$ (defined by $\phi(n)$ = number of $i$, $1 \leq i < n$, $(i, n) = 1$) the inequality $\phi(n) \geq 2\sqrt{n}$ for odd $n$, which is easily proved. No condition on $q$ is needed when $r_i \neq r_j$ for all $i \neq j$, except when some $r_i = 3$ and $q = 2$. It actually turns out that for $l = 2$, $q = 2$, and $\tau = -(1, 2, 3)$, the twisted Coxeter torus $T_\sigma$ contains no regular elements in the simply connected group $G_{un}$. But in the corresponding $G_{\mathrm{ad}} = PGL_3(\mathbb{F}^2)$

one can show that the twisted Coxeter torus $T_\sigma$ (with $\sigma$ as above) does contain regular elements. For elements in general position in anisotropic maximal tori in $G_\sigma/C_d$, $d\,|\,(q+1, l+1)$, $d \neq 1$, one has to set up an argument similar to that used for ${}^1A_l$. This leads to somewhat long computations. For all connectedness types of ${}^2A_2(q^2)$ the torus $T_\sigma$ with $\sigma^* = -q$ contains regular elements for all $q$, and elements in general position for $q > 2$. For the adjoint ${}^2A_2(2^2)$ it can be shown by direct computation that the torus $T_\sigma$ with $\tau = -1$ contains no elements in general position, nor does the twisted Coxeter torus ($\tau = -(1, 2, 3)$). So the adjoint ${}^2A_2(2^2)$ contains no elements in general position at all.

## 11. Type ${}^2D_l$

Now $\tau$ is a product of an odd number of negative cycles, hence the arguments are the same as for ${}^1D_l$ provided the cases $l$ even and $l$ odd are interchanged. Since $\sigma_0$ interchanges the kernels of the two half-spin groups, there are no ${}^2D_l$-forms of these groups, so we only have to deal with $G_{un}$, $SO_{2l}$ and $G_{\text{ad}}$.

## 12. The Suzuki groups ${}^2C_2(q^2)$

Let $G$ be of type $C_2$ over the field $\mathbb{F}_2$, $T_0$ a maximal torus, $X = \langle \omega_1, \omega_2 \rangle$ its character group, and $\pm 2\omega_i$, $\pm\omega_1 \pm \omega_2$ the roots. Consider the endomorphism $\sigma_0$ with $\sigma_0{}^* = q\tau_0$, where $q = 2^{n+\frac{1}{2}}$, $n = 0, 1, 2, \ldots$, and $\tau_0$ the reflection in the bissectrix of $\omega_1$ and $\omega_1 + \omega_2$, i.e., $\tau_0\alpha_1 = 2^{-1/2}\alpha_2$, $\tau_0\alpha_2 = 2^{1/2}\alpha_1$ for the simple roots $\alpha_1 = \omega_1 - \omega_2$, $\alpha_2 = 2\omega_2$. Representatives of the conjugacy classes in $W\sigma_0{}^*$ under the action of $W$ are the following $\sigma_i{}^* = q\tau_i$ (matrices with respect to the basis $\omega_1, \omega_2$, see [11, p. 307]):

$$\sigma_0{}^* = 2^n\begin{pmatrix} 1 & 1 \\ 1 & -1 \end{pmatrix}, \qquad \sigma_1{}^* = 2^n\begin{pmatrix} 1 & -1 \\ 1 & 1 \end{pmatrix}, \qquad \sigma_2{}^* = 2^n\begin{pmatrix} -1 & -1 \\ 1 & -1 \end{pmatrix}.$$

Only $\sigma_1{}^*$ and $\sigma_2{}^*$ yield anisotropic tori $T_\sigma$, as one sees by verifying that these $\tau_i$ have no eigenvalue 1 whereas $\tau_0$ has. $T_{\sigma_1}$ consists of the elements $t = (s^{q\cdot 2^{1/2}-1}, s)$ with $s^{q^2 - q\cdot 2^{1/2}+1} = 1$. All elements $t \neq 1$ turn out to be regular, so the number of regular elements is $q^2 - q\sqrt{2}$, which is $> 0$ for all $q > \sqrt{2}$ (i.e., $n \geq 1$), whereas for $q = \sqrt{2}$ there are no regular elements in $T_{\sigma_1}$.

$T_{\sigma_2}$ consists of the elements $t = (s, s^{-q\cdot 2^{1/2}-1})$ with $s^{q^2 + q\cdot 2^{1/2}+1} = 1$. All $t \neq 1$ are regular, so there are $q^2 + q\sqrt{2}$ regular elements, which is $> 0$ for all $q$.

In the same way it is easily verified that in $T_{\sigma_0}$, which is not anisotropic, all elements $\neq 1$ are regular, which means that there are regular elements for $q > \sqrt{2}$, but none for $q = \sqrt{2}$.

Since there is only one connectedness type for $G$ in characteristic 2, regular elements are in general position. Thus we have proved the following proposition.

***Proposition 1*** *In the Suzuki group* $^2C_2(q^2)$, $q = 2^{n+\frac{1}{2}}$ $(n \geq 0)$, *there are two anisotropic tori (up to conjugacy), viz.,* $T_{\sigma_1}$ *and* $T_{\sigma_2}$ *as given above. In both tori, all elements* $\neq 1$ *are in general position.* $T_{\sigma_1}$ *contains elements in general position for* $q > 2^{1/2}$, $T_{\sigma_2}$ *for all* $q$.

## 13. Groups of type $^3D_4$

Let $G_{un}$ be simply connected of type $D_4$, $T_0$ a maximal torus in $G_{un}$, and $\sigma_0$ the triality endomorphism with $\sigma_0 = q\tau_0$, $q$ a power of the characteristic $p$ of the ground field, and $\tau_0$ the diagram automorphism: $\alpha_1 \mapsto \alpha_3 \mapsto \alpha_4 \mapsto \alpha_1$, $\alpha_2 \mapsto \alpha_2$ ($\alpha_i$ as in Section 3, $D_l$ for $l = 4$). The character group $X$ of $T_0$ is generated by $\mu_1 = \omega_1$, $\mu_2 = \omega_2$, $\mu_3 = \omega_3$, $\mu_4 = \frac{1}{2}(\omega_1 + \omega_2 + \omega_3 + \omega_4)$ ($\omega_i$ as in Section 3). With respect to this basis the roots (up to sign) are

$$\mu_i \pm \mu_j \qquad (1 \leq i < j \leq 3),$$

$$-2\mu_4 + \mu_i + \mu_j \qquad (1 \leq i < j \leq 3),$$

$$-2\mu_4 + 2\mu_i + \mu_{i+1} + \mu_{i+2} \qquad (1 \leq i \leq 3, \text{ indices mod } 3, \text{ equal to } 1, 2, \text{ or } 3).$$

If $q$ is odd, the elements $\neq 1$ of the center $C$ of $G$ ($C \cong \mathbb{Z}_2 \times \mathbb{Z}_2$) are permuted cyclically by $\sigma_0$, so $SO_8$ and the two half-spin groups have no endomorphism corresponding to $\sigma_0$. In fact, $\sigma_0$ induces isomorphisms between these three groups. Furthermore, $C_\sigma = 1$, so the projection of $G_{un,\sigma}$ on $G_{\text{ad},\sigma}$ is an isomorphism (of abstract groups). Hence it suffices to consider $G_{un,\sigma}$, and there regular is equivalent to being in general position for elements of maximal tori. For even $q$, $C = 1$, so the above conclusions hold a fortiori.

The conjugacy classes of $W\tau_0$ under the action of $W$ can be computed in the same way as we did this for the Ree group $^2F_4$ in [11, §8]. Our results agree with those of Gager [4], who computed them in a different way. In Table 3 we list for each class a representative $\tau_i$, its order $|\tau_i|$, the number of conjugates under $W$ ($= |W|$ divided by $|W_{\tau_i}|$, where $W_{\tau_i}$ is the centralizer of $\tau_i$ in $W$), its characteristic polynomial $\chi_{\tau_i}$, and the elementary divisors of the $\mathbb{Z}$-matrix $q\tau_i - 1$. In the last column we indicate whether the corresponding maximal torus $T_{\sigma_i}$, $\sigma_i{}^* = q\tau_i$, is anisotropic or not, i.e., whether or not $\tau_i$ has an eigenvalue 1. The centralizer $W_{\tau_i}$ has been determined by using the eigenvectors and eigenvalues of $\tau_i$ (cf. [11, §8]). All matrices are with respect to the basis $\mu_1, \mu_2, \mu_3, \mu_4$ of $X$.

*Table 3*

*The Conjugacy Classes of $W\tau_0$ Under $W$ for ${}^3D_4$*

| $\tau_i$ | $\lvert\tau_i\rvert$ | Number of conjugates | $\chi_{\tau_i}$ | Elementary divisors of $q\tau_i - 1$ | $T_\sigma$ anisotropic? |
|---|---|---|---|---|---|
| $\tau_0 = \begin{bmatrix} 1 & 0 & 0 & 1 \\ 1 & 0 & -1 & 0 \\ 1 & -1 & 0 & 0 \\ -1 & 1 & 1 & 0 \end{bmatrix}$ | 3 | 16 | $(X^3 - 1)(X - 1)$ | $q^3 - 1; q - 1$ | No |
| $\tau_1 = \begin{bmatrix} -1 & 0 & 0 & -1 \\ -1 & 0 & 1 & 0 \\ -1 & 1 & 0 & 0 \\ 1 & -1 & -1 & 0 \end{bmatrix} = -\tau_0$ | 6 | 16 | $(X^3 + 1)(X + 1)$ | $q^3 + 1; q + 1$ | Anisotropic |
| $\tau_2 = \begin{bmatrix} 1 & 0 & -1 & 0 \\ 1 & 0 & 0 & 1 \\ 1 & -1 & 0 & 0 \\ -1 & 1 & 1 & 0 \end{bmatrix}$ | 6 | 48 | $(X^3 + 1)(X - 1)$ | $(q^3 + 1)(q - 1)$ | No |
| $\tau_3 = \begin{bmatrix} -1 & 0 & 1 & 0 \\ -1 & 0 & 0 & -1 \\ -1 & 1 & 0 & 0 \\ 1 & -1 & -1 & 0 \end{bmatrix} = -\tau_2$ | 6 | 48 | $(X^3 - 1)(X + 1)$ | $(q^3 - 1)(q + 1)$ | No |
| $\tau_4 = \begin{bmatrix} 1 & -1 & 0 & 0 \\ 1 & 0 & 0 & 1 \\ 1 & 0 & -1 & 0 \\ -1 & 1 & 1 & 0 \end{bmatrix}$ | 12 | 48 | $X^4 - X^2 + 1$ | $q^4 - q^2 + 1$ | Anisotropic: twisted Coxeter element |
| $\tau_5 = \begin{bmatrix} 0 & 0 & -1 & 0 \\ 0 & -1 & -1 & -1 \\ 1 & 0 & -1 & 0 \\ -1 & 1 & 1 & 0 \end{bmatrix}$ | 3 | 8 | $(X^2 + X + 1)^2$ | $q^2 + q + 1$ $(2\times)$ | Anisotropic |
| $\tau_6 = \begin{bmatrix} 0 & 0 & 1 & 0 \\ 0 & 1 & 1 & 1 \\ -1 & 0 & 1 & 0 \\ 1 & -1 & -1 & 0 \end{bmatrix} = -\tau_5$ | 3 | 8 | $(X^2 - X + 1)^2$ | $q^2 - q + 1$ $(2\times)$ | Anisotropic |

Regarding regular elements in anisotropic maximal tori we have the following result.

***Proposition 2*** *In a group of type ${}^3D_4(q^3)$ the anisotropic maximal tori $T_{\sigma_i}$, $i = 1, 4, 5, 6$, all contain regular elements (= elements in general position), except when $i = 1$ or $6$ and $q = 2$. In the twisted Coxeter torus $T_{\sigma_4}$ all elements $\neq 1$ are regular.*

*Proof* (i) $\sigma_1{}^* = q\tau_1$. The elements of $T_{\sigma_1}$ are

$$t = (\zeta^i, \zeta^{-i+(q^2-q+1)j}, \zeta^{(q^2-q)i-(q^2-q+1)j}, \zeta^{-qi}),$$

where $\zeta$ is a primitive $(q^3+1)$th root of unity in $\mathbb{F}_{q^6}$, $0 \le i < q^3+1$, $0 \le j < q+1$. For $\alpha$ a root, $t^\alpha = \zeta^{l(\alpha)}$, where $l(\alpha)$ is a homogeneous linear form in $i$ and $j$ whose coefficients are polynomials in $q$. For the roots $\alpha$ as given above one can easily compute the $l(\alpha)$; e.g., $l(\mu_1+\mu_2) = (q^2-q+1)j$, $l(\mu_1-\mu_2) = 2i-(q^2-q+1)j$. To find a regular $t$ one has to choose $i$ and $j$ such that all $l(\alpha) \not\equiv 0 \bmod q^3+1$. For $q>2$, $i=1$ and $j=2$ does the job, so $t = (\zeta, \zeta^{2q^2-2q+1}, \zeta^{-q^2+q-2}, \zeta^{-q})$ is regular. For $q=2$ a straightforward verification shows that no such $i$ and $j$ exist, so no regular elements.

(ii) $\sigma_4{}^* = q\tau_4$ (twisted Coxeter element). The elements of $T_{\sigma_4}$ are

$$t = (\zeta^{q^3i}, \zeta^i, \zeta^{(-q^3+q^2+q-1)i}, \zeta^{qi}),$$

with $\zeta$ a primitive $(q^4-q^2+1)$th root of unity in $\mathbb{F}_{q^{12}}$, and $0 \le i < q^4-q^2+1$. For $\alpha$ a root, $t^\alpha = \zeta^{p(\alpha,q)i}$, where $p(\alpha, q)$ is a polynomial in $q$ for each $\alpha$. All $p(\alpha, q)$ are easily seen to have g.c.d. 1 with $q^4-q^2+1$ for all values of $q$. For instance, if $\alpha = \mu_1+\mu_2$, then $p(\alpha, q) = q^3+1$, and

$$\begin{aligned}(q^3+1, q^4-q^2+1) &= (q^3+1, -q^2-q+1)\\ &= (-q^2+q+1, -q^2-q+1)\\ &= (2q, -q^2-q+1) = 1,\end{aligned}$$

by the Euclidean algorithm. This implies that $t^\alpha \ne 1$ for all $t \ne 1$, i.e., all $t \ne 1$ are regular. The number of such $t$ is $q^4-q^2>0$ for all $q$.

(iii) $\sigma_5{}^* = q\tau_5$. Here we count the number of regular elements. This has the advantage that for $\sigma_6{}^* = -q\tau_5$ we only have to replace $q$ by $-q$. The elements of $T_{\sigma_5}$ are $t = (\varepsilon^{-q-1}\eta^q, \varepsilon, \eta, \varepsilon^{-q})$ with $\varepsilon, \eta \in \mathbb{F}_{q^3}$, $\varepsilon^{q^2+q+1} = \eta^{q^2+q+1} = 1$. We now can write down the equations $t^\alpha = 1$ for all roots $\alpha$ as given above, and count the number of solutions, counting common solutions of distinct equations precisely once, of course. For instance,

$$\alpha = \mu_1+\mu_2, \qquad t^\alpha = \varepsilon^{-q}\eta^q = 1,$$

hence $\varepsilon = \eta$, $t = (\varepsilon^{-1}, \varepsilon, \varepsilon, \varepsilon^{-q})$. This yields $q^2+q+1$ irregular elements.

$$\alpha = \mu_1-\mu_2, \qquad t^\alpha = \varepsilon^{-q-2}\eta^q = 1.$$

Taking the $(q+1)$th power, we get $\eta = \varepsilon^{-2q-1}$, since $\eta^{q^2+q} = \eta^{-1}$ and $\varepsilon^{q^2+q+1} = 1$. So $t = (\varepsilon, \varepsilon, \varepsilon^{-2q-1}, \varepsilon^{-q})$. This coincides with a solution of the previous equation (for $\alpha = \mu_1+\mu_2$) if and only if $\varepsilon = \varepsilon^{-1}$, so $\varepsilon = \pm 1$. Thus we find $q^2+q-1$ new irregular elements if $q$ is odd, and $q^2+q$ ones if $q$ is even. Proceeding in this way through the whole list of roots $\alpha$, we finally find for the number of regular elements in $T_\sigma$:

$$R_\sigma = q^4 + 2q^3 - q^2 - 2q + 2 \qquad \text{if } q \text{ is odd,}$$

$$R_\sigma = q^4 + 2q^3 - q^2 - 2q \qquad \text{if } q \text{ is even.}$$

Both expressions are $> 0$ for all relevant $q$.

(iv) $\sigma_6{}^* = q\tau_6 = -q\tau_5$. Replacing $q$ by $-q$ in the above expressions, one finds

$$R_\sigma = q^4 - 2q^3 - q^2 + 2q + 2 \qquad \text{if } q \text{ is odd,}$$

$$R_\sigma = q^4 - 2q^3 - q^2 + 2q \qquad \text{if } q \text{ is even.}$$

Hence $R_\sigma > 0$ for $q > 2$, $R_\sigma = 0$ for $q = 2$.

## 14. Conjugacy classes in the Weyl group of $F_4$

For groups of type ${}^1F_4$ we need a good description of the conjugacy classes in the Weyl group $W(F_4)$. We shall use the description given by Gager in his thesis [4]; the author is very indebted to Dr. Gager for showing his results prior to publication.

***Lemma 2*** *Let $W$ be a Weyl group acting in the linear space $V$ spanned by the roots, $\tau$ a bijective linear transformation in $V$ of finite order which normalizes $W$, and $W^* = \langle W, \tau \rangle$, the group generated by $W$ and $\tau$.*

(i) *$W\tau$ is a union of conjugacy classes of $W^*$.*

(ii) *If two elements in $W\tau$ are conjugate under $W^*$, then so they are under $W$. That is, the $W^*$-conjugacy classes coincide with the $W$-conjugacy classes in $W\tau$.*

*Proof* (i) Immediate from the fact that $\tau$ normalizes $W$.

(ii) For $w \in W$, $\tau(w\tau)\tau^{-1} = w^{-1}(w\tau)w$.

So conjugacy in $W\tau$ under $W\tau^i$ implies conjugacy under $W$.

Let $V$ be the real linear space spanned by the root system of $F_4$, that is, $\Sigma(F_4)$. In $V$ we have the inclusions of root systems

$$\Sigma(F_4) \supset \Sigma(B_4) \supset \Sigma(D_4).$$

As a basis for $V$ we take $\omega_1, \omega_2, \omega_3, \omega_4$. The Weyl group $W(B_4)$ consists of the transformations

$$\omega_i \mapsto e_i\omega_{\pi i}, \qquad \pi \in S_4, \quad e_i = \pm 1,$$

and $W(D_4)$ of those with $\prod_{i=1}^4 e_i = 1$. Hence $W(B_4)$ is generated by $W(D_4)$ and the diagram automorphism $\rho$ of $\Sigma(D_4)$ of order 2 given by $\rho\omega_i = \omega_i$ for

$i = 1, 2, 3$, $\rho\omega_4 = -\omega_4$, which normalizes $W(D_4)$. Let $\tau_0$ be the diagram automorphism of $\Sigma(D_4)$ of order 3 we used in the previous section. It normalizes $W(D_4)$ and $\rho\tau_0\rho^{-1} = \tau_0^2$, hence $\rho$ normalizes $\langle W(D_4), \tau_0\rangle$. Coset representatives for $W(F_4)/W(D_4)$ are

$$1, \quad \tau_0, \quad \tau_0^2, \quad \rho, \quad \tau_0\rho, \quad \tau_0^2\rho.$$

By applying Lemma 2 repeatedly, we find the conjugacy classes in $W(F_4)$:

(a) $C \cup \tau_0 C\tau_0^{-1} \cup \tau_0^2 C\tau_0^{-2}$, with $C$ being a conjugacy class of $W(B_4)$ contained in $W(D_4)$. The union is not necessarily disjoint, but the whole $W(F_4)$-class is contained in $W(D_4)$;

(b) $C \dot{\cup} \rho C\rho^{-1}$, where $C$ is a $W(D_4)$-conjugacy class in $W(D_4)\tau_0$;

***Table 4***

*The Conjugacy Classes of $W(F_4)$*

| | Symbol | $\tau$ | $\chi_\tau$ | Anisotropic $T_\sigma$? |
|---|---|---|---|---|
| (a) | $\phi$ | $1^+ 1^+ 1^+ 1^+$ | $(x-1)^4$ | No |
| | $2A_1$ | $1^+ 1^+ 1^- 1^-$ or $2^+ 2^+$ | $(x-1)^2(x+1)^2$ | No |
| | $4A_1$ | $1^- 1^- 1^- 1^-$ | $(x+1)^4$ | Anisotropic |
| | $A_1$ | $2^+ 1^+ 1^+$ | $(x^2-1)(x-1)^2$ | No |
| | $3A_1$ | $2^+ 1^- 1^-$ | $(x^2-1)(x+1)^2$ | No |
| | $A_3$ | $2^- 1^- 1^+$ or $4^+$ | $(x^2+1)(x+1)(x-1)$ | No |
| | $A_2$ | $3^+ 1^+$ | $(x^3-1)(x-1)$ | No |
| | $D_4$ | $3^- 1^-$ | $(x^3+1)(x+1)$ | Anisotropic |
| | $D_4(a_1)$ | $2^- 2^-$ | $(x^2+1)^2$ | Anisotropic |
| (b) | $\tilde{A}_2$ | $\tau_0$ | $(x^3-1)(x-1)$ | No |
| | $C_3 + A_1$ | $\tau_1$ | $(x^3+1)(x+1)$ | Anisotropic |
| | $C_3$ | $\tau_2$ | $(x^3+1)(x-1)$ | No |
| | $\tilde{A}_2 + A_1$ | $\tau_3$ | $(x^3-1)(x+1)$ | No |
| | $F_4$ | $\tau_4$ | $x^4 - x^2 + 1$ | Anisotropic (Coxeter torus) |
| | $A_2 + \tilde{A}_2$ | $\tau_5$ | $(x^2+x+1)^2$ | Anisotropic |
| | $F_4(a_1)$ | $\tau_6$ | $(x^2-x+1)^2$ | Anisotropic |
| (c) | $\tilde{A}_1$ | $1^- 1^+ 1^+ 1^+$ | $(x+1)(x-1)^3$ | No |
| | $2A_1 + \tilde{A}_1$ | $1^- 1^- 1^- 1^+$ | $(x+1)^3(x-1)$ | No |
| | $A_1 + \tilde{A}_1$ | $2^+ 1^+ 1^-$ | $(x-1)^2(x+1)^2$ | No |
| | $B_2$ | $2^- 1^+ 1^+$ | $(x^2+1)(x-1)^2$ | No |
| | $A_3 + \tilde{A}_1$ | $2^- 1^- 1^-$ | $(x^2+1)(x+1)^2$ | Anisotropic |
| | $A_2 + \tilde{A}_1$ | $3^+ 1^-$ | $(x^3-1)(x+1)$ | No |
| | $B_3$ | $3^- 1^+$ | $(x^3+1)(x-1)$ | No |
| | $B_4$ | $4^-$ | $x^4+1$ | Anisotropic |
| | $B_2 + A_1$ | $2^+ 2^-$ | $(x^2-1)(x^2+1)$ | No |

(c) $C \dot{\cup} \tau_0 C \tau_0^{-1} \dot{\cup} \tau_0{}^2 C \tau_0^{-2}$, where $C$ is a $W(D_4)$-conjugacy class in $W(D_4)\rho$, hence $C$ is also a $W(B_4)$-conjugacy class.

For the conjugacy classes $C$ as in (b) we use Table 3. The classes $C$ in (a) and (c), which are $W(B_4)$-conjugacy classes, are denoted by their signed cycle type. We list a representative $\tau$ of each conjugacy class in Table 4, together with the characteristic polynomials $\chi_\tau$ and whether or not it gives rise to an anisotropic torus (cf. also [4]).

## 15. Groups of type ${}^1F_4$

For these groups we have the following result.

***Proposition 3*** *For groups of type ${}^1F_4(q)$ the following is an exhaustive list of the values of $q$ for which each anisotropic maximal torus contains regular (= in general position) elements (so for the other values of $q$ there are no regular elements):*

| | | | | | |
|---|---|---|---|---|---|
| $4A_1$ | $q > 9,$ | $C_3 + A_1$ | $q > 2,$ | $F_4(a_1)$ | $q > 3,$ |
| $D_4$ | $q > 2,$ | $F_4(\text{Coxeter})$ | all $q$, | $A_3 + \tilde{A}_1$ | $q > 2,$ |
| $D_4(a_1)$ | $q > 3,$ | $A_2 + \tilde{A}_2$ | $q > 2,$ | $B_4$ | all $q$. |

*In a Coxeter torus ($F_4$) only 1 is irregular, in a torus of type $B_4$ only 1 and $-1$ are irregular if $q$ is odd, and only 1 if $q$ is even.*

*Proof* As a basis for the character group $X$ of a maximal torus, we choose $\mu_1 = \omega_1$, $\mu_2 = \omega_2$, $\mu_3 = \omega_3$, $\mu_4 = \frac{1}{2}(\omega_1 + \omega_2 + \omega_3 + \omega_4)$. Then the roots (up to signs) are

$$\begin{array}{ll}
\mu_i & (1 \le i \le 4) \\
\mu_i \pm \mu_j & (1 \le i < j \le 3) \\
\mu_4 - \mu_i & (1 \le i \le 3) \\
\mu_4 - \mu_i - \mu_j & (1 \le i < j \le 3) \\
\mu_4 - \mu_1 - \mu_2 - \mu_3 & \\
2\mu_4 - \mu_i - \mu_j & (1 \le i < j \le 3) \\
2\mu_4 - 2\mu_i - \mu_{i+1} - \mu_{i+2} & (1 \le i \le 3, \text{ coefficients mod } 3) \\
2\mu_4 - \mu_1 - \mu_2 - \mu_3 . &
\end{array}$$

The elements of a maximal torus are written as $t = (t_1, t_2, t_3, t_4)$ with $t^{\mu_i} = t_i$.

(i) $4A_1$. $\sigma^* = -q$. Elements of $T_\sigma$: $(\varepsilon^i, \varepsilon^j, \varepsilon^k, \varepsilon^l)$, $\varepsilon$ a primitive $(q+1)$th root of unity in $\mathbb{F}_{q^2}$. For $q \geq 11$, the element $(\varepsilon, \varepsilon^2, \varepsilon^3, \varepsilon^7)$ is regular. If $q < 8$, the torus $T_\sigma$ contains no regular elements in ${}^1B_4(q)$, in which it is contained, hence a fortiori not in ${}^1F_4(q)$. For $q = 8$ and 9 a straightforward verification shows the absence of regular elements.

(ii) $D_4$. $\sigma^* = q\tau$, $\tau = (1, 2, 3)^-(4)^-$ (with respect to the basis $\omega_1, \omega_2, \omega_3, \omega_4$ of $X \otimes \mathbb{R}$). Elements of $T_\sigma$: $t = (\zeta^i, \zeta^{-q^2i}, \zeta^{-qi}, \zeta^{-q^2i+(q^2-q+1)j})$, where $\zeta$ is a primitive $(q^3+1)$th root of unity in $\mathbb{F}_{q^6}$, and $0 \leq i < q^3 + 1$, $0 \leq j < q + 1$. For $\alpha$ a root, $t^\alpha = \zeta^{l(\alpha)}$, where $l(\alpha)$ is a homogeneous linear form in $i$ and $j$ whose coefficients are polynomials in $q$. For $i = 1, j = 2$, none of these $l(\alpha) \equiv 0 \bmod q^3 + 1$, provided $q > 2$. For $q = 2$ one easily verifies that at least one of the forms $l(\alpha)$ for $\alpha = \mu_1 + \mu_2, \mu_4 - \mu_2, \mu_4 - \mu_1 - \mu_3$, $2\mu_4 - \mu_1 - \mu_2 - \mu_3$ must be $\equiv 0 \bmod q^3 + 1 = 9$, whatever $i$ and $j$ are, so there are no regular elements in this case.

(iii) $D_4(a_1)$. $\sigma^* = q\tau$ with $\tau = (1, 2)^-(3, 4)^-$. Elements of $T_\sigma$: $t = (\zeta^{qi}, \zeta^i, \zeta^{i+(q+1)j}, \zeta^{i+1})$ with $0 \leq i, j < q^2 + 1$ and $\zeta$ a primitive $(q^2+1)$th root of unity in $\mathbb{F}_{q^4}$. In the same way as in case (ii) one shows that the $t$ with $i = j = 1$ is regular for $q > 3$. For $q = 2$ or 3 there are no regular elements as one easily checks.

(iv) $C_3 + A_1$. $\sigma^* = q\tau_1$ with $\tau_1$ as in Table 3. Elements of $T_\sigma$ (see proof of Proposition 2, (i)): $t = (\zeta^i, \zeta^{-i+(q^2-q+1)j}, \zeta^{(q^2-q)i-(q^2-q+1)j}, \zeta^{-qi})$, with $\zeta$ a primitive $(q^3+1)$th root of unity in $\mathbb{F}_{q^6}$, $0 \leq i < q^3 + 1$, $0 \leq j < q + 1$. As before one shows that, e.g., $i = 1, j = 2$ yields a regular element for $q > 2$. For $q = 2$ there can be no regular element, since this was already the case in ${}^3D_4$ as we saw in Proposition 2.

(v) $F_4$. $\sigma^* = q\tau_4$, $\tau_4$ as in Table 3. In the same way as in the proof of Proposition 2, (ii), one shows that all elements $\neq 1$ of $T_\sigma$ are regular.

(vi) $A_2 + \tilde{A}_2$. $\sigma^* = q\tau_5$ with $\tau_5$ as in Table 3, (cf. proof of Proposition 2, (iii)). Elements of $T_\sigma$: $(\zeta^{-(q+1)i+qj}, \zeta^i, \zeta^j, \zeta^{-qi})$ with $0 \leq i, j < q^2 + q + 1$, $\zeta$ a primitive $(q^2+q+1)$th root of unity in $\mathbb{F}_{q^3}$. With $i = 1$ and $j = 3$ we get a regular element for $q > 2$. If $q = 2$, there is no regular element.

(vii) $F_4(a_1)$. $\sigma^* = q\tau_6$, $\tau_6$ as in Table 3. Elements of $T_\sigma$: replace $q$ by $-q$ in case (vi). With $i = 1, j = 2$ we get a regular element if $q > 3$. For $q = 2$ and 3 a straightforward verification shows the nonexistence of regular elements in $T_\sigma$.

(viii) $A_3 + \tilde{A}_1$. $\sigma^* = q\tau$ with $\tau = (1, 2)^-(3)^-(4)^-$. Elements of $T_\sigma$: $(\zeta^{(q+1)i}, \zeta^{-(q^2+q)i}, \zeta^{(q^2+1)j}, \zeta^i)$, with $\zeta$ a primitive $(q+1)(q^2+1)$th root of unity in $\mathbb{F}_{q^4}$, $0 \leq i < (q+1)(q^2+1)$, $0 \leq j < q + 1$. The element with $i = 1$, $j = 2$ turns out to be regular for $q > 2$. For $q = 2$ none of the possible values of $i$ and $j$ yields a regular element.

(ix) $B_4$. $\sigma^* = q\tau$, $\tau = (1, 2, 3, 4)^-$. Elements of $T_\sigma$: $t = (\zeta^{(q^3+1)i}, \zeta^{(-q^3+q^2)i}, \zeta^{(-q^2+q)i}, \zeta^i)$, where $0 \leq i < q^4 + 1$ and $\zeta$ is a primitive $(q^4+1)$th root of unity in $\mathbb{F}_{q^8}$. Write $t^\alpha = \zeta^{l(\alpha, q)i}$. All g.c.d.'s $(l(\alpha, q), q^4 + 1)$ are 1 if $q$ is even, and 1 or 2 if $q$ is odd. So the only irregular $t$ in $T_\sigma$ are 1 if $q$ is even, 1 and $-1$ if $q$ is odd.

## 16. The Ree groups ${}^2F_4$

Let $G$ be the split algebraic group of type $F_4$ over the field $\mathbb{F}_2$. In $G$ we choose a maximal torus $T_0$. We have the root system, characters, and Weyl group $W$ as before. Let $\sigma_0$ be the endomorphism of $G$ with

$$\sigma_0{}^* = 2^n \begin{bmatrix} -2 & 0 & 0 & -1 \\ -1 & 1 & 1 & 0 \\ -1 & 1 & -1 & -1 \\ 2 & 0 & 0 & 2 \end{bmatrix}$$

with respect to the basis $\mu_1, \mu_2, \mu_3, \mu_4$ of $X \otimes \mathbb{R}$. The conjugacy classes of $W\sigma_0{}^*$ under the action of $W$ have been determined in our previous paper [11, Table 1], where they are written with respect to the basis $\omega_1, \omega_2, \omega_3, \omega_4$. We only consider the classes with no eigenvalue equal to 1, which correspond to anisotropic tori $T_\sigma$, that are orbits numbered in [11] as 2, 5, 6, 8, 9, 10, and 11. Since the class number 6 is minus the class number 5, as one sees from the eigenvalues, we have replaced the representative of orbit number 6 by minus that of number 5. In the proof of the following proposition we give the matrices of the representatives with respect to the basis $\mu_1, \mu_2, \mu_3, \mu_4$, multiplied by $r = 2^n$, $n = 0, 1, 2, \ldots$. The value of $q$ is $2^{n+\frac{1}{2}} = r\sqrt{2}$.

***Proposition 4*** *Let $G_\sigma$ be the Ree group ${}^2F_4(q^2)$, with $q = 2^{n+\frac{1}{2}}$, $n \geq 0$. In the following list of anisotropic tori we give the values of $n$ for which each torus contains regular elements (so there are no regular elements for the other values of $n$):*

| | | | | | | | |
|---|---|---|---|---|---|---|---|
| 2 | $n \geq 1$, | 6 | $n \geq 0$, | 9 | $n \geq 2$, | 11 | $n \geq 1$. |
| 5 | $n \geq 1$, | 8 | $n \geq 1$, | 10 | $n \geq 2$, | | |

*In the maximal tori of orbits 5 and 6, all elements $\neq 1$ are regular; in the maximal tori of orbit 2, there are only 3 irregular elements, namely the elements of $T_\sigma$ with $n = 0$, i.e., $q = \sqrt{2}$.*

*Proof* (i) Orbit 2.

$$\sigma^* = r \begin{bmatrix} 0 & -2 & 0 & -1 \\ -1 & -1 & 1 & -1 \\ -1 & -1 & -1 & -2 \\ 0 & 2 & 0 & 2 \end{bmatrix}, \qquad \text{with} \quad r = 2^n.$$

$T_\sigma$ consists of the elements

$$t = (\zeta^{-(2r+1)i}, \zeta^{(-4r^3+2r+1)i}, \zeta^i, \zeta^{(-4r^3-2r^2+1)i}),$$
$$0 \le i < 4r^4 - 2r^2 + 1 = q^4 - q^2 + 1,$$

where $\zeta$ is a primitive $(q^4 - q^2 + 1)$th root of unity in $\mathbb{F}_{q^{12}} = \mathbb{F}_{2^{12n+6}}$. Then $t^\alpha = \zeta^{l(\alpha, r)i}$. The polynomials $l(\alpha, r)$ have g.c.d. = 1 or 3 with $4r^4 - 2r^2 + 1$, both possibilities occurring for every value of $n$. Hence the only irregular elements are the three $t$ with $i$ divisible by $\frac{1}{3}(q^4 - q^2 + 1)$. For $n = 0$, these are all elements of $T_\sigma$, but for $n > 0$, $T_\sigma$ contains other elements, which are therefore regular.

(ii) Orbit 5.

$$\sigma^* = r\begin{bmatrix} 0 & -2 & 0 & -1 \\ 1 & -1 & -1 & -1 \\ 1 & -1 & 1 & 0 \\ 0 & 2 & 0 & 2 \end{bmatrix}, \qquad r = 2^n.$$

Elements of $T_\sigma$:

$$t = (\zeta^{(2r-1)i}, \zeta^{(4r^3-4r^2+2r-1)i}, \zeta^i, \zeta^{(4r^3-2r^2+2r-1)i}),$$
$$0 \le i < 4r^4 - 4r^3 + 2r^2 - 2r + 1 = q^4 - q^3\sqrt{2} + q^2 - q\sqrt{2} + 1,$$

with $\zeta$ a primitive $(q^4 - q^3\sqrt{2} + q^2 - q\sqrt{2} + 1)$th root of unity over $\mathbb{F}_2$. The polynomials $l(\alpha, r)$ with $t^\alpha = \zeta^{l(\alpha, r)i}$ all have g.c.d. = 1 with $4r^4 - 4r^3 + 2r^2 - 2r + 1$, for all $r = 2^n$, hence all elements $\neq 1$ are regular. For $n = 0$, $|T_\sigma| = 1$, whereas $|T_\sigma| > 1$ for $n > 0$.

(iii) Orbit 6. Replace $r$ by $-r$ in the proof of the previous case. This yields, again, that 1 is the only irregular element, but now $|T_\sigma| > 1$ for all $n$.

(iv) Orbit 8.

$$\sigma^* = r\begin{bmatrix} 2 & 0 & 0 & 1 \\ 1 & -1 & 1 & 0 \\ 1 & -1 & -1 & -1 \\ -2 & 0 & 0 & 0 \end{bmatrix}, \qquad r = 2^n.$$

Elements of $T_\sigma$:

$$t = (\zeta^{(2r^2+2r+1)i}, \zeta^{(-4r^3+2r^2-1)j}, \zeta^{(2r^2-2r+1)j}, \zeta^{(2r^3+2r^2+r)i-(2r^3-2r^2+r)j}),$$

with

$$0 \le i < 2r^2 - 2r + 1 = q^2 - q\sqrt{2} + 1,$$
$$0 \le j < 2r^2 + 2r + 1 = q^2 + q\sqrt{2} + 1,$$

where $\zeta$ is a $(q^2 - q\sqrt{2} + 1)(q^2 + q\sqrt{2} + 1)$th $(= (q^4 + 1)\text{th})$ root of unity over $\mathbb{F}_2$. Write $t^\alpha = \zeta^{l(\alpha)}$, $l(\alpha)$ a linear form in $i$ and $j$. It turns out that for $i = j = 1$, for instance, no $l(\alpha)$ is divisible by the order of $\zeta$ if $n \geq 1$, so this yields a regular $t$. If $n = 0$, i.e., $r = 1$, $\zeta$ has order 5, and $l(\mu_1) = 5i$, hence all $t$ are irregular.

(v) Orbit 9.

$$\sigma^* = r \begin{bmatrix} 0 & 2 & 0 & 1 \\ -1 & 1 & -1 & 0 \\ -1 & 1 & -1 & -1 \\ 0 & -2 & 2 & 0 \end{bmatrix}, \qquad r = 2^n.$$

Elements of $T_\sigma$: $t = (\zeta^i, \zeta^{(2r-1)i-2rj}, \zeta^{i+2rj}, \zeta^j)$, with $0 \leq i, j < 2r^2 + 1 = q^2 + 1$, and $\zeta$ a primitive $(q^2+1)$th root of unity in $\mathbb{F}_{q^4} = \mathbb{F}_{2^{4n+2}}$. For $i = 1$, $j = 2$ we find a regular $t$, provided $n \geq 2$. If $n = 0$ or 1, then $T_\sigma$ is contained in the torus of type $4A_1$ in ${}^1F_4(q)$ with $q = 2$ or 8, respectively (see proof of Proposition 3, (i)), which does not contain regular elements.

(vi) Orbit 10.

$$\sigma^* = r \begin{bmatrix} 2 & 0 & 0 & 1 \\ 1 & 1 & -1 & 0 \\ 1 & 1 & 1 & 1 \\ -2 & 0 & 0 & 0 \end{bmatrix}, \qquad r = 2^n.$$

Elements of $T_\sigma$: $t = (\zeta^i, \zeta^{(2r-1)j}, \zeta^j, \zeta^{ri+rj})$, where $0 \leq i, j < 2r^2 - 2r + 1 = q^2 - q\sqrt{2} + 1$, and $\zeta$ a primitive $(q^2 - q\sqrt{2} + 1)$th root of unity in $\mathbb{F}_{q^8} = \mathbb{F}_{2^{8n+4}}$. For $i = 1$, $j = 2$, e.g., $t$ is regular, provided $n \geq 2$. For $n = 0$, $|T_\sigma| = 1$, and for $n = 1$, $T_\sigma$ is contained in the torus $4A_1$ in ${}^1F_4(q)$ with $q = 4$, which contains no regular elements.

(vii) Orbit 11. Replace $r$ by $-r$ in the previous case. For $i = 1, j = 2$ we find a regular $t$ in $T_\sigma$, if $n > 0$. If $n = 0$, $T_\sigma$ is contained in the torus $4A_1$ in ${}^1F_4(q)$ with $q = 4$, which has no regular elements.

## 17. Groups of type ${}^1G_2$

Let $G$ be of type $G_2$, and $\sigma_0$ an endomorphism of $G$ such that $G_{\sigma_0}$ is finite of type ${}^1G_2(q)$. Choose $T_0$ and $B_0$ as usual. The positive roots with respect to $T_0$ are $\alpha_1, \alpha_2, \alpha_1 + \alpha_2, 2\alpha_1 + \alpha_2, 3\alpha_1 + \alpha_2, 3\alpha_1 + 2\alpha_2$; $\alpha_1$ is short and $\alpha_2$ is long. They generate the character group of $T_0$. The conjugacy classes in the Weyl group $W$ are given in Table 5 (cf. [1, 2]).

The action of $\sigma_0$ on $X \otimes \mathbb{R}$ is $\sigma_0{}^* = q \cdot 1$. For the anisotropic tori we have the following results about regular elements.

***Proposition 5*** *Let $G_\sigma$ be of type ${}^1G_2(q)$, $q = p^t$ ($p$ prime). The anisotropic tori of $G$ contain regular elements if and only if $q$ satisfies the conditions*

$$G_2 \quad q \geq 3, \qquad A_2 \quad \text{all } q, \qquad A_1 + \tilde{A}_1 \quad q \geq 5.$$

*Table 5*

*The Conjugacy Classes in $W(G_2)$*

| Type | Conjugacy class | Characteristic polynomial | $T_\sigma$ anisotropic? |
|---|---|---|---|
| $A_1$ | Reflection in $\alpha^\perp$, $\alpha$ short root | $X^2 - 1$ | No |
| $\tilde{A}_1$ | Reflection in $\alpha^\perp$, $\alpha$ long root | $X^2 - 1$ | No |
| $G_2$ | Rotation over $\pm\pi/3$ (Coxeter class) | $X^2 - X + 1$ | Anisotropic |
| $A_2$ | Rotation over $\pm 2\pi/3$ | $X^2 + X + 1$ | Anisotropic |
| $A_1 + \tilde{A}_1$ | $-1$ | $(X + 1)^2$ | Anisotropic |
| $\varnothing$ | 1 | $(X - 1)^2$ | No |

*In the Coxeter tori (type $G_2$), 1 is the only irregular element if $3 \nmid q + 1$, i.e., if $p = 3$, or $3 \mid p - 1$, or $t$ even and $3 \mid p + 1$. If $t$ is odd and $3 \mid p + 1$, i.e., if $3 \mid q + 1$, then the irregular elements form a 3-cyclic subgroup.*

*In the tori of type $A_2$, 1 is the only irregular element if $3 \nmid q - 1$, i.e., if either $p = 3$ or $t$ odd and $3 \mid p + 1$, whereas the irregular elements form a 3-cyclic subgroup if $3 \mid q - 1$, i.e., if either $3 \mid p - 1$ or $t$ even and $3 \mid p + 1$.*

*Proof* The matrices for elements of the Weyl group will be written with respect to the basis $\alpha_1, \alpha_2$ of $X \otimes \mathbb{R}$. Elements of $T_0$ are $t = (t_1, t_2)$, with $t_i = t^{\alpha_i}$.

(i) $G_2$. Take

$$\sigma^* = q\begin{pmatrix} 2 & -3 \\ 1 & 1 \end{pmatrix}.$$

Elements of $T_\sigma$: $t = (\zeta^i, \zeta^{-(q+1)i})$, $0 \le i < q^2 - q + 1$, $\zeta$ a primitive $(q^2 - q + 1)$th root of unity in $\mathbb{F}_{q^6}$. For a root $\alpha$, $t^\alpha = \zeta^{l(\alpha, q)i}$, where each $l(\alpha, q)$ is a polynomial in $q$. The greatest common divisor of the $l(\alpha, q)$ with $q^2 - q + 1$ is either 1 or g.c.d. $(q + 1, 3)$, which is 1 if $p = 3$, or $3 \mid p - 1$, or $t$ even and $3 \mid p + 1$, and it is 3 if $t$ odd and $3 \mid p + 1$. In the former case, 1 is the only irregular element; in the latter case, the irregular elements are $t = (\zeta^i, 1)$ with $(q^2 - q + 1)/3$ dividing $i$, so they form a 3-cyclic group. $|T_\sigma| = 3$ if $q = 2$, so then there are no regular elements; for $q > 2$, $|T_\sigma| = q^2 - q + 1 > 3$.

(ii) $A_2$. Choose

$$\sigma^* = q\begin{pmatrix} 1 & -3 \\ 1 & -2 \end{pmatrix}.$$

Elements of $T_\sigma$: $t = (\zeta^i, \zeta^{-(q+2)i})$, $0 \le i < q^2 + q + 1$, $\zeta$ a primitive $(q^2 + q + 1)$th root of unity in $\mathbb{F}_{q^3}$. As in the previous case one shows that 1 is the only irregular element if $3 \nmid q-1$, and that the irregular elements form a 3-cyclic group if $3 \mid q-1$. Since $|T_\sigma| = q^2+q+1 > 3$ for all $q$, there always exist regular elements in $T_\sigma$.

(iii) $A_1 + \tilde{A}_1$. Now $\sigma^* = -q$, so $T_\sigma$ consists of the elements $t = (\zeta^i, \zeta^j)$, with $0 \le i, j < q+1$ and $\zeta$ a primitive $(q+1)$th root of unity in $\mathbb{F}_{q^2}$. If $i = j = 1$, $t$ is regular, provided $q \ge 5$. For $q = 2$, 3, or 4 it is easily checked that no $t$ is regular.

## 18. Groups of type ${}^2G_2$

In a group of type $G_2$ over $\mathbb{F}_3$ with a fixed maximal torus $T_0$ we now consider the endomorphism $\sigma_0$ such that $\sigma_0{}^*$ acts on the simple roots $\alpha_1$ (short) and $\alpha_2$ (long) by

$$\sigma_0\colon \alpha_1 \mapsto 3^n\alpha_2, \quad \alpha_2 \mapsto 3^{n+1}\alpha_1 \qquad (n \ge 0).$$

With $q = 3^{n+\frac{1}{2}}$, we have $\sigma_0{}^* = q\tau_0$, $\tau_0$ an isometry of $X \otimes \mathbb{R}$. The conjugacy classes of $W\tau_0$ under the action of $W$ are the following, together with their characteristic polynomials (cf. [11, §7]):

| | |
|---|---|
| All reflections in a bissectrix of a long and a short root | $X^2 - 1$, |
| Rotations over $\pm\pi/6$ | $X^2 - X\sqrt{3} + 1$, |
| Rotations over $\pm\pi/2$ | $X^2 + 1$, |
| Rotations over $\pm 5\pi/6$ | $X^2 + X\sqrt{3} + 1$. |

The last three classes give rise to anisotropic tori $T_\sigma$.

***Proposition 6*** *Let $G_\sigma$ be of type ${}^2G_2(q^2)$ with $q = 3^{n+\frac{1}{2}}$, $n \ge 0$. The anisotropic tori of $G_\sigma$ contain regular elements in the following cases (and only in these):*

(i) *$\tau$ a rotation over $\pm\pi/6$: $n > 0$,*
(ii) *$\tau$ a rotation over $\pm\pi/2$: $n > 0$,*
(iii) *$\tau$ a rotation over $\pm 5\pi/6$: $n \ge 0$.*

*In cases* (i) *and* (iii), *all elements $\ne 1$ are regular (but in case* (i), *$T_\sigma = \langle 1 \rangle$ if $n = 0$); in case* (ii) *there are 4 irregular elements, which form a direct product $C_2 \times C_2$.*

*Proof* (i) Take

$$\sigma^* = r\begin{pmatrix} 3 & -3 \\ 1 & 0 \end{pmatrix}, \qquad r = 3^n,$$

with respect to the basis $\alpha_1, \alpha_2$. Elements of $T_\sigma$: $t = (\zeta^i, \zeta^{-3ri})$, $0 \le i < 3r^2 - 3r + 1 = q^2 - q\sqrt{3} + 1$, $\zeta$ a primitive $(3r^2 - 3r + 1)$th root of unity over $\mathbb{F}_3$.

For $\alpha$ a root, $t^\alpha = \zeta^{l(\alpha, r)i}$, where the $l(\alpha, r)$ are polynomials in $r$ which all have g.c.d. $= 1$ with $3r^2 - 3r + 1$, so 1 is the only irregular element.

(ii) Take

$$\sigma^* = r\begin{pmatrix} 3 & -6 \\ 2 & -3 \end{pmatrix}, \qquad r = 3^n.$$

Elements of $T_\sigma$: $t = (\zeta^{\frac{1}{4}(3r^2+1)i+\frac{1}{2}(r-1)j}, \zeta^j)$ with $i = 0$ or $1, 0 \leq j < \frac{1}{2}(3r^2 + 1)$, and $\zeta$ a primitive $\frac{1}{2}(3r^2 + 1)$th root of unity in $\mathbb{F}_{3^{4n+2}}$. A straightforward computation shows that $(1, 1)$, $(1, -1)$, $(-1, 1)$, and $(-1, -1)$ are the only irregular elements. For $n = 0$, $|T_\sigma| = 4$; otherwise, $|T_\sigma| > 4$.

(iii) Now

$$\sigma^* = r\begin{pmatrix} 0 & -3 \\ 1 & -3 \end{pmatrix}, \quad r = 3^n.$$

Elements of $T_\sigma$: $t = (\zeta^{ri}, \zeta^i)$, $0 \leq i < 3r^2 + 3r + 1$, $\zeta$ a primitive $(3r^2 + 3r + 1)$th root of unity in $\mathbb{F}_3$. Precisely as in case (i) it is shown that 1 is the only irregular element.

## 19. Endomorphisms $\sigma$ with cyclic centralizer in $W$

So far we have been able to determine the existence of elements in general position in anisotropic tori in exceptional groups in a very direct way. In the case of groups of type $E_{6,7,8}$, however, this leads to unwieldy computations. It is for this reason that we introduce a different method which will allow us to handle at least some tori in these groups. We consider endomorphisms $\sigma$ whose centralizer in the Weyl group $W$ is cyclic, say $W_\sigma = \langle w_\sigma \rangle$. If $\sigma = qw_1$ for $w_1 \in W$, $w_1$ is a power of $w_\sigma$, hence there exists a common set of eigenvectors in $V_\mathbb{C} = X \otimes \mathbb{C}$ for $w_\sigma$ and $w_1$. In the case $\sigma = q\tau$ with $\tau \notin W$, we shall make the assumption that a similar property holds, i.e., that $\tau$ and $w_\sigma$ have a common set of eigenvectors in $V_\mathbb{C}$. This assumption is fulfilled if $w_\sigma$ is a power of $\tau$, as will always be the case in the applications.

If $A$ is a linear transformation in $V_\mathbb{C}$ of finite order $r$, its eigenvalues can be written as $\exp(2\pi i p_j r^{-1})$, $1 \leq j \leq l = \dim V_\mathbb{C}$. We call $p_1, p_2, \ldots, p_l$ the *exponents* of $A$.

For the rest, we keep notations and definitions as before.

***Proposition 7*** *Let G and $\sigma$ be as usual, $T$ a $\sigma$-invariant maximal torus. Assume that the centralizer of $\sigma$ in $W$ is cyclic: $W_\sigma = \langle w_\sigma \rangle$ for some $w_\sigma \in W$. Let $r$ be the order of $w_\sigma$, and $p_1, \ldots, p_l$ its exponents. Write $\sigma = q\tau$, and assume* *that $w_\sigma$ and $\tau$ have common eigenvectors in $V_{\mathbb{C}}$. Denote the order of $\tau$ by $t$ and its exponents by $m_1, \ldots, m_l$. For each $d$ dividing $r$, define the polynomial $g_d$ by*

$$g_d(T) = \prod_{d^{-1}r \mid p_j} (T - \exp(2\pi i m_j t^{-1})), \qquad \textit{if} \quad \mu(d^{-1}r) = 1,$$

$$g_d(T) = \prod_{d^{-1}r \mid p_j} (T - \exp(2\pi i m_j t^{-1})) \prod_{d^{-1} \nmid p_j} (1 - \exp(2\pi i p_j\, d r^{-1})),$$
$$\textit{if} \quad \mu(d^{-1}r) = -1,$$

*where $\mu$ is the Möbius function. Then the number of elements in general position in $T_\sigma$, denoted by $P_\sigma$, satisfies*

$$P_\sigma \geq \sum_{d \mid r} \mu(d^{-1}r) \, |g_d(q)| .$$

*Proof* An element of $T_\sigma$ is not in general position if and only if it belongs to $(T_w)_\sigma$ for some $w \in W_\sigma$, $w \neq 1$, where $T_w$ denotes the set of $w$-fixed elements in $T$ (cf. [7, 6.10, 6.12, 6.14]). Now assume $W_\sigma = \langle w_\sigma \rangle$, $r =$ order of $w_\sigma$. Then $T_{w_\sigma{}^i} = T_{w_\sigma{}^d}$ with $d =$ g.c.d.$(i, r)$ for $0 \leq i < r$, and for $d$ and $e$ dividing $r$,

$$T_{w_\sigma{}^d} \subseteq T_{w_\sigma{}^e} \Leftrightarrow d \mid e,$$

$$T_{w_\sigma{}^d} \cap T_{w_\sigma{}^e} = T_{w_\sigma{}^f} \qquad \text{where} \quad f = \text{g.c.d.}(d, e).$$

The maximal $T_{w_\sigma{}^d}$, $d \mid r$, are those for which $d^{-1}r$ is prime. So we can count the number of irregular elements in $T_\sigma$ by adding the orders of the $(T_{w_\sigma{}^d})_\sigma$ with $d^{-1}r$ prime, then subtracting the orders of their intersections, adding those of their triple intersections, etc. Noticing that $T_{w_\sigma{}^r} = T$, we find for the number of elements in general position

$$P_\sigma = \sum_{d \mid r} \mu(d^{-1}r) \, |(T_{w_\sigma{}^d})_\sigma| , \tag{1}$$

where $\mu$ is the Möbius function: $\mu(n) = (-1)^t$ if $n$ is a product of $t$ distinct primes, and $\mu(n) = 0$ otherwise.

Let $e_1, e_2, \ldots, e_l$ be a basis of eigenvectors of $w_\sigma$ and of $\tau$ in $V_{\mathbb{C}}$. These are also eigenvectors of $w_\sigma{}^d$ with eigenvalues $\exp(2\pi i p_j\, d r^{-1})$, $j = 1, 2, \ldots, l$. Assume $d \mid r$. Let $V_1$ be the subspace spanned by the $e_j$ such that $d^{-1}r$ divides $p_j$, i.e., $\exp(2\pi i p_j\, d r^{-1}) = 1$, and $V_2$ the subspace spanned by the remaining $e_j$. Then

$$V_{\mathbb{C}} = V_1 \oplus V_2 , \quad V_1 = \ker(w_\sigma{}^d - 1), \qquad V_2 = (w_\sigma{}^d - 1)V_{\mathbb{C}} = (w_\sigma{}^d - 1)V_2 .$$

Clearly, $(w_\sigma{}^d - 1)X \subseteq X \cap V_2$. Choose a basis $x_1, x_2, \ldots, x_l$ of $X$ such that $\alpha_{t+1} x_{t+1}, \ldots, \alpha_l x_l$ form a basis of $(w_\sigma{}^d - 1)X$ for suitable integers $\alpha_j$. Then $x_{t+1}, \ldots, x_l$ form a basis of $V_2$, since $V_2$ is spanned by $(w_\sigma{}^d - 1)X$.

We write $(T_{w_\sigma{}^d})^0{}_\sigma$ for $((T_{w_\sigma{}^d})^0)_\sigma$. It is immediate that

$$|(T_{w_\sigma{}^d})^0{}_\sigma| \leq |(T_{w_\sigma{}^d})_\sigma| \leq |(T_{w_\sigma{}^d})^0{}_\sigma| \cdot [T_{w_\sigma{}^d} : (T_{w_\sigma{}^d})^0]. \tag{2}$$

Now the character group of $(T_{w_\sigma^d})^0$ is the torsion-free part of $X/(w_\sigma{}^d - 1)X$. Let $\chi_{\tau|V_1}$ be the characteristic polynomial of the restriction of $\tau$ to $V_1$. It is also the characteristic polynomial of the transformation induced by $\tau$ on $V_{\mathbb{C}}/V_2$, hence also of the transformation that $\tau$ induces on the torsion-free part of $X/(w_\sigma{}^d - 1)X$. Therefore,

$$|(T_{w_\sigma^d})^0{}_\sigma| = |\chi_{\tau|V_1}(q)| = \left| \prod_{d^{-1}r|p_j} (q - \exp(2\pi i m_j t^{-1})) \right|. \tag{3}$$

Now the character group of $T_{w_\sigma^d}/(T_{w_\sigma^d})^0$ is the torsion part of $X/(w_\sigma{}^d - 1)X$, which is $X \cap V_2/(w_\sigma{}^d - 1)X$, hence

$$[T_{w_\sigma^d}: (T_{w_\sigma^d})^0] = |\alpha_{t+1} \cdot \cdots \cdot \alpha_l| = |\det(w_\sigma{}^d | V_2 - 1)|$$

$$= |\chi_{w_\sigma^d|V_2}(1)| = \left| \prod_{d^{-1}r \nmid p_j} (1 - \exp(2\pi i p_j\, dr^{-1})) \right|.$$

Combining this relation with (1)–(3), one gets the inequality for $P_\sigma$.

The inequality for $P_\sigma$ given in the above proposition suffices in most cases to prove the existence of elements in general position. But in a few cases it is too rough; for instance, the right hand side may become negative for some (usually very small) values of $q$. In those cases we must refine it by making use of more detailed information about $|(T_{w_\sigma^d})_\sigma|$. For instance, it can be shown for certain values of $q$ that $(T_{w_\sigma^d})_\sigma = (T_{w_\sigma^d})^0{}_\sigma$ for all $d\,|\,r$. Then we get for $P_\sigma$ the equality

$$P_\sigma = \sum_{d|r} \mu(d^{-1}r) \left| \prod_{d^{-1}r|p_j} (q - \exp(2\pi i m_j t^{-1})) \right|.$$

In other cases we still get an inequality for $P_\sigma$, but a reasoning as above for certain values of $d$ makes it sharper than the inequality given in the proposition.

In the following sections we consider only certain elements in $W$ or $W\sigma_0$ with cyclic centralizers. These are the so-called *regular* elements, as treated in [8], since these are the only elements we have sufficient information about. On the other hand, they include the Coxeter and twisted Coxeter elements. We consider only cases where $T_\sigma$ is anisotropic.

## 20. Groups of type ${}^1E_6$

Let $G$ be of type $E_6$. For $\sigma_0$ we take the Frobenius endomorphism with $\sigma_0{}^* = q \cdot 1$. For $\sigma$ with $\sigma^* = qw$, $w \in W$, we get a cyclic centralizer $W_\sigma$ in the Weyl group in three cases of regular elements $w$, two of which give rise to anisotropic tori. In these cases, $W_\sigma = \langle w \rangle$, so we can take $w_\sigma = w$, hence

$p_j = m_j$ for all $j$, and $r = t$. From [8, 5.4], we get the orders $t$ and the characteristic polynomials, hence the exponents.

**Proposition 8** *Let $G_\sigma$ be of type ${}^1E_6(q)$. The anisotropic maximal tori $T_\sigma$ of type $E_6(a_1)$ and $E_6$ (Coxeter torus) contain elements in general position for all possible $q$.*

*Proof* (i) $E_6(a_1)$. Order $r = t = 9$, exponents $p_j = 1, 2, 4, 5, 7, 8$. The $d \mid r$ with $\mu(r^{-1} d) \neq 0$ are $d = 3$ and $9$:

$d = 3$: $\mu(d^{-1}r) = -1$, $m_j$ with $d^{-1}r \mid p_j$: none, $g_3(q) = 27$;

$d = 9$: $\mu(d^{-1}r) = 1$, $m_j$ with $d^{-1}r \mid p_j$: all, $g_9(q) = q^6 + q^3 + 1$.

Hence by Proposition 7,

$$P_\sigma \geq q^6 + q^3 + 1 - 27 = q^6 + q^3 - 26 > 0 \qquad \text{for all} \quad q \geq 2.$$

(ii) $E_6$. Order $t = 12$, exponents $p_j = 1, 4, 5, 7, 8, 11$. Application of Proposition 7 leads to

$$P_\sigma \geq q^6 + q^5 - q^3 - 16q^2 - 15q - 41.$$

The right hand side is $> 0$ for $q \geq 3$, but for $q = 2$ it is $< 0$. So we inspect the case $q = 2$ more precisely.

For $d = 6$, $p_2 = 4$ and $p_5 = 8$ are divisible by $d^{-1}r = 2$, hence

$$[T_{w^6} : (T_{w^6})^0] = \prod_{p_j = 1, 5, 7, 11} (1 - \exp(2\pi i p_j\, d r^{-1})) = 16.$$

Similarly one computes $[T_{w^d} : (T_{w^d})^0] = 27$ for $d = 4$, and $= 3$ for $d = 2$. Now $|T_\sigma| = |\chi_\sigma(2)| = 91$, which is not divisible by 2 or 3. So all elements of $(T_{w^d})_\sigma$ must already be in $(T_{w^d})^0{}_\sigma$. Hence for $q = 2$,

$$P_\sigma = \left| \sum_{d|12} \mu(12d^{-1}) \prod_{12d^{-1}|p_j} (2 - \exp(2\pi i m_j t^{-1})) \right| = 84.$$

For arbitrary $q$, the inequality for $P_\sigma$ can be sharpened by observing that $(T_{w^6})_\sigma = (T_{w^6})^0{}_\sigma$, since $|T_\sigma| = (q^2 + q + 1)(q^4 - q^2 + 1)$ is odd for all $q$.

## 21. Groups of type ${}^2E_6$

Take in a group $G$ of type $E_6$ the endomorphism $\sigma_0$ with $\sigma_0{}^* = -q$. The classes of $W\tau_0 = -W$ are minus the conjugacy classes of $W$. So we can take classes with cyclic centralizers in $W$, and multiply these by $-1$.

**Proposition 9** *Let $G_\sigma$ be of type ${}^2E_6(q^2)$. The anisotropic maximal tori of type $-E_6(a_1)$ (twisted Coxeter torus) and $-E_6$ contain elements in general position for all possible values of $q$. In the twisted Coxeter torus $-E_6(a_1)$, the only irregular element is 1.*

*Proof* (i) The twisted Coxeter element $\tau$ of $W\tau_0$ is of type $-E_6(a_1)$ (cf. [8, §7, and Table 8 in §6.12]). The order $\tau$ is $t = 18$, and its exponents are $m_j = 1, 5, 7, 11, 13, 17$. As a generator of $W_\sigma$ we can take $w_\sigma = \tau^2$ (which is of type $E_6(a_1)$, i.e., $\tau^2$ is conjugate to $-\tau$, as one can check, e.g., by considering the characteristic polynomials); its order is $r = 9$, and its exponents are $p_j = 2, 1, 5, 4, 8, 7$, respectively. The only relevant divisors of $r$ are $d = 3$ and 9. One finds, using the fact that $[T_{w_\sigma d} : (T_{w_\sigma d})^0] = 1$ for both $d$'s,

$$P_\sigma = q^6 - q^3,$$

which is $> 0$ for all $q \geq 2$. Actually, $P_\sigma = |T_\sigma| - 1$.

(ii) $-E_6$. Here $t = r = 12$, the exponents of $\tau$ are $m_j = 7, 10, 11, 1, 2, 5$, those of $w_\sigma = -\tau$ are $p_j = 1, 4, 5, 7, 8, 11$, respectively. With Proposition 7 we get

$$P_\sigma \geq q^6 - q^5 + q^3 - 16q^2 + 15q - 41.$$

This is $> 0$ for $q \geq 3$ but $< 0$ for $q = 2$. So in the latter case we have to be a little bit more careful.

For $d = 6$, we have $[T_{w_\sigma 6} : (T_{w_\sigma 6})^0] = 16$. For $q = 2$, $|T_\sigma| = \chi_\tau(2) = 39$, so no element of $T_\sigma$ has even order. It follows that $(T_{w_\sigma 6})_\sigma = (T_{w_\sigma 6})^0{}_\sigma$. Taking this into account for the estimate of $P_\sigma$ (so leaving it the same for the other divisors $d$ of $r = 12$), we find

$$P_\sigma \geq (q^4 - q^2)(q^2 - q + 1) - 26 = 10$$

for $q = 2$.

## 22. Groups of type ${}^1E_7$

Let $G$ be of type $E_7$, and $\sigma_0$ the Frobenius endomorphism with $\sigma_0{}^* = q \cdot 1$. There are four regular elements with cyclic centralizers in $W$ [8, 5.4], two of which correspond to anisotropic tori.

**Proposition 10** *In a group $G_\sigma$ of type ${}^1E_7(q)$ the anisotropic maximal tori $T_\sigma$ of type $E_7$ (Coxeter torus) and $E_7(a_1)$ contain elements in general position for all $q$.*

*Proof* (i) $E_7$. Take $\sigma^* = qc$, $c$ a Coxeter element in $W$. The characteristic polynomial of $c$ is $(X^6 - X^3 + 1)(X + 1)$. Orders $r = t = 18$, and exponents $m_j = p_j = 1, 5, 7, 9, 11, 13, 17$. With Proposition 7 we get

$$P_\sigma \geq q^7 + q^6 - q^4 - q^3 - 26q - 153,$$

hence $P_\sigma > 0$ for $q \geq 3$. For $q = 2$, one has $|T_\sigma| = (2^6 - 2^3 + 1) \times (2 + 1) = 3^2 \cdot 19$. Now for $d = 9$, $[T_{c^d} : (T_{c^d})^0] = 2^7$, prime to $|T_\sigma|$, hence $(T_{c^d})_\sigma = (T_{c^d})^0{}_\sigma$. Thus one finds

$$P_\sigma \geq q^7 + q^6 - q^4 - q^3 - 26q - 26 = 90 \quad \text{for} \quad q = 2.$$

(ii) $E_7(a_1)$. Here $\sigma^* = qw$, characteristic polynomial of $w$: $X^7 + 1$. The centralizer $W_\sigma = \langle w \rangle$. Orders $r = t = 14$, exponents $m_j = p_j = 1, 3, 5, 7, 9, 11, 13$. Then

$$P_\sigma \geq q^7 - 7q - 133,$$

which is $> 0$ for $q \geq 3$. For $q = 2$,

$$P_\sigma = q^7 - q = 126.$$

## 23. Groups of type ${}^1E_8$

Let $G$ be of type $E_8$, and $\sigma_0$ the Frobenius endomorphism: $\sigma_0{}^* = q \cdot 1$.

***Proposition 11*** *In a group $G_\sigma$ of type ${}^1E_8(q)$ the anisotropic tori of type $E_8$ (Coxeter torus), $E_8(a_1)$, $E_8(a_2)$, and $E_8(a_5)$ contain elements in general position for all values of q. More precisely, 1 is the only irregular element of $E_8$, $E_8(a_1)$, and $E_8(a_5)$. In $E_8(a_2)$ there are at most 25 irregular elements; if 5 divides q, $q + 1$, or $q - 1$, then 1 is the only irregular element in $E_8(a_2)$.*

*Proof* (i) $E_8$. $\sigma^* = qc$, where $c$ is a Coxeter element in $W(E_8)$. The characteristic polynomial of $c$ is $\Phi_{30} = X^8 + X^7 - X^5 - X^4 - X^3 + X + 1$. The order of $c$ is $t = r = 30$, the exponents are $m_j = p_j = 1, 7, 11, 13, 17, 19, 23, 29$. For $d = 10$, $[T_{c^d} : (T_{c^d})^0] = 3^4$, for $d = 15$, $[T_{c^d} : (T_{c^d})^0] = 2^8$, in all other cases it is 1. Now $|T_\sigma| = q^8 + q^7 - q^5 - q^4 - q^3 + q + 1$ is not divisible by 2 or 3, hence $(T_{c^d})_\sigma = (T_{c^d})^0{}_\sigma$ for all $d$ dividing $r$. Hence 1 is the only irregular element, so $P_\sigma = |T_\sigma| - 1 > 0$ for all $q$.

(ii) $E_8(a_1)$. Write $\sigma^* = qw$. The characteristic polynomial of $w$ is $\Phi_{24} = X^8 - X^4 + 1$, its centralizer is $\langle w \rangle$. Now $t = r = 24$, the exponents are $m_j = p_j = 1, 5, 7, 11, 13, 17, 19, 23$. $[T_{w^d} : (T_{w^d})^0]$ is $3^4$ for $d = 8$, and $2^8$ for $d = 12$, in the other cases it is 1. But $|T_\sigma|$ does not contain 2 and 3 as prime factors, hence 1 is the only irregular element again.

(iii) $E_8(a_2)$. Writing $\sigma^* = qw$, we find as characteristic polynomial of $w$: $\Phi_{20} = X^8 - X^6 + X^4 - X^2 + 1$, and as centralizer $\langle w \rangle$. The order is $t = r = 20$, the exponents are 1, 3, 7, 9, 11, 13, 17, 19. $[T_{w^d} : (T_{w^d})^0]$ is $5^2$ for $d = 4$, $2^8$ for $d = 10$, and 1 for $d = 2$ or 20. Now $|T_\sigma| = q^8 - q^6 + q^4 - q^2 + 1$ is odd for all $q$, and divisible by 5 for $q = 5l \pm 2$. Hence the irregular elements are contained in $(T_{w^4})_\sigma$, which has order 25. For $q \neq 5l \pm 2$, 1 is the only irregular element.

(iv) $E_8(a_5)$. Here $\sigma = qw$, where $w = c^2$, $c$ a Coxeter element of $W(E_8)$ (cf. (i)). The characteristic polynomial of $w$ is $\Phi_{15} = X^8 - X^7 + X^5 - X^4 + X^3 - X + 1$, its centralizer is $\langle c \rangle$. The order of $w$ is $t = 15$, its exponents are $m_j = 1, 7, 11, 13, 2, 4, 8, 14$, and $w_\sigma = c$ has order $r = 30$ and exponents $p_j = 1, 7, 11, 13, 17, 19, 23, 29$, respectively. $[T_{w^d} : (T_{w^d})^0]$ is $3^4$ for $d = 10$, $2^8$

for $d = 15$ and 1 for the other divisors $d$ of $r = 30$. But $|T_\sigma| = q^8 - q^7 + q^5 - q^4 + q^3 - q + 1$ is neither divisible by 2 nor by 3. Therefore, 1 is the only irregular element in $T_\sigma$.

## References

1. R. W. Carter, Conjugacy classes in the Weyl group, *in* "Seminar on Algebraic Groups and Related Finite Groups" (A. Borel et al., eds.) (Lect. Notes in Math. **131**). Springer-Verlag, Berlin–Heidelberg–New York, 1970.
2. R. W. Carter, Conjugacy classes in the Weyl group, *Compos. Math.* **25** (1972), 1–59.
3. P. Deligne and G. Lusztig, Representations of reductive groups over finite fields, *Ann. of Math.* **103** (1976), 103–161.
4. P. C. Gager, Maximal tori in finite groups of Lie type, Ph.D. thesis, Univ. of Warwick, 1973.
5. Séminaire Chevalley 1956/1958: Classification des groupes de Lie algébriques. Paris 1958.
6. J.-P. Serre, "Existence d'éléments réguliers sur les corps finis," S.G.A. 3, Schémas en Groupes II, App. d'exposé 14 (Lect. Notes in Math. **152**), Springer-Verlag, Berlin–Heidelberg–New York, 1970.
7. T. A. Springer, On the characters of certain finite groups, *in* "Lie Groups and Their Representations" (Summer School Budapest 1971) (I. M. Gelfand, ed.). Hilger, London, 1975.
8. T. A. Springer, Regular elements of finite reflection groups, *Invent. Math.* **25** (1974), 159–198.
9. T. A. Springer and R. Steinberg, Conjugacy classes, *in* "Seminar on Algebraic Groups and Related Finite Groups" (A. Borel *et al.*, eds.) (Lect. Notes in Math. **131**). Springer-Verlag, Berlin–Heidelberg–New York, 1970.
10. R. Steinberg, Endomorphisms of linear algebraic groups, *Mem. Amer. Math. Soc.* **80** (1968).
11. F. D. Veldkamp, Roots and maximal tori in finite forms of semisimple algebraic groups, *Math. Ann.* **207** (1974), 301–314.
12. F. D. Veldkamp, Regular characters and regular elements, *Comm. Algebra* (to appear).

Part of the work on this paper was done while the author was visiting at The Ohio State University, and he is very grateful for the great hospitality he enjoyed there.

AMS (MOS) 1970 subject classification: 20G40

A
B 7
C 8
D 9
E 0
F 1
G 2
H 3
I 4
J 5